Nonlinear Potential Theory of Degenerate Elliptic Equations

Juha Heinonen
University of Michigan

Tero Kilpeläinen
University of Jyvaskyla

Olli Martio
University of Helsinki

Preface to the Dover Edition
by the Authors

Dover Publications, Inc.
Mineola, New York

Bibliographical Note

This Dover edition, first published in 2006, is an unabridged republication of
the work originally published in 1993 by Oxford University Press, Inc., New York.
For the Dover edition, the authors have provided a new Preface, Corrigenda, and
Epilogue consisting of other new material.

Library of Congress Cataloging-in-Publication Data

Heinonen, Juha.
    Nonlinear potential theory of degenerate elliptic equations / Juha Heinonen,
Tero Kilpeläinen, and Olli Martio.
        p. cm.
    Rev. ed. of: Oxford ; New York : Clarendon Press ; New York : Oxford
University Press, 1993.
    Includes bibliographical references and index.
    ISBN 0-486-45050-3 (pbk.)
    1. Potential theory (Mathematics) 2. Differential equations, Elliptic. 3.
Calculus of variations. I. Kilpeläinen,Tero. II. Martio, O. (Olli) III. Title.

QA404.7.H45 2006
515'.96—dc22

                                                                        2006040276

Manufactured in the United States of America
Dover Publications, Inc., 31 East 2nd Street, Mineola, N.Y. 11501

Contents

Preface to the Dover Edition

The first eighteen chapters of this Dover edition of our book constitute an identical copy of the original monograph published by the Oxford University Press in 1993. In the ensuing **Corrigenda**, we list all the errors that have come to our attention since the publication of the book.

Four chapters have been added specifically for this edition. In Chapter 19, we give a self-contained treatment of the John-Nirenberg lemma in a form that suffices for the theory as developed in this book; our original proof had a flaw in it. In Chapters 20, 21, and 22, we review some important developments that have taken place since the publication of the Oxford edition, and that are closely connected with the main themes of the book. In Chapter 20, we discuss the defining axioms for p-admissible weights as well as some new examples. In Chapter 21, we discuss the Riesz measure associated with an $\mathcal{A}$-superharmonic function; in particular, we discuss the Wolff potential estimate and its use in the boundary regularity for the Dirichlet problem and in the behavior of superharmonic functions. In the final Chapter 22, we briefly review recent more general nonlinear potential theories. The addition has its own relatively short bibliography. Many more articles on nonlinear potential theory have appeared since the publication of the Oxford edition.

We thank Anders Björn, Jana Björn, and Juha Kinnunen for carefully reading the four new chapters. The authors have been supported by the US National Science Foundation and the Academy of Finland.

Finally, we wish to thank Dover Publications for publishing our monograph as well as for their friendly and flexible cooperation in this matter.

The Authors
Ann Arbor, Jyväskylä, and Helsinki
June 2006

Corrigenda:

Here we list and correct all the errors in the main text that we are aware of; the only exception is the John-Nirenberg lemma, which will be treated separately in Chapter 19. In addition to actual corrections, we have listed a few clarifications. We thank Anders Björn, Donatella Danielli, and Juha Kinnunen for pointing out some of these errors.

Page 5, line 1.   "prize" should read "price".

Page 5.   Add the following four items to **Notation:**

$\overline{B}(x, r) = \{y \in \mathbf{R}^n : |x - y| \leq r\}$.

By a *ball* we mean an open ball unless otherwise specifically stated.

χ_E stands for the characteristic function of a set E; that is, if $E \subset X$, then $\chi_E : X \to \mathbf{R}$ is defined by $\chi_E(x) = 1$ if $x \in E$ and $\chi_E(x) = 0$ if $x \notin E$.

We denote by L_{loc}^p various local Lebesgue spaces. For example, a function u belongs to $L_{loc}^p(\Omega; \mu)$ if $u \in L^p(D; \mu)$ for every $D \Subset \Omega$.

Page 13, lines 13–18.   The definition of the space $L_0^{1,p}(\Omega; \mu)$ is given somewhat ambiguously. Following the lines 17–18, $L_0^{1,p}(\Omega; \mu)$ is by definition the set of all functions $u \in L^{1,p}(\Omega; \mu)$ for which there exists a sequence $\varphi_j \in C_0^\infty(\Omega)$ such that $\nabla \varphi_j \to \nabla u$ in $L^p(\Omega; \mu)$.

We do not mean the space $L_0^{1,p}(\Omega; \mu)$ to be the abstract norm closure of $C_0^\infty(\Omega)$ with respect to the norm

$$(0.1) \qquad p(u) = \left(\int_\Omega |\nabla u|^p \, d\mu \right)^{1/p},$$

nor is such an interpretation used later in the text. This closure is not, in general, a space of functions. (See, for example, p. 307 in *Function Spaces and Potential Theory* by D. R. Adams and L. I. Hedberg, Springer-Verlag, 1996.) Also note that p in (0.1) is actually a norm, not only a seminorm, for functions in $C_0^\infty(\Omega)$, and that p is a bad choice for notation here as it is also used for the exponent.

Page 20, lines 2–5.   These lines should be corrected to: "... then $\max(u, \lambda) \in L^{1,p}(\Omega; \mu)$ and

$$\nabla \max(u, \lambda) = \begin{cases} \nabla u & \text{if } u > \lambda \\ 0 & \text{if } u \leq \lambda. \end{cases}$$

If, in addition, $\max(u, \lambda) \in L^p(\Omega; \mu)$, then $\max(u, \lambda) \in H^{1,p}(\Omega; \mu)$. A similar conclusion holds...".

Page 30. The definition for the sets K_i in the last paragraph is incorrect and should be replaced by

$$K_i = \{x \in K : \operatorname{dist}(x, \complement E_i) \geq \delta\},$$

where

$$\delta = \min\{\operatorname{dist}(K, \complement \bigcup_{i=1}^{k} E_i), \operatorname{dist}(C_1, \complement F_1), \operatorname{dist}(C_2, \complement F_2), \ldots, \operatorname{dist}(C_k, \complement F_k)\}$$

and $\operatorname{dist}(X, Y) = \inf\{|x - y| : x \in X, y \in Y\}$ denotes the distance between two sets X and Y in $\mathbf{R}^n$.

Page 36, line 17. The definition of u should read

$$u(x) = \begin{cases} \dfrac{\int_{|x-x_0|}^{R} t^{(1-n)/(p-1)} \, dt}{\int_{r}^{R} t^{(1-n)/(p-1)} \, dt} & \text{if } r < |x - x_0| < R \\ 1 & \text{if } 0 \leq |x - x_0| \leq r \end{cases}$$

Page 68. In the formulation of Theorem 3.41, replace the phrase "locally bounded" by "locally essentially bounded" (four times).

Page 69. In the formulation of the John-Nirenberg lemma 3.46, replace the assumption $B \Subset \Omega$ by $2B \subset \Omega$ (two times).

Page 73, line -4. Replace η by η_j.

Page 76, line -4. Replace $\subset$ by $\Subset$.

Page 76, lines -1, -2; page 77, lines 2–3. Replace each m_r by m_{2r}.

Page 81, line 16. Replace $h \mapsto \mathcal{A}(x, \xi)$ by $h \mapsto \mathcal{A}(x, h)$.

Page 87, line 12. Replace Ω by $\mathbf{R}^n$.

Page 148, line 4. The estimate on line 4 should be

$$\frac{1}{c}\left(\frac{\mathrm{cap}_{p,\mu}\left(B(x_0,r),B\right)}{\mathrm{cap}_{p,\mu}\left(\frac{1}{2}B,B\right)}\right)^{1/(1-p)} \leq h(x) \leq c\left(\frac{\mathrm{cap}_{p,\mu}\left(B(x_0,r),B\right)}{\mathrm{cap}_{p,\mu}\left(\frac{1}{2}B,B\right)}\right)^{1/(1-p)}$$

Page 148. Delete line 11, and line 13 should be

$$\frac{1}{m_r} = \frac{h(x_1)}{m_r} \geq \omega_r(x_1) \geq \frac{1}{c}\left(\frac{\mathrm{cap}_{p,\mu}\left(B(x_0,r),B\right)}{\mathrm{cap}_{p,\mu}\left(\frac{1}{2}B,B\right)}\right)^{1/(p-1)}.$$

Page 159, line 8. Replace $\inf_{E_k} s_k$ by $\inf_{E_k} s$.

Page 200, lines -4, -5. It should read: "... for a given $\mathcal{A}$-polar G_δ-set...".

Page 221, line -7. Replace x_0 by 0.

Page 265, line -9. Delete: $= [a,b]$.

Page 268, line 14. Replace A by $\mathcal{A}$.

Page 275. Radó's theorem for quasiregular mappings (Theorem 14.47) is a special case of Theorems 13.17 and 13.18 for harmonic morphisms. In particular, the proof of 13.17 shows that the set $f^{-1}(C) \cap D$ is of n-capacity zero whenever D is a component of Ω. Therefore f is quasiregular in Ω by Theorem 14.46.

Page 307, line -2. Replace $=$ by $\leq$.

Page 318, line -12. It should read: "... is called a *Brelot harmonic space* if the following two conditions are satisfied: ...".

Page 333, line -1.   Replace the second $||\varphi||$ by $||\varphi||^{p-1}$.

Page 336 ff.   The proof of the John–Nirenberg lemma is not correct. The flaw occurs in the discussion of a Calderón-Zygmund type decomposition in 18.4; specifically, the inequality $|B_{i,x}| \le c(n)|B_{i+1,x}|$ on line 6 of the last paragraph on p. 337 may not be true for $i = 0$. A new and self-contained argument for a slightly different statement is given in Chapter 19.

Introduction

The solutions to a second order quasilinear elliptic equation

$$\text{(1)} \qquad -\operatorname{div}\mathcal{A}(x, \nabla u(x)) = 0$$

have many features in common with harmonic functions. Most notably, the Dirichlet solutions are order preserving: if u and v are two solutions in a bounded open set Ω in $\mathbf{R}^n$ with $u \leq v$ on the boundary $\partial\Omega$, then $u \leq v$ in the interior of Ω. We call this property of solutions the *comparison principle*. Roughly speaking, this principle makes it possible to develop a genuine potential theory without having a linear solution space. The purpose of this book is to present such a theory.

To illustrate the kind of potential theory we have in mind, consider the following *p-Laplace equation*

$$\text{(2)} \qquad -\operatorname{div}(|\nabla u|^{p-2}\nabla u) = 0, \quad 1 < p < \infty.$$

When $p = 2$ equation (2) reduces to the Laplace equation $\Delta u = 0$ whose solutions, harmonic functions, are the primary object of study in classical potential theory. When $p \neq 2$ equation (2) is nonlinear and degenerates at the zeros of the gradient of u. Consequently, in this case the solutions, commonly referred to as *p-harmonic functions*, need not be smooth, nor even C^2, and equation (2) must be understood in a weak sense. To motivate the study of equation (2) as a prototype of equation (1), we mention the following three facts: (i) equation (2) is the Euler equation for the variational integral

$$\int |\nabla u|^p \, dx,$$

which in a sense is the simplest variational functional of nonquadratic growth; (ii) solutions of (2) are naturally associated with the first order Sobolev space $H^{1,p}$, where they play the role of functions with extremal properties; (iii) when $p = n$, the dimension of the underlying space, equation (2) and its solutions are central to the theory of quasiconformal and quasiregular mappings.

Equation (1) itself can be viewed as a measurable perturbation of (2). In particular, our theory covers equations of the form

$$\text{(3)} \qquad -\operatorname{div}\left((\theta(x)\nabla u \cdot \nabla u)^{(p-2)/2}\theta(x)\nabla u\right) = 0, \quad 1 < p < \infty,$$

1

where $\theta : \mathbf{R}^n \to GL(n, \mathbf{R})$ is a measurable matrix function satisfying, for some $\lambda > 0$, the ellipticity condition

$$\lambda^{-1}|\xi|^2 \le \theta(x)\xi \cdot \xi \le \lambda|\xi|^2$$

for $x, \xi \in \mathbf{R}^n$. When $p = 2$ in (3), we recover linear elliptic equations with measurable coefficients:

$$-\sum_{i,j=1}^{n} \partial_j(\theta_{j,i}(x)\partial_i u(x)) = 0 \,.$$

Generalizing one step further, we impose a growth condition on $\mathcal{A}$ replacing the requirement

$$\mathcal{A}(x,\xi) \cdot \xi \approx |\xi|^p \,,$$

with the weaker condition

$$\mathcal{A}(x,\xi) \cdot \xi \approx w(x)|\xi|^p \,,$$

where w is a nonnegative locally integrable function in $\mathbf{R}^n$, called a *weight*, and $1 < p < \infty$. Then our prototype equation is the *weighted p-Laplacian*

$$-\operatorname{div}(w(x)|\nabla u|^{p-2}\nabla u) = 0 \,.$$

To obtain standard regularity results for solutions of (1) in the weighted situation, certain restrictions on the weight are necessary. Recently, the search for admissible weights has been intensive and, although a complete characterization remains to be found, several partial results indicate that the required conditions are not overly severe. Indeed, it suffices to have weighted versions of the Poincaré and Sobolev inequalities, together with a doubling property of the measure $w(x)dx$. We have chosen to axiomatize these properties. The class of weights satisfying the given requirements is large; it includes the A_p-weights of Muckenhoupt and certain powers of the Jacobian of a quasiconformal mapping.

This book consists of sixteen chapters. First we develop in detail the theory of weighted Sobolev spaces $H^{1,p}$. Our treatment includes a study of the so-called refined Sobolev spaces, that is spaces of Sobolev functions which are defined up to a set of zero capacity. We define the Sobolev space to be the closure of smooth functions with respect to the weighted norm

$$\| u \|_{1,p} = \| u \|_p + \| \nabla u \|_p \,.$$

Chapter 2 contains the theory of weighted variational capacity. Therein we collect and prove all the basic properties of variational capacity; the only exception is Choquet's capacitability theorem, which we quote without proof.

In Chapters 3 and 6 we investigate solutions and supersolutions of equation (1). We follow the now classical scheme using the Moser iteration method and prove that supersolutions can be chosen to be lower semicontinuous functions. Consequently, we arrive at the (Hölder) continuity of solutions. The particular case when equation (1) is the Euler equation of a variational integral is treated in detail in Chapter 5; there we offer a quick proof for the existence of solutions to variational problems with homogeneous kernels.

After this preparation we are ready for the potential theory. In Chapter 7 we define $\mathcal{A}$-superharmonic functions via the comparison principle: a lower semicontinuous function $u : \Omega \to (-\infty, \infty]$ is $\mathcal{A}$-superharmonic if for each open $D \Subset \Omega$ and $\mathcal{A}$-harmonic $h \in C(\overline{D})$ the inequality $h \leq u$ on ∂D implies $h \leq u$ in D. Here the term $\mathcal{A}$-harmonic refers to continuous solutions of (1). An intriguing fact is that the above definition, which requires no a priori regularity, leads to the existence and the integrability of the gradient of a (locally bounded) $\mathcal{A}$-superharmonic function. This enables us to use variational methods in potential theoretic problems, and vice versa. The main idea involved here is the obstacle method, which we explore in Chapters 3, 5, and 7. These methods also show that the above definition has a local character and hence $\mathcal{A}$-superharmonic functions constitute a sheaf.

Chapters 8–14 form the core of the nonlinear potential theory of $\mathcal{A}$-superharmonic functions. We study balayage, Perron's method, polar sets, harmonic measure, the fine topology, and harmonic morphisms. A large portion of this study (Chapter 14) is dedicated to quasiconformal and quasiregular mappings and their role in nonlinear potential theory. Today the potential theory of $\mathcal{A}$-harmonic functions is an indispensable part of quasiconformal theory; similarly, numerous examples show that function theoretic methods are significant in studying geometric properties of solutions to partial differential equations. In Chapter 15 we establish the admissibility of certain weights, namely A_p-weights and powers of Jacobians of quasiconformal mappings. Our exposition of A_p-weights is self-contained; we prove the important open-ended property and Muckenhoupt's theorem. In the last chapter we briefly examine an axiomatic nonlinear potential theory. In the appendices we discuss the existence of solutions and establish a weighted version of the John–Nirenberg lemma.

By now nonlinear potential theory is more than twenty years old. This field grew from the necessity to understand better various function spaces frequently encountered in the theory of partial differential equations. Initially it consisted primarily of a study of nonlinear potentials, and foundational contributions were made by Maz'ya and Khavin (1972), Meyers (1970, 1975), Reshetnyak (1969), Adams and Meyers (1972, 1973), Fuglede (1971a), Hedberg (1972), and others. The theory of nonlinear potentials and function spaces is explored in the forthcoming monograph by Adams and Hedberg. The approach based on solutions and supersolutions of quasilinear elliptic equations has its origins in the papers by Granlund, Lindqvist, and Martio in the early 1980s, and in the later work of Heinonen and Kilpeläinen. See Granlund *et al.* (1982, 1983, 1985, 1986), Lindqvist and Martio (1985, 1988), Lindqvist (1986), Heinonen and Kilpeläinen (1988a, b, c), Kilpeläinen (1989). Our aim here is to present a unified nonlinear potential theory based on this latter approach. Although some of the topics can be viewed as purely potential theoretic, we hope that the inclusion of quasiconformal mappings, A_p-weights, and the regularity theory of (super)solutions will make the book appealing to readers with different backgrounds and interests. This in mind, an effort has been made to keep the presentation self-contained; the only exception is Chapter 14, where many of the basic properties of quasiregular mappings are quoted without proof. We always give a precise reference whenever we invoke a result not proved in the text.

Bibliographic notes at the end of each chapter provide additional, necessarily incomplete, information on the subject, which despite its short history has already evolved into quite an elaborate theory, with connections to numerous branches of analysis.

Ostensibly, the addition of weights makes most results in this book new. However, they often represent fairly straightforward extensions of the unweighted results and thus are probably known to specialists. Some innovations deserve special attention. These include the boundary regularity theorems and regularity theorems for weighted variational inequalities. Some of the topics have not been treated before in this generality. Moreover, we feel that many proofs presented here simplify those previously found in the literature, even for linear elliptic equations.

Weighted nonlinear potential theory from somewhat different points of view has been studied by Adams (1986) and Vodop'yanov (1990).

Although the treatment is reasonably self-contained, we assume that the reader is familiar with real analysis slightly beyond the level of standard graduate courses. In particular, some acquaintance with the usual Sobolev spaces is necessary in reading the first chapter. On the other hand, the reader interested only in the unweighted theory may safely skip most of Chapter 1 and rely, for example, on the excellent monographs by Maz'ya (1985) and Ziemer (1989).

Finally, a few words for those who may be acquainted with the unweighted nonlinear potential theory developed earlier by the authors and others.

Most of the theory goes through with weights but there is a prize to be paid. Although some proofs are new and simpler, many things that are trivial or easy in the nonweighted situation now require extra care. For instance, the possibility that, while working with a fixed equation, we can have both points of zero capacity and points of positive capacity sometimes causes new technical trouble. However, such cases are usually easily located and the reader who is interested only in the unweighted theory may proceed without much concern.

Acknowledgements

Many friends and colleagues have contributed to this book by providing useful comments and advice. We wish to thank all of them. Especially we want to mention Jorgen Harmse, Pekka Koskela, Jan Malý, Juan Manfredi, Jean McKemie, Bruce Palka, Karen E. Smith, and Susan Staples, who patiently read various early manuscripts and whose suggestions led to substantial improvements in the text. Also we thank Eira Henriksson who carefully typed part of the manuscript and Ari Lehtonen who helped us with $\mathcal{A}_{\mathcal{M}}\mathcal{S}$-TEX problems.

The first author was supported in part by grants from the National Science Foundation.

Notation

Here we introduce the basic notation which will be observed throughout this book.

$\mathbf{R}$ – the real numbers.

$\mathbf{R}^n$ – the real Euclidean n-space, $n \geq 2$. Unless otherwise stated, all the topological notions are taken with respect to $\mathbf{R}^n$.

$\overline{\mathbf{R}}^n = \mathbf{R}^n \cup \{\infty\}$ – the one-point compactification of $\mathbf{R}^n$.

Ω – an open nonempty subset of $\mathbf{R}^n$; by a *domain* we mean an open connected set.

The Euclidean norm of a point $x = (x_1, x_2, \ldots, x_n) \in \mathbf{R}^n$ is denoted by $|x| = (x_1^2 + x_2^2 + \cdots + x_n^2)^{1/2}$.

$B(x, r) = \{y \in \mathbf{R}^n : |x - y| < r\}$.

If $B = B(x, r)$ and $\lambda > 0$, then $\lambda B = B(x, \lambda r)$.

If $E \subset \mathbf{R}^n$, the boundary, the closure, and the complement of E with respect to $\mathbf{R}^n$ are denoted by ∂E, $\overline{E}$, and $\complement E = \mathbf{R}^n \setminus E$, respectively; diam E is the Euclidean diameter of E.

$E \Subset F$ means that $\overline{E}$ is a compact subset of F.

$|E|$ – the Lebesgue n-measure of a measurable $E \subset \mathbf{R}^n$.

ω_{n-1} – the surface measure of the boundary of the unit ball in $\mathbf{R}^n$.

If ν is a measure in Ω, $Y = \mathbf{R}$ or $Y = \mathbf{R}^n$, and $q > 0$, then $L^q(\Omega; \nu; Y)$ is the space of all ν-a.e. on Ω defined ν-measurable functions u with values in Y such that

$$||u||_q = (\int_\Omega |u|^q \, d\nu)^{1/q} < \infty.$$

We often write $L^q(X; \nu; Y) = L^q(X; \nu)$. Furthermore, $L^\infty(\Omega; \nu; Y)$ denotes the space of ν-essentially bounded ν-measurable functions u with

$$||u||_\infty = \operatorname{ess\,sup} |u| < \infty.$$

For a sequence of points (x_j) or functions (φ_j) we drop the parentheses and denote them simply as x_j and φ_j.

If $u : E \to \mathbf{R}$ is a function, then

$$\operatorname{osc}(u, E) = \sup_E u - \inf_E u$$

is the *oscillation* of u in E.

If X is a topological space, $C(X)$ is the set of all continuous functions $u : X \to \mathbf{R}$. Moreover, spt u is the smallest closed set such that u vanishes outside spt u.

$C^k(\Omega) = \{\varphi : \Omega \to \mathbf{R} : \text{ the } k\text{th-derivative of } \varphi \text{ is continuous}\}$

$C_0^k(\Omega) = \{\varphi \in C^k(\Omega) : \operatorname{spt} \varphi \Subset \Omega\}$

$C^\infty(\Omega) = \bigcap_{k=1}^\infty C^k(\Omega)$

$C_0^\infty(\Omega) = \{\varphi \in C^\infty(\Omega) : \operatorname{spt} \varphi \Subset \Omega\}.$

For a function $\varphi \in C^\infty(\Omega)$ we write

$$\nabla \varphi = (\partial_1 \varphi, \partial_2 \varphi, \ldots, \partial_n \varphi)$$

for the gradient of φ.

Throughout, c will denote a positive constant whose value is not necessarily the same at each occurrence; it may vary even within a line. $c(a, b, \cdots)$ is a constant that depends only on $a, b, \cdots$. Occasionally, when there is no danger of ambiguity, we use the expression $A \approx B$ meaning that there is a constant c such that

$$\frac{1}{c} A \leq B \leq c A.$$

1

Weighted Sobolev spaces

In this first chapter we introduce the weighted Sobolev spaces $H^{1,p}(\Omega; \mu)$ and investigate their basic properties which are needed in chapters to come. Although many features of the unweighted theory are retained, a somewhat different approach is mandatory.

We do not try to characterize those weights or measures which are admissible for our purposes. Instead, we elude the characterization problem in a customary way: the basic inequalities which are necessary for the development of the theory are included in the definition. The class of weights satisfying the given requirements is by no means restricted.

Throughout this book Ω will denote an open subset of $\mathbf{R}^n$, $n \geq 2$, and $1 < p < \infty$.

1.1. p-admissible weights

Let w be a locally integrable, nonnegative function in $\mathbf{R}^n$. Then a Radon measure μ is canonically associated with the *weight* w,

$$(1.2) \qquad \mu(E) = \int_E w(x)\, dx\,.$$

Thus $d\mu(x) = w(x)\, dx$, where dx is the n-dimensional Lebesgue measure. In what follows the weight w and the measure μ are identified via (1.2). We say that w (or μ) is *p-admissible* if the following four conditions are satisfied:

I $0 < w < \infty$ almost everywhere in $\mathbf{R}^n$ and the measure μ is *doubling*, i.e. there is a constant $C_\mathbf{I} > 0$ such that

$$\mu(2B) \leq C_\mathbf{I}\, \mu(B)$$

whenever B is a ball in $\mathbf{R}^n$.

II If D is an open set and $\varphi_i \in C^\infty(D)$ is a sequence of functions such that $\int_D |\varphi_i|^p\, d\mu \to 0$ and $\int_D |\nabla\varphi_i - v|^p\, d\mu \to 0$ as $i \to \infty$, where v is a vector-valued measurable function in $L^p(D; \mu; \mathbf{R}^n)$, then $v = 0$.

III There are constants $\varkappa > 1$ and $C_\mathbf{III} > 0$ such that

$$(\frac{1}{\mu(B)} \int_B |\varphi|^{\varkappa p}\, d\mu)^{1/\varkappa p} \leq C_\mathbf{III}\, r\, (\frac{1}{\mu(B)} \int_B |\nabla\varphi|^p\, d\mu)^{1/p}$$

whenever $B = B(x_0, r)$ is a ball in $\mathbf{R}^n$ and $\varphi \in C_0^\infty(B)$.

IV There is a constant $C_{IV} > 0$ such that

$$\int_B |\varphi - \varphi_B|^p \, d\mu \leq C_{IV} \, r^p \int_B |\nabla\varphi|^p \, d\mu$$

whenever $B = B(x_0, r)$ is a ball in $\mathbf{R}^n$ and $\varphi \in C^\infty(B)$ is bounded. Here

$$\varphi_B = \frac{1}{\mu(B)} \int_B \varphi \, d\mu \, .$$

Convention. *From now on, unless otherwise stated, we assume that μ is a p-admissible measure and $d\mu(x) = w(x) \, dx$.*

Let us make some remarks on conditions **I–IV**. It follows immediately from condition **I** that the measure μ and Lebesgue measure dx are mutually absolutely continuous, i.e. they have the same zero sets; so there is no need to specify the measure when using the ubiquitous expressions *almost everywhere* and *almost every*, both abbreviated *a.e.* Moreover, it easily follows from the doubling property that $\mu(\mathbf{R}^n) = \infty$.

Condition **II** guarantees that the gradient of a Sobolev function is well defined, a conclusion that cannot be expected in general (Fabes *et al.* 1982a, pp. 91–92).

Condition **III** is the *weighted Sobolev embedding theorem* or the *weighted Sobolev inequality* and condition **IV** is the *weighted Poincaré inequality*. The validity of these inequalities is crucial to the theory in this book.

The Lebesgue differentiation theorem holds: if $f \in L^1_{loc}(\mathbf{R}^n; \mu)$, then for a.e. x in $\mathbf{R}^n$

$$(1.3) \qquad \lim_{r \to 0} \frac{1}{\mu(B(x,r))} \int_{B(x,r)} f(y) \, d\mu(y) = f(x) \, .$$

For a proof, see Ziemer (1989, p. 14).

In general, if ν is a measure and f is a ν-integrable function on a set E with $0 < \nu(E) < \infty$, we write the integral average of f on E as

$$\fint_E f \, d\nu = \frac{1}{\nu(E)} \int_E f \, d\nu \, .$$

For example, (1.3) is usually written as

$$\lim_{r \to 0} \fint_{B(x,r)} f(y) \, d\mu(y) = f(x) \, .$$

The weighted Sobolev inequality **III** implies the following Poincaré type inequality. With an obvious abuse of terminology, in this book both condition **IV** and inequality (1.5) are referred to as the *Poincaré inequality*.

1.4. Poincaré inequality. *If Ω is bounded, then*

$$(1.5) \qquad \int_\Omega |\varphi|^p \, d\mu \le C_{\text{III}}^p (\text{diam } \Omega)^p \int_\Omega |\nabla\varphi|^p \, d\mu$$

for $\varphi \in C_0^\infty(\Omega)$.

PROOF: Let $x_0 \in \Omega$ and write $B = B(x_0, \text{diam } \Omega)$. If $\varphi \in C_0^\infty(\Omega)$, then the Hölder inequality and **III** imply

$$\left(\fint_B |\varphi|^p \, d\mu \right)^{1/p} \le \left(\fint_B |\varphi|^{\varkappa p} \, d\mu \right)^{1/\varkappa p}$$

$$\le C_{\text{III}} \, \text{diam } \Omega \left(\fint_B |\nabla\varphi|^p \, d\mu \right)^{1/p},$$

and the lemma follows. □

NOTATION. Qualitatively, many properties of μ depend only on the constants which appear in conditions **I**, **III**, and **IV**. For short we write

$$c_\mu = (C_{\text{I}}, \varkappa, C_{\text{III}}, C_{\text{IV}}).$$

Thus, saying that something depends on c_μ means it depends on the above constants associated with μ.

1.6. Examples of p-admissible weights

Next we give some examples of p-admissible weights and show that a p-admissible weight is also q-admissible for all q greater than p.

The first example is the usual case when $w = 1$ and μ is Lebesgue measure. Then **I** is obvious, **II** is easy, and **III** is the ordinary Sobolev inequality which holds with

$$\varkappa = \begin{cases} \dfrac{n}{n-p} & \text{if } p < n \\ 2 & \text{if } p \ge n. \end{cases}$$

Moreover, for $p < n$ we have that

$$(1.7) \qquad \left(\int_\Omega |\varphi|^{np/(n-p)} \, dx \right)^{(n-p)/np} \le c(n,p) \left(\int_\Omega |\nabla\varphi|^p \, dx \right)^{1/p}$$

for $\varphi \in C_0^\infty(\Omega)$.

Condition **IV** is the classical Poincaré inequality; see, for instance, Chapter 7 in Gilbarg and Trudinger (1983).

For the second example consider the Muckenhoupt class A_p which consists of all nonnegative locally integrable functions w in $\mathbf{R}^n$ such that

$$\sup \left(\fint_B w \, dx \right) \left(\fint_B w^{1/(1-p)} \, dx \right)^{p-1} = c_{p,w} < \infty \,,$$

where the supremum is taken over all balls B in $\mathbf{R}^n$. If w belongs to A_p, then w is p-admissible; we emphasize that the index p is the same. The weight w is said to be in A_1 if there is a constant c such that

$$\fint_B w(x) \, dx \le c \, \operatorname*{ess\,inf}_B w$$

for all balls B in $\mathbf{R}^n$. Since $A_1 \subset A_p$ whenever $p > 1$, an A_1-weight is p-admissible for every $p > 1$.

We give the basic theory of A_p-weights in Chapter 15, where we also establish their p-admissibility.

The third example arises from the theory of quasiconformal mappings: if $f : \mathbf{R}^n \to \mathbf{R}^n$ is a K-quasiconformal mapping and $J_f(x)$ the determinant of its Jacobian matrix, then

$$w(x) = J_f(x)^{1-p/n}$$

is p-admissible for $1 < p < n$. This weight need not be in A_p. For instance, the function $|x|^\delta$ is in A_p if and only if $-n < \delta < n(p-1)$, but for the quasiconformal mapping

$$f(x) = x|x|^\gamma \,, \quad \gamma > -1 \,,$$

$J_f(x)^{1-p/n}$ is comparable to $|x|^{\gamma(n-p)}$. Thus, if $p < n$, the function $w(x) = |x|^\delta$ satisfies I–IV whenever $\delta > -n$. The constants for μ depend only on n and δ. It follows from Theorem 1.8 that $w(x) = |x|^\delta$, $\delta > -n$, is a p-admissible weight for all $p > 1$.

The above facts about quasiconformal mappings and admissible weights are proved in Chapter 15.

This discussion does not exhaust the body of admissible weights; there is a rapidly growing literature on weighted Sobolev and Poincaré inequalities. See Notes to this chapter.

1.8. Theorem. *Suppose that w is a p-admissible weight and $q > p$. Then w is q-admissible.*

PROOF: Condition **I** is trivial. Condition **II** follows by observing that

$$\int_D |\varphi_i|^q \, d\mu \to 0 \qquad \text{and} \qquad \int_D |\nabla \varphi_i - v|^q \, d\mu \to 0$$

implies

$$\int_G |\varphi_i|^p \, d\mu \to 0 \qquad \text{and} \qquad \int_G |\nabla\varphi_i - v|^p \, d\mu \to 0$$

for each $G \Subset D$ by Hölder's inequality.

Next we prove **III**. Let $\psi \in C_0^\infty(B)$, $B = B(x_0, r)$, and $\varphi = \max(0, \psi)$. Then let

$$s = \frac{q}{p} > 1$$

and note that the p-type inequality **III** holds for the function φ^s; this follows by approximation (see the proof of Lemma 1.11). Moreover, it suffices to verify the q-type inequality **III** for φ. To do so, we combine

$$\left(\fint_B (\varphi^s)^{\varkappa p} \, d\mu\right)^{1/\varkappa p} \le C_{\text{III}} \, s \, r \left(\fint_B |\nabla\varphi|^p \varphi^{(s-1)p} \, d\mu\right)^{1/p}$$

$$\le C_{\text{III}} \, s \, r \left(\fint_B |\nabla\varphi|^{sp} \, d\mu\right)^{1/sp} \left(\fint_B \varphi^{sp} \, d\mu\right)^{(s-1)/sp}$$

and

$$\left(\fint_B \varphi^{sp} \, d\mu\right)^{1/p} \le \left(\fint_B \varphi^{\varkappa sp} \, d\mu\right)^{1/\varkappa p}$$

to obtain

$$\left(\fint_B \varphi^{\varkappa q} \, d\mu\right)^{1/\varkappa q} \le c \, \frac{q}{p} \, r \left(\fint_B |\nabla\varphi|^q \, d\mu\right)^{1/q},$$

as desired.

To verify inequality **IV** with p replaced by q, let $\varphi \in C^\infty(B)$ be bounded. It suffices to find constants γ and C such that

$$\int_B |\varphi - \gamma|^q \, d\mu \le C \, r^q \int_B |\nabla\varphi|^q \, d\mu;$$

this is due to the fact that

$$\int_B |\varphi - \varphi_B|^q \, d\mu \le 2^q \int_B |\varphi - \gamma|^q \, d\mu.$$

Again let $s = q/p > 1$ and write

$$v = \max(\varphi - \gamma, 0)^s - \max(\gamma - \varphi, 0)^s,$$

where γ is chosen so that

$$\int_B v \, d\mu = 0.$$

It is easily demonstrated (cf. Lemma 1.11) that the p-Poincaré inequality holds for v, that is

$$\int_B |v|^p \, d\mu \le C_{IV} \, r^p \int_B |\nabla v|^p \, d\mu \, .$$

Since

$$|\nabla v| = s |\nabla \varphi| \, |v|^{(s-1)/s}$$

and $q = sp$, Hölder's inequality yields

$$\int_B |v|^p \, d\mu \le C_{IV} \, s^p \, r^p \, \Big(\int_B |\nabla \varphi|^q \, d\mu \Big)^{1/s} \Big(\int_B |v|^p \, d\mu \Big)^{(s-1)/s} \, .$$

Finally, because $|v|^p = |\varphi - \gamma|^q$, it follows that

$$\int_B |\varphi - \gamma|^q \, d\mu \le C_{IV}^{q/p} \, s^q \, r^q \int_B |\nabla \varphi|^q \, d\mu \, ,$$

as desired.   □

1.9. Sobolev spaces

For a function $\varphi \in C^\infty(\Omega)$ we let

$$\| \varphi \|_{1,p} = \Big(\int_\Omega |\varphi|^p \, d\mu \Big)^{1/p} + \Big(\int_\Omega |\nabla \varphi|^p \, d\mu \Big)^{1/p} \, ,$$

where, we recall, $\nabla \varphi = (\partial_1 \varphi, \ldots, \partial_n \varphi)$ is the gradient of φ. The *Sobolev space* $H^{1,p}(\Omega; \mu)$ is defined to be the completion of

$$\{\varphi \in C^\infty(\Omega) \colon \|\varphi\|_{1,p} < \infty\}$$

with respect to the norm $\| \cdot \|_{1,p}$. In other words, a function u is in $H^{1,p}(\Omega; \mu)$ if and only if u is in $L^p(\Omega; \mu)$ and there is a vector-valued function v in $L^p(\Omega; \mu) = L^p(\Omega; \mu; \mathbf{R}^n)$ such that for some sequence $\varphi_i \in C^\infty(\Omega)$

$$\int_\Omega |\varphi_i - u|^p \, d\mu \to 0$$

and

$$\int_\Omega |\nabla \varphi_i - v|^p \, d\mu \to 0$$

as $i \to \infty$. The function v is called the *gradient of u in* $H^{1,p}(\Omega; \mu)$ and denoted by $v = \nabla u$. Condition **II** implies that ∇u is a uniquely defined function in $L^p(\Omega; \mu)$.

The space $H_0^{1,p}(\Omega;\mu)$ is the closure of $C_0^\infty(\Omega)$ in $H^{1,p}(\Omega;\mu)$. It is clear that $H^{1,p}(\Omega;\mu)$ and $H_0^{1,p}(\Omega;\mu)$ are Banach spaces under the norm $\|\cdot\|_{1,p}$. Moreover, the norm $\|\cdot\|_{1,p}$ is uniformly convex and therefore the Sobolev spaces $H^{1,p}(\Omega;\mu)$ and $H_0^{1,p}(\Omega;\mu)$ are reflexive (Yosida 1980, p. 127).

The corresponding local space $H_{loc}^{1,p}(\Omega;\mu)$ is defined in the obvious manner: a function u is in $H_{loc}^{1,p}(\Omega;\mu)$ if and only if u is in $H^{1,p}(\Omega';\mu)$ for each open set $\Omega' \Subset \Omega$. Note that for a function $u \in H_{loc}^{1,p}(\Omega;\mu)$ the gradient ∇u is a well-defined function in $L_{loc}^p(\Omega;\mu)$.

We alert the reader that the symbol ∇u stands for the gradient of u in a Sobolev space $H_{loc}^{1,p}(\Omega;\mu)$; even for a C^1-function u it is not a priori obvious that ∇u coincides with the usual gradient of u. We shall show later that they are equal (Lemma 1.11).

We also repeatedly invoke the *Dirichlet spaces* $L^{1,p}(\Omega;\mu)$ and $L_0^{1,p}(\Omega;\mu)$:

$$L^{1,p}(\Omega;\mu) = \{u \in H_{loc}^{1,p}(\Omega;\mu) \colon \nabla u \in L^p(\Omega;\mu)\}$$

and $L_0^{1,p}(\Omega;\mu)$ is the closure of $C_0^\infty(\Omega)$ with respect to the seminorm

$$p(u) = (\int_\Omega |\nabla u|^p \, d\mu)^{1/p}.$$

That is, $L_0^{1,p}(\Omega;\mu)$ is the set of all functions $u \in L^{1,p}(\Omega;\mu)$ for which there exists a sequence $\varphi_j \in C_0^\infty(\Omega)$ such that $\nabla\varphi_j \to \nabla u$ in $L^p(\Omega;\mu)$.

As opposed to the standard Sobolev space $H^{1,p}(\Omega;dx)$, an element in $H^{1,p}(\Omega;\mu)$ may have some peculiar features. For instance, a function in $H^{1,p}(\Omega;\mu)$ need not be locally integrable with respect to Lebesgue measure. To display a particular example, fix $p > 1$ and let $w(x) = |x|^{p(n+1)}$; then w is a p-admissible weight as discussed in Section 1.6. Now the function $u(x) = |x|^{-n}$ is in $H_{loc}^{1,p}(\mathbf{R}^n;\mu)$ and $\nabla u(x) = -nx|x|^{-n-2}$, but u is not locally integrable. In particular, there is no distribution in $\mathbf{R}^n$ that agrees with $|x|^{-n}$ in $\mathbf{R}^n \setminus \{0\}$. The gradient of $|x|^{-n}$ above can be computed by using Lemma 1.11 and a truncation argument.

Sometimes a weighted Sobolev space is defined as the set of all locally Lebesgue integrable functions u such that u and its distributional gradient both belong to $L^p(\Omega;\mu)$. Equipped with the norm $\|u\|_{1,p}$ this produces a normed space which is not necessarily Banach as the example above shows. Consequently, this definition does not lead to the space $H^{1,p}(\Omega;\mu)$. However, if the weight w is in A_p, it can be shown that these two definitions give the same space (Kilpeläinen 1992b).

If we impose a mild additional condition on the weight w, each Sobolev function is a distribution. More precisely, if $w^{1/(1-p)} \in L_{loc}^1(\Omega;dx)$, in particular if $w \in A_p$, then every Sobolev function u in $H^{1,p}_{loc}(\Omega; \mu)$ is a distribution and ∇u is the distributional gradient of u; that is, u is locally Lebesgue integrable in Ω and

$$\int_\Omega u \partial_i \varphi \, dx = -\int_\Omega \partial_i u \, \varphi \, dx$$

for all $\varphi \in C^\infty_0(\Omega)$ and $i = 1, 2, \ldots, n$. Here $\partial_i u$ is the ith coordinate of ∇u. To see this, first apply the Hölder inequality to $u \in L^p(D; \mu)$, $D \Subset \Omega$, and obtain

$$\int_D |u| \, dx = \int_D |u| \, w^{1/p} \, w^{-1/p} \, dx$$
$$\leq \left(\int_D w^{1/(1-p)} \, dx \right)^{(p-1)/p} \left(\int_D |u|^p \, d\mu \right)^{1/p}.$$

This implies that $L^p(D; \mu)$ is continuously embedded in $L^1(D; dx)$. Thus if $\varphi_j \in C^\infty(\Omega)$ converges in $H^{1,p}(\Omega; \mu)$ to u, then the sequences φ_j and $\partial_i \varphi_j$ converge to u and $\partial_i u$, respectively, in $L^1(D; dx)$ for all $D \Subset \Omega$, $i = 1, 2, \ldots, n$. We have for all $\varphi \in C^\infty_0(\Omega)$

$$\left| \int_\Omega u \, \partial_i \varphi + \varphi \, \partial_i u \, dx \right| = \left| \int_{\operatorname{spt} \varphi} (u - \varphi_j) \, \partial_i \varphi + (\partial_i u - \partial_i \varphi_j) \, \varphi \, dx \right| \to 0$$

as $j \to \infty$. This proves that ∇u is the distributional gradient of u.

We prove in Lemma 1.11 that if u is a locally Lipschitz function in $H^{1,p}(\Omega; \mu)$, then ∇u is the distributional gradient of u.

1.10. Basic properties of Sobolev spaces

In the following few pages we demonstrate the basic properties of the Sobolev space $H^{1,p}(\Omega; \mu)$. The first fundamental fact to observe is that the Sobolev and Poincaré inequalities **III**, **IV**, and (1.5) hold for functions in $H^{1,p}_0(B; \mu)$, $H^{1,p}(B; \mu)$, and $H^{1,p}_0(\Omega; \mu)$, respectively.

Before proceeding we recall the usual regularization procedure. Let η be a nonnegative function in $C^\infty_0(\mathbf{R}^n)$ such that

$$\int_{\mathbf{R}^n} \eta \, dx = 1.$$

Such a function η is called a *mollifier*. For example, we can take

$$\eta(x) = \begin{cases} c \exp\{-1/(1 - |x|^2)\} & \text{if } |x| < 1 \\ 0 & \text{if } |x| \geq 1. \end{cases}$$

Next write
$$\eta_j(x) = j^n\eta(jx), \quad j = 1, 2, \ldots,$$
and recall that, for $u \in L^1_{loc}(\mathbf{R}^n; dx)$, the convolution

$$u_j(x) = \eta_j * u(x) = \int_{\mathbf{R}^n} \eta_j(x - y)u(y)\, dy$$

enjoys the following properties:

(i) $u_j \in C^\infty(\mathbf{R}^n)$ and $\partial_i u_j = (\partial_i \eta_j) * u$.

(ii) $u_j(x) \to u(x)$ whenever x is a Lebesgue point for u. If u is continuous, then $u_j \to u$ locally uniformly on $\mathbf{R}^n$.

(iii) If $u \in L^q(\mathbf{R}^n; dx)$, $1 \le q \le \infty$, then

$$\|u_j\|_q \le \|u\|_q,$$

where $\|v\|_q$ is the $L^q(\mathbf{R}^n; dx)$-norm of v. Moreover, $u_j \to u$ in $L^q(\mathbf{R}^n; dx)$ if $q < \infty$.

(iv) If u has a distributional derivative $D_i u \in L^1_{loc}(\mathbf{R}^n; dx)$, then

$$D_i u_j = \eta_j * D_i u.$$

For these, see e.g. Ziemer (1989, 1.6.1 and 2.1.3).

Recall that a function $u: E \to \mathbf{R}$ is *Lipschitz on* $E \subset \mathbf{R}^n$, if there is $L > 0$ such that

$$|u(x) - u(y)| \le L\,|x - y|$$

for all $x, y \in E$. Moreover, u is *locally Lipschitz on* E, if u is Lipschitz on each compact subset of E. It is well known that a locally Lipschitz function on $\mathbf{R}^n$ is a.e. differentiable; this is Rademacher's theorem (Federer 1969, 3.1.6).

1.11. Lemma. *Let* $u: \Omega \to \mathbf{R}$ *be a locally Lipschitz function. Then* $u \in H^{1,p}_{loc}(\Omega; \mu)$ *and* $\nabla u = (\partial_1 u, \partial_2 u, \ldots, \partial_n u)$ *is the usual gradient of* u.

PROOF: Let $D \Subset \Omega$ be open. Multiplying u by a cut-off function $\psi \in C^\infty_0(\Omega)$, $0 \le \psi \le 1$ with $\psi = 1$ in D, we may assume that u is Lipschitz and bounded in $\mathbf{R}^n$. Choose a mollifier $\eta \in C^\infty_0(B(0,1))$ and let $\eta_j(x) = j^n\eta(jx)$ and $u_j = \eta_j * u$ be as above. Then $u_j \in C^\infty(\mathbf{R}^n)$,

$$\|u_j\|_\infty \le \|u\|_\infty,$$

and $u_j \to u$ uniformly on D. Since

$$\partial_i u_j = \eta_j * \partial_i u \to \partial_i u$$

a.e. and since

$$\|\partial_i u_j\|_\infty \leq \|\partial_i u\|_\infty,$$

we obtain

$$\int_D |u_j - u|^p \, d\mu \to 0$$

and

$$\int_D |\nabla u_j - (\partial_1 u, \partial_2 u, \ldots, \partial_n u)|^p \, d\mu \to 0$$

by the Lebesgue dominated convergence theorem. The lemma follows. □

We have the following relationship between weighted and unweighted Sobolev spaces.

1.12. Lemma. *Suppose that* $1 < s \leq \infty$, $s' = s/(s-1)$ *if* $s < \infty$, $s' = 1$ *if* $s = \infty$, *and that* $w \in L^s(\Omega; dx)$. *If* $u \in H^{1,s'p}(\Omega; dx)$, *then* $u \in H^{1,p}(\Omega; \mu)$ *and the gradient of* u *in* $H^{1,p}(\Omega; \mu)$ *is the distributional gradient of* u. *Moreover, the embedding*

$$H^{1,s'p}(\Omega; dx) \subset H^{1,p}(\Omega; \mu)$$

is continuous. In particular, if w *is bounded, then* $H^{1,p}(\Omega; dx)$ *is a subspace of* $H^{1,p}(\Omega; \mu)$.

PROOF: Hölder's inequality implies

$$\left(\int_\Omega |f|^p \, d\mu\right)^{1/p} \leq \left(\int_\Omega |f|^{s'p} \, dx\right)^{1/s'p} \left(\int_\Omega w^s \, dx\right)^{1/sp}$$

whenever $f \in L^{s'p}(\Omega; dx)$. The lemma follows easily from this. □

1.13. Lemma. *Suppose that* $1 < s \leq p$ *and that* $w^{1/(1-s)} \in L^1(\Omega; dx)$. *If* $u \in H^{1,p}(\Omega; \mu)$, *then* $u \in H^{1,p/s}(\Omega; dx)$ *and the gradient of* u *in* $H^{1,p}(\Omega; \mu)$ *is the distributional gradient of* u. *Moreover, the embedding*

$$H^{1,p}(\Omega; \mu) \subset H^{1,p/s}(\Omega; dx)$$

is continuous.

PROOF: The assertion again follows from the Hölder inequality since

$$\left(\int_\Omega |f|^{p/s} \, dx\right)^{s/p} \leq \left(\int_\Omega |f|^p \, d\mu\right)^{1/p} \left(\int_\Omega w^{1/(1-s)} \, dx\right)^{(s-1)/p}.$$

□

It is also easy to see that if $w^{-1} \in L^\infty(\Omega)$, then $H^{1,p}(\Omega; \mu)$ is continuously embedded in $H^{1,p}(\Omega; dx)$. Moreover, it is trivial that if both w and w^{-1} are bounded, then $H^{1,p}(\Omega; \mu) = H^{1,p}(\Omega; dx)$.

The usual Sobolev embedding theorem (Gilbarg and Trudinger 1983, Theorem 7.10) says that if $p > n$, then $H_0^{1,p}(\Omega; dx)$ is continuously embedded into $C(\overline{\Omega})$. Hence we have

1.14. Theorem. *If $1 < s < p/n$ and $w^{1/(1-s)} \in L_{loc}^1(\Omega; dx)$, then each function in $H^{1,p}(\Omega; \mu)$ has a continuous representative.*

The next lemma shows that

$$H^{1,p}(\Omega; \mu) = L^p(\Omega; \mu) \cap L^{1,p}(\Omega; \mu).$$

1.15. Lemma. *If $u \in H_{loc}^{1,p}(\Omega; \mu)$ and $u, \nabla u \in L^p(\Omega; \mu)$, then $u \in H^{1,p}(\Omega; \mu)$.*

PROOF: Choose open sets $\Omega_j \Subset \Omega_{j+1} \Subset \Omega$, $j \geq 1$, such that $\bigcup_j \Omega_j = \Omega$. Let Ψ be a partition of unity of Ω, subordinate to the covering $\Omega_{j+1} \setminus \overline{\Omega}_{j-1}$; see e.g. Ziemer (1989, Lemma 2.3.1). Let ψ_j denote the (finite) sum of those $\psi \in \Psi$ for which spt $\psi \subset \Omega_{j+1} \setminus \overline{\Omega}_{j-1}$ and spt $\psi \not\subset \complement\overline{\Omega}_j$. Thus $\psi_j \in C_0^\infty(\Omega_{j+1} \setminus \overline{\Omega}_{j-1})$ and $\sum_j \psi_j \equiv 1$ in Ω (here we define $\Omega_0 = \emptyset$). Fix $\varepsilon > 0$ and for each $j = 1, 2, \ldots$ choose $\varphi_j \in C_0^\infty(\Omega_{j+1} \setminus \overline{\Omega}_{j-1})$ such that $\| \varphi_j - \psi_j u \|_{1,p} < \varepsilon 2^{-j}$. Then $\varphi = \sum_j \varphi_j \in C^\infty(\Omega)$, and

$$\| \varphi - u \|_{1,p} = \| \sum_j \varphi_j - \sum_j \psi_j u \|_{1,p}$$
$$\leq \sum_j \| \varphi_j - \psi_j u \|_{1,p} < \varepsilon.$$

Thus u is in $H^{1,p}(\Omega; \mu)$ as required. $\qquad\qquad\qquad\qquad\qquad\qquad\square$

Lemma 1.15 and its proof can be used to show that if

$$u \in H^{1,p}(\Omega_1; \mu) \cap H^{1,p}(\Omega_2; \mu),$$

then $u \in H^{1,p}(\Omega_1 \cup \Omega_2; \mu)$; we leave this for the reader. Consequently, $u \in H_{loc}^{1,p}(\Omega; \mu)$ if and only if each $x \in \Omega$ has a neighborhood D such that $u \in H^{1,p}(D; \mu)$.

1.16. Lemma. *Let Ω be a domain and $u \in H^{1,p}(\Omega; \mu)$. If $\nabla u = 0$, then u is constant.*

PROOF: Choose a sequence of functions $\varphi_j \in C^\infty(\Omega)$ converging to u in $H^{1,p}(\Omega; \mu)$. Let $B \Subset \Omega$ be a ball. Since $\nabla \varphi_j \to 0$ in $L^p(B; \mu)$, we obtain from the Poincaré inequality **IV** that $\varphi_j \overset{\bullet}{\sim} \varphi_{j,B} \to 0$ in $L^p(B;\mu)$, where

$$\varphi_{j,B} = \fint_B \varphi_j \, d\mu \, .$$

Hence passing to a subsequence, if necessary, we have that

$$\varphi_j \to c = \lim_{j \to \infty} \varphi_{j,B}$$

a.e. in B. Hence u is constant in B, and since Ω is connected, u is constant in Ω.   □

1.17. Lemma. *If $u \in H_0^{1,p}(\Omega;\mu)$ with $\nabla u = 0$, then $u = 0$.*

PROOF: Since u can be regarded as a function in $H_0^{1,p}(\mathbf{R}^n;\mu)$ whose gradient vanishes, Lemma 1.16 implies that u is constant, say c, in $\mathbf{R}^n$. This constant must be zero because it follows from the doubling property that $\mu(\mathbf{R}^n) = \infty$, while

$$|c|^p \mu(\mathbf{R}^n) = \int_{\mathbf{R}^n} |u|^p \, d\mu < \infty \, .$$

                                                                        □

1.18. Theorem. *Suppose that $f \in C^1(\mathbf{R})$, f' is bounded, and $u \in H^{1,p}(\Omega;\mu)$. Then $f \circ u \in L^{1,p}(\Omega;\mu)$ and*

$$\nabla(f \circ u) = f'(u)\nabla u \, .$$

If, in addition, $f \circ u \in L^p(\Omega;\mu)$, then $f \circ u \in H^{1,p}(\Omega;\mu)$.

PROOF: Suppose that $\varphi_i \in C^\infty(\Omega)$ is a sequence such that $\varphi_i \to u$ in $H^{1,p}(\Omega;\mu)$ and pointwise a.e. Then $f \circ \varphi_i$ is locally Lipschitz; hence $f \circ \varphi_i \in L^{1,p}(\Omega;\mu)$ and

$$\nabla(f \circ \varphi_i) = f'(\varphi_i)\nabla \varphi_i$$

by Lemma 1.11. Moreover, since

$$|f(t) - f(s)| \le \sup |f'| \, |t - s| \, ,$$

we have

$$\int_\Omega |f \circ \varphi_i - f \circ u|^p \, d\mu \le \sup |f'|^p \int_\Omega |\varphi_i - u|^p \, d\mu \to 0 \, .$$

Further, by the dominated convergence theorem

$$\left(\int_\Omega |f'(\varphi_i)\nabla\varphi_i - f'(u)\nabla u|^p \, d\mu \right)^{1/p}$$

$$\le \sup |f'|\left(\int_\Omega |\nabla\varphi_i - \nabla u|^p \, d\mu \right)^{1/p} + \left(\int_\Omega |\nabla u|^p \, |f'(\varphi_i) - f'(u)|^p \, d\mu \right)^{1/p}$$

$$\to 0$$

because $f'(\varphi_i) \to f'(u)$ a.e. in Ω. This proves the theorem.   □

We use the notation

$$u^+ = \max(u,0) \quad \text{and} \quad u^- = \min(u,0).$$

1.19. Lemma. *If* $u \in H^{1,p}(\Omega;\mu)$, *then* $u^+ \in H^{1,p}(\Omega;\mu)$ *and*

$$\nabla u^+ = \begin{cases} \nabla u & \text{if } u > 0 \\ 0 & \text{if } u \leq 0. \end{cases}$$

PROOF: Let g be the characteristic function of the interval $(0,\infty)$, i.e. $g(t) = 0$ if $t \leq 0$ and $g(t) = 1$ if $t > 0$. Write for $j = 1, 2, \ldots$

$$f_j(t) = \begin{cases} 0 & \text{if } t \leq 0 \\ \frac{j}{j+1} t^{\frac{j+1}{j}} & \text{if } 0 \leq t \leq 1 \\ t - \frac{1}{j+1} & \text{if } t \geq 1. \end{cases}$$

Then $f_j \in C^1(\mathbf{R})$ and

$$f_j'(t) = \begin{cases} 0 & \text{if } t \leq 0 \\ t^{1/j} & \text{if } 0 < t < 1 \\ 1 & \text{if } t \geq 1. \end{cases}$$

Moreover, f_j' is bounded and increases to g pointwise, and $0 \leq f_j(t) \leq t^+$ for $t \in \mathbf{R}$. Thus it follows from Theorem 1.18 that $f_j \circ u \in H^{1,p}(\Omega;\mu)$ and

$$\nabla(f_j \circ u) = f_j'(u)\nabla u.$$

Now $f_j \circ u$ increases pointwise to u^+ a.e. and, moreover,

$$\nabla(f_j \circ u) = f_j'(u)\nabla u \to \begin{cases} \nabla u & \text{if } u > 0 \\ 0 & \text{if } u \leq 0 \end{cases}$$

pointwise a.e. in Ω. Let v be the vector-valued function on the right. Since

$$|\nabla(f_j \circ u)| \leq |\nabla u|,$$

from the dominated convergence theorem we obtain $\nabla(f_j \circ u) \to v$ in $L^p(\Omega;\mu)$, and the monotone convergence theorem implies $f_j \circ u \to u^+$ in $L^p(\Omega;\mu)$. Thus $f_j \circ u$ is a Cauchy sequence, hence convergent in $H^{1,p}(\Omega;\mu)$. It follows that $f_j \circ u \to u^+$ in $H^{1,p}(\Omega;\mu)$ and

$$\nabla u^+ = v = \begin{cases} \nabla u & \text{if } u > 0 \\ 0 & \text{if } u \leq 0. \end{cases}$$

$\square$

It follows from Lemmas 1.15 and 1.19 that $H^{1,p}(\Omega;\mu)$ is closed under truncation; more precisely, if $u \in H^{1,p}(\Omega;\mu)$ and $\lambda \in \mathbf{R}$, then $\max(u,\lambda) \in H^{1,p}(\Omega;\mu)$ and

$$\nabla \max(u,\lambda) = \begin{cases} \nabla u & \text{if } u > \lambda \\ 0 & \text{if } u \leq \lambda. \end{cases}$$

A similar conclusion holds for $\min(u,\lambda)$ whenever $\lambda \in \mathbf{R}$. From this we deduce that a function $u \in H^{1,p}(\Omega;\mu)$ can be approximated by its truncations in $H^{1,p}(\Omega;\mu)$. Indeed, for $j = 1, 2, \ldots$ the functions $\min(u^+,j)$ and $\max(u^-,-j)$ belong to $H^{1,p}(\Omega;\mu)$ and clearly converge to u^+ and u^-, respectively. Hence

$$\min(u^+,j) + \max(u^-,-j) \to u$$

in $H^{1,p}(\Omega;\mu)$.

It is important that the first order Sobolev space $H^{1,p}(\Omega;\mu)$ is a lattice.

1.20. Theorem. If u and v are in $H^{1,p}(\Omega;\mu)$, then $\max(u,v)$ and $\min(u,v)$ are in $H^{1,p}(\Omega;\mu)$ with

$$\nabla \max(u,v)(x) = \begin{cases} \nabla u(x) & \text{if } u(x) \geq v(x) \\ \nabla v(x) & \text{if } v(x) \geq u(x) \end{cases}$$

and

$$\nabla \min(u,v)(x) = \begin{cases} \nabla u(x) & \text{if } u(x) \leq v(x) \\ \nabla v(x) & \text{if } v(x) \leq u(x). \end{cases}$$

In particular, $|u| = u^+ - u^-$ belongs to $H^{1,p}(\Omega;\mu)$.

PROOF: Since $\max(u,v) = (u-v)^+ + v$ and $\min(u,v) = u - (u-v)^+$, the assertions follow from Lemma 1.19. $\qquad\qquad\square$

1.21. Corollary. If $u \in H^{1,p}_{loc}(\Omega;\mu)$ and λ is a real number, then $\nabla u = 0$ a.e. on the set $\{x \in \Omega : u(x) = \lambda\}$.

1.22. Lemma. If $u_j, v_j \in H^{1,p}(\Omega;\mu)$ are such that $u_j \to u$ and $v_j \to v$ in $H^{1,p}(\Omega;\mu)$, then $\min(u_j,v_j) \to \min(u,v)$ and $\max(u_j,v_j) \to \max(u,v)$ in $H^{1,p}(\Omega;\mu)$.

PROOF: It suffices to show that if u_j converge to u in $H^{1,p}(\Omega;\mu)$, then u_j^+ converge to u^+ in $H^{1,p}(\Omega;\mu)$. Since $|u_j^+ - u^+| \leq |u_j - u|$, it is clear that $u_j^+ \to u^+$ in $L^p(\Omega;\mu)$. To establish the convergence of the gradients, let g be the characteristic function of the interval $(0,\infty)$. Then

$$\left(\int_\Omega |\nabla u_j^+ - \nabla u^+|^p \, d\mu\right)^{1/p} = \left(\int_\Omega |g(u_j)\nabla u_j - g(u)\nabla u|^p \, d\mu\right)^{1/p}$$

$$\leq \left(\int_\Omega |\nabla u_j - \nabla u|^p \, d\mu\right)^{1/p} + \left(\int_\Omega |\nabla u|^p |g(u_j) - g(u)|^p \, d\mu\right)^{1/p} \to 0,$$

for the first integral trivially converges to 0 and the second can be treated by means of subsequences and the Lebesgue dominated convergence theorem; note that $\nabla u = 0$ a.e. on the set where $u = 0$. □

Next we show that the space $H_0^{1,p}(\Omega; \mu)$ is also a lattice.

1.23. Lemma. *If $u, v \in H_0^{1,p}(\Omega; \mu)$, then $\min(u, v)$ and $\max(u, v)$ are in $H_0^{1,p}(\Omega; \mu)$. Moreover, if $u \in H_0^{1,p}(\Omega; \mu)$ is nonnegative, then there is a sequence of nonnegative functions $\varphi_j \in C_0^\infty(\Omega)$ converging to u in $H^{1,p}(\Omega; \mu)$.*

PROOF: Let $\varphi \in C_0^\infty(\Omega)$. That φ^+ can be approximated in $H^{1,p}(\Omega; \mu)$ by nonnegative functions in $C_0^\infty(\Omega)$ was actually established in the proof of Lemma 1.11. The lemma follows since $\varphi_j^+ \to u^+$ in $H^{1,p}(\Omega; \mu)$ provided that $\varphi_j \to u$ in $H^{1,p}(\Omega; \mu)$ (Lemma 1.22). □

1.24. Theorem. *Suppose that u and v are bounded and belong to $H^{1,p}(\Omega; \mu)$. Then*

(i) *$uv \in H^{1,p}(\Omega; \mu)$ and $\nabla(uv) = v\nabla u + u\nabla v$.*

(ii) *If, in addition, $u \in H_0^{1,p}(\Omega; \mu)$, then $uv \in H_0^{1,p}(\Omega; \mu)$.*

PROOF: (i) We are free to assume that $|u|, |v| \leq 1$. Let $\tilde{u}_j, \tilde{v}_j \in C^\infty(\Omega) \cap H^{1,p}(\Omega; \mu)$ be such that $\tilde{u}_j \to u$, $\tilde{v}_j \to v$ both in $H^{1,p}(\Omega; \mu)$ and pointwise a.e. in Ω. Write $u_j = \max(-1, \min(\tilde{u}_j, 1))$ and $v_j = \max(-1, \min(\tilde{v}_j, 1))$. Then $u_j v_j$ is a locally Lipschitz function and therefore belongs to $H_{loc}^{1,p}(\Omega; \mu)$ by Lemma 1.11. Moreover,

$$\nabla(u_j v_j) = v_j \nabla u_j + u_j \nabla v_j \, .$$

Thus it follows from Lemma 1.15 that $u_j v_j \in H^{1,p}(\Omega; \mu)$. Then $u_j \to u$ and $v_j \to v$ both in $H^{1,p}(\Omega; \mu)$ and pointwise a.e. (Lemma 1.22). Therefore, since $|u_j|, |v_j| \leq 1$, we have

$$\left(\int_\Omega |v_j u_j - vu|^p \, d\mu \right)^{1/p}$$

$$\leq \left(\int_\Omega |v_j|^p |u_j - u|^p \, d\mu \right)^{1/p} + \left(\int_\Omega |u|^p |v_j - v|^p \, d\mu \right)^{1/p} \to 0$$

as $j \to \infty$. Furthermore,

$$\left(\int_\Omega |(u_j \nabla v_j + v_j \nabla u_j) - (u\nabla v + v\nabla u)|^p \, d\mu \right)^{1/p}$$

$$\leq \left(\int_\Omega |u_j \nabla v_j - u\nabla v|^p \, d\mu \right)^{1/p} + \left(\int_\Omega |v_j \nabla u_j - v\nabla u|^p \, d\mu \right)^{1/p}$$

$$\leq \left(\int_\Omega |u_j|^p |\nabla v_j - \nabla v|^p \, d\mu \right)^{1/p} + \left(\int_\Omega |\nabla v|^p |u_j - u|^p \, d\mu \right)^{1/p}$$

$$+ \left(\int_\Omega |v_j|^p |\nabla u_j - \nabla u|^p \, d\mu \right)^{1/p} + \left(\int_\Omega |\nabla u|^p |v_j - v|^p \, d\mu \right)^{1/p}$$

which tends to 0 by the dominated convergence theorem. Hence (i) is proved.

Assertion (ii) follows from the argument used in the proof of (i) by requiring that $\tilde{u}_j$ have compact support in Ω; cf. (i) in the next lemma. $\qquad\square$

Observe that if $v \in H^{1,p}(\Omega; \mu)$ and if both v and ∇v are bounded, the argument in Theorem 1.24 gives that

$$\max(-j, \min(u, j)) \to uv$$

in $H^{1,p}(\Omega; \mu)$, and hence $uv \in H^{1,p}(\Omega; \mu)$, whenever $u \in H^{1,p}(\Omega; \mu)$. Moreover,

$$\nabla(uv) = u\nabla v + v\nabla u$$

in this case.

1.25. Lemma. *Suppose that a function v belongs to $H^{1,p}(\Omega; \mu)$.*

(i) *If v has compact support, then $v \in H_0^{1,p}(\Omega; \mu)$.*

(ii) *If $u \in H_0^{1,p}(\Omega; \mu)$ and if $0 \le v \le u$ a.e. in Ω, then $v \in H_0^{1,p}(\Omega; \mu)$.*

(iii) *If $u \in H_0^{1,p}(\Omega; \mu)$ and if $|v| \le |u|$ a.e. in $\Omega \setminus K$, where K is a compact subset of Ω, then $v \in H_0^{1,p}(\Omega; \mu)$.*

PROOF: For the proof of (i), let $\psi \in C_0^\infty(\Omega)$ be such that $\psi = 1$ on the support of v. If a sequence $\psi_j \in C^\infty(\Omega)$ converges to v in $H^{1,p}(\Omega; \mu)$, then $\psi \psi_j \in C_0^\infty(\Omega)$ converges to $\psi v = v$ in $H_0^{1,p}(\Omega; \mu)$.

As to assertion (ii), let $\varphi_j \in C_0^\infty(\Omega)$ be a nonnegative approximating sequence for u in $H^{1,p}(\Omega; \mu)$. Then $\min(v, \varphi_j)$ has compact support and hence belongs to $H_0^{1,p}(\Omega; \mu)$. Moreover, since $\min(v, \varphi_j)$ converges to $\min(v, u) = v$ in $H^{1,p}(\Omega; \mu)$ (Lemma 1.22), we have $v \in H_0^{1,p}(\Omega; \mu)$.

To prove (iii), let $\eta \in C_0^\infty(\Omega)$, $0 \le \eta \le 1$, be such that $\eta = 1$ on K. Then

$$\tilde{u} = (1 - \eta)|u| + \eta v^+ \in H_0^{1,p}(\Omega; \mu)$$

(see the remark after Theorem 1.24) and $0 \le v^+ \le \tilde{u}$ a.e. in Ω. Hence $v^+ \in H_0^{1,p}(\Omega; \mu)$. Similarly we see that $v^- \in H_0^{1,p}(\Omega; \mu)$, and the proof is complete. $\qquad\square$

Roughly speaking, a function $u \in H^{1,p}(\Omega; \mu)$ belongs to $H_0^{1,p}(\Omega; \mu)$ if u vanishes on the boundary of Ω. The following easy but often used lemma displays this phenomenon; a sharp form of the lemma appears in Theorem 4.5. The assumption that Ω is bounded is not essential; it is assumed for the sake of a very simple proof. However, it is not too difficult to prove the lemma directly for a general Ω.

1.26.   Lemma. *If Ω is bounded and $u \in H^{1,p}(\Omega; \mu)$ is such that* $\lim_{x \to y} u(x) = 0$ *for all* $y \in \partial\Omega$, *then* $u \in H_0^{1,p}(\Omega; \mu)$.

PROOF: Recalling that $u = u^+ + u^-$ we may assume that u is nonnegative. The function $u_\varepsilon = \max(u - \varepsilon, 0)$ is in $H^{1,p}(\Omega; \mu)$ for $\varepsilon > 0$ and has compact support in Ω. Thus $u_\varepsilon \in H_0^{1,p}(\Omega; \mu)$ and the lemma follows because $u_\varepsilon \to u$ in $H^{1,p}(\Omega; \mu)$ as $\varepsilon \to 0$. $\square$

If $\complement\Omega$ is small enough, the spaces $H^{1,p}(\Omega; \mu)$ and $H_0^{1,p}(\Omega; \mu)$ coincide; this means that functions from $H^{1,p}(\Omega; \mu)$ can be approximated in $H^{1,p}(\Omega; \mu)$ by functions from $C_0^\infty(\Omega)$. In Chapter 2 we characterize those Ω for which $H^{1,p}(\Omega; \mu) = H_0^{1,p}(\Omega; \mu)$. The next theorem is the first result in this direction.

1.27.   Theorem. $H^{1,p}(\mathbf{R}^n; \mu) = H_0^{1,p}(\mathbf{R}^n; \mu)$.

PROOF: Let $u \in H^{1,p}(\mathbf{R}^n; \mu)$. For $j = 1, 2, \ldots$ let $A_j = B(0, j+1) \setminus \overline{B}(0, j-1)$ and choose functions $\varphi_j \in C_0^\infty(A_j)$, $0 \le \varphi_j \le 1$, such that

$$\sum_{j=1}^\infty \varphi_j(x) = 1$$

for each $x \in \mathbf{R}^n$; i.e. φ_j is a partition of unity subordinate to the covering A_j. Then $u\varphi_j \in H_0^{1,p}(A_j; \mu)$. Fix $\varepsilon > 0$ and choose $\psi_j \in C_0^\infty(A_j)$ with

$$\|\psi_j - u\varphi_j\|_{1,p} < 2^{-(j+1)}\varepsilon.$$

Then $\psi = \sum_j \psi_j$ is in $C^\infty(\mathbf{R}^n)$ and

$$\|\psi - u\|_{1,p} \le \|\sum_j \psi_j - u \sum_j \varphi_j\|_{1,p} < \varepsilon/2.$$

Now choose j_0 such that

$$\left(\int_{\complement B(0, j_0)} |\psi|^p \, d\mu\right)^{1/p} + \left(\int_{\complement B(0, j_0)} |\nabla\psi|^p \, d\mu\right)^{1/p} < \varepsilon/2.$$

Then $\eta = \sum_{k=1}^{j_0} \psi_k$ is in $C_0^\infty(\mathbf{R}^n)$ and

$$\|\eta - \psi\|_{1,p} \le \left(\int_{\complement B(0, j_0)} |\psi|^p \, d\mu\right)^{1/p} + \left(\int_{\complement B(0, j_0)} |\nabla\psi|^p \, d\mu\right)^{1/p} < \varepsilon/2.$$

Hence

$$\|\eta - u\|_{1,p} < \varepsilon$$

which shows that $u \in H_0^{1,p}(\mathbf{R}^n; \mu)$. The theorem is proved. $\square$

1.28. Weak compactness in Sobolev spaces

We discuss the important fact that $H^{1,p}(\Omega;\mu)$ and $H_0^{1,p}(\Omega;\mu)$ are sequentially weakly compact. The assumption $p > 1$, which is necessary for most of this book, is used here.

Recall that a sequence $u_j \in L^p(\Omega;\mu)$ *converges weakly* in $L^p(\Omega;\mu)$ to a function $u \in L^p(\Omega;\mu)$ if

$$\int_\Omega v\,u_j\,d\mu \to \int_\Omega v\,u\,d\mu$$

whenever $v \in L^{p/(p-1)}(\Omega;\mu)$. The weak convergence of vector-valued functions in $L^p(\Omega;\mu;\mathbf{R}^n)$ has an obvious interpretation in terms of the coordinate functions. For more on weak convergence, see e.g. the monographs by Hewitt and Stromberg (1965) and Yosida (1980).

Next we record Mazur's lemma (Yosida 1980, p. 120).

1.29. The Mazur lemma. *If X is a normed space and x_j converges weakly in X to x, then there exists a sequence $\tilde{x}_j$ of convex combinations of x_j,*

$$\tilde{x}_k = \sum_{j=1}^k \lambda_{k,j} x_j\,, \quad \lambda_{k,j} \ge 0\,, \quad \sum_{j=1}^k \lambda_{k,j} = 1\,,$$

such that $\tilde{x}_j \to x$ in the norm topology of X.

1.30. Theorem. *Suppose that $\mathcal{K}$ is a convex and closed subset of $H^{1,p}(\Omega;\mu)$. If $u_j \in \mathcal{K}$ is a sequence and if $u \in L^p(\Omega;\mu)$ and $v \in L^p(\Omega;\mu;\mathbf{R}^n)$ are functions such that $u_j \to u$ weakly in $L^p(\Omega;\mu)$ and $\nabla u_j \to v$ weakly in $L^p(\Omega;\mu)$, then $u \in \mathcal{K}$ and $v = \nabla u$.*

PROOF: First observe that $(u_j, \nabla u_j)$ converges to (u,v) weakly in the normed space $L^p(\Omega;\mu) \times L^p(\Omega;\mu;\mathbf{R}^n)$. The Mazur lemma implies that there is a sequence of convex combinations

$$\tilde{u}_k = \sum_{j=1}^k \lambda_{k,j}(u_j, \nabla u_j)\,, \quad \lambda_{k,j} \ge 0\,, \text{ and } \sum_{j=1}^k \lambda_{k,j} = 1\,,$$

which converges to (u,v) in $L^p(\Omega;\mu) \times L^p(\Omega;\mu;\mathbf{R}^n)$. In particular,

$$\tilde{v}_k = \sum_{j=1}^k \lambda_{k,j} u_j \in \mathcal{K}$$

is a Cauchy sequence in $H^{1,p}(\Omega;\mu)$, and hence there exists a limit function

$$\tilde{u} = \lim \tilde{v}_k \in \mathcal{K}.$$

It follows that $\tilde{u} = u$ and $v = \nabla \tilde{u} = \nabla u$ as required. $\square$

1.31. Theorem. *Suppose that u_j is a bounded sequence in $H^{1,p}(\Omega;\mu)$.
Then there is a subsequence u_{j_i} and a function $u \in H^{1,p}(\Omega;\mu)$ such that
$u_{j_i} \to u$ weakly in $L^p(\Omega;\mu)$ and $\nabla u_{j_i} \to \nabla u$ weakly in $L^p(\Omega;\mu)$. Moreover,
if $u_j \in H_0^{1,p}(\Omega;\mu)$, then $u \in H_0^{1,p}(\Omega;\mu)$.*

PROOF: The existence of weakly convergent subsequences u_{j_i} and ∇u_{j_i}
is a standard theorem of functional analysis (Yosida 1980, Theorem 1, p.
126). An appeal to Theorem 1.30 completes the proof.   □

1.32. Theorem. *Suppose that u_j is a bounded sequence in $H^{1,p}(\Omega;\mu)$
such that $u_j \to u$ pointwise a.e. Then $u \in H^{1,p}(\Omega;\mu)$, $u_j \to u$ weakly
in $L^p(\Omega;\mu)$, and $\nabla u_j \to \nabla u$ weakly in $L^p(\Omega;\mu)$. Furthermore, if $u_j \in
H_0^{1,p}(\Omega;\mu)$, then $u \in H_0^{1,p}(\Omega;\mu)$.*

PROOF: That u_j converges to u weakly in $L^p(\Omega;\mu)$ is standard (Hewitt and
Stromberg 1965, Theorem 13.44). Thus, by Theorems 1.30 and 1.31, $u \in
H^{1,p}(\Omega;\mu)$ (or $u \in H_0^{1,p}(\Omega;\mu)$ if $u_j \in H_0^{1,p}(\Omega;\mu)$) and a subsequence ∇u_{j_i}
converges weakly to ∇u in $L^p(\Omega;\mu)$. Since the weak limit is independent
of the choice of the subsequence, it follows that $\nabla u_j \to \nabla u$ weakly in
$L^p(\Omega;\mu)$, as desired.   □

The next lemma concerning weak compactness in $L_0^{1,p}(\Omega;\mu)$ is sometimes
useful.

1.33. Lemma. *Suppose that $u_j \in L_0^{1,p}(\Omega;\mu)$ is a sequence converging
to u a.e. If the sequence ∇u_j is bounded in $L^p(\Omega;\mu)$ and if for each open
$D \Subset \Omega$ the sequence u_j is bounded in $H^{1,p}(D;\mu)$, then $u \in L_0^{1,p}(\Omega;\mu)$ and
$\nabla u_j \to \nabla u$ weakly in $L^p(\Omega;\mu)$.*

PROOF: Theorem 1.32 implies that $u \in H_{loc}^{1,p}(\Omega;\mu)$ and that ∇u_j converges
weakly to ∇u in $L^p(D;\mu)$ whenever $D \Subset \Omega$. But the sequence ∇u_j is
bounded in $L^p(\Omega;\mu)$ and hence it has a weakly converging subsequence.
This subsequence is easily seen to converge weakly to ∇u which also yields
that $u \in L^{1,p}(\Omega;\mu)$. Moreover, the weak limit being independent of the
subsequence, we infer that $\nabla u_j \to \nabla u$ weakly in $L^p(\Omega;\mu)$. That u belongs
to $L_0^{1,p}(\Omega;\mu)$ follows from the Mazur lemma: we may pick a sequence
$v_j \in L_0^{1,p}(\Omega;\mu)$ of convex combinations of u_j such that $\nabla v_j \to \nabla u$ in
$L^p(\Omega;\mu)$. Then for $\varepsilon > 0$ we choose an index j and a function $\varphi \in C_0^\infty(\Omega)$
such that

$$\int_\Omega |\nabla v_j - \nabla u|^p \, d\mu < (\frac{\varepsilon}{2})^p$$

and

$$\int_\Omega |\nabla v_j - \nabla\varphi|^p \, d\mu < (\frac{\varepsilon}{2})^p \, ;$$

hence
$$(\int_\Omega |\nabla\varphi - \nabla u|^p \, d\mu)^{1/p} < \varepsilon ,$$

as desired.   [

NOTES TO CHAPTER 1. The weighted Poincaré and Sobolev inequalities have been studied rather extensively in recent years. For an earlier work, see Fabes *et al.* (1982a), Maz'ya (1985), Stredulinsky (1984), and the references therein. The problem of characterizing admissible weights continues to be under active research and for the latest developments we refer to Chanillo and Wheeden (1985, In press), Franchi and Serapioni (1987), Chiarenza *et al.* (1989), Sawyer and Wheeden (In press), Franchi (1991), and the numerous references in these articles. Much of the recent work concerns inequalities involving two weight functions.

The idea to impose the four conditions **I–IV** on a weight w arises from the paper by Fabes *et al.* (1982a), where it was observed that these properties suffice in employing Moser's iteration technique to obtain local regularity for solutions.

There are many textbooks where Sobolev spaces have been treated. These include Adams (1975), Gilbarg and Trudinger (1983), Kufner *et al.* (1977), Maz'ya (1985), Morrey (1966), Reshetnyak (1989), Stein (1970), and Ziemer (1989). Weighted Sobolev spaces are studied in Kufner (1985) and Stredulinsky (1984); see also Miller (1982) and Chabrowski (1991).

2
Capacity

The concept of capacity is indispensable to an understanding of the local behavior of functions in a Sobolev space. In a sense, capacity takes the place of measure in Egorov and Lusin type theorems for Sobolev functions, as we shall see in Chapter 4. Various capacity estimates also play a decisive role in studies of solutions to partial differential equations. In the present chapter we develop the theory of variational capacity connected with the weighted Sobolev space $H^{1,p}(\Omega; \mu)$.

We conform to the notation of Chapter 1. In particular, we assume that w is a p-admissible weight and that the measure μ is obtained by $d\mu(x) = w(x)\,dx$.

DEFINITION. Suppose that K is a compact subset of Ω. Let

$$W(K, \Omega) = \{u \in C_0^\infty(\Omega) : u \geq 1 \text{ on } K\}$$

and define

$$\operatorname{cap}_{p,\mu}(K, \Omega) = \inf_{u \in W(K,\Omega)} \int_\Omega |\nabla u|^p \, d\mu.$$

Further, if $U \subset \Omega$ is open, set

$$\operatorname{cap}_{p,\mu}(U, \Omega) = \sup_{K \subset U \text{ compact}} \operatorname{cap}_{p,\mu}(K, \Omega),$$

and, finally, for an arbitrary set $E \subset \Omega$

$$\operatorname{cap}_{p,\mu}(E, \Omega) = \inf_{\substack{E \subset U \subset \Omega \\ U \text{ open}}} \operatorname{cap}_{p,\mu}(U, \Omega).$$

The number $\operatorname{cap}_{p,\mu}(E, \Omega) \in [0, \infty]$ is called the *(variational)* (p, μ)-*capacity* of the *condenser* (E, Ω).

Clearly, $\operatorname{cap}_{p,\mu}(E, \Omega) < \infty$ if $E \Subset \Omega$. There is no ambiguity in having two different definitions for the (p, μ)-capacity of a condenser (K, Ω) when K is compact, for we soon show that they give the same number $\operatorname{cap}_{p,\mu}(K, \Omega)$; see Theorem 2.2(i) and (iv).

Using an approximation, we observe that the set $W(K, \Omega)$ in the definition above can be replaced by a larger set

$$W_0(K, \Omega) = \{u \in H_0^{1,p}(\Omega; \mu) \cap C(\Omega) : u \geq 1 \text{ on } K\}$$

without affecting $\text{cap}_{p,\mu}(K,\Omega)$. Indeed, let $u \in W_0(K,\Omega)$; we may clearly assume that $u = 1$ in a neighborhood $U \Subset \Omega$ of K. Then choose a cut-off function $\eta \in C^\infty(\mathbf{R}^n)$ such that $\eta = 1$ on $\Omega \setminus U$ and $\eta = 0$ in a neighborhood of K. Now, if $\varphi_j \in C_0^\infty(\Omega)$ is a sequence converging to u in $H_0^{1,p}(\Omega;\mu)$, then

$$\psi_j = 1 - \eta(1 - \varphi_j)$$

belongs to $W(K,\Omega)$, and it also converges to u in $H_0^{1,p}(\Omega;\mu)$. This establishes the assertion.

It is also useful to observe that if $\psi \in H^{1,p}(\Omega;\mu)$ is such that $\varphi - \psi \in H_0^{1,p}(\Omega \setminus K;\mu)$ for some $\varphi \in W_0(K,\Omega)$, then

$$\text{cap}_{p,\mu}(K,\Omega) \le \int_\Omega |\nabla\psi|^p \, d\mu \, .$$

The functions in $W_0(K,\Omega)$ are said to be *admissible* for the condenser (K,Ω). It is also clear that the same capacity is achieved if one requires that admissible functions u satisfy $0 \le u \le 1$; when convenient, we tacitly assume this extra condition.

If μ is Lebesgue measure, we call the (p,μ)-capacity simply the *p-capacity* and write

$$\text{cap}_{p,dx}(E,\Omega) = \text{cap}_p(E,\Omega) \, .$$

2.1. Basic properties of (p,μ)-capacity

In abstract theories, a capacity is a monotone, subadditive set function. The following theorem expresses, among other things, that this is true for the (p,μ)-capacity.

2.2. Theorem. *The set function* $E \mapsto \text{cap}_{p,\mu}(E,\Omega)$, $E \subset \Omega$, *enjoys the following properties:*

  (i) *If* $E_1 \subset E_2$, *then* $\text{cap}_{p,\mu}(E_1,\Omega) \le \text{cap}_{p,\mu}(E_2,\Omega)$.

  (ii) *If* $\Omega_1 \subset \Omega_2$ *are open and* $E \subset \Omega_1$, *then*

$$\text{cap}_{p,\mu}(E,\Omega_2) \le \text{cap}_{p,\mu}(E,\Omega_1) \, .$$

  (iii) *If* K_1 *and* K_2 *are compact subsets of* Ω, *then*

$$\text{cap}_{p,\mu}(K_1 \cup K_2,\Omega) + \text{cap}_{p,\mu}(K_1 \cap K_2,\Omega) \le \text{cap}_{p,\mu}(K_1,\Omega) + \text{cap}_{p,\mu}(K_2,\Omega) \, .$$

  (iv) *If* K_i *is a decreasing sequence of compact subsets of* Ω *with* $K = \bigcap_i K_i$, *then*

$$\text{cap}_{p,\mu}(K,\Omega) = \lim_{i \to \infty} \text{cap}_{p,\mu}(K_i,\Omega) \, .$$

(v) If $E_1 \subset E_2 \subset \cdots \subset \bigcup_i E_i = E \subset \Omega$, then

$$\operatorname{cap}_{p,\mu}(E, \Omega) = \lim_{i \to \infty} \operatorname{cap}_{p,\mu}(E_i, \Omega).$$

(vi) If $E = \bigcup_i E_i \subset \Omega$, then

$$\operatorname{cap}_{p,\mu}(E, \Omega) \leq \sum_i \operatorname{cap}_{p,\mu}(E_i, \Omega).$$

PROOF: Properties (i) and (ii) are immediate consequences of the definition.

To prove (iii), we note that if $u_1 \in W(K_1, \Omega)$ and $u_2 \in W(K_2, \Omega)$, then

$$\int_\Omega |\nabla \max(u_1, u_2)|^p \, d\mu + \int_\Omega |\nabla \min(u_1, u_2)|^p \, d\mu$$
$$= \int_\Omega |\nabla u_1|^p \, d\mu + \int_\Omega |\nabla u_2|^p \, d\mu$$

(Theorem 1.20). Since $\max(u_1, u_2)$ and $\min(u_1, u_2)$ are admissible for the condensers $(K_1 \cup K_2, \Omega)$ and $(K_1 \cap K_2, \Omega)$, respectively, we have that

$$\operatorname{cap}_{p,\mu}(K_1 \cup K_2, \Omega) + \operatorname{cap}_{p,\mu}(K_1 \cap K_2, \Omega) \leq \int_\Omega |\nabla u_1|^p \, d\mu + \int_\Omega |\nabla u_2|^p \, d\mu.$$

By taking the infimum over all admissible functions u_1, u_2, we arrive at (iii).

To prove (iv), fix a small $\varepsilon > 0$ and pick a function u from $W(K, \Omega)$ such that

$$\int_\Omega |\nabla u|^p \, d\mu < \operatorname{cap}_{p,\mu}(K, \Omega) + \varepsilon.$$

When i is large, the sets K_i lie in the compact set $\{ u \geq 1 - \varepsilon \}$; therefore

$$b = \lim_{i \to \infty} \operatorname{cap}_{p,\mu}(K_i, \Omega) \leq \operatorname{cap}_{p,\mu}(\{ u \geq 1 - \varepsilon \}, \Omega)$$
$$\leq (1 - \varepsilon)^{-p} \int_\Omega |\nabla u|^p \, d\mu.$$

Letting $\varepsilon \to 0$ yields $b \leq \operatorname{cap}_{p,\mu}(K, \Omega)$, whence (iv) follows because obviously $b \geq \operatorname{cap}_{p,\mu}(K, \Omega)$.

To prove (v) and (vi) we require the following lemma.

2.3. Lemma. *Suppose that $E_1, \ldots, E_k \subset \Omega$. Then*

(2.4)

$$\text{cap}_{p,\mu}\left(\bigcup_{i=1}^{k} E_i, \Omega\right) - \text{cap}_{p,\mu}\left(\bigcup_{i=1}^{k} F_i, \Omega\right) \leq \sum_{i=1}^{k}\left(\text{cap}_{p,\mu}(E_i, \Omega) - \text{cap}_{p,\mu}(F_i, \Omega)\right)$$

whenever $F_i \subset E_i$, $i = 1, 2, \ldots, k$, and $\text{cap}_{p,\mu}\left(\bigcup_{i=1}^{k} F_i, \Omega\right) < \infty$.

PROOF: First note that if C, K, and F are compact subsets of Ω with $C \subset K$, then it follows from (i) and (iii) of Theorem 2.2 that

$$\text{cap}_{p,\mu}(K \cup F, \Omega) + \text{cap}_{p,\mu}(C, \Omega)$$
$$\leq \text{cap}_{p,\mu}(K \cup (C \cup F), \Omega) + \text{cap}_{p,\mu}(K \cap (C \cup F), \Omega)$$
$$\leq \text{cap}_{p,\mu}(K, \Omega) + \text{cap}_{p,\mu}(C \cup F, \Omega)$$

or

$$\text{cap}_{p,\mu}(K \cup F, \Omega) - \text{cap}_{p,\mu}(C \cup F, \Omega) \leq \text{cap}_{p,\mu}(K, \Omega) - \text{cap}_{p,\mu}(C, \Omega).$$

Repeating this we obtain (2.4) in the case where $E_i = K_i$ and $F_i = C_i$ are compact. Indeed, by induction

$$\text{cap}_{p,\mu}\left(\bigcup_{i=1}^{k} K_i, \Omega\right) - \text{cap}_{p,\mu}\left(\bigcup_{i=1}^{k} C_i, \Omega\right)$$
$$= \text{cap}_{p,\mu}\left(\bigcup_{i=1}^{k-1} K_i \cup K_k, \Omega\right) - \text{cap}_{p,\mu}\left(\bigcup_{i=1}^{k-1} C_i \cup K_k, \Omega\right)$$
$$\quad + \text{cap}_{p,\mu}\left(K_k \cup \bigcup_{i=1}^{k-1} C_i, \Omega\right) - \text{cap}_{p,\mu}\left(C_k \cup \bigcup_{i=1}^{k-1} C_i, \Omega\right)$$
$$\leq \sum_{i=1}^{k-1}\left(\text{cap}_{p,\mu}(K_i, \Omega) - \text{cap}_{p,\mu}(C_i, \Omega)\right)$$
$$\quad + \text{cap}_{p,\mu}(K_k, \Omega) - \text{cap}_{p,\mu}(C_k, \Omega)$$
$$= \sum_{i=1}^{k}\left(\text{cap}_{p,\mu}(K_i, \Omega) - \text{cap}_{p,\mu}(C_i, \Omega)\right)$$

as required.

To establish (2.4) for open sets E_i and F_i, we need the following elementary fact: *if $K \subset \cup_{i=1}^{k} E_i$ and $C_i \subset F_i$ are compact sets with $\cup_{i=1}^{k} C_i \subset K$, then the compact set*

$$K_i = K \setminus \bigcup_{\substack{j=1 \\ j \neq i}}^{k} E_j$$

is a subset of E_i and it contains C_i. Moreover, $K = \cup_{i=1}^{k} K_i$. Using this result we can easily conclude that (2.4) for open sets follows from (2.4) for compacta. Similarly, once established for open sets, estimate (2.4) follows for arbitrary sets $E_i \supset F_i$. This proves Lemma 2.3.   □

Now we are ready to prove property (v) of Theorem 2.2. By the monotonicity in (i) only the inequality

$$\mathrm{cap}_{p,\mu}(E, \Omega) \leq \lim_{i \to \infty} \mathrm{cap}_{p,\mu}(E_i, \Omega)$$

requires a proof. Moreover, we are free to assume that $\mathrm{cap}_{p,\mu}(E_i, \Omega)$ is finite for each i. Fix $\varepsilon > 0$ and choose open U_i such that $E_i \subset U_i \subset \Omega$ and

$$\mathrm{cap}_{p,\mu}(U_i, \Omega) \leq \mathrm{cap}_{p,\mu}(E_i, \Omega) + \varepsilon\, 2^{-i}.$$

Since $\mathrm{cap}_{p,\mu}(\bigcup_{i=1}^{k} E_i, \Omega) = \mathrm{cap}_{p,\mu}(E_k, \Omega) < \infty$ for each k, it follows from Lemma 2.3 that

$$\mathrm{cap}_{p,\mu}(\bigcup_{i=1}^{k} U_i, \Omega) - \mathrm{cap}_{p,\mu}(\bigcup_{i=1}^{k} E_i, \Omega) \leq \sum_{i=1}^{k} \varepsilon\, 2^{-i} < \varepsilon.$$

Thus if $K \subset \bigcup_{i=1}^{\infty} U_i$ is compact, then $K \subset \bigcup_{i=1}^{k} U_i$ for some k, and we have

$$\mathrm{cap}_{p,\mu}(K, \Omega) \leq \mathrm{cap}_{p,\mu}(\bigcup_{i=1}^{k} U_i, \Omega) \leq \mathrm{cap}_{p,\mu}(\bigcup_{i=1}^{k} E_i, \Omega) + \varepsilon$$

$$\leq \lim_{k \to \infty} \mathrm{cap}_{p,\mu}(E_k, \Omega) + \varepsilon.$$

Hence

$$\mathrm{cap}_{p,\mu}(E, \Omega) \leq \mathrm{cap}_{p,\mu}(\bigcup_{i=1}^{\infty} U_i, \Omega)$$

$$= \sup_{K} \mathrm{cap}_{p,\mu}(K, \Omega)$$

$$\leq \lim_{k \to \infty} \mathrm{cap}_{p,\mu}(E_k, \Omega) + \varepsilon,$$

where the supremum is taken over all compact sets $K \subset \bigcup_{i=1}^{\infty} U_i$. So (v) is proved.

It remains to establish that (vi) is a consequence of (v). To this end, note that Lemma 2.3 implies a finite version of (vi):

$$\mathrm{cap}_{p,\mu}(\bigcup_{i=1}^{k} E_i, \Omega) \leq \sum_{i=1}^{k} \mathrm{cap}_{p,\mu}(E_i, \Omega).$$

Since $\bigcup_{i=1}^{k} E_i$ increases to $\bigcup_{i=1}^{\infty} E_i$, an application of (v) yields (vi). Theorem 2.2 is proved.   □

Any set function which is defined in the family of all subsets of Ω and which satisfies the properties (i), (iv), and (v) in Theorem 2.2 is called a *Choquet capacity* (relative to Ω). We may thus invoke an important capacitability theorem of Choquet and state

2.5. Theorem. *The set function $E \mapsto \mathrm{cap}_{p,\mu}(E, \Omega)$, $E \subset \Omega$, is a Choquet capacity. In particular, all Borel (in fact, all analytic) subsets E of Ω are capacitable, i.e.*

$$\mathrm{cap}_{p,\mu}(E, \Omega) = \sup_{\substack{K \subset E \text{ compact}}} \mathrm{cap}_{p,\mu}(K, \Omega).$$

We do not present the proof of Choquet's theorem here but refer to the discussion in Doob (1984, Appendix II) or Helms (1969, p. 149).

If $G \Subset \Omega$ is an open set, then it is easy to show by examples that in general $\mathrm{cap}_{p,\mu}(G, \Omega)$ differs from $\mathrm{cap}_{p,\mu}(\overline{G}, \Omega)$. However, if B is a ball with $\overline{B} \subset \Omega$, then

$$\mathrm{cap}_{p,\mu}(\overline{B}, \Omega) = \mathrm{cap}_{p,\mu}(B, \Omega).$$

We show in Chapter 6 that this equality follows from a Wiener type boundary estimate; see (6.40).

On the other hand, the definition of the (p, μ)-capacity readily implies that

$$\mathrm{cap}_{p,\mu}(K, \Omega) = \mathrm{cap}_{p,\mu}(\partial K, \Omega)$$

whenever K is a compact set in Ω.

The following property of capacities is often useful.

2.6. Theorem. *If $E_1 \subset \Omega_1 \subset E_2 \subset \Omega_2 \subset \cdots \subset \Omega = \cup_i \Omega_i$, then*

$$\mathrm{cap}_{p,\mu}(E_1, \Omega) \leq \Big(\sum_{i=1}^{\infty} \mathrm{cap}_{p,\mu}(E_i, \Omega_i)^{1/(1-p)} \Big)^{1-p}.$$

PROOF: There is no loss of generality in assuming that $\mathrm{cap}_{p,\mu}(E_1, \Omega_1) < \infty$. Fix an integer j. Then fix $\varepsilon > 0$ and choose an open set $U \subset \Omega_1$ such that $E_1 \subset U$ and

$$\mathrm{cap}_{p,\mu}(U, \Omega_1) \leq \mathrm{cap}_{p,\mu}(E_1, \Omega_1) + \frac{\varepsilon}{2}.$$

Let $K_1 \subset U$ be compact and let $u_1 \in W(K_1, \Omega_1)$ be such that

$$\int_{\Omega_1} |\nabla u_1|^p \, d\mu \leq \mathrm{cap}_{p,\mu}(K_1, \Omega_1) + \frac{\varepsilon}{2}.$$

Choose functions $u_i \in C_0^\infty(\Omega)$, $i = 2, 3, \ldots, j$, inductively such that $u_i \in W(K_i, \Omega_i)$, where $K_i = \operatorname{spt} u_{i-1}$, and that

$$\int_{\Omega_i} |\nabla u_i|^p \, d\mu \leq \operatorname{cap}_{p,\mu}(K_i, \Omega_i) + \frac{\varepsilon}{2}.$$

Let a_i be a sequence of nonnegative numbers with $\sum_{i=1}^{j} a_i = 1$ and put

$$v = \sum_{i=1}^{j} a_i u_i.$$

Then $v \in W(K_1, \Omega)$, and hence

$$\operatorname{cap}_{p,\mu}(K_1, \Omega) \leq \int_\Omega |\nabla v|^p \, d\mu = \sum_{i=1}^{j} a_i^p \int_{\Omega_i} |\nabla u_i|^p \, d\mu;$$

note that the sets where $\nabla u_i \neq 0$ are pairwise disjoint. Since $K_i \subset \Omega_{i-1} \subset E_i$, for $i = 2, 3, \ldots$, we have

$$\operatorname{cap}_{p,\mu}(K_1, \Omega) \leq \sum_{i=1}^{j} a_i^p \operatorname{cap}_{p,\mu}(K_i, \Omega_i) + \frac{\varepsilon}{2}$$

$$\leq a_1^p \operatorname{cap}_{p,\mu}(U, \Omega_1) + \sum_{i=2}^{j} a_i^p \operatorname{cap}_{p,\mu}(E_i, \Omega_i) + \frac{\varepsilon}{2}$$

$$\leq \sum_{i=1}^{j} a_i^p \operatorname{cap}_{p,\mu}(E_i, \Omega_i) + \varepsilon.$$

It follows that

$$\operatorname{cap}_{p,\mu}(E_1, \Omega) \leq \sum_{i=1}^{j} a_i^p \operatorname{cap}_{p,\mu}(E_i, \Omega_i).$$

If $\operatorname{cap}_{p,\mu}(E_i, \Omega_i) > 0$ for all $i = 1, 2, \ldots, j$, choose

$$a_i = \operatorname{cap}_{p,\mu}(E_i, \Omega_i)^{1/(1-p)} \Big(\sum_{k=1}^{j} \operatorname{cap}_{p,\mu}(E_k, \Omega_k)^{1/(1-p)} \Big)^{-1}$$

and obtain

$$\operatorname{cap}_{p,\mu}(E_1, \Omega) \leq \Big(\sum_{i=1}^{j} \operatorname{cap}_{p,\mu}(E_i, \Omega_i)^{1/(1-p)} \Big)^{1-p}.$$

If $\operatorname{cap}_{p,\mu}(E_i, \Omega_i) = 0$ for some i, then $\operatorname{cap}_{p,\mu}(E_1, \Omega) = 0$ as well, and the estimate above holds trivially. The conclusion follows by letting $j \to \infty$.

□

2.7. Sets of (p, μ)-capacity zero

A set E in $\mathbf{R}^n$ is said to be of (p, μ)-*capacity zero*, or to *have* (p, μ)-*capacity zero*, if

$$\mathrm{cap}_{p,\mu}(E \cap \Omega, \Omega) = 0$$

for all open $\Omega \subset \mathbf{R}^n$. In this case we write $\mathrm{cap}_{p,\mu} E = 0$. If E is not of (p, μ)-capacity zero, we write $\mathrm{cap}_{p,\mu} E > 0$. We make the simple observation that any set E of (p, μ)-capacity zero is contained in a Borel set $\tilde{E}$ of (p, μ)-capacity zero; in fact, we can choose $\tilde{E}$ to be a G_δ-set, i.e. a countable intersection of open sets.

We also say that a property holds (p, μ)-*quasieverywhere*, often abbreviated (p, μ)-q.e. or simply q.e., if it holds except on a set of (p, μ)-capacity zero.

In potential theory the sets of zero capacity are in many ways negligible, and therefore it is important to know whether or not a given set E has (p, μ)-capacity zero. The following lemma is an immediate consequence of Theorem 2.2(vi); it shows that we may always assume E to be bounded if we want to show that $\mathrm{cap}_{p,\mu} E = 0$.

2.8. Lemma. *A countable union of sets of (p, μ)-capacity zero has (p, μ)-capacity zero.*

The next lemma shows that, if E is bounded, one needs to test only a single bounded open set Ω containing E in showing that E has zero (p, μ)-capacity.

2.9. Lemma. *Suppose that E is bounded and that there is a bounded neighborhood Ω of E with $\mathrm{cap}_{p,\mu}(E, \Omega) = 0$. Then E is of (p, μ)-capacity zero.*

PROOF: Let Ω' be an open set. Since there is a G_δ-set $\tilde{E} \subset \Omega$ such that $E \subset \tilde{E}$ and $\mathrm{cap}_{p,\mu}(\tilde{E}, \Omega) = \mathrm{cap}_{p,\mu}(E, \Omega) = 0$, we may assume that E itself is a Borel set. Thus by invoking Choquet's capacitability theorem 2.5, we may assume that $E \cap \Omega'$ is compact. If $u \in W(E \cap \Omega', \Omega)$ and $v \in W(E \cap \Omega', \Omega')$, then $uv \in W(E \cap \Omega', \Omega')$, and hence

$$\mathrm{cap}_{p,\mu}(E \cap \Omega', \Omega') \leq 2^p \max |v|^p \int_\Omega |\nabla u|^p \, d\mu + 2^p \max |\nabla v|^p \int_\Omega |u|^p \, d\mu$$
$$\leq c \int_\Omega |\nabla u|^p \, d\mu \, ;$$

the Poincaré inequality (1.5) is also used here. Since the last integral can be chosen to be as small as we please, we have $\mathrm{cap}_{p,\mu}(E \cap \Omega', \Omega') = 0$ as desired.   □

It easily follows from the Poincaré inequality that each set of (p, μ)-capacity zero has μ-measure zero, hence Lebesgue measure zero; the converse is not true since, for example, sets of (p, μ)-capacity zero do not separate the space $\mathbf{R}^n$ (see Lemma 2.46).

2.10. Lemma. *If* $\mathrm{cap}_{p,\mu} E = 0$, *then* $\mu(E) = 0$ *and hence* $|E| = 0$.

PROOF: Let Ω be a bounded open set and fix $\varepsilon > 0$. Since $\mathrm{cap}_{p,\mu}(E \cap \Omega, \Omega) = 0$, we find an open neighborhood $U \subset \Omega$ of $E \cap \Omega$ such that $\mathrm{cap}_{p,\mu}(U, \Omega) < \varepsilon$. Let $K \subset U$ be a compact set and choose a function $\varphi \in W(K, \Omega)$ such that

$$\int_\Omega |\nabla\varphi|^p \, d\mu < \mathrm{cap}_{p,\mu}(K, \Omega) + \varepsilon < 2\varepsilon \, .$$

It follows from the Poincaré inequality that

$$\mu(K) \leq \int_\Omega |\varphi|^p \, d\mu \leq c \int_\Omega |\nabla\varphi|^p \, d\mu < c\varepsilon \, ,$$

and hence $\mu(U) \leq c\varepsilon$, where $c = c(p, c_\mu, \mathrm{diam}\,\Omega) > 0$. In conclusion, $\mu(E) = 0$ as desired. $\qquad\qquad\square$

2.11. Estimates for capacities

Only a few capacities can be computed explicitly. We next do this for the spherical condenser $(\overline{B}(x_0, r), B(x_0, R))$ in the unweighted case $d\mu = dx$.

2.12. EXAMPLE. If $0 < r < R < \infty$, then

(2.13)
$$\mathrm{cap}_p(\overline{B}(x_0, r), B(x_0, R)) = \begin{cases} \omega_{n-1}\left(\frac{|n-p|}{p-1}\right)^{p-1} |R^{\frac{p-n}{p-1}} - r^{\frac{p-n}{p-1}}|^{1-p} & p \neq n \\ \omega_{n-1}(\log\frac{R}{r})^{1-n} & p = n \, . \end{cases}$$

In particular,

$$\mathrm{cap}_p(\overline{B}(x_0, r), B(x_0, 2r)) = c_1(n, p) r^{n-p}$$

for all $p > 1$,

$$\mathrm{cap}_p(\overline{B}(x_0, r), \mathbf{R}^n) = c_2(n, p) r^{n-p}$$

for $1 < p < n$, and

$$\mathrm{cap}_p(\{x_0\}, B(x_0, r)) = c_3(n, p) r^{n-p}$$

for $p > n$. It is also clear from (2.13) that

$$\text{cap}_p(B(x_0, r), B(x_0, R)) = \text{cap}_p(\overline{B}(x_0, r), B(x_0, R)).$$

To prove (2.13), pick $u \in C_0^\infty(B(x_0, R))$ such that $u = 1$ on $\overline{B}(x_0, r)$. Then for each $y \in \partial B(0, 1)$,

$$1 \leq \int_r^R |\frac{d}{ds} u(sy)| \, ds \leq \int_r^R |\nabla u(sy)| \, ds$$

$$\leq (\int_r^R s^{(1-n)/(p-1)} \, ds)^{(p-1)/p} (\int_r^R |\nabla u(sy)|^p s^{n-1} \, ds)^{1/p}$$

by the Hölder inequality. This implies

$$1 \leq A^{p-1} \int_r^R |\nabla u(sy)|^p s^{n-1} \, ds,$$

where

$$A = \begin{cases} |\frac{p-1}{n-p}| |R^{(p-n)/(p-1)} - r^{(p-n)/(p-1)}| & p \neq n \\ \log \frac{R}{r} & p = n. \end{cases}$$

By integrating with respect to y, we obtain

$$\omega_{n-1} \leq A^{p-1} \int_{\partial B(0,1)} \int_r^R |\nabla u(sy)|^p s^{n-1} \, ds \, dy$$

$$= A^{p-1} \int_{B(x_0, R)} |\nabla u(x)|^p \, dx,$$

and thus

$$\text{cap}_p(\overline{B}(x_0, r), B(x_0, R)) \geq \omega_{n-1} A^{1-p}.$$

To establish the reverse inequality, we notice that the function

$$u(x) = \begin{cases} \dfrac{\int_{|x|}^R t^{(1-n)/(p-1)} \, dt}{\int_r^R t^{(1-n)/(p-1)} \, dt} & \text{if } r < |x| < R \\ 1 & \text{if } 0 \leq |x| \leq r \end{cases}$$

is admissible. A straightforward computation then yields

$$\text{cap}_p(\overline{B}(x_0, r), B(x_0, R)) \leq \int_{B(x_0, R)} |\nabla u(x)|^p \, dx = \omega_{n-1} A^{1-p}$$

as required. This completes the proof of (2.13).

For weighted capacities $\text{cap}_{p,\mu}(B(x_0, r), B(x_0, R))$ no explicit expression similar to (2.13) is known. However, the following rough estimates are usually sufficient for applications.

2.14. Lemma.   *There is a positive constant c, depending only on n, p, and c_μ, such that*

(2.15)
$$\frac{1}{c}\,\mu(B(x_0,r))r^{-p} \leq \text{cap}_{p,\mu}\left(B(x_0,r), B(x_0,2r)\right) \leq c\,\mu(B(x_0,r))r^{-p}.$$

PROOF: Let first $u \in C_0^\infty(B(x_0,2r))$ be a function such that $u = 1$ on $\overline{B}(x_0,r)$ and $|\nabla u| \leq 2/r$; such a function u is easily constructed by the aid of mollifiers (see Section 1.10). Then

$$\text{cap}_{p,\mu}\left(B(x_0,r), B(x_0,2r)\right) \leq \int_{B(x_0,2r)} |\nabla u|^p\,d\mu$$
$$\leq cr^{-p}\mu(B(x_0,2r)) \leq cr^{-p}\mu(B(x_0,r)),$$

where the final step follows from the doubling property. On the other hand, if $0 < r' < r$ and u is admissible for the condenser $(\overline{B}(x_0,r'), B(x_0,2r))$, then we have by the Poincaré inequality that

$$\mu(B(x_0,r')) \leq \int_{B(x_0,2r)} |u|^p\,d\mu \leq c\,r^p \int_{B(x_0,2r)} |\nabla u|^p\,d\mu.$$

Taking the infimum over all such u and letting $r' \to r$, we arrive at the left inequality of (2.15).   □

2.16. Lemma.   *If $E \subset B(x_0,r)$ and $0 < r \leq s \leq 2r$, then*

(2.17)
$$\frac{1}{c}\,\text{cap}_{p,\mu}\left(E, B(x_0,2r)\right) \leq \text{cap}_{p,\mu}\left(E, B(x_0,2s)\right) \leq \text{cap}_{p,\mu}\left(E, B(x_0,2r)\right),$$

where c depends only on n, p, and c_μ.

PROOF: Since the second inequality is trivial, it suffices to verify the first inequality in the extremal case $s = 2r$,

$$\text{cap}_{p,\mu}\left(E, B(x_0,2r)\right) \leq c\,\text{cap}_{p,\mu}\left(E, B(x_0,4r)\right).$$

To achieve this, we may assume that E is compact. Then choose a cut-off function $\eta \in C_0^\infty(B(x_0,2r))$, $0 \leq \eta \leq 1$, such that $\eta = 1$ in $B(x_0,r)$ and $|\nabla \eta| \leq 2/r$. If $u \in C_0^\infty(B(x_0,4r))$ is admissible for the condenser

$(E, B(x_0, 4r))$, then ηu is admissible for the condenser $(E, B(x_0, 2r))$, and we find

$$
\begin{aligned}
&\mathrm{cap}_{p,\mu}\left(E, B(x_0, 2r)\right) \\
&\leq \int_{B(x_0, 2r)} |\nabla(\eta u)|^p \, d\mu \\
&\leq 2^p \int_{B(x_0, 2r)} |\nabla u|^p |\eta|^p \, d\mu + 2^p \int_{B(x_0, 2r)} |u|^p |\nabla \eta|^p \, d\mu \\
&\leq 2^p \int_{B(x_0, 4r)} |\nabla u|^p \, d\mu + c \, r^{-p} \int_{B(x_0, 4r)} |u|^p \, d\mu \\
&\leq c \int_{B(x_0, 4r)} |\nabla u|^p \, d\mu,
\end{aligned}
$$

where the last step follows from the Poincaré inequality. The lemma follows by taking the infimum over all such functions u.   □

2.18. Theorem. *If $0 < r < R$, then*

$$
\begin{aligned}
&\mathrm{cap}_{p,\mu}\left(B(x_0, r), B(x_0, R)\right) \\
&\geq \omega_{n-1}^p \Big(\int_{A(x_0;r,R)} |x - x_0|^{(1-n)p/(p-1)} w(x)^{1/(1-p)} \, dx \Big)^{1-p},
\end{aligned}
$$

where $A(x_0; r, R)$ is the annulus $B(x_0, R) \setminus B(x_0, r)$.

PROOF: Let $u \in C_0^\infty(B(x_0, R))$ be an admissible function for the condenser $(\overline{B}(x_0, r), B(x_0, R))$ such that $u = 1$ on $\overline{B}(x_0, r)$. The standard representation theorem, see e.g. Gilbarg and Trudinger (1983, Lemma 7.14), yields

$$
u(y) = \omega_{n-1}^{-1} \int_{\mathbf{R}^n} \frac{\nabla u(x) \cdot (y - x)}{|x - y|^n} \, dx
$$

for all $y \in \mathbf{R}^n$. Hence

$$
\begin{aligned}
\omega_{n-1}^p = \left(\omega_{n-1} u(x_0)\right)^p &= \Big(\int_{A(x_0;r,R)} \frac{\nabla u(x) \cdot (x_0 - x)}{|x - x_0|^n} \, dx \Big)^p \\
&\leq \int_{B(x_0, R)} |\nabla u(x)|^p \, d\mu \Big(\int_{A(x_0;r,R)} |x - x_0|^{(1-n)p/(p-1)} w(x)^{1/(1-p)} \, dx \Big)^{p-1}.
\end{aligned}
$$

After dividing and taking the infimum over all such u, we arrive, by continuity of the integral, at the desired result.   □

If $w^{1/(1-p)}$ is not integrable in the annulus $A(x_0; r, R)$, the estimate in Theorem 2.18 is trivial; if $w \in A_p$, then it is of right magnitude:

2.19. Theorem. *Suppose that $0 < 2r \leq R$. If $w \in A_p$, then there is a positive constant c, depending only on n, p, and the A_p constant $c_{p,w}$ of w, such that*

$$(2.20) \quad \begin{aligned} &\operatorname{cap}_{p,\mu}\big(B(x_0, r), B(x_0, R)\big) \\ &\leq c \Big(\int_{A(x_0; r, R)} |x - x_0|^{(1-n)p/(p-1)} w(x)^{1/(1-p)} \, dx \Big)^{1-p}, \end{aligned}$$

where $A(x_0; r, R)$ is the annulus $B(x_0, R) \setminus B(x_0, r)$.

Conversely, if (2.20) holds for all $x_0 \in \mathbf{R}^n$ and $R = 2r > 0$, then $w \in A_p$.

PROOF: Suppose first that $w \in A_p$. Then we have by (2.15) that

$$(2.21) \quad \begin{aligned} &\operatorname{cap}_{p,\mu}\big(B(x_0, r), B(x_0, 2r)\big) \\ &\leq c\, r^{n-p} \fint_{B(x_0, 2r)} w(x) \, dx \\ &\leq c\, r^{n-p} \Big(\fint_{B(x_0, 2r)} w(x)^{1/(1-p)} \, dx \Big)^{1-p} \\ &\leq c\, r^{p(n-1)} \Big(\int_{A(x_0; r, 2r)} w(x)^{1/(1-p)} \, dx \Big)^{1-p} \\ &\leq c \Big(\int_{A(x_0; r, 2r)} |x - x_0|^{(1-n)p/(p-1)} w(x)^{1/(1-p)} \, dx \Big)^{1-p}. \end{aligned}$$

Since it follows from Lemma 2.6 that

$$\begin{aligned} &\operatorname{cap}_{p,\mu}\big(B(x_0, r), B(x_0, 2^k r)\big) \\ &\leq \Big(\sum_{j=1}^{k} \operatorname{cap}_{p,\mu}\big(B(x_0, 2^{j-1} r), B(x_0, 2^j r)\big)^{1/(1-p)} \Big)^{1-p} \end{aligned}$$

for all integers $k \geq 2$, we have by (2.21) that

$$\begin{aligned} &\operatorname{cap}_{p,\mu}\big(B(x_0, r), B(x_0, 2^k r)\big) \\ &\leq c \Big(\sum_{j=1}^{k} \int_{A(x_0; 2^{j-1} r, 2^j r)} |x - x_0|^{(1-n)p/(p-1)} w(x)^{1/(1-p)} \, dx \Big)^{1-p} \\ &= c \Big(\int_{A(x_0; r, 2^k r)} |x - x_0|^{(1-n)p/(p-1)} w(x)^{1/(1-p)} \, dx \Big)^{1-p} \\ &\leq c \Big(\int_{A(x_0; r, R)} |x - x_0|^{(1-n)p/(p-1)} w(x)^{1/(1-p)} \, dx \Big)^{1-p} \end{aligned}$$

whenever $k \geq 2$ is an integer with $R < 2^k r$. Now the desired estimate follows since if we fix the integer $k \geq 2$ such that $2^{k-1}r \leq R < 2^k r$, then Lemma 2.16 implies

$$\text{cap}_{p,\mu}\left(B(x_0,r), B(x_0,R)\right) \leq c \, \text{cap}_{p,\mu}\left(B(x_0,r), B(x_0, 2^k r)\right).$$

To prove the second assertion, suppose that estimate (2.20) holds for all radii r and $R = 2r$. Let $B = B(y, r/2)$ be a ball. Choose $x_0 \in \mathbf{R}^n$ with $\text{dist}(x_0, B) = r$. Then the annulus $A(x_0; r, 2r)$ contains B. Consequently, by (2.15) and (2.20) we have

$$\mu\left(B(x_0,r)\right)r^{-p} \leq c \, \text{cap}_{p,\mu}\left(B(x_0,r), B(x_0,2r)\right)$$

$$\leq c \Big(\int_{A(x_0;r,2r)} |x-x_0|^{(1-n)p/(p-1)} w(x)^{1/(1-p)} \, dx\Big)^{1-p}$$

$$\leq c \, r^{p(n-1)} \Big(\int_{A(x_0;r,2r)} w(x)^{1/(1-p)} \, dx\Big)^{1-p}$$

$$\leq c \, r^{p(n-1)} \Big(\int_{B} w(x)^{1/(1-p)} \, dx\Big)^{1-p}.$$

The doubling property of μ now implies

$$\fint_{B} w(x) \, dx \leq c\,\mu\left(B(x_0,2r)\right) r^{-n} \leq c\,\mu(B(x_0,r))r^{-n}$$

$$\leq c \Big(\fint_{B} w(x)^{1/(1-p)} \, dx\Big)^{1-p},$$

and hence $w \in A_p$.   □

It may seem that an estimate of type (2.20) yields a new characterization of A_p-weights. However, a look at the proof reveals the a priori assumption that w be doubling and gives rise to a Poincaré type inequality (1.5).

2.22. EXAMPLE. As a final example we consider the weight $w(x) = |x|^\delta$ for $\delta > -n$. Then w is p-admissible for all $p > 1$ (see Section 1.6). It is clear by Example 2.12 that if $x_0 \neq 0$, then $\text{cap}_{p,\mu}\{x_0\} = 0$ if and only if $1 < p \leq n$. Next we examine when $\text{cap}_{p,\mu}\{0\} = 0$. This will depend not only on p but also on δ.

The most convenient way to estimate the capacities

$$\text{cap}_{p,\mu}(\overline{B}(r), B), \qquad 0 < r < 1,$$

where $d\mu = |x|^\delta dx$, $B(r) = B(0,r)$, and $B = B(0,1)$, is to use some results from the ensuing chapters. More specifically, consider the following weighted p-Laplace equation

$$(2.23) \qquad \Delta_{p,\delta} u(x) = \text{div}(|x|^\delta |\nabla u(x)|^{p-2} \nabla u(x)) = 0.$$

Then a direct computation shows that the functions

$$u_{p,\delta}(x) = |x|^{(p-n-\delta)/(p-1)} - 1, \qquad p - n - \delta \neq 0,$$

and

$$u_{p,\delta}(x) = -\log|x|, \qquad p - n - \delta = 0,$$

are solutions to (2.23) outside the origin. In particular, the functions

$$v_{p,\delta}(x) = \frac{u_{p,\delta}(x)}{u_{p,\delta}(re)}, \qquad |e| = 1,$$

are solutions to (2.23) in $B \setminus \overline{B}(r)$ with $\lim v_{p,\delta}(x) \to 1$ as $|x| \to r$, $\lim v_{p,\delta}(x) \to 0$ as $|x| \to 1$. It follows from Theorem 8.6 that

$$\text{cap}_{p,\mu}(\overline{B}(r), B) = \int_B |\nabla v_{p,\delta}|^p \, d\mu.$$

A computation now gives

$$\text{cap}_{p,\mu}(\overline{B}(r), B) = c(n, p, \delta)(1 - r^{(p-n-\delta)/(p-1)})^{1-p}, \qquad p - n - \delta \neq 0,$$

and

$$\text{cap}_{p,\mu}(\overline{B}(r), B) = \omega_{n-1}\left(\log\frac{1}{r}\right)^{1-n}, \qquad p - n - \delta = 0.$$

Since $\text{cap}_{p,\mu}(\overline{B}(r), B) \to \text{cap}_{p,\mu}(\{0\}, B)$, we observe that $\text{cap}_{p,\mu}\{0\} = 0$ if and only if $p - n - \delta \leq 0$.

2.24. Hausdorff measure and capacity

In this section we examine the relationship between Hausdorff measures and (p, μ)-capacities, mainly in the unweighted case..

Let $s > 0$ and $0 < \delta \leq \infty$. For a set E in $\mathbf{R}^n$ we define

$$\Lambda_s^\delta(E) = \inf \sum_i r_i^s,$$

where the infimum is taken over all coverings of E by balls B_i with diameter r_i not exceeding δ. The number $\Lambda_s^\infty(E)$ is often called the *s-content* of E. Clearly Λ_s^δ is an outer measure, i.e. for all sets $E_1, E_2, \cdots \subset \mathbf{R}^n$ it holds that

(i) $$\Lambda_s^\delta(\emptyset) = 0,$$

(ii) $$\Lambda_s^\delta(E_1) \leq \Lambda_s^\delta(E_2) \quad \text{if } E_1 \subset E_2, \text{ and}$$

(iii) $$\Lambda_s^\delta(\cup_i E_i) \leq \sum_i \Lambda_s^\delta(E_i).$$

However, Λ_s^δ is not an additive measure on any reasonable σ-algebra in $\mathbf{R}^n$ because, in general, it fails to be additive on families of disjoint compact sets. Therefore we define the *s-Hausdorff measure* of E by

$$\Lambda_s(E) = \sup_{\delta > 0} \Lambda_s^\delta(E) = \lim_{\delta \to 0} \Lambda_s^\delta(E).$$

The measure Λ_s is a Borel regular measure; that is, it is an additive measure on Borel sets of $\mathbf{R}^n$ and for each $E \subset \mathbf{R}^n$ there is a Borel set G such that $E \subset G$ and $\Lambda_s(E) = \Lambda_s(G)$.

It is immediate from the definition that $\Lambda_s(E) < \infty$ implies $\Lambda_r(E) = 0$ for all $r > s$. The smallest $t \geq 0$ that satisfies $\Lambda_r(E) = 0$ for all $r > t$ is called the *Hausdorff dimension* of E.

If E is a bounded set, then $\Lambda_s^\delta(E)$ is finite, but $\Lambda_s(E) = \infty$ for example if E has nonempty interior and $0 < s < n$. Nevertheless, Λ_s and Λ_s^δ have the same zero sets.

2.25. Lemma. *Suppose that $0 < \delta \leq \infty$. Then $\Lambda_s(E) = 0$ if and only if $\Lambda_s^\delta(E) = 0$.*

PROOF: Since

$$\Lambda_s(E) \geq \Lambda_s^\delta(E),$$

it suffices to show that $\Lambda_s(E) = 0$ if $\Lambda_s^\delta(E) = 0$. So assume that $\Lambda_s^\delta(E) = 0$ and fix $\varepsilon > 0$. Choose a covering of E by balls B_i with diameter r_i not exceeding δ such that

$$\sum_i r_i^s < \varepsilon^s.$$

Then $r_i < \varepsilon$ and hence

$$\Lambda_s^\varepsilon(E) \leq \sum_i r_i^s < \varepsilon^s.$$

Letting $\varepsilon \to 0$ we obtain

$$\Lambda_s(E) = \lim_{\varepsilon \to 0} \Lambda_s^\varepsilon(E) = 0,$$

as required. $\square$

There are many excellent sources on Hausdorff measures including Falconer (1985), Federer (1969), Mattila (1986), and Rogers (1970). No deep properties of Hausdorff measures are needed in what follows. We require, however, a slightly more general concept than the *s*-Hausdorff measure.

Let h be a real-valued, increasing function on $[0,1)$ with

$$\lim_{t \to 0} h(t) = h(0) = 0 \,.$$

Define the h-*Hausdorff measure* of E by

$$\Lambda_h(E) = \sup_{\delta > 0} \inf \sum_i h(r_i) \,,$$

where, as above, the infimum is taken over all coverings of E by balls B_i with diameter r_i not exceeding δ. The measures Λ_h are still Borel regular measures in $\mathbf{R}^n$ (Federer 1969, p. 170); the choice $h(t) = t^s$ gives the s-Hausdorff measure Λ_s.

Our goal is to prove the following two theorems. Recall the convention that if $d\mu = dx$, then μ is dropped from the terminology; for example, the (p,μ)-capacity of a condenser (E, Ω) is denoted simply by $\text{cap}_p(E, \Omega)$.

2.26. Theorem. *Suppose that $1 < p \leq n$ and that E is a set in $\mathbf{R}^n$ of p-capacity zero. Then the Hausdorff dimension of E is at most $n - p$.*

2.27. Theorem. *Let $h(t) = t^{n-p}$ when $1 < p < n$ and $h(t) = (\log 1/t)^{1-n}$ when $p = n$. Then $\Lambda_h(E) < \infty$ implies $\text{cap}_p E = 0$ for every set E in $\mathbf{R}^n$.*

The proof of Theorem 2.26 hinges on three real analytic lemmas. Before proving these, recall the following covering theorem (Stein 1970, pp. 9–10; Ziemer 1989, Theorem 1.3.1):

2.28. Covering theorem. *Let $\mathcal{B}$ be any family of balls in $\mathbf{R}^n$, with uniformly bounded diameter. Then there is a countable pairwise disjoint subfamily $\{B_1, B_2, \dots\}$ of $\mathcal{B}$ such that*

$$\bigcup_{B \in \mathcal{B}} B \subset \bigcup_{i=1}^{\infty} 5B_i \,.$$

For $s > 0$ define a maximal function

$$M_{s,p}f(y) = \sup \left(r^{-s} \int_B |f|^p \, dx \right)^{1/p} \,,$$

where the supremum is taken over all balls B containing y with radius r.

2.29. Lemma. *For $f \in L^p(\mathbf{R}^n; dx)$ and $t > 0$ we have*

$$\Lambda_s^{\infty}(\{y \in \mathbf{R}^n : M_{s,p}f(y) > t\}) < \frac{10^s}{t^p} \int_{\mathbf{R}^n} |f|^p \, dx \,.$$

PROOF: Fix $t > 0$ and write $E_t = \{y \in \mathbf{R}^n : M_{s,p}f(y) > t\}$. Then for each $y \in E_t$ we can find a ball B_y containing y with radius r_y such that

$$\int_{B_y} |f|^p \, dx > t^p r_y^s \, .$$

Thus

$$E_t \subset \bigcup_{y \in E_t} B_y \, ,$$

and because $f \in L^p(\mathbf{R}^n; dx)$, the radii r_y have an upper bound which does not depend on y. The covering theorem 2.28 implies that we can find a subfamily $B_1, B_2, \ldots$ of the family $\{B_y\}$ such that the balls B_i are pairwise disjoint and that

$$E_t \subset \bigcup 5B_i \, .$$

Therefore, if we write r_i for the radius of B_i, we have that

$$\Lambda_s^\infty(E_t) \le 10^s \sum_i r_i^s < \frac{10^s}{t^p} \sum \int_{B_i} |f|^p \, dx \le \frac{10^s}{t^p} \int_{\mathbf{R}^n} |f|^p \, dx \, ,$$

as required.   □

2.30. Lemma.   Let $u \in C_0^\infty(B(y_0, R))$ and $0 \le n - s < p \le n$. Then there is a constant $c = c(n, p, s) > 0$ such that

$$|u(y)| \le c R^{1-n/p+s/p} M_{s,p} \nabla u(y)$$

for all $y \in \mathbf{R}^n$.

PROOF: We may assume that $y_0 = 0$. With fixed $y \in \mathbf{R}^n$ the function

$$f(t) = \fint_{B(y,t)} u(x) \, dx$$

is continuously differentiable on $(0, \infty)$, $f(0) = u(y)$, and $f(t)$ tends to 0 as t tends to ∞. Hence

$$\int_0^\infty f'(t) \, dt = -u(y)$$

and after a change of variables we arrive at

$$u(y) = -\int_0^\infty \left(\frac{d}{dt} \fint_{B(0,1)} u(y+tz) \, dz\right) dt = -\int_0^\infty \fint_{B(0,1)} z \cdot \nabla u(y+tz) \, dz \, dt \, .$$

Consequently, we have

$$|u(y)| \leq \int_0^\infty \fint_{B(0,1)} |\nabla u(y + tz)| \, dz \, dt \, .$$

Making the change of variables $(z, t) \mapsto (x, r)$, $x = y + tz$, $r = t$, we obtain for $y \in B(0, R)$ the estimate

$$
\begin{aligned}
|u(y)| &\leq \int_0^\infty \fint_{B(y,r)} |\nabla u| \, dx \, dr \\
&\leq \int_0^{3R} (\fint_{B(y,r)} |\nabla u|^p \, dx)^{1/p} \, dr + \int_{3R}^\infty \fint_{B(y,r)} |\nabla u| \, dx \, dr \\
&\leq c \Big(\int_0^{3R} (r^{s-n} r^{-s} \int_{B(y,r)} |\nabla u|^p \, dx)^{1/p} \, dr + \int_{B(0,R)} |\nabla u| \, dx \int_{3R}^\infty r^{-n} \, dr \Big) \\
&\leq c \Big(M_{s,p} \nabla u(y) \int_0^{3R} r^{(s-n)/p} \, dr + R^{1-n} \int_{B(0,R)} |\nabla u| \, dx \Big) \\
&\leq c \Big(M_{s,p} \nabla u(y) R^{(s-n+p)/p} + R (\fint_{B(0,R)} |\nabla u|^p \, dx)^{1/p} \Big) \\
&\leq c \Big(M_{s,p} \nabla u(y) R^{(s-n+p)/p} + R^{1+(s-n)/p} (R^{-s} \int_{B(0,R)} |\nabla u|^p \, dx)^{1/p} \Big) \\
&\leq c M_{s,p} \nabla u(y) R^{1-n/p+s/p} \, ,
\end{aligned}
$$

as required. $\qquad\qquad\qquad\qquad\qquad\qquad\qquad\qquad\qquad\qquad\qquad\qquad\qquad\quad\square$

2.31. Lemma. Let $u \in C_0^\infty(B(y_0, R))$ and $0 \leq n - s < p \leq n$. Then there is a constant $c = c(n, p, s) > 0$ such that

$$\Lambda_s^\infty(\{y \in B(y_0, R) \colon |u(y)| > t\}) \leq c R^{p+s-n} t^{-p} \int_{B(y_0,R)} |\nabla u|^p \, dx$$

for all $t > 0$.

PROOF: If $|u(y)| > t$, then by Lemma 2.30 we have the pointwise estimate

$$M_{s,p} \nabla u(y) > c R^{n/p-s/p-1} t$$

for the maximal function. Hence the assertion follows from the weak type estimate in Lemma 2.29. $\qquad\qquad\qquad\qquad\qquad\qquad\qquad\qquad\qquad\quad\square$

PROOF OF THEOREM 2.26: Since $\text{cap}_p E = 0$, E is contained in a G_δ-set of p-capacity zero, and hence using the Choquet capacitability theorem 2.5 we may as well assume that E is compact. Further, it suffices to show that $\Lambda_s^\infty(E) = 0$ for all s with $n - p < s \le n$ (see Lemma 2.25). To this end, fix an open ball B containing E and choose a sequence of functions $u_j \in C_0^\infty(B)$, admissible for the condenser (E, B) with

$$\int_B |\nabla u_j|^p \, dx \to 0$$

as $j \to \infty$. Then Lemma 2.31 implies

$$\Lambda_s^\infty(E) \le \Lambda_s^\infty(\{y \in B : |u_j(y)| > 1/2\}) \le c \int_B |\nabla u_j|^p \, dx,$$

where c is independent of j. Hence $\Lambda_s^\infty(E) = 0$ and the theorem is proved.
□

Theorem 2.26 yields interesting information about weighted capacities as well:

2.32. Theorem. *Suppose that $1 < q < p$ and that w is a p-admissible weight with $w^{1/(1-q)} \in L_{loc}^1(\mathbf{R}^n; dx)$. Let E be a nonempty set of (p, μ)-capacity zero. Then $p \le qn$ and the p/q-capacity of E is zero. In particular, the Hausdorff dimension of E is at most $n - p/q$.*

PROOF: We may assume that E is compact and contained in a ball B. For each function $\varphi \in C_0^\infty(B)$ we have that

$$\int_B |\nabla \varphi|^{p/q} \, dx \le \left(\int_B |\nabla \varphi|^p \, d\mu \right)^{1/q} \left(\int_B w^{1/(1-q)} \, dx \right)^{(q-1)/q}.$$

Thus $\text{cap}_{p,\mu}(E, B) = 0$ implies $\text{cap}_{p/q} E = 0$. In particular, it follows that $p/q \le n$, and hence the Hausdorff dimension of E does not exceed $n - p/q$ by Theorem 2.26.
□

2.33. Corollary. *Let $w \in A_p$ and define $p_0 = \inf\{q : w \in A_q\}$. If $\text{cap}_{p,\mu} E = 0$ for a nonempty set E, then $p \le p_0 n$ and the p/p_0-capacity of E is zero. In particular, the Hausdorff dimension of E is at most $n - p/p_0$.*

PROOF: In Chapter 15 we prove that any A_p-weight is in A_q for some $q < p$ (see Theorem 15.13) so that $p_0 < p$. Now the assertion follows from Theorem 2.32 because $w^{1/(1-q)}$ is locally integrable for all $q > p_0$.
□

We point out that Theorem 2.26 remains true also for the value $p = 1$; the 1-capacity is defined analogously and the proof is the same. Consequently, in Theorem 2.32 we could allow for q the range $1 < q \leq p$.

The corollary, in turn, has an interesting consequence: if the weight w belongs to A_1, then $\text{cap}_{p,\mu} E = 0$ implies $\text{cap}_{p,dx} E = 0$ and the Hausdorff dimension of E is at most $n - p$. Moreover, if w is any p-admissible weight such that $w^{1/(1-q)}$ is locally integrable for some $1 < q \leq p$, then for $p > nq$ the only set of (p, μ)-capacity zero is the empty set.

Now we turn to the proof of Theorem 2.27. The general formulation with weights in the following lemma is not employed in this book but it has some independent interest.

2.34. Lemma. *Let E be a compact set in $\mathbf{R}^n$. If there is a constant $M < \infty$ such that*

$$\text{cap}_{p,\mu}(E, \Omega) \leq M < \infty$$

for all open sets Ω containing E, then $\text{cap}_{p,\mu} E = 0$.

PROOF: Choose a descending sequence of bounded open sets

$$\Omega_1 \ni \Omega_2 \ni \cdots \ni \bigcap_i \Omega_i = E$$

and choose $\varphi_i \in C_0^\infty(\Omega_i)$, $0 \leq \varphi_i \leq 1$, with $\varphi = 1$ on E and

$$\int_{\Omega_i} |\nabla \varphi_i|^p \, d\mu \leq M + 1.$$

By the Poincaré inequality,

$$\int_{\Omega_i} |\varphi_i|^p \, d\mu \leq c < \infty,$$

where c is independent of i. Because φ_i converges pointwise to a function ψ which is 0 in $\Omega_1 \setminus E$ and 1 on E, we have by the weak compactness of $H_0^{1,p}(\Omega_1; \mu)$ that $\psi \in H_0^{1,p}(\Omega_1; \mu)$ and that $\nabla \psi = 0$ a.e. in Ω_1 (see Theorem 1.32). If $\varphi \in C_0^\infty(\Omega_1)$ is such that $\varphi = 1$ on E, then $\varphi - \varphi_i \in H_0^{1,p}(\Omega_1 \setminus E; \mu)$ and the weak compactness of the space $H_0^{1,p}(\Omega_1 \setminus E; \mu)$ guarantees that $\varphi - \psi \in H_0^{1,p}(\Omega_1 \setminus E; \mu)$. Consequently,

$$\text{cap}_{p,\mu}(E, \Omega_1) \leq \int_{\Omega_1} |\nabla \psi|^p \, d\mu = 0,$$

and the lemma follows.   □

PROOF OF THEOREM 2.27: Since Λ_h is a Borel regular measure, we may assume that E is a Borel set and furthermore, in light of the Choquet capacitability theorem 2.5, we may assume that E is compact. Let Ω be a bounded open set containing E and denote by δ the distance from E to the complement of Ω. Fix $0 < \varepsilon < 1$ such that $\varepsilon \leq \delta^2$; then $r < \varepsilon$ implies $\log \delta/r \geq \frac{1}{2} \log 1/r$. Next cover E by open balls $B(x_i, r_i)$ such that $r_i < \varepsilon/2$. Since we may assume that the balls $B(x_i, r_i)$ intersect E, we have $B(x_i, \delta/2) \subset \Omega$. Now using the subadditivity and monotonicity properties of the capacity (Theorem 2.2) we obtain

$$\mathrm{cap}_p(E, \Omega) \leq \sum_i \mathrm{cap}_p(B(x_i, r_i), \Omega)$$

$$\leq \sum_i \mathrm{cap}_p(B(x_i, r_i), B(x_i, \delta/2)) \leq c(n,p) \sum_i h(r_i),$$

where in the last step we also used formula (2.13) for p-capacities of spherical condensers together with our choice of ε. Taking the infimum over all such coverings and letting $\varepsilon \to 0$, we conclude

$$\mathrm{cap}_p(E, \Omega) \leq c(n,p) \Lambda_h(E) < \infty.$$

Since Ω was an arbitrary bounded open set containing E, we have by Lemma 2.34 that

$$\mathrm{cap}_p E = 0,$$

as desired.   □

2.35. Sobolev capacity
Next we present a variant of capacity which is sometimes more convenient in use than the relative (p, μ)-capacity $\mathrm{cap}_{p,\mu}(E, \Omega)$.

DEFINITION. For a set $E \subset \mathbf{R}^n$ define

$$C_{p,\mu}(E) = \inf_u \int_{\mathbf{R}^n} (|u|^p + |\nabla u|^p) \, d\mu,$$

where u runs through the set

$$S(E) = \{u \in H^{1,p}(\mathbf{R}^n; \mu) : u = 1 \text{ in an open set containing } E\}.$$

If $S(E)$ is empty, we set $C_{p,\mu}(E) = \infty$.

The number $C_{p,\mu}(E)$ is called the *Sobolev (p, μ)-capacity* of E. If $d\mu = dx$, then we drop μ and speak about the Sobolev p-capacity of E.

The advantages over the variational (p, μ)-capacity of a condenser are clear: no underlying set Ω is needed. On the other hand, the extremal functions for $C_{p,\mu}$ are much more difficult to determine.

It is evident that the same number $C_{p,\mu}(E)$ is obtained if the infimum in the definition is taken over $u \in S(E)$ with $0 \leq u \leq 1$. Moreover:

2.36. Lemma. *If K is compact, then*

$$C_{p,\mu}(K) = \inf_{u \in S_0(K)} \int_{\mathbf{R}^n} \left(|u|^p + |\nabla u|^p \right) d\mu,$$

where $S_0(K) = S(K) \cap C_0^\infty(\mathbf{R}^n)$.

PROOF: Let $u \in S(K)$. Since $H^{1,p}(\mathbf{R}^n; \mu) = H_0^{1,p}(\mathbf{R}^n; \mu)$ (Theorem 1.27), we may choose a sequence of functions $\varphi_j \in C_0^\infty(\mathbf{R}^n)$ converging to u in $H^{1,p}(\mathbf{R}^n; \mu)$. Let U be an open neighborhood of K such that $u = 1$ in U. Let $\psi \in C^\infty(\mathbf{R}^n)$, $0 \le \psi \le 1$, be such that $\psi = 1$ in $\complement U$ and $\psi = 0$ in an open neighborhood of K. Then it is easily seen that the functions $\psi_j = 1 - (1 - \varphi_j)\psi$ converge to u in $H^{1,p}(\mathbf{R}^n; \mu)$. This establishes the assertion since $\psi_j \in S_0(K)$. $\qquad \square$

We can apply the proof of Theorem 2.2 almost verbatim to conclude:

2.37. Theorem. *The set function $E \mapsto C_{p,\mu}(E)$, $E \subset \mathbf{R}^n$, is a Choquet capacity. In particular:*

(i) *If $E_1 \subset E_2$, then $C_{p,\mu}(E_1) \le C_{p,\mu}(E_2)$.*
(ii) *If $E = \bigcup_i E_i$, then*

$$C_{p,\mu}(E) \le \sum_i C_{p,\mu}(E_i).$$

We have introduced two different capacities, and it is next shown that for many practical purposes they are essentially equivalent. Suppose that K is a compact subset of a ball $B = B(x_0, r)$. Let $\eta \in C_0^\infty(B(x_0, 2r))$ be a cut-off function with $0 \le \eta \le 1$, $\eta = 1$ on $\overline{B}$, and $|\nabla \eta| \le 2/r$. Then, if $u \in S(K)$, we have

$$\mathrm{cap}_{p,\mu}\left(K, B(x_0, 2r)\right) \le \int_{B(x_0, 2r)} |\nabla(\eta u)|^p \, d\mu$$

$$\le 4^p \left(1 + r^{-p}\right) \int_{\mathbf{R}^n} \left(|u|^p + |\nabla u|^p \right) d\mu.$$

On the other hand, if $u \in C_0^\infty(B(x_0, 2r))$ is such that $u = 1$ in an open neighborhood of K, we have by the Poincaré inequality that

$$\int_{\mathbf{R}^n} \left(|u|^p + |\nabla u|^p \right) d\mu \le (1 + c\, r^p) \int_{B(x_0, 2r)} |\nabla u|^p \, d\mu,$$

where $c = c(p, C_{\mathrm{III}})$. By combining these inequalities, we arrive at

$$(1 + c\, r^p)^{-1}\, C_{p,\mu}(K) \le \mathrm{cap}_{p,\mu}\left(K, B(x_0, 2r)\right) \le 4^p(1 + r^{-p})\, C_{p,\mu}(K).$$

Since both $\mathrm{cap}_{p,\mu}\left(\cdot, B(x_0, 2r)\right)$ and $C_{p,\mu}(\cdot)$ are Choquet capacities, we obtain:

2.38. Theorem. *If $E \subset B(x_0, r)$, then*

$$(1 + c\, r^p)^{-1}\, C_{p,\mu}(E) \le \mathrm{cap}_{p,\mu}\left(E, B(x_0, 2r)\right) \le 4^p(1 + r^{-p})\, C_{p,\mu}(E),$$

where the constant c depends only on p and C_{III}.

2.39. Corollary. *For $E \subset \mathbf{R}^n$ we have $C_{p,\mu}(E) = 0$ if and only if $\mathrm{cap}_{p,\mu} E = 0$.*

2.40. Corollary. *There is a constant c, depending only on p and C_{III}, such that*

$$\frac{1}{c}\, C_{p,\mu}\left(B(x_0, r)\right) \le \mathrm{cap}_{p,\mu}\left(B(x_0, r), B(x_0, 2)\right) \le c\, C_{p,\mu}\left(B(x_0, r)\right)$$

whenever $r \le 1$.

PROOF: This follows from Theorem 2.38, because $B(x_0, r) \subset B(x_0, 1)$ for each $r \le 1$. ☐

Finally if $d\mu = dx$, we have the following estimate:

2.41. Corollary. *If $0 < r \le 1$, then*

$$C_p(B(x_0, r)) \approx \begin{cases} r^{n-p} & \text{if } 1 < p < n \\ (\log \frac{2}{r})^{1-n} & \text{if } p = n, \end{cases}$$

where the constants in $\approx$ depend only on n and p.

PROOF: Combine Corollary 2.40 with (2.13). ☐

2.42. Exceptional sets in Sobolev spaces
In the theory of partial differential equations it is desirable to know when a set is negligible for a Sobolev space. The ensuing two theorems are useful when removability questions are studied for solutions.

If there is an isometric isomorphism between two normed spaces X and Y we write $X = Y$. In particular, if E is a relatively closed subset of Ω, then by

$$H_0^{1,p}(\Omega; \mu) = H_0^{1,p}(\Omega \setminus E; \mu)$$

we mean that each function $u \in H_0^{1,p}(\Omega; \mu)$ can be approximated in $\| \ \|_{1,p}$-norm by functions from $C_0^\infty(\Omega \setminus E)$. Similarly,

$$H^{1,p}(\Omega; \mu) = H^{1,p}(\Omega \setminus E; \mu)$$

means that $\mu(E) = 0$ and that each function $u \in H^{1,p}(\Omega \setminus E; \mu)$ can be approximated in $\| \ \|_{1,p}$-norm by functions from $C^\infty(\Omega)$.

2.43. Theorem. *Suppose that E is a relatively closed subset of Ω. Then*

$$H_0^{1,p}(\Omega;\mu) = H_0^{1,p}(\Omega \setminus E;\mu)$$

if and only if E is of (p,μ)-capacity zero.

PROOF: Suppose that $C_{p,\mu}(E) = 0$. Let $\varphi \in C_0^\infty(\Omega)$ and choose a sequence $u_j \in H^{1,p}(\mathbf{R}^n;\mu)$, $0 \le u_j \le 1$, such that $u_j = 1$ in a neighborhood of E and $u_j \to 0$ in $H^{1,p}(\mathbf{R}^n;\mu)$. Then $(1 - u_j)\varphi$ has compact support in $\Omega \setminus E$ and hence it belongs to $H_0^{1,p}(\Omega \setminus E;\mu)$. Moreover, $(1 - u_j)\varphi$ converges to φ in $H_0^{1,p}(\Omega \setminus E;\mu)$ so that $\varphi \in H_0^{1,p}(\Omega \setminus E;\mu)$. Hence

$$H_0^{1,p}(\Omega;\mu) \subset H_0^{1,p}(\Omega \setminus E;\mu),$$

and since the reverse inclusion is trivial, the sufficiency is established.

For the only if part, let $K \subset E$ be compact. It suffices to show that $C_{p,\mu}(K) = 0$. Choose $\varphi \in C_0^\infty(\Omega)$ with $\varphi = 1$ in a neighborhood of K. Since $H_0^{1,p}(\Omega;\mu) = H_0^{1,p}(\Omega \setminus E;\mu)$, we may choose a sequence of functions $\varphi_j \in C_0^\infty(\Omega \setminus E)$ such that $\varphi_j \to \varphi$ in $H^{1,p}(\Omega;\mu)$. Consequently

$$C_{p,\mu}(K) \le \lim_{j\to\infty} \int_\Omega \left(|\varphi - \varphi_j|^p + |\nabla\varphi - \nabla\varphi_j|^p\right) d\mu = 0,$$

and the theorem follows. $\square$

2.44. Theorem. *Suppose that E is a relatively closed subset of Ω. If E is of (p,μ)-capacity zero, then*

$$H^{1,p}(\Omega;\mu) = H^{1,p}(\Omega \setminus E;\mu).$$

PROOF: Since $\mu(E) = 0$, the inclusion

$$H^{1,p}(\Omega;\mu) \subset H^{1,p}(\Omega \setminus E;\mu)$$

is trivial. For the reverse inclusion, pick $u \in H^{1,p}(\Omega\setminus E;\mu)$. We may assume that u is bounded and it suffices to show that u can be approximated in $H^{1,p}(\Omega\setminus E;\mu)$ by functions from $H^{1,p}(\Omega;\mu)$. To this end, choose a sequence of functions $v_j \in H^{1,p}(\mathbf{R}^n;\mu)$, $0 \le v_j \le 1$, such that $v_j = 1$ in an open neighborhood U_j of E and $v_j \to 0$ in $H^{1,p}(\mathbf{R}^n;\mu)$. Then the function $u_j = (1 - v_j)u$ belongs to $H^{1,p}(\Omega\setminus E)$ (see Theorem 1.24 and Lemma 1.15). Moreover, it is clear that u_j also belongs to $H^{1,p}(U_j;\mu)$. Thus we conclude that $u_j \in H^{1,p}(\Omega;\mu)$ (see the remark after Lemma 1.15). Now it is easy to see that $u_j \to u$ in $H^{1,p}(\Omega \setminus E;\mu)$, whence $u \in H^{1,p}(\Omega;\mu)$ as desired. $\square$

If $w \in A_p$, Theorem 2.44 is far from being sharp. Indeed, the result then remains true if E is of $(n-1)$-measure zero. This is an easy consequence of the ACL-characterization of Sobolev functions if $w = 1$; the general A_p-case can be reduced to the unweighted situation as then $u \in H^{1,p}(\Omega; \mu)$ if and only if $u \in L^p(\Omega; \mu)$ and the first distributional derivatives of u belong to $L^p(\Omega; \mu)$. Note that, by Corollary 2.33, for $w \in A_p$ the Hausdorff dimension of a set of (p, μ)-capacity zero is always less than $n-1$. For a more thorough discussion of this topic, see Kilpeläinen (1992b).

The space $H_0^{1,p}(\Omega; \mu)$ consists of functions from $H^{1,p}(\Omega; \mu)$ that in some sense vanish on the boundary of Ω. The two spaces coincide if the functions in $H^{1,p}(\Omega; \mu)$ can be approximated in $H^{1,p}(\Omega; \mu)$ by functions that belong to $C_0^\infty(\Omega)$. This, however, requires that $\complement\Omega$ be rather small.

2.45. Theorem. $H^{1,p}(\Omega; \mu) = H_0^{1,p}(\Omega; \mu)$ *if and only if* $\complement\Omega$ *has zero* (p, μ)-*capacity.*

PROOF: If $\complement\Omega$ is of (p, μ)-capacity zero, then

$$H^{1,p}(\Omega; \mu) = H^{1,p}(\mathbf{R}^n \setminus \complement\Omega; \mu) = H^{1,p}(\mathbf{R}^n; \mu)$$
$$= H_0^{1,p}(\mathbf{R}^n; \mu) = H_0^{1,p}(\mathbf{R}^n \setminus \complement\Omega; \mu) = H_0^{1,p}(\Omega; \mu),$$

where we used Theorems 2.44, 1.27, and 2.43.

For the only if part we have by Theorem 1.27 that

$$H_0^{1,p}(\mathbf{R}^n; \mu) = H^{1,p}(\mathbf{R}^n; \mu) \subset H^{1,p}(\Omega; \mu) = H_0^{1,p}(\Omega; \mu) \subset H_0^{1,p}(\mathbf{R}^n; \mu).$$

Hence

$$H_0^{1,p}(\mathbf{R}^n; \mu) = H_0^{1,p}(\Omega; \mu) = H_0^{1,p}(\mathbf{R}^n \setminus \complement\Omega; \mu),$$

and $\complement\Omega$ is of (p, μ)-capacity zero by Theorem 2.43. $\square$

We close this chapter with the following lemma that expresses a topological property of sets of (p, μ)-capacity zero: they cannot separate $\mathbf{R}^n$. We recall that for many weights, e.g. for A_p-weights, this follows from Theorem 2.32.

2.46. Lemma. *Suppose that* $E \subset \mathbf{R}^n$ *is a set of* (p, μ)-*capacity zero. Then* $\complement E$ *is connected.*

PROOF: Suppose, on the contrary, that there is a partition $\complement E = U \cup V$, where U and V are disjoint, nonempty, and open relative to $\complement E$. Then $U \cap \overline{V} = \emptyset$ and $V \cap \overline{U} = \emptyset$. Let $F = \overline{U} \cap \overline{V}$. Then $F \subset E$ and hence

$\mathrm{cap}_{p,\mu} F = 0$. Fix a ball $B \subset \mathbf{R}^n$ such that B intersects both $\complement\overline{U}$ and $\complement\overline{V}$. Define a function u in $\complement F$ by

$$u(x) = \begin{cases} \mu(B \setminus \overline{U})^{-1} & \text{if } x \in \complement\overline{U} \\ -\mu(B \setminus \overline{V})^{-1} & \text{if } x \in \complement\overline{V}; \end{cases}$$

note that $\complement\overline{U} \cap \complement\overline{V} = \emptyset$ because E does not have interior points (Lemma 2.10). Then $u \in H^{1,p}(B \setminus F; \mu)$ and since F is removable for Sobolev functions (Theorem 2.44), we indeed have $u \in H^{1,p}(B; \mu)$. Moreover, because $\mu(F) = 0$,

$$u_B = \frac{1}{\mu(B)} \int_B u \, d\mu = \frac{1}{\mu(B)} \Big(\int_{B \setminus \overline{U}} u \, d\mu + \int_{B \setminus \overline{V}} u \, d\mu \Big) = 0.$$

Thus the Poincaré inequality yields

$$\int_B |u|^p \, d\mu \le c \int_B |\nabla u|^p \, d\mu = 0,$$

whence $u = 0$ a.e. in B, an obvious contradiction. Consequently, no partition of $\complement E$ is possible, and hence $\complement E$ is connected. □

NOTES TO CHAPTER 2. "Every analytic problem has its own capacity" and the essential capacity of this book is the relative first order variational (p, μ)-capacity together with its global counterpart. In classical potential theory the capacity based on the equilibrium measure is much used; see Helms (1969), Landkof (1972), or Doob (1984). The nonlinear theory in this book is mostly built on variational principles which are directly connected with (p, μ)-capacity integrals. The Choquet theory of capacities provides a standard approach to deal with general sets in $\mathbf{R}^n$; see Choquet (1953–54), Landkof (1972, Chapter II), and Doob (1984, Appendix II).

Another approach to nonlinear potential theory is based on Bessel or Riesz (or more general) potentials and their associated capacities. This program has been executed in the papers by Reshetnyak (1969), Meyers (1970), Maz'ya and Khavin (1972), Fuglede (1971a), and elsewhere. We refer to Ziemer (1989, Section 2) for a treatment of nonlinear Bessel and Riesz capacities; a more thorough exposition can be found in the forthcoming monograph by Adams and Hedberg.

In analytic problems good lower bounds for capacities are essential – upper bounds are usually easy to obtain directly from the definition. The simple bounds in Lemmas 2.14 and 2.16 suffice for the theory of this book.

In addition to (2.13) not many p-capacities of condensers are known; for the 2-capacity, see Landkof (1972, pp. 165–167, 172–173).

The study of the relationship between Hausdorff measures and capacities has a long history going back to Frostman and Nevanlinna; see Landkof (1972, Chapter III.4), and also Carleson (1967). For the subsequent development, see Reshetnyak (1969) or Maz'ya and Khavin (1972), and also Martio (1978/79). The proof of Theorem 2.26 is from Bojarski and Iwaniec (1983). The idea for the short proof of Theorem 2.27 is similar to that in Väisälä (1975).

The issue of identifying domains for which $H^{1,p}(\Omega; dx) = H_0^{1,p}(\Omega; dx)$ as domains whose complement has zero capacity is well known; see Adams (1975, Chapter III) or Maz'ya (1985, p. 396).

Most of the weighted results in this chapter are likely to be known to specialists. The observations concerning the size of sets of zero (p, μ)-capacity may be new. Weighted capacities and Hausdorff measures have also been studied by Nieminen (1991).

3
Supersolutions and the obstacle problem

This chapter deals with quasilinear elliptic equations of the form

$$(3.1) \qquad -\operatorname{div}\mathcal{A}(x,\nabla u)=0\,,$$

where $\mathcal{A}:\mathbf{R}^n\times\mathbf{R}^n\to\mathbf{R}^n$ is a mapping satisfying certain structural assumptions, given in (3.3)–(3.7) below. In particular, we impose the growth condition

$$\mathcal{A}(x,\xi)\cdot\xi\approx w(x)|\xi|^p\,,$$

where w is a p-admissible weight as defined in Chapter 1. It is advisable to think of equation (3.1) as a perturbed form of the *weighted p-Laplace equation*

$$(3.2) \qquad -\Delta_{p,w}u=-\operatorname{div}(w(x)|\nabla u|^{p-2}\nabla u)=0\,,$$

which is canonically attached to the weighted Sobolev space $H^{1,p}(\,\cdot\,;\mu)$. Solutions of (3.2) can be recognized as local minimizers of the weighted p-Dirichlet integral

$$\int|\nabla u(x)|^p w(x)\,dx\,.$$

More generally, Euler equations of variational integrals

$$\int F\bigl(x,\nabla u(x)\bigr)\,dx$$

are prime examples of the equations considered. The connection to the calculus of variations will be discussed in Chapter 5.

In this chapter we cover the basic regularity theory of nonlinear partial differential equations of type (3.1). We establish weak Harnack inequalities for supersolutions and, as a consequence, that a supersolution always has a lower semicontinuous representative. A large part of the treatment is devoted to proving various fundamental estimates, which ultimately lead to a proof of the continuity of the solution to the obstacle problem with continuous obstacle. Roughly speaking, a solution to the obstacle problem is a minimal supersolution of equation (3.1) above the obstacle. Solving the obstacle problem provides a basic tool in nonlinear potential theory.

55

As always, we assume that $1 < p < \infty$, that w is a p-admissible weight in $\mathbf{R}^n$, and denote by μ the measure

$$\mu(E) = \int_E w(x)\,dx\,.$$

Let $\mathcal{A} : \mathbf{R}^n \times \mathbf{R}^n \to \mathbf{R}^n$ be a mapping satisfying the following assumptions for some constants $0 < \alpha \le \beta < \infty$:

(3.3)   the mapping $x \mapsto \mathcal{A}(x,\xi)$ is measurable for all $\xi \in \mathbf{R}^n$ and
the mapping $\xi \mapsto \mathcal{A}(x,\xi)$ is continuous for a.e. $x \in \mathbf{R}^n$;

for all $\xi \in \mathbf{R}^n$ and a.e. $x \in \mathbf{R}^n$

(3.4) $\mathcal{A}(x,\xi) \cdot \xi \ge \alpha\,w(x)\,|\xi|^p$

(3.5) $|\mathcal{A}(x,\xi)| \le \beta\,w(x)\,|\xi|^{p-1}$

(3.6) $\bigl(\mathcal{A}(x,\xi_1) - \mathcal{A}(x,\xi_2)\bigr) \cdot \bigl(\xi_1 - \xi_2\bigr) > 0$

whenever $\xi_1, \xi_2 \in \mathbf{R}^n$, $\xi_1 \ne \xi_2$; and

(3.7) $\mathcal{A}(x,\lambda\xi) = \lambda\,|\lambda|^{p-2}\mathcal{A}(x,\xi)$

whenever $\lambda \in \mathbf{R}$, $\lambda \ne 0$.

Assumption (3.3) guarantees that the composed mapping $x \mapsto \mathcal{A}(x,g(x))$ is measurable whenever g is a measurable function. Assumption (3.4) describes the ellipticity degeneracy of equation (3.8). In view of (3.5) the map $u \mapsto Lu$ defines a continuous operator from $L^{1,p}(\Omega;\mu)$ into its dual via the pairing

$$\langle Lu, \varphi \rangle = \int_\Omega \mathcal{A}(x, \nabla u) \cdot \nabla\varphi\,dx\,;$$

indeed,

$$\left| \int_\Omega \mathcal{A}(x, \nabla u) \cdot \nabla\varphi\,dx \right| \le \beta \int_\Omega |\nabla u|^{p-1}|\nabla\varphi|\,d\mu$$

$$\le \beta \Bigl(\int_\Omega |\nabla u|^p\,d\mu \Bigr)^{(p-1)/p}\Bigl(\int_\Omega |\nabla\varphi|^p\,d\mu \Bigr)^{1/p}\,.$$

The monotoneity assumption (3.6) plays a crucial role in deriving the existence and the uniqueness of solutions to variational problems. Finally, the homogeneity assumption (3.7) ensures that the class of solutions is closed under a scalar multiplication.

Many results in this chapter are true in more generality. Assumptions (3.3)–(3.7) become essential later in this book.

DEFINITION. A function u in $H^{1,p}_{loc}(\Omega; \mu)$ is a (weak) solution of the equation

$$(3.8) \qquad - \operatorname{div} \mathcal{A}(x, \nabla u) = 0$$

in Ω if

$$(3.9) \qquad \int_\Omega \mathcal{A}(x, \nabla u(x)) \cdot \nabla \varphi(x)\, dx = 0$$

whenever $\varphi \in C_0^\infty(\Omega)$. We usually write the integral in (3.9) as

$$\int_\Omega \mathcal{A}(x, \nabla u) \cdot \nabla \varphi\, dx \,.$$

A function u in $H^{1,p}_{loc}(\Omega; \mu)$ is a supersolution of (3.8) in Ω if

$$- \operatorname{div} \mathcal{A}(x, \nabla u) \geq 0$$

weakly in Ω, i.e.

$$(3.10) \qquad \int_\Omega \mathcal{A}(x, \nabla u) \cdot \nabla \varphi\, dx \geq 0$$

whenever $\varphi \in C_0^\infty(\Omega)$ is nonnegative. A function v is a subsolution of (3.8) if $-v$ is a supersolution of (3.8). In other words, $v \in H^{1,p}_{loc}(\Omega; \mu)$ is a subsolution in Ω if (3.10) holds for all nonpositive $\varphi \in C_0^\infty(\Omega)$. We often use the word "solution" to mean a solution of (3.8), and similarly for super- and subsolutions.

Appealing to a partition of unity, it is evident that being a solution or a supersolution of (3.8) is a local property, i.e. a function u is a (super)solution in Ω if and only if Ω can be covered by open sets where u is a (super)solution.

We observe the following important property of solutions and supersolutions: if u is a solution (a supersolution) and $\lambda, \tau \in \mathbf{R}$ ($\lambda \geq 0$), then $\lambda u + \tau$ is again a solution (a supersolution).

It is often desirable to deal with a larger class of test functions than $C_0^\infty(\Omega)$ in (3.9). The following lemma expresses that this can be done, provided we know a priori that ∇u is in $L^p(\Omega; \mu)$.

3.11. Lemma. *If $u \in L^{1,p}(\Omega; \mu)$ is a solution (respectively, a supersolution) of (3.8) in Ω, then*

$$\int_{\Omega} \mathcal{A}(x, \nabla u) \cdot \nabla \varphi \, dx = 0 \quad \text{(respectively, } \geq 0\text{)}$$

for all $\varphi \in H_0^{1,p}(\Omega; \mu)$ (respectively, for all nonnegative $\varphi \in H_0^{1,p}(\Omega; \mu)$).

PROOF: Let $\varphi \in H_0^{1,p}(\Omega; \mu)$ and choose a sequence of functions $\varphi_i \in C_0^{\infty}(\Omega)$ such that $\varphi_i \to \varphi$ in $H^{1,p}(\Omega; \mu)$. If φ is nonnegative, pick nonnegative functions φ_i (see Lemma 1.23). Then by the boundedness assumption (3.5)

$$\left| \int_{\Omega} \mathcal{A}(x, \nabla u) \cdot \nabla \varphi \, dx - \int_{\Omega} \mathcal{A}(x, \nabla u) \cdot \nabla \varphi_i \, dx \right|$$

$$\leq \beta \int_{\Omega} |\nabla u|^{p-1} |\nabla \varphi - \nabla \varphi_i| \, d\mu$$

$$\leq \beta \left(\int_{\Omega} |\nabla u|^p \, d\mu \right)^{(p-1)/p} \left(\int_{\Omega} |\nabla \varphi - \nabla \varphi_i|^p \, d\mu \right)^{1/p}.$$

Because the last integral tends to zero as $i \to \infty$, we have

$$\int_{\Omega} \mathcal{A}(x, \nabla u) \cdot \nabla \varphi \, dx = \lim_{i \to \infty} \int_{\Omega} \mathcal{A}(x, \nabla u) \cdot \nabla \varphi_i \, dx \geq 0,$$

and the lemma follows. $\qquad \square$

The proof of Lemma 3.11 implies more than the actual statement. Indeed, if u is any (super)solution in Ω, then (3.10) holds for all (nonnegative) $\varphi \in H_0^{1,p}(\Omega; \mu)$ with compact support. Moreover, if $u \in L^{1,p}(\Omega; \mu)$ is a solution of (3.8), then

$$(3.12) \qquad \int_{\Omega} \mathcal{A}(x, \nabla u) \cdot \nabla \varphi \, dx = 0$$

for all $\varphi \in L_0^{1,p}(\Omega; \mu)$, and if $u \in L^{1,p}(\Omega; \mu)$ is a supersolution, then

$$\int_{\Omega} \mathcal{A}(x, \nabla u) \cdot \nabla \varphi \, dx \geq 0$$

whenever $\varphi \in L_0^{1,p}(\Omega; \mu)$ is such that there is a sequence of nonnegative functions $\varphi_i \in C_0^{\infty}(\Omega)$ with $\nabla \varphi_i \to \nabla \varphi$ in $L^p(\Omega; \mu)$.

A function u is a solution of (3.8) if and only if u is a supersolution and a subsolution. To see this, consider the positive and negative parts of a test function $\varphi \in C_0^{\infty}(\Omega)$. They belong to $H_0^{1,p}(\Omega; \mu)$ and have compact support, and it follows that the integral in (3.9) is nonnegative for all test functions φ; hence (3.9) holds.

3.13. Quasiminimizers

An important property of solutions of (3.8) is that they are *quasiminimizers* of the weighted p-Dirichlet integral

$$(3.14) \qquad \int_\Omega |\nabla u|^p \, d\mu.$$

To see what this means, let $u \in H^{1,p}(\Omega; \mu)$ be a solution of (3.8) and $u - \varphi \in H_0^{1,p}(\Omega; \mu)$. Then

$$\int_\Omega |\nabla u|^p \, d\mu \le \alpha^{-1} \int_\Omega \mathcal{A}(x, \nabla u) \cdot \nabla u \, dx$$

$$= \alpha^{-1} \int_\Omega \mathcal{A}(x, \nabla u) \cdot \nabla \varphi \, dx$$

$$\le \frac{\beta}{\alpha} \Big(\int_\Omega |\nabla u|^p \, d\mu \Big)^{(p-1)/p} \Big(\int_\Omega |\nabla \varphi|^p \, d\mu \Big)^{1/p}$$

and thus

$$(3.15) \qquad \int_\Omega |\nabla u|^p \, d\mu \le \Big(\frac{\beta}{\alpha}\Big)^p \int_\Omega |\nabla \varphi|^p \, d\mu.$$

In other words, amongst all functions φ having the "same boundary values" as u, i.e. $u - \varphi \in H_0^{1,p}(\Omega; \mu)$, the solution u has the least weighted p-Dirichlet integral, up to a factor $(\beta/\alpha)^p$. In particular, if (3.8) is the p-Laplace equation (3.2), then $\alpha = \beta = 1$ and u has the least weighted p-energy among all functions φ with $u - \varphi \in H_0^{1,p}(\Omega; \mu)$. Similarly, a supersolution $u \in H^{1,p}(\Omega; \mu)$ is a *quasisuperminimizer*, i.e. (3.15) holds for all functions φ with $u - \varphi \in H_0^{1,p}(\Omega; \mu)$ and $\varphi \ge u$.

Minimization problems are pursued in Chapter 5.

3.16. Existence and uniqueness of solutions

The next theorem tells us that the Dirichlet problem for equation (3.8) is solvable with *Sobolev boundary values*.

3.17. Theorem. *Suppose that Ω is bounded and that $\vartheta \in H^{1,p}(\Omega; \mu)$. There is a unique solution $u \in H^{1,p}(\Omega; \mu)$ of (3.8) in Ω with $u - \vartheta \in H_0^{1,p}(\Omega; \mu)$.*

Proving the existence requires the theory of monotone operators and its full treatment would be a major digression; we refer the reader to Appendix I. If, however, equation (3.8) arises from a variational integral, a direct proof is given in Chapter 5. The uniqueness follows from the next lemma which is the first version of the comparison principle.

3.18. Lemma. *Let $u \in H^{1,p}(\Omega; \mu)$ be a supersolution and $v \in H^{1,p}(\Omega; \mu)$ a subsolution of (3.8) in Ω. If $\eta = \min(u - v, 0) \in H_0^{1,p}(\Omega; \mu)$, then $u \geq v$ a.e. in Ω.*

PROOF: Since

$$0 \leq \int_\Omega \mathcal{A}(x, \nabla v) \cdot \nabla \eta \, dx - \int_\Omega \mathcal{A}(x, \nabla u) \cdot \nabla \eta \, dx$$

$$= -\int_{\{u<v\}} \left(\mathcal{A}(x, \nabla v) - \mathcal{A}(x, \nabla u) \right) \cdot \left(\nabla v - \nabla u \right) dx \leq 0,$$

we have $\nabla \eta = 0$ a.e. in Ω by (3.6). Because $\eta \in H_0^{1,p}(\Omega; \mu)$, $\eta = 0$ a.e. in Ω (Lemma 1.17). The lemma follows. □

3.19. Obstacle problem

Suppose that Ω is bounded, that ψ is any function in Ω with values in the extended reals $[-\infty, \infty]$, and that $\vartheta \in H^{1,p}(\Omega; \mu)$. Let

$$\mathcal{K}_{\psi,\vartheta} = \mathcal{K}_{\psi,\vartheta}(\Omega) = \{ v \in H^{1,p}(\Omega; \mu) : v \geq \psi \text{ a.e. in } \Omega, \, v - \vartheta \in H_0^{1,p}(\Omega; \mu) \}.$$

If $\psi = \vartheta$, we write $\mathcal{K}_{\psi,\psi}(\Omega) = \mathcal{K}_\psi(\Omega)$.

The problem is to find a function u in $\mathcal{K}_{\psi,\vartheta}$ such that

$$(3.20) \qquad \int_\Omega \mathcal{A}(x, \nabla u) \cdot \nabla(v - u) \, dx \geq 0$$

whenever $v \in \mathcal{K}_{\psi,\vartheta}$. The function ψ is called an *obstacle*.

DEFINITION. A function u in $\mathcal{K}_{\psi,\vartheta}(\Omega)$ that satisfies (3.20) for all $v \in \mathcal{K}_{\psi,\vartheta}(\Omega)$ is called a *solution to the obstacle problem with obstacle ψ and boundary values ϑ* or a *solution to the obstacle problem in $\mathcal{K}_{\psi,\vartheta}(\Omega)$.*

If u is a solution to the obstacle problem in $\mathcal{K}_{\psi,u}(\Omega)$, then u is called a *solution to the obstacle problem with obstacle ψ.*

Observe the following convention throughout the book: the notation $\mathcal{K}_{\psi,\vartheta}(\Omega)$ tacitly assumes that Ω is bounded, for the obstacle problem is defined in bounded open sets only. Otherwise, we retain the standing assumption that Ω is any open subset of $\mathbf{R}^n$.

Since $u + \varphi \in \mathcal{K}_{\psi,\vartheta}(\Omega)$ for all nonnegative $\varphi \in C_0^\infty(\Omega)$, the solution u to the obstacle problem is always a supersolution of (3.8) in Ω. Conversely, a supersolution is always a solution to the obstacle problem in $\mathcal{K}_{u,u}(D)$ for all open $D \Subset \Omega$. It is also worth while to observe the fact that a solution u to equation (3.8) in an open set Ω is a solution to the obstacle problem in $\mathcal{K}_{-\infty,u}(D)$ for all open sets $D \Subset \Omega$. Similarly, a solution to the obstacle problem in $\mathcal{K}_{-\infty,u}(\Omega)$ is a solution to equation (3.8).

In the case of variational integrals the solution u to the obstacle problem in $\mathcal{K}_{\psi,\vartheta}(\Omega)$ minimizes the corresponding variational integral among the functions in $\mathcal{K}_{\psi,\vartheta}(\Omega)$; in other words, u has the minimal "energy" among all the functions that lie above the obstacle ψ and have boundary values ϑ (see Chapter 5). In the general case the solution u to the obstacle problem in $\mathcal{K}_{\psi,\vartheta}(\Omega)$ quasiminimizes the weighted p-Dirichlet integral in $\mathcal{K}_{\psi,\vartheta}(\Omega)$, that is,

$$\int_\Omega |\nabla u|^p \, d\mu \le \left(\frac{\beta}{\alpha}\right)^p \int_\Omega |\nabla v|^p \, d\mu$$

whenever $v \in \mathcal{K}_{\psi,\vartheta}(\Omega)$. This is easily established in the same manner as (3.15).

The following simple fact is constantly used in what follows. A function $u \in H^{1,p}(\Omega;\mu)$ is a solution to the obstacle problem in $\mathcal{K}_{\psi,u}(\Omega)$ if and only if u is a solution to the obstacle problem in $\mathcal{K}_{\psi,u}(D)$ whenever $D \subset \Omega$ is open. Indeed, let $v \in \mathcal{K}_{\psi,u}(D)$. Then the function $u - v$ has a zero extension to a function in $H_0^{1,p}(\Omega;\mu)$ and hence the function

$$\tilde{u} = \begin{cases} u & \text{in } \Omega \setminus D \\ v & \text{in } D \end{cases}$$

belongs to $\mathcal{K}_{\psi,u}(\Omega)$. Therefore we have

$$0 \le \int_\Omega \mathcal{A}(x, \nabla u) \cdot \nabla(\tilde{u} - u) \, dx = \int_D \mathcal{A}(x, \nabla u) \cdot \nabla(v - u) \, dx$$

which shows that u is a solution to the obstacle problem in $\mathcal{K}_{\psi,u}(D)$.

3.21. Theorem. *If $\mathcal{K}_{\psi,\vartheta}(\Omega)$ is nonempty, there is a unique solution u to the obstacle problem in $\mathcal{K}_{\psi,\vartheta}(\Omega)$.*

The existence is established in Appendix I; see also Chapter 5 for the case of variational integrals. The uniqueness follows from the next comparison lemma which tells us that the solution to the obstacle problem in $\mathcal{K}_{\psi,\vartheta}$ is the smallest supersolution of (3.8) in $\mathcal{K}_{\psi,\vartheta}$.

3.22. Lemma. *Suppose that u is a solution to the obstacle problem in $\mathcal{K}_{\psi,\vartheta}(\Omega)$. If $v \in H^{1,p}(\Omega;\mu)$ is a supersolution of (3.8) in Ω such that $\min(u,v) \in \mathcal{K}_{\psi,\vartheta}(\Omega)$, then $v \ge u$ a.e. in Ω.*

PROOF: The function $u - \min(u,v)$ belongs to $H_0^{1,p}(\Omega;\mu)$ and is nonnegative. Since u is a solution to the obstacle problem and v is a supersolution, we have by Lemma 3.11 that

$$0 \le \int_\Omega \big(\mathcal{A}(x, \nabla v) - \mathcal{A}(x, \nabla u)\big) \cdot \nabla\big(u - \min(u,v)\big) \, dx$$

$$= \int_\Omega \big(\mathcal{A}(x, \nabla \min(u,v)) - \mathcal{A}(x, \nabla u)\big) \cdot \big(\nabla u - \nabla \min(u,v)\big) \, dx \le 0.$$

Using the monotoneity assumption (3.6) and Lemma 1.17 we infer that $u = \min(u, v)$ as required.   □

3.23. Theorem. *If u and v are two supersolutions of (3.8) in Ω, then so is $\min(u, v)$.*

PROOF: Fix an open set $G \Subset \Omega$ and write $s = \min(u, v)$. It suffices to show that s is a supersolution in G. If s_0 is the solution to the obstacle problem in $\mathcal{K}_s(G)$, then $s_0 \geq s$ a.e. in G. Moreover, since $\min(u, s_0)$ and $\min(v, s_0)$ belong to $\mathcal{K}_s(G)$, Lemma 3.22 implies that $u \geq s_0$ and $v \geq s_0$ a.e. in G. Thus $s = s_0$ is a supersolution in G.   □

3.24. Theorem. *If u is a solution to the obstacle problem in $\mathcal{K}_{\psi,\vartheta}(\Omega)$, where ψ and ϑ are essentially bounded above by a constant M, then u is essentially bounded above by M in Ω. In particular, if u is a solution to (3.8) in Ω with boundary values ϑ, then*

$$\operatorname*{ess\,inf}_{\Omega} \vartheta \leq u \leq \operatorname*{ess\,sup}_{\Omega} \vartheta$$

a.e. in Ω.

PROOF: Since the function $\min(u, M)$ is a supersolution and belongs to $\mathcal{K}_{\psi,\vartheta}(\Omega)$, the first assertion follows from Lemma 3.22.

To prove the second assertion we only need to note that u is then a solution to the obstacle problem in $\mathcal{K}_{-\infty,\vartheta}(\Omega)$ and $-u$ in $\mathcal{K}_{-\infty,-\vartheta}(\Omega)$; the claim then follows from the first assertion.   □

3.25. Lemma. *Suppose that u is a solution to the obstacle problem in $\mathcal{K}_{\psi,u}(\Omega)$ and that $D \subset \Omega$ is open. If there is a subsolution v of (3.8) in D with $\psi \leq v \leq u$ a.e. in D, then u is a solution of (3.8) in D.*

PROOF: Let $G \Subset D$ be open and let h be the solution of equation (3.8) in G with $h - u \in H_0^{1,p}(G; \mu)$. We show that $u = h$ in G, which implies the desired assertion. To this end, note that the comparison lemma 3.18 implies that $h \leq u$ in G. On the other hand,

$$\min(h - v, 0) = \min(u - v, u - h) + (h - u) \in H_0^{1,p}(G; \mu),$$

and appealing again to the comparison lemma 3.18, we deduce that $h \geq v$ in G. Hence h belongs to $\mathcal{K}_{\psi,u}(G)$. Because u is a solution to the obstacle problem also in $\mathcal{K}_{\psi,u}(G)$, it is the smallest supersolution in $\mathcal{K}_{\psi,u}(G)$ (Lemma 3.22) and so $u \leq h$ in G. We conclude that $u = h$ in G, and the lemma follows.   □

Lemma 3.25 implies in particular that if u is a solution to the obstacle problem in $\mathcal{K}_{\psi,u}(\Omega)$ and if there is a constant c such that $\psi \le c \le u$ in an open set D, then u is a solution of (3.8) in D.

3.26. Basic estimates

Next we establish several basic estimates for supersolutions and solutions to obstacle problems. In particular, we show that a solution to (3.8) always becomes continuous after a redefinition in a set of measure zero; the same is true for a solution to the obstacle problem with continuous obstacle.

We begin with *Caccioppoli type estimates*. Roughly speaking these estimates provide upper bounds locally for the L^p-norm of ∇u in terms of the L^p-norm of the solution u.

3.27. Lemma. *Suppose that $\eta \in C_0^\infty(\Omega)$ is nonnegative and $q \ge 0$.*

(i) *If u is a solution to the obstacle problem in $\mathcal{K}_{\psi,u}(\Omega)$ with nonpositive obstacle ψ, then*

$$(3.28) \qquad \int_\Omega |u^+|^q |\nabla u^+|^p \eta^p \, d\mu \le c \int_\Omega |u^+|^{p+q} |\nabla \eta|^p \, d\mu.$$

(ii) *If u is a supersolution of (3.8) in Ω, then*

$$(3.29) \qquad \int_\Omega |u^-|^q |\nabla u^-|^p \eta^p \, d\mu \le c \int_\Omega |u^-|^{p+q} |\nabla \eta|^p \, d\mu.$$

Here $c = p^p(\beta/\alpha)^p$.

PROOF: We prove only (i); the proof of (ii) is analogous and left to the reader. Without loss of generality we may assume that $0 \le \eta \le 1$. Let $\varphi = -u^+ \eta^p$. Then φ is in $H_0^{1,p}(\Omega; \mu)$ and has compact support in Ω. Since $u + \varphi \in \mathcal{K}_{\psi,u}(\Omega)$, we have

$$
\begin{aligned}
0 &\le \int_\Omega \mathcal{A}(x, \nabla u) \cdot \nabla \varphi \, dx \\
&= \int_\Omega \mathcal{A}(x, \nabla u) \cdot (-\nabla u^+ \eta^p - p u^+ \eta^{p-1} \nabla \eta) \, dx \\
&\le -\alpha \int_\Omega |\nabla u^+|^p \eta^p \, d\mu + p\beta \int_\Omega |u^+| \, |\nabla \eta| \, |\nabla u^+|^{p-1} \eta^{p-1} \, d\mu,
\end{aligned}
$$

and therefore

$$
\begin{aligned}
\int_\Omega |\nabla u^+|^p \eta^p \, d\mu &\le p \frac{\beta}{\alpha} \int_\Omega |u^+| \, |\nabla \eta| \, |\nabla u^+|^{p-1} \eta^{p-1} \, d\mu \\
&\le p \frac{\beta}{\alpha} \Big(\int_\Omega |u^+|^p |\nabla \eta|^p \, d\mu \Big)^{1/p} \Big(\int_\Omega |\nabla u^+|^p \eta^p \, d\mu \Big)^{(p-1)/p}.
\end{aligned}
$$

From this and the fact that $\int_\Omega |\nabla u^+|^p \eta^p \, d\mu < \infty$ we obtain

$$\int_\Omega |\nabla u^+|^p \eta^p \, d\mu \leq c \int_\Omega |u^+|^p |\nabla \eta|^p \, d\mu,$$

where $c = p^p(\beta/\alpha)^p$. This is (3.28) for $q = 0$.

Next we observe that the function $u - t$, where t is a real constant, is a solution to the obstacle problem in $\mathcal{K}_{\psi-t, u-t}$; hence for $t > 0$

$$(3.30) \qquad \int_\Omega |\nabla(u - t)^+|^p \eta^p \, d\mu \leq c \int_\Omega |(u - t)^+|^p |\nabla \eta|^p \, d\mu.$$

To complete the proof, we make use of a standard representation theorem: *if f is a nonnegative ν-measurable function in a measure space X, ν is a measure in X, then for $0 < q < \infty$*

$$(3.31) \qquad \int_X f^q \, d\nu = q \int_0^\infty t^{q-1} \nu(\{x : f(x) > t\}) \, dt.$$

If $q = 1$, equation (3.31) is immediate for simple functions and then it follows from the monotone convergence theorem for measurable functions. Thereafter we obtain (3.31) for all $0 < q < \infty$ by changing the variables. Another proof can be based on the Fubini theorem (Bauer 1990, p. 160).

Now the desired estimate follows by applying (3.31). Indeed, we have by (3.31) and (3.30) that

$$\begin{aligned}
\int_\Omega |u^+|^q |\nabla u^+|^p \eta^p \, d\mu &= q \int_0^\infty t^{q-1} \int_{\{u>t\}} |\nabla u^+|^p \eta^p \, d\mu \, dt \\
&= q \int_0^\infty t^{q-1} \int_{\{u>t\}} |\nabla(u - t)^+|^p \eta^p \, d\mu \, dt \\
&\leq cq \int_0^\infty t^{q-1} \int_{\{u>t\}} |(u - t)^+|^p |\nabla \eta|^p \, d\mu \, dt \\
&\leq cq \int_0^\infty t^{q-1} \int_{\{u>t\}} |u^+|^p |\nabla \eta|^p \, d\mu \, dt \\
&= c \int_\Omega |u^+|^{q+p} |\nabla \eta|^p \, d\mu
\end{aligned}$$

as desired. $\qquad\qquad\qquad\qquad\qquad\qquad\qquad\qquad\qquad\qquad\qquad\qquad$ $\square$

For solutions Lemma 3.27 yields the following estimate.

3.32. Lemma. *If u is a solution of equation (3.8) in Ω, then*

$$(3.33) \qquad \int_\Omega |u|^q |\nabla u|^p \eta^p \, d\mu \le c \int_\Omega |u|^{p+q} |\nabla \eta|^p \, d\mu$$

whenever $\eta \in C_0^\infty(\Omega)$ is nonnegative and $q \ge 0$. Here $c = p^p(\beta/\alpha)^p$.

Next we employ the celebrated Moser iteration technique in establishing two fundamental estimates for supersolutions.

3.34. Theorem. *Suppose that $B = B(x_0, r)$ is a ball, $0 < \lambda < 1$, and $0 < q < \infty$.*

(i) *If u is a solution to the obstacle problem in $\mathcal{K}_{\psi,u}(B)$ with nonpositive obstacle ψ, then*

$$(3.35) \qquad \operatorname*{ess\,sup}_{\lambda B} u^+ \le c(1-\lambda)^{-\xi} \Big(\fint_B |u^+|^q \, d\mu\Big)^{1/q}.$$

(ii) *If u is a supersolution of (3.8) in B, then*

$$(3.36) \qquad \operatorname*{ess\,sup}_{\lambda B} |u^-| \le c(1-\lambda)^{-\xi} \Big(\fint_B |u^-|^q \, d\mu\Big)^{1/q}.$$

Here $\xi = \max(p\varkappa/q(\varkappa - 1), \varkappa/(\varkappa - 1))$, $\varkappa$ is the number in the Sobolev inequality **III**, and $c = c(p, q, \beta/\alpha, c_\mu)$ is a positive constant.

PROOF: We prove case (i) only, the proof for (ii) being similar. Let

$$r_\ell = \lambda + (1 - \lambda)2^{-\ell}, \qquad \ell = 0, 1, 2, \ldots;$$

thus $r_0 = 1$ and $r_{\ell+1}$ is the mean value of r_ℓ and λ. Let $\eta_\ell \in C_0^\infty(r_\ell B)$ be a nonnegative function such that $\eta_\ell = 1$ in $r_{\ell+1}B$ and that $|\nabla \eta| \le 4(1-\lambda)^{-1}2^\ell r^{-1}$. Fix $t \ge 0$ and let

$$\omega_\ell = |u^+|^{1+t/p}\eta_\ell.$$

By using the Caccioppoli estimate (3.28), we obtain

$$\Big(\int_{r_\ell B} |\nabla \omega_\ell|^p \, d\mu\Big)^{1/p}$$

$$\le \Big(\frac{p+t}{p}\Big)\Big(\int_{r_\ell B} |\nabla u^+|^p |u^+|^t \eta_\ell^p \, d\mu\Big)^{1/p} + \Big(\int_{r_\ell B} |u^+|^{p+t} |\nabla \eta_\ell|^p \, d\mu\Big)^{1/p}$$

$$\le c(p+t)\Big(\int_{r_\ell B} |u^+|^{p+t} |\nabla \eta_\ell|^p \, d\mu\Big)^{1/p}$$

$$\le c(1-\lambda)^{-1} 2^\ell r^{-1} (p+t)\Big(\int_{r_\ell B} |u^+|^{p+t} \, d\mu\Big)^{1/p},$$

where $c = c(p, \beta/\alpha)$. The weighted Sobolev inequality **III** implies

$$\left(\fint_{r_\ell B} |\omega_\ell|^{\varkappa p} \, d\mu\right)^{1/\varkappa p} \le c \, r_\ell \, r \left(\fint_{r_\ell B} |\nabla\omega_\ell|^p \, d\mu\right)^{1/p}$$

$$\le c_0 \, (1-\lambda)^{-1} \, r_\ell \, 2^\ell \, (p + t) \left(\fint_{r_\ell B} |u^+|^{p+t} \, d\mu\right)^{1/p},$$

where c_0 depends only on p, β/α, and c_μ, and $\varkappa > 1$ is from **III**. Now by setting $\kappa = p + t$ and using the doubling property of μ we obtain

$$\left(\fint_{r_{\ell+1} B} |u^+|^{\kappa\varkappa} \, d\mu\right)^{1/\kappa\varkappa} \le c_1^{1/\kappa} \, (1-\lambda)^{-p/\kappa} \, 2^{p\ell/\kappa} \, \kappa^{p/\kappa} \left(\fint_{r_\ell B} |u^+|^\kappa \, d\mu\right)^{1/\kappa},$$

where c_1 depends only on p, β/α, and c_μ. Because this estimate holds for all $\kappa > p$, it can be applied with $\kappa = \kappa_\ell = p\varkappa^\ell$ for all $\ell = 0, 1, 2, \ldots$. By iterating we arrive at the desired estimate for $q = p$:

$$\operatorname*{ess\,sup}_{\lambda B} u^+ \le \lim_{\ell\to\infty} \left(\fint_{r_\ell B} |u^+|^{\kappa_\ell \varkappa} \, d\mu\right)^{1/\kappa_\ell \varkappa}$$

$$(3.37) \qquad \le c \, (1-\lambda)^{-\sum_{i=0}^\infty \varkappa^{-\ell}} \prod_{\ell=0}^\infty 2^{\ell\varkappa^{-\ell}} \prod_{\ell=0}^\infty (p\varkappa^\ell)^{\varkappa^{-\ell}} \left(\fint_B |u^+|^p \, d\mu\right)^{1/p}$$

$$\le c \, (1-\lambda)^{-\varkappa/(\varkappa-1)} \left(\fint_B |u^+|^p \, d\mu\right)^{1/p}.$$

It is important to observe that (3.37) does not hold only in the ball $B = B(x_0, r)$ but also in each ball inside B; this is because the constants c and $\varkappa$ are independent of x_0 and r. In the following we use the estimate for the balls $B' \subset B$ that are concentric with B. An extrapolation argument then shows that the exponent p in (3.37) can be replaced by any positive number q. This is formulated as an individual lemma, which concludes the proof of Theorem 3.34 because, by Hölder's inequality, there is no loss of generality in assuming that $q \le p$.

In the next lemma we interpret

$$\left(\fint_E v^s \, d\mu\right)^{1/s} = \operatorname*{ess\,sup}_E v$$

for $s = \infty$.

3.38. Lemma. Suppose that $0 < q < p < s \le \infty$, $\xi \in \mathbf{R}$, and that $B = B(x_0, r)$ is a ball. If a nonnegative function $v \in L^p(B; \mu)$ satisfies

$$(3.39) \qquad \left(\fint_{\lambda B'} v^s \, d\mu\right)^{1/s} \le c_1 (1-\lambda)^\xi \left(\fint_{B'} v^p \, d\mu\right)^{1/p}$$

for each ball $B' = B(x_0, r')$ with $r' \leq r$ and for all $0 < \lambda < 1$, then

$$(\fint_{\lambda B} v^s \, d\mu)^{1/s} \leq c\,(1 - \lambda)^{\xi/\theta}(\fint_B v^q \, d\mu)^{1/q}$$

for all $0 < \lambda < 1$. Here $c = c(p, q, s, \xi, c_1, c_\mu) > 0$ and $\theta \in (0, 1)$ is such that

$$\frac{1}{p} = \frac{\theta}{q} + \frac{1 - \theta}{s}.$$

PROOF: Since

$$(\fint_{\lambda B} v^s \, d\mu)^{1/s}$$

increases to ess $\sup_{\lambda B} v$ as s grows to ∞, there is no loss of generality in assuming that $s < \infty$. Then let

$$\Phi(q) = \sup_{\frac{1}{2} < \lambda < 1} (1 - \lambda)^{\hat{q}} (\fint_{\lambda B} v^p \, d\mu)^{1/p},$$

where $\hat{q} = \xi(\theta - 1)/\theta$. Writing $\lambda' = \frac{1}{2}(1 + \lambda)$ for each $\lambda \in (0, 1)$ we have by (3.39) that

$$(3.40) \quad (1 - \lambda)^{\hat{q}/(1-\theta)} (\fint_{\lambda B} v^s \, d\mu)^{1/s} \leq c\,(1 - \lambda')^{\hat{q}} (\fint_{\lambda' B} v^p \, d\mu)^{1/p} \leq c\,\Phi(q),$$

where $c = c(c_1, \xi, \theta) > 0$. Fix $\delta > 0$ and choose $\lambda' \in (\frac{1}{2}, 1)$ such that

$$\Phi(q) \leq (1 - \lambda')^{\hat{q}} (\fint_{\lambda' B} v^p \, d\mu)^{1/p} + \delta.$$

Next we apply *Young's inequality*:

$$ab \leq \varepsilon a^r + \varepsilon^{-1/(r-1)} b^{r/(r-1)}, \quad \varepsilon > 0 \text{ and } r > 1.$$

Since

$$1 = \theta \frac{p}{q} + (1 - \theta)\frac{p}{s},$$

we have

$$\Phi(q) \leq (1 - \lambda')^{\hat{q}} (\fint_{\lambda' B} v^p \, d\mu)^{1/p} + \delta$$

$$= (1 - \lambda')^{\hat{q}} (\fint_{\lambda' B} v^{p\theta} v^{p(1-\theta)} \, d\mu)^{1/p} + \delta$$

$$\leq (1 - \lambda')^{\hat{q}} (\fint_{\lambda' B} v^q \, d\mu)^{\theta/q} (\fint_{\lambda' B} v^s \, d\mu)^{(1-\theta)/s} + \delta$$

$$\leq c(\theta, \varepsilon) (\fint_{\lambda' B} v^q \, d\mu)^{1/q} + \varepsilon\,(1 - \lambda')^{\hat{q}/(1-\theta)} (\fint_{\lambda' B} v^s \, d\mu)^{1/s} + \delta$$

$$= c(\theta, q, \varepsilon, c_\mu) (\fint_B v^q \, d\mu)^{1/q} + \varepsilon\,(1 - \lambda')^{\hat{q}/(1-\theta)} (\fint_{\lambda' B} v^s \, d\mu)^{1/s} + \delta$$

$$\leq c(\theta, q, \varepsilon, c_\mu) (\fint_B v^q \, d\mu)^{1/q} + \varepsilon\,c_0\,\Phi(q) + \delta,$$

where the last two inequalities follow from the doubling property of μ and (3.40). Choosing $\varepsilon = (2c_0)^{-1}$ and letting δ tend to zero we obtain

$$\Phi(q) \leq c \left(\fint_B v^q \, d\mu \right)^{1/q}.$$

Hence it follows from (3.40) that

$$(1 - \lambda)^{\hat{q}/(1-\theta)} \left(\fint_{\lambda B} v^s \, d\mu \right)^{1/s} \leq c \left(\fint_B v^q \, d\mu \right)^{1/q},$$

where $c = c(c_1, \xi, \theta, q, c_\mu) > 0$. This is the desired estimate and Lemma 3.38 follows.   □

Theorem 3.34 leads to the following theorem.

3.41. Theorem.
  (i) *Each supersolution of (3.8) in Ω is locally bounded below.*
  (ii) *Each solution of (3.8) in Ω is locally bounded.*
  (iii) *A solution to the obstacle problem in Ω is locally bounded above provided the obstacle is locally bounded above.*

From Theorem 3.34 it follows that if v is a nonnegative subsolution in a ball B, then

$$(3.42) \qquad\qquad \operatorname*{ess\,sup}_{\frac{1}{2}B} v \leq c \left(\fint_B v^q \, d\mu \right)^{1/q}$$

for all $q > 0$, where the constant c depends on p, q, β/α, and c_μ. Next we proceed to show that a nonnegative supersolution u in a ball B satisfies an important *weak Harnack inequality*:

$$\left(\fint_B u^s \, d\mu \right)^{1/s} \leq c \operatorname*{ess\,inf}_{\frac{1}{2}B} u,$$

where c and s are positive constants that depend only on the data. For solutions this together with (3.42) yields the Harnack inequality, which is discussed later in Chapter 6.

We require some lemmas. The first is a special case of a more general convexity theorem 7.5.

3.43. Lemma. *If u is a supersolution of (3.8) in Ω and $u \geq \varepsilon > 0$, then $v = -1/u$ is also a supersolution in Ω.*

PROOF: If $q < 0$, then u^{-q} belongs to $H^{1,p}_{loc}(\Omega; \mu)$ and $\nabla u^{-q} = -qu^{-q-1}\nabla u$ (Theorem 1.18). Let $\varphi \in C^\infty_0(\Omega)$ be nonnegative and write $\psi = u^{2(1-p)}\varphi$. Then $\psi \in H^{1,p}_0(\Omega; \mu)$ with compact support and since $\psi \geq 0$, the homogeneity assumption (3.7) yields

$$0 \leq \int_\Omega \mathcal{A}(x, \nabla u) \cdot \nabla \psi \, dx$$

$$= 2(1-p) \int_\Omega u^{1-2p}\varphi \mathcal{A}(x, \nabla u) \cdot \nabla u \, dx + \int_\Omega u^{2(1-p)} \mathcal{A}(x, \nabla u) \cdot \nabla \varphi \, dx$$

$$\leq \int_\Omega \mathcal{A}(x, u^{-2}\nabla u) \cdot \nabla \varphi \, dx = \int_\Omega \mathcal{A}(x, \nabla v) \cdot \nabla \varphi \, dx$$

as desired. □

3.44. Lemma. *Suppose that u is a nonnegative supersolution of (3.8) in a ball B. Then*

$$(3.45) \qquad \underset{\lambda B}{\mathrm{ess\,inf}}\, u \geq c\,(1-\lambda)^\xi \left(\fint_B u^{-q}\, d\mu \right)^{-1/q}$$

whenever $0 < \lambda < 1$ and $0 < q < \infty$, where $\xi > 0$ is as in Theorem 3.34 and $c = c(p, q, \beta/\alpha, c_\mu)$ is a positive constant.

PROOF: Let $u_\varepsilon = u + \varepsilon$ for $\varepsilon > 0$. Then $-1/u_\varepsilon$ is a supersolution by Lemma 3.43. Hence (3.45) for u_ε follows from Theorem 3.34 and letting $\varepsilon \to 0$ completes the proof. □

In order to replace the negative exponent $-q$ in (3.45) with a positive number we need a weighted version of the John–Nirenberg lemma; its proof is offered in Appendix II.

3.46. John–Nirenberg lemma. *Suppose that v is a locally μ-integrable function in Ω with*

$$\sup_{\substack{B \subseteq \Omega \\ B \text{ ball}}} \fint_B |v - v_B|\, d\mu \leq c_0,$$

where

$$v_B = \fint_B v\, d\mu.$$

Then there are positive constants c_1 and c_2 depending on c_0, n, and c_μ such that

$$\sup_{\substack{B \subseteq \Omega \\ B \text{ ball}}} \fint_B e^{c_1|v - v_B|}\, d\mu \leq c_2.$$

Next we show that if u is a positive supersolution, then $v = \log u$ satisfies the hypothesis of the John–Nirenberg lemma.

3.47. Lemma. *Let $u \geq \varepsilon > 0$ be a supersolution of (3.8) in Ω. Then $\log u \in H^{1,p}_{loc}(\Omega; \mu)$ and $\nabla \log u = \nabla u/u$. Moreover, if $E \subset \Omega$ is measurable, then*

$$(3.48) \qquad \int_E |\nabla \log u|^p \, d\mu \leq c \operatorname{cap}_{p,\mu}(E, \Omega),$$

where $c = (p\beta/(p-1)\alpha)^p$.

PROOF: The first assertion is clear (see Theorem 1.18) and it suffices to establish estimate (3.48) for a compact set $E \subset \Omega$. Let $\varphi \in C^\infty_0(\Omega)$ be nonnegative with $\varphi = 1$ in E. Then the function $\eta = \varphi^p u^{1-p}$ belongs to $H^{1,p}_0(\Omega; \mu)$. Since η is nonnegative and has compact support, we have

$$0 \leq \int_\Omega \mathcal{A}(x, \nabla u) \cdot \nabla \eta \, dx$$
$$= \int_\Omega \mathcal{A}(x, \nabla u) \cdot (p\varphi^{p-1} u^{1-p} \nabla\varphi - (p-1)u^{-p}\varphi^p \nabla u) \, dx$$

and hence

$$\int_{\text{spt } \varphi} |\nabla u|^p u^{-p} \varphi^p \, d\mu \leq c \int_{\text{spt } \varphi} |\nabla u|^{p-1} u^{1-p} |\nabla\varphi| \varphi^{p-1} \, d\mu$$
$$\leq c \left(\int_{\text{spt } \varphi} |\nabla u|^p u^{-p} \varphi^p \, d\mu \right)^{(p-1)/p} \left(\int_\Omega |\nabla\varphi|^p \, d\mu \right)^{1/p},$$

where $c = p\beta/(p-1)\alpha$. Since φ was an arbitrary function, admissible for the condenser (E, Ω), the desired estimate follows. $\qquad\square$

By combining estimate (3.48) with the capacity estimate (2.15),

$$\operatorname{cap}_{p,\mu}(B, 2B) \approx \mu(B)(\operatorname{diam}(B))^{-p},$$

we obtain:

3.49. Corollary. *Let $u \geq \varepsilon > 0$ be a supersolution of (3.8) in Ω. Then*

$$(3.50) \qquad \int_B |\nabla \log u|^p \, d\mu \leq c\,\mu(B)\,r^{-p}$$

whenever $B = B(x, r)$ is a ball with $2B \subset \Omega$; here $c = c(p, \beta/\alpha, c_\mu)$.

Now we can invoke the John–Nirenberg lemma to obtain a weak Harnack inequality for supersolutions.

3.51. Theorem. *Suppose that u is a nonnegative supersolution of (3.8) in Ω. Then there is a constant $s_0 = s_0(n, p, \beta/\alpha, c_\mu) > 0$ such that for all $0 < s < s_0$*

$$(3.52) \qquad \operatorname*{ess\,inf}_{\lambda B} u \geq c\,(1 - \lambda)^\xi \left(\fint_B u^s\, d\mu \right)^{1/s}$$

whenever B is a ball with $2B \subset \Omega$ and $0 < \lambda < 1$. Here $c = c(n, p, \beta/\alpha, c_\mu)$ and $\xi = \max(p\varkappa/s(\varkappa - 1), \varkappa/(\varkappa - 1))$.

PROOF: Let $\varepsilon > 0$ and write $v = \log(u + \varepsilon)$. Then the Poincaré inequality and Corollary 3.49 yield

$$\fint_B |v - v_B|^p\, d\mu \leq c\, r^p \fint_B |\nabla v|^p\, d\mu \leq c,$$

where r is the radius of B. It follows from the John–Nirenberg lemma 3.46 that there are constants c_1 and c_2 such that

$$\left(\fint_B e^{c_1 v}\, d\mu \right)\left(\fint_B e^{-c_1 v}\, d\mu \right) = \left(\fint_B e^{c_1(v - v_B)}\, d\mu \right)\left(\fint_B e^{c_1(v_B - v)}\, d\mu \right)$$

$$\leq \left(\fint_B e^{c_1|v - v_B|}\, d\mu \right)^2 \leq c_2^2.$$

Hence by letting $\varepsilon \to 0$ we have that

$$\left(\fint_B u^{c_1}\, d\mu \right)^{1/c_1} \leq c \left(\fint_B u^{-c_1}\, d\mu \right)^{-1/c_1}.$$

Then for all $0 < s < c_1$ it holds that

$$\left(\fint_B u^s\, d\mu \right)^{1/s} \leq c \left(\fint_B u^{-s}\, d\mu \right)^{-1/s} \leq c\,(1 - \lambda)^{-\xi} \operatorname*{ess\,inf}_{\lambda B} u,$$

where the last inequality follows from (3.45) with the desired ξ. □

Now we can improve Lemma 3.47:

3.53. Theorem. *Let u be a positive supersolution of (3.8) in Ω. Then $\log u \in H^{1,p}_{loc}(\Omega; \mu)$ and $\nabla \log u = \nabla u/u$. Moreover, if $E \subset \Omega$ is measurable, then*

$$(3.54) \qquad \int_E |\nabla \log u|^p\, d\mu \leq c\, \mathrm{cap}_{p,\mu}(E, \Omega),$$

where $c = (p\beta/(p-1)\alpha)^p$. In particular,

$$(3.55) \qquad \int_B |\nabla \log u|^p \, d\mu \leq c \, \mu(B) \, r^{-p}$$

whenever $B = B(x, r)$ is a ball with $2B \subset \Omega$; here $c = c(p, \beta/\alpha, c_\mu)$.

PROOF: As in Lemma 3.47 we may assume that E is compact. Let $\varepsilon > 0$ and choose an open set $D \Subset \Omega$ such that $E \subset D$ and

$$\mathrm{cap}_{p,\mu}(E, D) \leq \mathrm{cap}_{p,\mu}(E, \Omega) + \varepsilon \,.$$

By Theorem 3.51 $u \geq \delta > 0$ in D for some $\delta > 0$ and so Lemma 3.47 implies that $\log u \in H^{1,p}(D; \mu)$, $\nabla \log u = \nabla u / u$, and

$$\int_E |\nabla \log u|^p \, d\mu \leq c \, \mathrm{cap}_{p,\mu}(E, D) \leq c \, \mathrm{cap}_{p,\mu}(E, \Omega) + c\varepsilon \,.$$

The theorem follows. $\qquad\qquad\qquad\qquad\qquad\qquad\qquad\qquad\qquad\quad\square$

3.56. A sharp form of the weak Harnack inequality

It is often desirable to know how large the exponent s in the weak Harnack inequality (3.52) can be. Next we show that any positive s which is less than $\varkappa(p-1)$ will do. Our proof uses a weak reverse Hölder inequality for supersolutions. We start with the following crucial estimate that extends Lemma 3.47.

3.57. Lemma. *Suppose that u is a positive supersolution in Ω. If $\eta \in C_0^\infty(\Omega)$ and $\varepsilon > 0$, then*

$$\int_\Omega |\nabla u|^p u^{-1-\varepsilon} |\eta|^p \, d\mu \leq c \int_\Omega u^{p-1-\varepsilon} |\nabla \eta|^p \, d\mu,$$

where $c = (p\beta/\varepsilon\alpha)^p$.

PROOF: We may assume that η is nonnegative. For a positive integer j write $u_j = u + 1/j$. Then $v = u_j^{-\varepsilon} \eta^p$ is a nonnegative function in $H_0^{1,p}(\Omega; \mu)$ with compact support. Hence

$$0 \leq \int_\Omega \mathcal{A}(x, \nabla u) \cdot \nabla v \, dx$$

$$= -\varepsilon \int_\Omega \mathcal{A}(x, \nabla u) \eta^p u_j^{-1-\varepsilon} \cdot \nabla u_j \, dx + p \int_\Omega \mathcal{A}(x, \nabla u) \eta^{p-1} u_j^{-\varepsilon} \cdot \nabla \eta \, dx$$

and using the structural assumptions together with the Hölder inequality we obtain

$$\alpha \varepsilon \int_\Omega |\nabla u|^p u_j^{-1-\varepsilon} \eta^p \, d\mu \le p\beta \int_\Omega |\nabla u|^{p-1} u_j^{-\varepsilon} \eta^{p-1} |\nabla \eta| \, d\mu$$

$$\le p\beta \left(\int_\Omega |\nabla u|^p u_j^{-1-\varepsilon} \eta^p \, d\mu \right)^{(p-1)/p} \left(\int_\Omega u_j^{p-1-\varepsilon} |\nabla \eta|^p \, d\mu \right)^{1/p}.$$

Consequently,

$$\int_\Omega |\nabla u|^p u_j^{-1-\varepsilon} \eta^p \, d\mu \le c \int_\Omega u_j^{p-1-\varepsilon} |\nabla \eta|^p \, d\mu,$$

and the lemma follows. $\qquad \square$

This estimate implies a weak reverse Hölder inequality for supersolutions.

3.58. Theorem. *Suppose that u is a positive supersolution in $2B$, where B is a ball. If*

$$0 < q \le s < \varkappa(p-1),$$

then

$$\left(\fint_B u^s \, d\mu \right)^{1/s} \le c \left(\fint_{2B} u^q \, d\mu \right)^{1/q},$$

where $c = c(n, p, \beta/\alpha, q, s, c_\mu) > 0$.

PROOF: We may assume that $u \ge \varepsilon > 0$ in $2B$. Let $B = B(x_0, r)$ and let $k \ge 1$ be the smallest integer such that $\varkappa^{-k} s \le q$. For $j = 1, \ldots, k$ write $B_j = (1 + j/k)B$ and let $\eta_j \in C_0^\infty(B_j)$ be a cut-off function such that $0 \le \eta_j \le 1$, $\eta_j = 1$ on $\overline{B}_{j-1}$, and $|\nabla \eta_j| \le 4k/r$, where $B_0 = B$. Write

$$\varepsilon_j = p - 1 - \varkappa^{-j} s > 0$$

and

$$v_j = \eta_j \, u^{1-(1+\varepsilon_j)/p}.$$

Now it follows from the Sobolev inequality **(III)** and Lemma 3.57 that

$$\left(\fint_{B_j} v_j^{\varkappa p} \, d\mu \right)^{1/\varkappa p} \le cr \left(\fint_{B_j} |\nabla v_j|^p \, d\mu \right)^{1/p}$$

$$\le cr \left(\fint_{B_j} u^{p-1-\varepsilon_j} |\nabla \eta|^p \, d\mu \right)^{1/p}$$

$$\le c \left(\fint_{B_j} u^{\varkappa^{-j} s} \, d\mu \right)^{1/p}$$

$$= c \left(\fint_{B_j} v_{j+1}^{\varkappa p} \, d\mu \right)^{1/p}$$

$$\le c \left(\fint_{B_{j+1}} v_{j+1}^{\varkappa p} \, d\mu \right)^{1/p}.$$

Because $B_k = 2B$ and $\varkappa^{-k}s \leq q$, we have after k steps that

$$\left(\fint_B u^s \, d\mu \right)^{1/s} \leq c \left(\fint_{B_1} v_1^{\varkappa p} \, d\mu \right)^{1/s}$$

$$\leq c \left(\fint_{B_k} u^{\varkappa^{-k}s} \, d\mu \right)^{\varkappa^k/s}$$

$$\leq c \left(\fint_{2B} u^q \, d\mu \right)^{1/q},$$

where $c = c(n, p, \beta/\alpha, q, s, c_\mu) > 0$.   □

Combining the weak Harnack inequality (Theorem 3.51) with the weak reverse Hölder inequality (Theorem 3.58), we obtain the following sharp form of Theorem 3.51.

3.59. Theorem. *Suppose that u is a nonnegative supersolution of (3.8) in Ω. If $0 < s < \varkappa(p-1)$, then*

$$(3.60) \qquad \left(\fint_B u^s \, d\mu \right)^{1/s} \leq c \operatorname*{ess\,inf}_B u$$

whenever B is a ball with $4B \subset \Omega$. Here $c = c(n, p, \beta/\alpha, c_\mu, s) > 0$.

In general, the bound $s < \varkappa(p-1)$ in Theorem 3.59 is the best possible; see Section 7.42 (in particular Example 7.47) where we discuss the local integrability of $\mathcal{A}$-superharmonic functions. In the unweighted case $\varkappa = n/(n-p)$ if $p < n$ and we may choose any $\varkappa < \infty$ if $p \geq n$; thus (3.60) holds for all positive s with $s(n-p) < n(p-1)$. In particular, it is true for $s = 1$ if $p > 2n/(n+1)$.

As an interesting consequence of Theorem 3.59 we mention that if u is a positive supersolution of (3.8) in $\mathbf{R}^n$, if $w = 1$, and if $p > 2n/(n+1)$, then u is an A_1-weight; in particular, u is q-admissible for all $q > 1$. See Chapter 15.

3.61. Lower semicontinuous functions

Before we examine the pointwise behavior of supersolutions it is appropriate to recall some basic facts about lower semicontinuous functions.

A function $u \colon A \to \mathbf{R} \cup \{\infty\}$, defined on a set $A \subset \mathbf{R}^n$, is *lower semicontinuous* if the set $\{x \in A : u(x) > \lambda\}$ is open in A for every $\lambda \in \mathbf{R}$. Equivalently, u is lower semicontinuous in A if

$$\liminf_{y \to x} u(y) \geq u(x).$$

for every $x \in A$, where

$$\liminf_{y \to x} u(y) = \lim_{r \to 0} \inf\{u(y) \colon y \in B(x,r) \cap (A \setminus \{x\})\}.$$

A function $u \colon A \to \mathbf{R} \cup \{-\infty\}$ is *upper semicontinuous* if $-u$ is lower semicontinuous.

Clearly, a real-valued function u is continuous in A if and only if u is both lower and upper semicontinuous in A.

It is important to observe that a lower semicontinuous function u attains its infimum on any compact set and hence it is bounded below on compact sets.

If u_i, $i \in I$, are lower semicontinuous functions in A, then $u = \sup_{i \in I} u_i$ is lower semicontinuous because

$$\{u > \lambda\} = \bigcup_{i \in I} \{u_i > \lambda\}.$$

In particular, if $u_1 \leq u_2 \leq \ldots$ are lower semicontinuous, so is $u = \lim u_i$.

We shall constantly invoke the following approximation of lower semicontinuous functions by continuous functions: if u is lower semicontinuous in Ω, then there is an increasing sequence f_j of continuous functions in Ω with

$$u = \lim_{j \to \infty} f_j$$

in Ω. To establish this, let

$$\mathcal{F} = \{f \in C(\Omega) \colon f \leq u \text{ in } \Omega\}.$$

Choose an exhaustion of Ω by compact sets $K_1 \subset K_2 \subset \cdots \subset \Omega$, $\cup K_j = \Omega$. Since u is bounded below on K_j, it is easy to see that $\mathcal{F} \neq \emptyset$ and, moreover, that

$$u = \sup \mathcal{F}.$$

Let

$$\{B_k \colon k = 1, 2, \ldots\}$$

be an enumeration of all balls $B \Subset \Omega$ with rational center and radius. For each $i = 1, 2, \ldots$ and $k = 1, 2, \ldots$ choose a function $f_{i,k} \in \mathcal{F}$ such that

$$f_{i,k} = \begin{cases} \inf_{B_k} u - 1/i & \text{if } \inf_{B_k} u < \infty \\ i & \text{if } \inf_{B_k} u = \infty \end{cases}$$

in $\frac{1}{2}B_k$. Since u is lower semicontinuous, it is easy to see that

$$u = \sup_{i,k} f_{i,k}$$

in Ω. Hence

$$f_j = \max_{1 \le i,k \le j} f_{i,k}$$

is the desired sequence.

By using a smoothing procedure (see Section 1.10), the approximating functions f_j above can be chosen to be C^∞-smooth; moreover, replacing f_j by $f_j - 1/j$, we have that $f_j < f_{j+1} < u$ for all $j = 1, 2, \ldots$

Next, for a function u defined in Ω we let

$$\text{ess}\liminf_{y \to x} u(y) = \lim_{r \to 0} \text{ess} \inf_{B(x,r)} u.$$

Then for a lower semicontinuous function

$$u(x) \le \liminf_{y \to x} u(y) \le \text{ess}\liminf_{y \to x} u(y),$$

and it is easily checked that if a function $u \colon \Omega \to \mathbf{R} \cup \{\infty\}$ satisfies

$$u(x) = \text{ess}\liminf_{y \to x} u(y)$$

for all x in Ω, then u is lower semicontinuous in Ω.

3.62. Pointwise behavior of solutions

In this section we show that each supersolution has a lower semicontinuous representative. Moreover, a solution to the obstacle problem with continuous obstacle has a continuous representative.

We begin with the following theorem:

3.63. Theorem. *Suppose that u is a supersolution of (3.8) in Ω. Then u is locally essentially bounded below and there is a lower semicontinuous representative of u such that*

$$(3.64) \qquad u(x) = \text{ess}\liminf_{y \to x} u(y)$$

for each $x \in \Omega$.

PROOF: Since u is locally bounded below (Theorem 3.41), by working locally we may assume that u is nonnegative. It suffices to find a representative of u which satisfies (3.64), for such a function is lower semicontinuous. To this end, suppose first that $M = \sup_\Omega u < \infty$ and fix $x \in \Omega$. Write $B_r = B(x,r)$ and $m_r = \text{ess}\inf_{B_r} u$ for r so small that $2B_r \subset \Omega$. The weak Harnack inequality (3.52) yields for some $0 < s \le 1$

$$m_{\frac{r}{2}} - m_r \ge c \Big(\fint_{B_r} (u - m_r)^s \, d\mu \Big)^{1/s}$$

$$\ge c \, (M - m_r)^{(s-1)/s} \Big(\fint_{B_r} (u - m_r) \, d\mu \Big)^{1/s}$$

or

$$0 \leq \fint_{B_r} u \, d\mu - m_r \leq c \, (M - m_r)^{1-s} \, (m_{\frac{r}{2}} - m_r)^s$$
$$\leq c \, M^{1-s} \, (m_{\frac{r}{2}} - m_r)^s \,,$$

and since the right hand side tends to zero as $r \to 0$, we have

$$(3.65) \qquad \underset{y \to x}{\text{ess lim inf}} \, u(y) = \lim_{r \to 0} \fint_{B(x,r)} u \, d\mu$$

for each $x \in \Omega$. By the Lebesgue differentiation theorem (see (1.3)) the right hand side of (3.65) equals $u(x)$ a.e., and the proof is complete if u is bounded above.

The general case follows from the first case, since the functions $u_k = \min(u, k)$ are supersolutions of (3.8) and since a locally μ-integrable lower semicontinuous function can be redefined in a set of measure zero so that (3.64) holds. Indeed, because $u = \lim_{k \to \infty} \min(u, k)$, we may assume that u is lower semicontinuous, and then by letting $\hat{u}(x) = \text{ess lim inf}_{y \to x} \, u(y)$, we have for a.e. x that

$$u(x) \leq \liminf_{y \to x} u(y) \leq \hat{u}(x)$$
$$\leq \liminf_{r \to 0} \fint_{B(x,r)} u(y) \, d\mu(y) = u(x) \,,$$

where the last equality follows from the Lebesgue differentiation theorem. Thus $\hat{u} = u$ a.e. and $\hat{u}(x) = \text{ess lim inf}_{y \to x} \, \hat{u}(y)$. Theorem 3.63 is then proved. $\qquad \square$

3.66. Theorem. *Let $p > 1 + 1/\varkappa$, where $\varkappa > 1$ is from the Sobolev inequality* **III**. *Then each supersolution u of (3.8) has a lower semicontinuous representative satisfying*

$$u(x) = \underset{y \to x}{\text{ess lim inf}} \, u(y) = \lim_{r \to 0} \fint_{B(x,r)} u(y) \, d\mu(y)$$

for all $x \in \Omega$.

PROOF: We proceed as in the proof of Theorem 3.63. Fix $x \in \Omega$ and write $m_r = \text{ess inf}_{B(x,r)} u$ for r so small that $B(x, 4r) \subset \Omega$. By Theorem 3.63 we may assume that $m_r \leq M < \infty$. Because $\varkappa(p - 1) > 1$, the weak Harnack inequality (Theorem 3.59) holds with exponent $s = 1$; thus

$$\fint_{B(x,r)} (u - m_{4r}) \, d\mu \leq c \, (m_r - m_{4r}) \,.$$

Since the right hand side tends to zero as $r \to 0$, we have

$$\operatorname{ess\,lim\,inf}_{y \to x} u(y) = \lim_{r \to 0} \fint_{B(x,r)} u \, d\mu$$

for each $x \in \Omega$. The assertion follows from Lebesgue's differentiation theorem. $\qquad\qquad\square$

In the unweighted case the hypothesis of Theorem 3.66 is satisfied for $p > 2n/(n+1)$, because then the Sobolev inequality **III** holds with

$$\varkappa = \begin{cases} \frac{n}{n-p} & \text{if } p < n \\ 2 & \text{if } p \geq n \,. \end{cases}$$

We do not know whether Theorem 3.66 remains true if $p \leq 1 + 1/\varkappa$. However, it follows from (3.65) that if u is a locally upper bounded supersolution, then u has a lower semicontinuous representative that possesses Lebesgue points everywhere.

Now we prove that a solution to the obstacle problem is continuous if the obstacle itself is continuous. This property of solutions has an important bearing on the subsequent development of the theory. In obstacle problems the set

$$\{x \in \Omega : u(x) = \psi(x)\}$$

is called the *coincidence set*. The next theorem also demonstrates the fact that a solution to the obstacle problem is actually a solution to equation (3.8) in each open set that does not meet the coincidence set.

3.67. Theorem. *Let Ω be bounded and let $\psi \colon \Omega \to [-\infty, \infty)$ be continuous. If $\vartheta \in H^{1,p}(\Omega; \mu)$ is such that $\vartheta \geq \psi$ a.e. in Ω, then there is a unique continuous solution u to the obstacle problem in $\mathcal{K}_{\psi,\vartheta}(\Omega)$.*

Moreover, u is a solution of (3.8) in the open set $\{x \in \Omega : u(x) > \psi(x)\}$.

PROOF: The existence and the uniqueness of the solution u in $\mathcal{K}_{\psi,\vartheta}$ have already been established (Theorem 3.21). Let u be the lower semicontinuous solution given by Theorem 3.63 so that

$$(3.68) \qquad\qquad u(x) = \operatorname{ess\,lim\,inf}_{y \to x} u(y)$$

for all $x \in \Omega$. We show that

$$(3.69) \qquad\qquad \operatorname{ess\,lim\,sup}_{y \to x} u(y) \leq u(x)$$

whenever $x \in \Omega$. Since u is locally bounded (Theorem 3.41), it then follows that u is real valued and continuous in Ω.

To achieve (3.69), fix $x \in \Omega$ and $\varepsilon > 0$. Since ψ is continuous,

$$u(x) = \operatorname{ess\,lim\,inf}_{y \to x} u(y) \geq \operatorname{ess\,lim\,inf}_{y \to x} \psi(y) = \psi(x)$$

for all $x \in \Omega$, and we may choose a ball $B = B(x, r) \Subset \Omega$ such that

$$\sup_B \psi \leq u(x) + \varepsilon = \gamma_0$$

and that

$$\inf_B u > u(x) - \varepsilon = \gamma_1,$$

because u is lower semicontinuous. Then estimate (3.35) yields

$$\operatorname{ess\,sup}_{\frac{1}{2}B}(u - \gamma_0) \leq c \fint_B (u - \gamma_0)^+ \, d\mu,$$

where c does not depend on r, the radius of B. On the other hand,

$$\fint_B (u - \gamma_0)^+ \, d\mu = \fint_B (u - \min(u, \gamma_0)) \, d\mu$$

$$\leq \fint_B (u - \gamma_1) \, d\mu = \fint_B u \, d\mu - u(x) + \varepsilon.$$

Since u is locally bounded,

$$\lim_{r \to 0} \fint_{B(x,r)} u \, d\mu = u(x)$$

(see the remark after Theorem 3.66), and the inequalities above imply

$$\operatorname{ess\,lim\,sup}_{y \to x} u(y) \leq \gamma_0 + c\varepsilon = u(x) + c\varepsilon.$$

Finally, we arrive at (3.69) by letting $\varepsilon \to 0$. Thus u is continuous in Ω.

To prove the second assertion, pick a point x_0 with $u(x_0) > \psi(x_0)$. Then there are a constant λ and an open neighborhood D of x_0 such that

$$u \geq \lambda \geq \psi$$

in D. Since the constant function $v \equiv \lambda$ is a subsolution, u is a solution to equation (3.8) in the open set $\{x \in \Omega : u(x) > \psi(x)\}$ by Lemma 3.25, and the proof of Theorem 3.67 is complete. $\qquad\square$

Because each solution u of equation (3.8) is a solution to an obstacle problem with continuous obstacle $-\infty$, we know from Theorem 3.67 that u can be redefined in a set of measure zero so that it becomes continuous. In fact, solutions of (3.8) are locally Hölder continuous, as will be shown in Chapter 6.

3.70. Theorem.   *If u is a solution of equation (3.8) in Ω, then there is a continuous function v in Ω such that $u = v$ a.e.*

The next theorem characterizes the solution to the obstacle problem in $\mathcal{K}_{\psi,\vartheta}$, with continuous obstacle, among all supersolutions in $\mathcal{K}_{\psi,\vartheta}$.

3.71. Theorem.   *Let Ω be bounded. Suppose that $\psi: \Omega \to \mathbf{R}$ is continuous, that $\vartheta \in H^{1,p}(\Omega; \mu)$, and that $u \in \mathcal{K}_{\psi,\vartheta}$ is a continuous supersolution in Ω. Then u is a solution to the obstacle problem in $\mathcal{K}_{\psi,\vartheta}$ if and only if u is a solution of (3.8) in the open set $\{u > \psi\}$.*

PROOF: The necessity was proved in Theorem 3.67. For the converse, let v be the continuous solution to the obstacle problem in $\mathcal{K}_{\psi,\vartheta}$. Since $\min(u,v) \in \mathcal{K}_{\psi,\vartheta}$, we have $u \geq v$ in Ω (Lemma 3.22). On the other hand, it is not difficult to see that $u - v \in H_0^{1,p}(\{u > \psi\}; \mu)$ (the easiest way to establish this would be the use of Theorem 4.5; however, it also follows rather straightforwardly from the definition of $H_0^{1,p}$). Now the comparison lemma 3.18 ensures that $u \leq v$, and hence $u = v$ is the solution to the obstacle problem.   □

3.72. Convergence properties of solutions
We close this chapter by establishing some convergence results which will be used in coming chapters.

3.73. Lemma.   *Suppose that the vector-valued functions u and u_i, $i = 1, 2, \ldots,$ belong to $L^p(\Omega; \mu)$. If*

$$\lim_{i \to \infty} \int_\Omega \big(\mathcal{A}(x, u(x)) - \mathcal{A}(x, u_i(x))\big) \cdot \big(u(x) - u_i(x)\big)\, dx = 0\,,$$

then u_i converges to u weakly in $L^p(\Omega; \mu)$ and $\mathcal{A}(\cdot, u_i)w^{-1/p}$ converges to $\mathcal{A}(\cdot, u)w^{-1/p}$ weakly in $L^{p/(p-1)}(\Omega; dx)$.

PROOF: Write

$$I_i = \int_\Omega \big(\mathcal{A}(x, u(x)) - \mathcal{A}(x, u_i(x))\big) \cdot \big(u(x) - u_i(x)\big)\, dx$$

for $i = 1, 2, \ldots,$ and pick a subsequence v_i of u_i such that

$$(3.74) \qquad \big(\mathcal{A}(x, v_i(x)) - \mathcal{A}(x, u(x))\big) \cdot \big(v_i(x) - u(x)\big) \to 0$$

for a.e. x in Ω; note that the integrand is nonnegative by (3.6). Let $x \in \Omega$ be such that both (3.74) and assumptions (3.3)–(3.6) are valid, that $0 < w(x) < \infty$, and that $|u(x)| < \infty$. Let $u_{i_j}(x)$ be any subsequence of v_i such that $u_{i_j}(x)$ converges to a point ξ in the extended space $\mathbf{R}^n \cup \{\infty\}$. Then $|\xi| < \infty$ since $|\xi| = \infty$ would violate (3.74) in view of the inequality

$$\big(\mathcal{A}\big(x, u_{i_j}(x)\big) - \mathcal{A}\big(x, u(x)\big)\big) \cdot \big(u_{i_j}(x) - u(x)\big)$$
$$\geq w(x)\,\big(\alpha\,|u_{i_j}(x)|^p - \beta\,\big(|u(x)|\,|u_{i_j}(x)|^{p-1} + |u(x)|^{p-1}|u_{i_j}(x)|\big)\big)$$
$$\geq w(x)\,\alpha\,\big(|u_{i_j}(x)|^p - c\,\big(|u_{i_j}(x)|^{p-1} + |u_{i_j}(x)|\big)\big),$$

where $c = \max(|u(x)|, |u(x)|^{p-1})\beta/\alpha$. From this we also see that

$$\alpha\,||u_{i_j}||^p \leq I_{i_j} + c\,(||u_{i_j}||^{p-1} + ||u_{i_j}||),$$

where

$$||f|| = (\int_\Omega |f|^p\,d\mu)^{1/p}$$

and $c = \beta \max(||u||, ||u||^{p-1})$. The above inequality shows that

$$||u_{i_j}|| \leq M < \infty,$$

where M is independent of the index i_j.

Next, since the mapping $h \mapsto \mathcal{A}(x, \xi)$ is continuous, we have

$$\big(\mathcal{A}(x, \xi) - \mathcal{A}(x, u(x))\big) \cdot \big(\xi - u(x)\big) = 0,$$

and hence $u(x) = \xi$ by (3.6). It follows that

$$\lim_{j \to \infty} u_{i_j}(x) = u(x)$$

and

$$\lim_{j \to \infty} \mathcal{A}\big(x, u_{i_j}(x)\big)w^{-1/p}(x) = \mathcal{A}\big(x, u(x)\big)w^{-1/p}(x)$$

for a.e. x in Ω. Since the $L^p(\Omega; \mu)$-norms of u_{i_j} are uniformly bounded, we have that the $L^{p/(p-1)}(\Omega; dx)$-norms of $\mathcal{A}(x, u_{i_j})w^{-1/p}$ are also uniformly bounded, and it follows from the properties of weak convergence that $u_{i_j} \to u$ weakly in $L^p(\Omega; \mu)$ and

$$\mathcal{A}(x, u_{i_j})w^{-1/p} \to \mathcal{A}(x, u)w^{-1/p}$$

weakly in $L^{p/(p-1)}(\Omega; dx)$.

We observe that the weak limits are independent of the choice of the subsequences v_i and u_{i_j}, and that the sequence u_i is bounded in $L^p(\Omega; \mu)$. Therefore $u_i \to u$ weakly in $L^p(\Omega; \mu)$ and

$$\mathcal{A}(x, u_i)w^{-1/p} \to \mathcal{A}(x, u)w^{-1/p}$$

weakly in $L^{p/(p-1)}(\Omega; dx)$, as required. $\square$

3.75. Theorem. *Suppose that u_i is an increasing and locally bounded sequence of supersolutions in Ω. Then $u = \lim u_i$ is a supersolution of (3.8) in Ω.*

PROOF: We may assume that each u_i is lower semicontinuous and satisfies the "ess lim inf" property (3.64). Fix open sets $D \Subset G \Subset \Omega$. Since u is locally bounded, we may clearly assume that $u < 0$ in G, and it follows from the Caccioppoli estimate (3.29) that the sequence ∇u_i is uniformly bounded in $L^p(G; \mu)$. Thus Theorem 1.32 yields that $u \in H^{1,p}(G; \mu)$ and that $\nabla u_i \to \nabla u$ weakly in $L^p(G; \mu)$. Choose $\eta \in C_0^\infty(G)$ such that $0 \leq \eta \leq 1$ and $\eta = 1$ in D. Write $\psi = \eta(u - u_i)$ and

$$I_i = \int_D \big(\mathcal{A}(x, \nabla u) - \mathcal{A}(x, \nabla u_i) \big) \cdot \big(\nabla u - \nabla u_i \big) \, dx \, .$$

Using ψ as a test function for the supersolution u_i and applying the Hölder inequality, we obtain

$$- \int_G \eta \mathcal{A}(x, \nabla u_i) \cdot (\nabla u - \nabla u_i) \, dx \leq \int_G (u - u_i) \mathcal{A}(x, \nabla u_i) \cdot \nabla \eta \, dx$$

$$\leq \beta \big(\int_G |u - u_i|^p |\nabla \eta|^p \, d\mu \big)^{1/p} \big(\int_G |\nabla u_i|^p \, d\mu \big)^{(p-1)/p}$$

$$\leq c \big(\int_G |u - u_i|^p \, d\mu \big)^{1/p} \to 0 \, ,$$

where the last conclusion follows from the dominated convergence theorem. Moreover, since

$$\eta \mathcal{A}(x, \nabla u) w^{-1} \in L^{p/(p-1)}(G; \mu) \, ,$$

the weak convergence implies

$$\int_G \eta \mathcal{A}(x, \nabla u) \cdot (\nabla u - \nabla u_i) \, dx \to 0 \, .$$

Then, since

$$\eta \big(\mathcal{A}(x, \nabla u) - \mathcal{A}(x, \nabla u_i) \big) \cdot \big(\nabla u - \nabla u_i \big) \geq 0$$

a.e. in G, we conclude

(3.76) $\lim_{i \to \infty} I_i = 0 \, .$

Now Lemma 3.73 implies that $\mathcal{A}(\cdot, \nabla u_i) w^{-1/p}$ converges to $\mathcal{A}(\cdot, \nabla u) w^{-1/p}$ weakly in $L^{p/(p-1)}(D; dx)$.

It is now an easy task to show that u is a supersolution of (3.8). Indeed, if $\varphi \in C_0^\infty(\Omega)$, $\varphi \geq 0$, is such that $\operatorname{spt} \varphi \subset D$, then $\nabla \varphi \in L^p(D; \mu)$ and $\nabla \varphi w^{1/p} \in L^p(D; dx)$. Hence

$$0 \leq \int_\Omega \mathcal{A}(x, \nabla u_i) \cdot \nabla \varphi \, dx = \int_D \mathcal{A}(x, \nabla u_i) w^{-1/p} \cdot \nabla \varphi w^{1/p} \, dx$$

$$\to \int_D \mathcal{A}(x, \nabla u) w^{-1/p} \cdot \nabla \varphi w^{1/p} \, dx = \int_\Omega \mathcal{A}(x, \nabla u) \cdot \nabla \varphi \, dx.$$

This concludes the proof of Theorem 3.75. $\square$

There is a similar result for decreasing sequences.

3.77. Theorem. *Suppose that u_i is a decreasing and locally bounded sequence of supersolutions in Ω. Then $u = \lim_{i \to \infty} u_i$ is a supersolution.*

PROOF: As in the proof of Theorem 3.75 it follows from the Caccioppoli estimate (3.29) that $u \in H_{loc}^{1,p}(\Omega; \mu)$. Pick an open set $G \Subset \Omega$ and let v be the solution to the obstacle problem in $\mathcal{K}_u(G)$. Then $v \geq u$ and Lemma 3.22 implies $v \leq u_i$ for each i. Thus $v = u$ a.e. in G and therefore u is a supersolution. $\square$

Minor modifications to the proofs of Theorems 3.75 and 3.77 reveal that the same conclusions hold if instead of requiring that the sequence u_i be locally bounded, one insists that the limit function u is in $H_{loc}^{1,p}(\Omega; \mu)$.

The reader is asked to mimic the proof of Theorem 3.75 to obtain the following theorem.

3.78. Theorem. *Let $u_i \in C(\Omega)$ be a sequence of solutions of (3.8) in Ω such that $u_i \to u$ locally uniformly in Ω. Then u is a solution of (3.8) in Ω.*

Another way to establish Theorem 3.78 is to modify the sequence by adding constants in such a way that it becomes increasing (see the proof of Theorem 16.10); hence the limit function u is a supersolution by Theorem 3.75 and a subsolution by Theorem 3.77.

The chapter is closed by two useful convergence theorems for obstacle problems.

3.79. Theorem. *Suppose that Ω is bounded and that $\psi_i \in H^{1,p}(\Omega; \mu)$ is a decreasing sequence such that $\psi_i \to \psi$ in $H^{1,p}(\Omega; \mu)$. Let $u_i \in H^{1,p}(\Omega; \mu)$ be a solution to the obstacle problem in $\mathcal{K}_{\psi_i}$. Then the sequence u_i is decreasing and the limit function $u = \lim u_i$ is a solution to the obstacle problem in $\mathcal{K}_\psi$.*

PROOF: It immediately follows from Lemma 3.22 that the sequence u_i is decreasing. Hence, by the dominated convergence theorem, $u_i \to u$ in

$L^p(\Omega; \mu)$. Next, because each u_i is the solution to the obstacle problem, the quasiminimizing property (3.15) implies

$$\sup_i \int_\Omega |\nabla u_i|^p \, d\mu \leq \sup_i c \int_\Omega |\nabla \psi_i|^p \, d\mu < \infty.$$

Hence $\nabla u_i \to \nabla u$ weakly in $L^p(\Omega; \mu)$ and $u - \psi \in H_0^{1,p}(\Omega; \mu)$ (Theorems 1.31 and 1.32). Because $u \geq \psi$ a.e. in Ω, it suffices to verify that

$$(3.80) \qquad \int_\Omega \mathcal{A}(x, \nabla u) \cdot \nabla \varphi \, dx \geq 0$$

whenever $\varphi \in H_0^{1,p}(\Omega; \mu)$ with $\varphi \geq \psi - u$ a.e. in Ω.

First consider $v_i = \psi_i - \psi$. Then $u + v_i - \psi_i \in H_0^{1,p}(\Omega; \mu)$ and since also $u + v_i \geq \psi_i$ a.e., we find that

$$\begin{aligned}
0 &\geq \int_\Omega \big(\mathcal{A}(x, \nabla u_i) - \mathcal{A}(x, \nabla u)\big) \cdot (\nabla u - \nabla u_i) \, dx \\
&= \int_\Omega \mathcal{A}(x, \nabla u_i) \cdot \nabla(u + v_i - u_i) \, dx - \int_\Omega \mathcal{A}(x, \nabla u_i) \cdot \nabla v_i \, dx \\
&\quad - \int_\Omega \mathcal{A}(x, \nabla u) \cdot (\nabla u - \nabla u_i) \, dx \\
&\geq - \int_\Omega \mathcal{A}(x, \nabla u_i) \cdot \nabla v_i \, dx - \int_\Omega \mathcal{A}(x, \nabla u) \cdot (\nabla u - \nabla u_i) \, dx.
\end{aligned}$$

Since $\nabla v_i \to 0$ in $L^p(\Omega; \mu)$ and since $\nabla u_i \to \nabla u$ weakly in $L^p(\Omega; \mu)$, the last two integrals tend to zero as $j \to \infty$. Hence

$$\int_\Omega \big(\mathcal{A}(x, \nabla u_i) - \mathcal{A}(x, \nabla u)\big) \cdot (\nabla u - \nabla u_i) \, dx \to 0,$$

and so

$$\int_\Omega \mathcal{A}(x, \nabla u_i) \cdot (\nabla u_i - \nabla u) \, dx \to 0.$$

Moreover, it follows from Lemma 3.73 that

$$\mathcal{A}(\cdot, \nabla u_i) w^{-1/p} \to \mathcal{A}(\cdot, \nabla u) w^{-1/p}$$

weakly in $L^{p/(p-1)}(\Omega; dx)$.

To complete the proof of (3.80), fix a function $\varphi \in H_0^{1,p}(\Omega; \mu)$ with $\varphi \geq \psi - u$ a.e. and write

$$\varphi_i = \varphi + u + v_i - u_i.$$

Then $\varphi_i \in H_0^{1,p}(\Omega; \mu)$ and $\varphi_i \geq \psi_i - u_i$ a.e. in Ω. Hence

$$\int_\Omega \mathcal{A}(x, \nabla u_i) \cdot \nabla \varphi_i \, dx \geq 0$$

and we conclude that

$$\int_\Omega \mathcal{A}(x, \nabla u_i) \cdot \nabla \varphi \, dx$$

$$= \int_\Omega \mathcal{A}(x, \nabla u_i) \cdot \nabla \varphi_i \, dx + \int_\Omega \mathcal{A}(x, \nabla u_i) \cdot (\nabla \varphi - \nabla \varphi_i) \, dx$$

$$\geq \int_\Omega \mathcal{A}(x, \nabla u_i) \cdot (\nabla u_i - \nabla u) \, dx - \int_\Omega \mathcal{A}(x, \nabla u_i) \cdot \nabla v_i \, dx.$$

Since the last two integrals tend to zero as $i \to \infty$, we obtain

$$\int_\Omega \mathcal{A}(x, \nabla u) \cdot \nabla \varphi \, dx$$

$$= \int_\Omega w^{-1/p} \mathcal{A}(x, \nabla u) \cdot \nabla \varphi \, w^{1/p} \, dx$$

$$= \lim_{i \to \infty} \int_\Omega w^{-1/p} \mathcal{A}(x, \nabla u_i) \cdot \nabla \varphi \, w^{1/p} \, dx$$

$$= \lim_{i \to \infty} \int_\Omega \mathcal{A}(x, \nabla u_i) \cdot \nabla \varphi \, dx \geq 0$$

as desired. The theorem is proved. □

3.81. Theorem. *Suppose that Ω is bounded. Let ψ_i and u_i be increasing sequences of functions in Ω such that $\psi_i \to \psi$ and that u_i is the solution to the obstacle problem in $\mathcal{K}_{\psi_i, u_i}$. Then the limit function $u = \lim u_i$ is the solution to the obstacle problem in $\mathcal{K}_{\psi, u}$ provided that $u \in H^{1,p}(\Omega; \mu)$.*

PROOF: Let v be the solution to the obstacle problem in $\mathcal{K}_{\psi, u}$. Since u is a supersolution (see Theorem 3.75 and the remark after Theorem 3.77), it follows from the comparison lemma 3.22 that $u \geq v$. The comparison lemma also implies that $v \geq u_i$, and hence $v \geq u$. We conclude that $u = v$, and the proof is complete. □

NOTES TO CHAPTER 3. The fundamental work on quasilinear elliptic equations is Serrin (1964). Serrin generalizes the iteration method in Moser (1961) which has become a standard technique in the theory of partial differential equations. Moser's idea is behind the proof of Lemma 3.34. The important works prior to Moser are De Giorgi (1957) and Nash (1958).

In this connection, see also the monographs Morrey (1966), Ladyzhenskaya and Ural'tseva (1968), Gilbarg and Trudinger (1983). Supersolutions were systematically studied in Trudinger (1967). The treatise by Kinderlehrer and Stampacchia (1980) covers the basic theory of variational inequalities. The existence of solutions to general nonlinear equations was first established in Leray and Lions (1965).

The continuity of solutions to nonlinear obstacle problems with minimal assumptions was first demonstrated in Michael and Ziemer (1986). They proved that indeed a weaker assumption than continuity on the obstacle ψ suffices; a converse to their result was provided in Heinonen and Kilpeläinen (1988c) and Kilpeläinen and Malý (1992b). Earlier works include Granlund et al. (1983), where the continuity was achieved in the borderline case $p = n$ by using a generalization of the concept of monotone functions.

The argument that leads to Theorems 3.63 and 3.67 is an adaptation of Michael and Ziemer (1986) to the weighted situation; see also Heinonen and Kilpeläinen (1988a). The sharp weak Harnack inequalities in the unweighted case were first proved in Trudinger (1967). The comparison lemmas and the convergence theorems are taken from Heinonen and Kilpeläinen (1988a,c). Lemma 3.38 appears e.g. in Di Benedetto and Trudinger (1984) and Iwaniec and Nolder (1985). An unweighted version of Lemma 3.73 is from Maz'ya (1976).

Weighted linear elliptic equations have been investigated in many papers. See Fabes et al. (1982a), Stredulinsky (1984), Chanillo and Wheeden (1986), Franchi and Serapioni (1987), Chiarenza et al. (1989), Franchi (1991), and the references therein.

4
Refined Sobolev spaces

The elements in the Sobolev space $H^{1,p}(\Omega; \mu)$ are equivalence classes of functions which agree a.e. in Ω. In the theory of partial differential equations it is often necessary to have a more accurate description of functions in $H^{1,p}(\Omega; \mu)$. By using a notion of capacity, the almost everywhere equivalence can be refined. Next we discuss this refinement.

Recall the definition for the Sobolev (p, μ)-capacity from Chapter 2: if E is a subset of $\mathbf{R}^n$, then

$$C_{p,\mu}(E) = \inf \int_{\mathbf{R}^n} (|u|^p + |\nabla u|^p) \, d\mu \, ,$$

where the infimum is taken over all $u \in H^{1,p}(\mathbf{R}^n; \mu)$ such that $u = 1$ (equivalently $u \geq 1$ a.e.) in a neighborhood of E.

Then $E \mapsto C_{p,\mu}(E)$ is a well-defined monotone, subadditive set function on the subsets of Ω; it is a Choquet capacity. In Chapter 2 we covered these and other properties of Sobolev (p, μ)-capacity.

The sets of zero (p, μ)-capacity appear naturally in connection with the theory of the Sobolev spaces $H^{1,p}(\Omega; \mu)$, as was shown in Section 2.42. Recall here that $C_{p,\mu}(E) = 0$ if and only if $\mathrm{cap}_{p,\mu} E = 0$. Also recall that $\mu(E) = |E| = 0$ if E is of (p, μ)-capacity zero (Lemma 2.10), while the converse is not true since a closed set of (p, μ)-capacity zero cannot separate $\mathbf{R}^n$ (Lemma 2.46).

4.1. Quasicontinuous functions

We show that for each $u \in H^{1,p}(\Omega; \mu)$ there is a function v such that $u = v$ a.e. and that v is quasicontinuous, i.e. v is continuous when restricted to a set whose complement has arbitrarily small Sobolev (p, μ)-capacity. Moreover, this quasicontinuous representative is unique up to a set of (p, μ)-capacity zero. To be more precise, we first state the definition.

DEFINITION. A function $u : \Omega \to \overline{\mathbf{R}}$ is (p, μ)-*quasicontinuous* in Ω if for every $\varepsilon > 0$ there is an open set G such that $C_{p,\mu}(G) < \varepsilon$ and the restriction of u to $\Omega \setminus G$ is finite valued and continuous.

A sequence of functions $\psi_j : \Omega \to \mathbf{R}$ converges (p, μ)-*quasiuniformly* in Ω to a function ψ if for every $\varepsilon > 0$ there is an open set G such that $C_{p,\mu}(G) < \varepsilon$ and $\psi_j \to \psi$ uniformly in $\Omega \setminus G$.

The sequence ψ_j converges *locally* (p, μ)-*quasiuniformly* if it converges (p, μ)-quasiuniformly in each open $D \Subset \Omega$.

4.2. Lemma. *A sequence ψ_j converges locally (p, μ)-quasiuniformly in Ω if and only if for every $\varepsilon > 0$ there is an open set $G \subset \Omega$ with $\mathrm{C}_{p,\mu}(G) < \varepsilon$ such that the sequence converges uniformly on every compact subset of $\Omega \setminus G$.*

PROOF: The condition is clearly sufficient. For the converse, fix $\varepsilon > 0$ and exhaust Ω by open sets $D_1 \Subset D_2 \Subset \cdots \Subset \Omega$ with $\cup D_i = \Omega$. For each $i = 1, 2, \ldots$, choose an open set G_i with $\mathrm{C}_{p,\mu}(G_i) < 2^{-i} \varepsilon$ such that ψ_j converges uniformly on $D_i \setminus G_i$. Let $G = \cup G_i$. Then $\mathrm{C}_{p,\mu}(G) < \varepsilon$ and if K is a compact subset of $\Omega \setminus G$, K is contained in D_i for some i. Because $K \subset D_i \setminus G \subset D_i \setminus G_i$, the sequence ψ_j converges uniformly on K. $\square$

For the following theorem recall that a property holds (p, μ)-quasievery-where, or simply q.e., if it holds except on a set of (p, μ)-capacity zero.

4.3. Theorem. *Let $\varphi_j \in C(\Omega) \cap H^{1,p}(\Omega; \mu)$ be a Cauchy sequence in $H^{1,p}(\Omega; \mu)$. Then there is a subsequence φ_k which converges locally (p, μ)-quasiuniformly in Ω to a function $u \in H^{1,p}(\Omega; \mu)$. In particular, u is (p, μ)-quasicontinuous and $\varphi_k \to u$ pointwise (p, μ)-quasieverywhere in Ω.*

PROOF: Suppose that a locally quasiuniformly convergent subsequence can be selected. Then it clearly converges (p, μ)-quasieverywhere to a (p, μ)-quasicontinuous function u. Moreover, $u \in H^{1,p}(\Omega; \mu)$ because $H^{1,p}(\Omega; \mu)$ is a Banach space.

Thus it suffices to show that a locally quasiuniformly convergent subsequence can be found.

Since φ_j is a Cauchy sequence in $H^{1,p}(\Omega; \mu)$, there is a subsequence, denoted again by φ_j, such that the series

$$\sum_{j=1}^{\infty} \int_{\Omega} 2^{jp} (|\varphi_j - \varphi_{j+1}|^p + |\nabla \varphi_j - \nabla \varphi_{j+1}|^p) \, d\mu$$

converges. Let $D \Subset \Omega$ be an open set. If $\psi \in C_0^{\infty}(\Omega)$ is such that $\psi = 1$ on D, then the series

$$\sum_{j=1}^{\infty} \int_{\Omega} 2^{jp} (|\psi(\varphi_j - \varphi_{j+1})|^p + |\nabla(\psi(\varphi_j - \varphi_{j+1}))|^p) \, d\mu$$

converges, and thus for every $\varepsilon > 0$ there is a j_ε such that

$$\sum_{j=j_\varepsilon}^{\infty} \int_{\Omega} 2^{jp} (|\psi(\varphi_j - \varphi_{j+1})|^p + |\nabla(\psi(\varphi_j - \varphi_{j+1}))|^p) \, d\mu < \varepsilon.$$

On the other hand, for the open set

$$E_j = \{ x \in D : |\varphi_j(x) - \varphi_{j+1}(x)| > 2^{-j} \}$$

we have

$$C_{p,\mu}(E_j) \leq \int_\Omega 2^{jp}(|\psi(\varphi_j - \varphi_{j+1})|^p + |\nabla(\psi(\varphi_j - \varphi_{j+1}))|^p)\, d\mu.$$

Put

$$E_\varepsilon = \bigcup_{j=j_\varepsilon}^{\infty} E_j.$$

Then we have by the subadditivity of the capacity that

$$C_{p,\mu}(E_\varepsilon) \leq \sum_{j=j_\varepsilon}^{\infty} C_{p,\mu}(E_j) < \varepsilon.$$

Moreover, for $j_\varepsilon \leq j \leq k$

$$|\varphi_j - \varphi_k| \leq \sum_{l=j}^{k-1} 2^{-l} \leq 2^{1-j}$$

in $D \setminus E_\varepsilon$, and this means that φ_j converges uniformly in $D \setminus E_\varepsilon$.

Next, fix $\varepsilon > 0$ and choose an exhaustion of Ω by open sets $D_1 \Subset D_2 \Subset \cdots \Subset \Omega$ with $\cup_{k=1}^{\infty} D_k = \Omega$. Let φ_j^0 denote the sequence φ_j. Then for each $k \geq 1$ we can choose a subsequence φ_l^k of φ_j^{k-1} such that φ_l^k converges uniformly in $D_k \setminus E_k$, where E_k is an open set with $C_{p,\mu}(E_k) < 2^{-k}\varepsilon$. Let $E = \cup E_k$. Then $C_{p,\mu}(E) < \varepsilon$, and if $F \subset \Omega \setminus E$ is a compact set, we have that $F \subset D_k \setminus E_k$ for all sufficiently large k. Thus the diagonal sequence φ_m^m converges uniformly on F, hence locally (p, μ)-quasiuniformly in Ω. $\square$

Theorem 4.3 implies:

4.4. Theorem. *Suppose that $u \in H^{1,p}(\Omega; \mu)$. Then there exists a (p, μ)-quasicontinuous function $v \in H^{1,p}(\Omega; \mu)$ such that $u = v$ a.e.*

Thus each $u \in H^{1,p}(\Omega; \mu)$ has a (p, μ)-quasicontinuous representative. This fact is important in the theory of Sobolev spaces and especially in the theory of partial differential equations, where it is often desirable to know the accurate pointwise behavior of solutions. We analyze quasicontinuous functions a bit more and show later (Theorem 4.12) that the quasicontinuous representative is essentially unique: any two quasicontinuous functions that coincide almost everywhere coincide quasieverywhere. This result is surprisingly difficult to establish. First we prove a useful theorem which characterizes functions in $H^{1,p}(\Omega; \mu)$ that can be approximated by $C_0^\infty(\Omega)$ functions in $H^{1,p}(\Omega; \mu)$.

4.5. Theorem. *Suppose that $u \in H^{1,p}(\Omega; \mu)$. Then $u \in H_0^{1,p}(\Omega; \mu)$ if and only if there is a (p, μ)-quasicontinuous function v in $\mathbf{R}^n$ such that $v = u$ a.e. in Ω and $v = 0$ q.e. in $\complement\Omega$.*

PROOF: Fix $u \in H_0^{1,p}(\Omega; \mu)$ and let $\varphi_j \in C_0^\infty(\Omega)$ be a sequence converging to u in $H^{1,p}(\Omega; \mu)$. By Theorem 4.3 there is a subsequence of φ_j which converges (p, μ)-quasieverywhere in $\mathbf{R}^n$ to a (p, μ)-quasicontinuous function v such that $v = u$ a.e. in Ω and $v = 0$ q.e. on $\complement\Omega$. Hence v is the desired function.

To prove the converse, assume first that Ω is bounded. Because the truncations of v converge to v in $H^{1,p}(\Omega; \mu)$, we may assume that v is bounded. Let

$$E = \{x \in \partial\Omega : v(x) \neq 0\}$$

and choose open sets G_j such that $E \subset G_j$,

$$C_{p,\mu}(G_j) \to 0,$$

and $v|_{\complement G_j}$ is continuous. Pick a sequence $\varphi_j \in H^{1,p}(\mathbf{R}^n; \mu)$ such that $0 \le \varphi_j \le 1$, $\varphi_j = 1$ everywhere in G_j, and

$$\int_{\mathbf{R}^n} (|\varphi_j|^p + |\nabla\varphi_j|^p)\, d\mu \to 0.$$

Then by Theorem 1.24

$$w_j = (1 - \varphi_j)v \in H^{1,p}(\Omega; \mu).$$

Moreover, $\lim_{x \to y} w_j(x) = 0$ for $y \in \partial\Omega$. Thus $w_j \in H_0^{1,p}(\Omega; \mu)$ (Lemma 1.26), and clearly $w_j \to v$ in $L^p(\Omega; \mu)$. By the dominated convergence theorem

$$\left(\int_\Omega |\nabla w_j - \nabla v|^p\, d\mu\right)^{1/p} = \left(\int_\Omega |\nabla(\varphi_j v)|^p\, d\mu\right)^{1/p}$$

$$\le \left(\int_\Omega |v\nabla\varphi_j|^p\, d\mu\right)^{1/p} + \left(\int_\Omega |\varphi_j \nabla v|^p\, d\mu\right)^{1/p} \to 0.$$

So $w_j \to v$ in $H^{1,p}(\Omega; \mu)$, and hence $v \in H_0^{1,p}(\Omega; \mu)$. The proof is complete in case Ω is bounded.

Assume that Ω is unbounded, let $A_j = B(0, j+1) \setminus \overline{B}(0, j-1)$, $j = 1, 2, \ldots$, and choose functions $\varphi_j \in C_0^\infty(A_j)$, $0 \le \varphi_j \le 1$ such that

$$\sum_{j=1}^\infty \varphi_j(x) = 1$$

for each $x \in \mathbf{R}^n$; i.e. φ_j is a partition of unity subordinate to the covering A_j. Then $v\varphi_j \in H_0^{1,p}(A_j \cap \Omega; \mu)$ by the previous part of the proof. Fix $\varepsilon > 0$ and choose $\psi_j \in C_0^\infty(A_j \cap \Omega)$ with

$$\|\psi_j - v\varphi_j\|_{1,p} < 2^{-(j+1)}\varepsilon.$$

Then $\psi = \sum_j \psi_j \in C^\infty(\mathbf{R}^n)$, $\psi = 0$ in $\complement\Omega$, and

$$\|\psi - v\|_{1,p} \le \|\sum_j \psi_j - v\sum_j \varphi_j\|_{1,p} < \varepsilon/2.$$

Now choose j such that

$$\left(\int_{\complement B(0,j)} |\psi|^p \, d\mu\right)^{1/p} + \left(\int_{\complement B(0,j)} |\nabla\psi|^p \, d\mu\right)^{1/p} < \varepsilon/2.$$

Then $\eta = \sum_{k=1}^j \psi_k \in C_0^\infty(\Omega)$ and

$$\|\eta - \psi\|_{1,p} \le \left(\int_{\complement B(0,j)} |\psi|^p \, d\mu\right)^{1/p} + \left(\int_{\complement B(0,j)} |\nabla\psi|^p \, d\mu\right)^{1/p} < \varepsilon/2.$$

Hence

$$\|\eta - v\|_{1,p} < \varepsilon$$

which shows that $v \in H_0^{1,p}(\Omega; \mu)$. The theorem is proved. $\qquad\square$

We discuss the uniqueness of a (p, μ)-quasicontinuous representative of a Sobolev function. A global version of Theorem 4.3 is presented first.

4.6. Theorem. *Suppose that $\varphi_j \in C(\mathbf{R}^n) \cap H^{1,p}(\mathbf{R}^n; \mu)$ is a Cauchy sequence in $H^{1,p}(\mathbf{R}^n; \mu)$. Then there is a subsequence of φ_j converging (p, μ)-quasiuniformly in $\mathbf{R}^n$ to a function $u \in H^{1,p}(\mathbf{R}^n; \mu)$.*

PROOF: The proof of Theorem 4.3 applies almost word for word; the only difference is that no exhaustion is needed: the set D and the function ψ are replaced by $\mathbf{R}^n$ and 1, respectively. $\qquad\square$

We denote by

$$Q^{1,p} = Q^{1,p}(\mathbf{R}^n; \mu)$$

the set of all functions $u \in H^{1,p}(\mathbf{R}^n; \mu)$ such that there is a sequence $\varphi_j \in C(\mathbf{R}^n) \cap H^{1,p}(\mathbf{R}^n; \mu)$ converging to u both in $H^{1,p}(\mathbf{R}^n; \mu)$ and (p, μ)-quasiuniformly in $\mathbf{R}^n$.

It is immediate that the functions in $Q^{1,p}$ are (p, μ)-quasicontinuous, and for each $v \in H^{1,p}(\mathbf{R}^n; \mu)$ there is $u \in Q^{1,p}$ such that $u = v$ a.e. (Theorem 4.6). We soon show that, conversely, each (p, μ)-quasicontinuous function v of $H^{1,p}(\mathbf{R}^n; \mu)$ belongs to $Q^{1,p}$.

4.7. Lemma.   Let $u \in Q^{1,p}$. If $u \geq 1$ (p, μ)-q.e. on E, then

$$C_{p,\mu}(E) \leq \int_{\mathbf{R}^n} (|u|^p + |\nabla u|^p) \, d\mu.$$

PROOF: Fix $\varepsilon \in (0, 1)$. Let $\varphi_j \in C(\mathbf{R}^n) \cap H^{1,p}(\mathbf{R}^n; \mu)$ be such that $\varphi_j \to u$ both in $H^{1,p}(\mathbf{R}^n; \mu)$ and (p, μ)-quasiuniformly in $\mathbf{R}^n$. Choose an open set G with $C_{p,\mu}(G) < \varepsilon$ and $\varphi_j \to u$ uniformly on $\complement G$. Let

$$E_1 = \{x \in E : u(x) \geq 1\}$$

and

$$G_j = \{x \in \mathbf{R}^n : \varphi_j(x) > 1 - \varepsilon\}.$$

Then G_j is open and

$$E_1 \setminus G \subset G_j \quad \text{for } j \geq j_\varepsilon.$$

Consequently, for $j \geq j_\varepsilon$

$$C_{p,\mu}(E) = C_{p,\mu}(E_1) \leq C_{p,\mu}(G_j) + C_{p,\mu}(G).$$

Since $\min(1, (1 - \varepsilon)^{-1} \varphi_j) = 1$ in G_j, we have

$$C_{p,\mu}(G_j) \leq (1 - \varepsilon)^{-p} \int_{\mathbf{R}^n} (|\varphi_j|^p + |\nabla \varphi_j|^p) \, d\mu,$$

and hence by letting $j \to \infty$ we obtain

$$C_{p,\mu}(E) \leq (1 - \varepsilon)^{-p} \int_{\mathbf{R}^n} (|u|^p + |\nabla u|^p) \, d\mu + \varepsilon.$$

The lemma follows. $\square$

Next we refine Theorem 4.6.

4.8. Lemma.   Suppose that $u_j \in Q^{1,p}$ converge in $H^{1,p}(\mathbf{R}^n; \mu)$ to a function $u \in Q^{1,p}$. Then there is a subsequence of u_j which converges to u (p, μ)-quasiuniformly in $\mathbf{R}^n$.

PROOF: The proof is very similar to that of Theorem 4.3. Passing to a subsequence we may assume that

$$\sum_{j=1}^{\infty} 2^{jp} \int_{\mathbf{R}^n} (|u_j - u|^p + |\nabla u_j - \nabla u|^p) \, d\mu < \infty.$$

If
$$E_j = \{x \colon |u_j(x) - u(x)| > 2^{-j}\}$$
and
$$\tilde{E}_k = \bigcup_{j \geq k} E_j,$$

we have by Lemma 4.7 that

$$C_{p,\mu}(\tilde{E}_k) \leq \sum_{j \geq k} C_{p,\mu}(E_j)$$

$$\leq \sum_{j \geq k} 2^{jp} \int_{\mathbf{R}^n} (|u_j - u|^p + |\nabla u_j - \nabla u|^p)\, d\mu \to 0$$

as $k \to \infty$. Since u_j converges uniformly to u outside $\tilde{E}_k$, the desired (p, μ)-quasiuniform convergence is established. $\square$

For the proof of the uniqueness of a quasicontinuous representative we still need a lemma; to prove it we use results from Chapter 3.

4.9. Lemma. *Suppose that G_j are open sets with $C_{p,\mu}(G_j) \to 0$. Then for (p, μ)-q.e. x there is an index $j = j(x)$ and a positive number $r_0 = r_0(x)$ such that*

$$\mu(G_j \cap B(x,r)) < \frac{1}{2}\mu(B(x,r))$$

for all $r \leq r_0$.

PROOF: Since the problem is local, there is no loss of generality in assuming that all G_j are contained in a ball B. Fix j and let u_j be a solution to the obstacle problem in $\mathcal{K}_{\chi_j, 0}(4B)$ for $\mathcal{A}(x, \xi) = w(x)|\xi|^{p-2}\xi$, where χ_j is the characteristic function of G_j (see Theorem 3.21). Choose a lower semicontinuous representative of u_j that satisfies

$$(4.10) \qquad \lim_{r \to 0} \int_{B(x,r)} u_j\, d\mu = u_j(x)$$

for each $x \in 4B$ (see Theorem 3.63 and equation (3.65)).

We first show that

$$(4.11) \qquad \int_{4B} |\nabla u_j|^p\, d\mu = \mathrm{cap}_{p,\mu}(G_j, 4B).$$

The inequality $\geq$ is immediate. To achieve the reverse inequality, exhaust G_j by compact sets $K_1 \Subset K_2 \Subset \cdots \Subset G_j$, $\cup K_k = G_j$, and let v_k be the solution to the obstacle problem in $\mathcal{K}_{f_k, 0}(4B)$ for $\mathcal{A}(x, \xi) = w(x)|\xi|^{p-2}\xi$, where

f_k is the characteristic function of K_k. Since v_k minimizes the weighted p-energy among all functions in $\mathcal{K}_{f_k,0}(4B)$ (see Section 3.13), we have

$$\int_{4B} |\nabla v_k|^p \, d\mu \leq \text{cap}_{p,\mu}(K_k, 4B) \leq \int_{4B} |\nabla u_j|^p \, d\mu \, .$$

By Theorem 3.81 we have that $v_k \to u_j$ a.e. in $4B$. So it follows that $\nabla v_k \to \nabla u_j$ weakly in $L^p(4B; \mu)$ (Theorem 1.32), and hence

$$\int_{4B} |\nabla u_j|^p \, d\mu \leq \liminf_{k \to \infty} \int_{4B} |\nabla v_k|^p \, d\mu$$
$$\leq \lim_{k \to \infty} \text{cap}_{p,\mu}(K_k, 4B) = \text{cap}_{p,\mu}(G_j, 4B)$$

by the weak lower semicontinuity of norms (see Remark 5.25) and by Theorem 2.2. So (4.11) is proved.

Let then $\eta \in C_0^\infty(3B)$, $0 \leq \eta \leq 1$, be a cut-off function with $\eta = 1$ on $2\overline{B}$. To show that $\eta u_j \in Q^{1,p}$, choose an increasing sequence $\psi_k \in C_0^\infty(4B)$ that converges pointwise to u_j. Let s_k be the continuous solution to the obstacle problem in $\mathcal{K}_{\psi_k}(4B)$ with $\mathcal{A}(x, \xi) = w(x)|\xi|^{p-2}\xi$ (Theorem 3.67). Since $\psi_k \leq s_k \leq u_j$ everywhere in $4B$ (Lemma 3.22), we have that s_k increases to u_j pointwise in $4B$. Since $u_j \in \mathcal{K}_{\psi_k}(4B)$, the minimizing property of solutions to obstacle problems again guarantees

$$\int_{4B} |\nabla s_k|^p \, d\mu \leq \int_{4B} |\nabla u_j|^p \, d\mu < \infty \, .$$

Hence s_k is a bounded sequence in $H^{1,p}(3B; \mu)$, and it follows that $\nabla s_k \to \nabla u_j$ weakly in $L^p(3B; \mu)$ (Theorem 1.32). Using the Mazur lemma 1.29 we easily select a sequence of convex combinations $\tilde{s}_i$ of s_k's such that $\tilde{s}_i \to u_j$ both in $H^{1,p}(3B; \mu)$ and pointwise in $3B$. Now $\eta\tilde{s}_i \in H^{1,p}(\mathbf{R}^n; \mu)$ is continuous and $\eta\tilde{s}_i \to \eta u_j$ both in $H^{1,p}(\mathbf{R}^n; \mu)$ and pointwise in $\mathbf{R}^n$. Consequently, $\eta u_j \in Q^{1,p}$ (Theorem 4.6).

Now we are ready to conclude the proof. Because of (4.11), we obtain from the equivalence of the two capacities (Theorem 2.38) and the Poincaré inequality that

$$\int_{\mathbf{R}^n} (|\eta u_j|^p + |\nabla(\eta u_j)|^p) \, d\mu \leq c \, \mathrm{C}_{p,\mu}(G_j) \, ,$$

where the constant c is independent of j. Hence a subsequence of ηu_j converges to zero (p, μ)-quasiuniformly (Lemma 4.8). So for (p, μ)-q.e. x

we may choose an index $j = j(x)$ such that $\eta u_j(x) < 1/2$. Hence by (4.10)

$$
\limsup_{r \to 0} \frac{\mu(G_j \cap B(x,r))}{\mu(B(x,r))} \leq \limsup_{r \to 0} \mu(B(x,r))^{-1} \int_{G_j \cap B(x,r)} u_j \, d\mu
$$
$$
\leq \lim_{r \to 0} \fint_{B(x,r)} \eta u_j \, d\mu
$$
$$
= \eta u_j(x) < \frac{1}{2}
$$

for (p,μ)-q.e. $x \in 4B$. On the other hand, $\eta u_j = 0$ outside $3B$. This completes the proof of the lemma.   □

4.12. Theorem. *If u and v are (p,μ)-quasicontinuous and $u = v$ a.e. in Ω, then $u = v$ (p,μ)-quasieverywhere in Ω.*

PROOF: We may clearly assume that $v \equiv 0$. Choose open sets G_j, $j = 1, 2, \ldots$, such that $u|_{\complement G_j}$ is continuous and that

$$
C_{p,\mu}(G_j) \to 0 .
$$

For (p,μ)-q.e. $x \in \Omega$ there is an index $j = j(x)$ such that

$$
\mu(G_j \cap B(x,r)) \leq \frac{1}{2}\mu(B(x,r))
$$

for all $r > 0$ small enough (Lemma 4.9). Fix such a point x. Then $x \notin G_j$ because G_j is open. Moreover, since $u = 0$ a.e., each neighborhood of x contains a point y from $\complement G_j$ with $u(y) = 0$. Since $u|_{\complement G_j}$ is continuous, we conclude $u(x) = 0$. The theorem is proved.   □

Combining Lemma 4.7 and Theorem 4.12 we obtain

4.13. Corollary. *Suppose that $E \subset \mathbf{R}^n$. Then*

$$
C_{p,\mu}(E) = \inf \int_{\mathbf{R}^n} (|u|^p + |\nabla u|^p) \, d\mu ,
$$

where the infimum is taken over all (p,μ)-quasicontinuous $u \in H^{1,p}(\mathbf{R}^n; \mu)$ such that $u = 1$ (p,μ)-q.e. on E.

Theorems 4.4 and 4.12 imply that each $u \in H^{1,p}(\Omega; \mu)$ has a "unique" quasicontinuous version.

4.14. Theorem. *Suppose that* $u \in H^{1,p}(\Omega; \mu)$. *Then there exists a* (p, μ)-*quasicontinuous function* v *such that* $u = v$ *a.e.*
Moreover, if $\tilde{v}$ *is another* (p, μ)-*quasicontinuous function such that* $u = \tilde{v}$ *a.e., then* $v = \tilde{v}$ (p, μ)-*quasieverywhere.*

We could define the *refined Sobolev space* $\tilde{H}^{1,p}(\Omega; \mu)$ which consists of equivalence classes of (p, μ)-quasicontinuous functions in $H^{1,p}(\Omega; \mu)$. The obstacle problem as discussed in Chapter 3 can be studied by using the refined space $\tilde{H}^{1,p}(\Omega; \mu)$. However, we do not pursue this.

NOTES TO CHAPTER 4. In the classical case $d\mu = dx$ and $p = 2$ the existence of quasicontinuous representatives is due to Deny (1950). The refined Sobolev spaces were later investigated by Deny (1954) and Deny and Lions (1953–54); instead of the Sobolev spaces $W^{1,p}(\mathbf{R}^n)$, so-called potential spaces $L^{1,p}(\mathbf{R}^n)$ were used. Calderón (1961) showed that these are the same spaces. Major contributions to the subject are made by Aronszajn and Smith (1956, 1961), by Federer and Ziemer (1972), and by Maz'ya and Khavin (1972). The refined properties of functions in the unweighted Sobolev space $W^{1,p}(\Omega)$ are studied in great detail in the book by Ziemer (1989); see also Evans and Gariepy (1992).

In general, the methods used to prove the capacitary properties of functions in $W^{1,p}(\Omega)$ apply rather easily to the weighted spaces $H^{1,p}(\Omega; \mu)$. The proof for Theorem 4.3 is taken from Frehse (1982); Theorem 4.5 in the unweighted case was originally proved by Bagby (1972); the proof for Theorem 4.12 follows the idea in Maz'ya and Khavin (1972).

5

Variational integrals

In this chapter we introduce a class of variational integrals whose Euler equations satisfy assumptions (3.3)–(3.7). Thus variational integrals provide abundant examples of the equations we consider in this book. We investigate the relationship between the minimization problem and the Euler equation and give a simple proof of the existence of solutions by applying direct methods of the calculus of variations. The results of this chapter are not used in the rest of the book; however, it makes the somewhat abstract approach of Chapter 3 easier to comprehend.

5.1. Variational integrals

Suppose, as always, that w is a p-admissible weight and that $d\mu(x) = w(x)dx$. Let $F : \mathbf{R}^n \times \mathbf{R}^n \to \mathbf{R}$ be a variational kernel satisfying the following assumptions for some constants $0 < \gamma \leq \delta < \infty$:

(5.2)   the mapping $x \mapsto F(x,\xi)$ is measurable for all $\xi \in \mathbf{R}^n$;

for a.e. $x \in \mathbf{R}^n$

(5.3) $$\gamma w(x)|\xi|^p \leq F(x,\xi) \leq \delta w(x)|\xi|^p, \quad \xi \in \mathbf{R}^n,$$

(5.4)   the mapping $\xi \mapsto F(x,\xi)$ is strictly convex and differentiable,

and

(5.5) $$F(x,\lambda\xi) = |\lambda|^p F(x,\xi), \ \lambda \in \mathbf{R}, \xi \in \mathbf{R}^n.$$

The *strict convexity* of $\xi \mapsto F(x,\xi)$ means that

$$F(x,t\xi_1 + (1-t)\xi_2) < tF(x,\xi_1) + (1-t)F(x,\xi_2)$$

whenever $0 < t < 1$ and $\xi_1 \neq \xi_2$. Note also that a convex function is differentiable if and only if it is continuously differentiable (Roberts and Varberg 1973, p. 117). Thus by (5.4) the mapping $\xi \mapsto F(x,\xi)$ is in fact continuously differentiable for a.e. x. The usual gradient of F with respect to the second variable, which exists for a.e. x, is denoted by $\nabla_\xi F(x,\cdot)$.

The convexity assumption (5.4) also implies a useful inequality.

97

5.6. Lemma. *For a.e.* $x \in \mathbf{R}^n$

$$(5.7) \qquad F(x, \xi_1) - F(x, \xi_2) > \nabla_\xi F(x, \xi_2) \cdot (\xi_1 - \xi_2)$$

whenever $\xi_1, \xi_2 \in \mathbf{R}^n$, $\xi_1 \neq \xi_2$.

PROOF: Fix $x \in \mathbf{R}^n$ such that $\xi \mapsto F(x, \xi)$ is strictly convex and differentiable. Then

$$F(x, \xi_2 + t(\xi_1 - \xi_2)) = F(x, (1-t)\xi_2 + t\xi_1) < (1-t)F(x, \xi_2) + tF(x, \xi_1)$$

for $0 < t < 1$, and setting $\xi = \xi_1 - \xi_2$ we obtain

$$F(x, \xi_2 + t\xi) - F(x, \xi_2) < t(F(x, \xi_2 + \xi) - F(x, \xi_2)).$$

Subtracting $\nabla_\xi F(x, \xi_2) \cdot (t\xi)$ from both sides and dividing by t we have

$$\frac{F(x, \xi_2 + t\xi) - F(x, \xi_2)}{t} - \nabla_\xi F(x, \xi_2) \cdot \xi$$
$$< F(x, \xi_2 + \xi) - F(x, \xi_2) - \nabla_\xi F(x, \xi_2) \cdot \xi.$$

The left hand side goes to zero as $t \to 0$, and the lemma follows. $\qquad\square$

Suppose that E is a measurable set and that $u \in H^{1,p}_{loc}(\Omega; \mu)$ for an open neighborhood Ω of E. Then the *variational integral*

$$(5.8) \qquad I_F(u, E) = \int_E F(x, \nabla u(x)) \, dx$$

is well defined. The value of $I_F(u, E)$ lies in the interval $[0, \infty]$ and it is finite if and only if $\nabla u \in L^p(E; \mu)$. For short, we write $F(x, \nabla u)$ for $F(x, \nabla u(x))$.

Associated with variational kernel $F(x, \xi) = w(x)|\xi|^p$ is the variational integral

$$I_F(u, E) = \int_E |\nabla u|^p \, d\mu,$$

called the *weighted p-Dirichlet integral* or the *weighted p-energy*. If μ is the Lebesgue measure, $d\mu = dx$, then $I_F(u, E)$ is simply called the p-Dirichlet integral; if, in addition, $p = 2$, it reduces to the classical Dirichlet integral

$$\int |\nabla u|^2 \, dx.$$

The (local) minimizers of the classical Dirichlet integral are exactly harmonic functions. We next pursue this analogue in a more general setting and establish that the variational integral in (5.8) gives rise to an equation of the type

$$- \operatorname{div} \mathcal{A}(x, \nabla u) = 0$$

as its Euler equation, where the mapping $\mathcal{A}(x, \xi) = \nabla_\xi F(x, \xi)$ satisfies the structural assumptions in Chapter 3.

5.9. Lemma. *Suppose that F is a variational kernel satisfying (5.2)– (5.5) and let $\mathcal{A}(x,\xi) = \nabla_\xi F(x,\xi)$. Then $\mathcal{A}$ satisfies assumptions (3.3)–(3.7) with $\alpha = \gamma$ and $\beta = 2^p \delta$.*

PROOF: We are free to define $\mathcal{A}(x,\xi)$ arbitrarily for those x for which $\xi \mapsto F(x,\xi)$ is not continuously differentiable. Let $x \in \mathbf{R}^n$ be such that F satisfies (5.3)–(5.5). Then the kth coordinate of $\mathcal{A}(x,\xi)$ equals

$$\lim_{i\to\infty} i(F(x,\xi + e_k/i) - F(x,\xi));$$

this demonstrates the measurability of the mapping $x \mapsto \mathcal{A}(x,\xi)$. Moreover, since $\xi \mapsto F(x,\xi)$ is continuously differentiable (see the remark after (5.5)), $\mathcal{A}$ satisfies (3.3).

To establish the remaining assumptions (3.4), (3.5), and (3.6), we make use of the basic convexity inequality from Lemma 5.6:

$$(5.10) \qquad F(x,\xi_1) - F(x,\xi) > \mathcal{A}(x,\xi) \cdot (\xi_1 - \xi), \quad \xi_1 \neq \xi.$$

Estimate (3.4) follows at once from this and (5.3) by choosing $\xi_1 = 0$. To obtain (3.5), fix ξ; we may assume that $\xi \neq 0$, since it follows from (5.3) and (5.4) that $\mathcal{A}(x,\xi) = 0$ if and only if $\xi = 0$. Write

$$v = \frac{\mathcal{A}(x,\xi)}{|\mathcal{A}(x,\xi)|}$$

and apply (5.10) with $\xi_1 = \xi + |\xi|v$ to conclude

$$|\xi||\mathcal{A}(x,\xi)| = \mathcal{A}(x,\xi) \cdot (\xi_1 - \xi) \leq F(x,\xi + |\xi|v) - F(x,\xi)$$
$$\leq F(x,\xi + |\xi|v) \leq 2^p \delta w(x)|\xi|^p.$$

The monotoneity inequality (3.6) is obtained by adding (5.10) to the inequality

$$F(x,\xi) - F(x,\xi_1) > \mathcal{A}(x,\xi_1) \cdot (\xi - \xi_1), \quad \xi \neq \xi_1,$$

which is (5.10) with the roles of ξ and ξ_1 interchanged.

Finally, to prove (3.7) note that differentiation of the equality

$$F(x,\lambda\xi) = |\lambda|^p F(x,\xi)$$

with respect to ξ yields

$$\lambda \nabla_\xi F(x,\lambda\xi) = |\lambda|^p \nabla_\xi F(x,\xi)$$

which takes the desired form

$$\mathcal{A}(x,\lambda\xi) = |\lambda|^{p-2} \lambda \mathcal{A}(x,\xi), \quad \lambda \neq 0.$$

The proof of Lemma 5.9 is now complete. $\qquad\square$

The homogeneity assumption (5.5) allows us to use Euler's theorem for homogeneous functions and conclude that for a.e. $x \in \mathbf{R}^n$

$$(5.11) \qquad \nabla_\xi F(x, \xi) \cdot \xi = p \, F(x, \xi)$$

for all $\xi \in \mathbf{R}^n$. Equality (5.11) is easily proved by differentiating equation (5.5) with respect to λ and then evaluating the result at $\lambda = 1$.

5.12. Minimization problems and the Euler inequality

We next show that minimizers of the variational integral $I_F(u, \Omega)$ are solutions to the corresponding Euler equation, and vice versa. In general obstacle problems it is advisable to treat the Euler equation as an inequality.

5.13. Theorem. *Suppose that $\mathcal{K} \subset L^{1,p}(\Omega; \mu)$ is a convex set and that $u \in \mathcal{K}$. Then*

$$(5.14) \qquad I_F(u, \Omega) = \min\{I_F(v, \Omega) \colon v \in \mathcal{K}\}$$

if and only if

$$(5.15) \qquad \int_\Omega \nabla_\xi F(x, \nabla u) \cdot (\nabla v - \nabla u) \, dx \geq 0$$

for all $v \in \mathcal{K}$.

PROOF: That (5.15) implies (5.14) is an easy consequence of the basic convexity inequality (5.7). Indeed, for $v \in \mathcal{K}$

$$\nabla_\xi F(x, \nabla u) \cdot \nabla(v - u) \leq F(x, \nabla v) - F(x, \nabla u)$$

a.e. in Ω. Integrating this we obtain from (5.15) that

$$0 \leq \int_\Omega \nabla_\xi F(x, \nabla u) \cdot \nabla(v - u) \, dx \leq I_F(v, \Omega) - I_F(u, \Omega)$$

and (5.14) follows.

The proof for the converse is slightly more difficult. Fix $v \in \mathcal{K}$ and set $\varphi = v - u$. Let $0 < \varepsilon \leq 1$. Then $u + \varepsilon\varphi = (1 - \varepsilon)u + \varepsilon v$ is in $\mathcal{K}$ since $\mathcal{K}$ is convex. Hence

$$I_F(u, \Omega) \leq I_F(u + \varepsilon\varphi, \Omega)$$

and therefore

$$(5.16) \qquad \int_\Omega \frac{F(x, \nabla u + \varepsilon\nabla\varphi) - F(x, \nabla u)}{\varepsilon} \, dx \geq 0 \,.$$

Because

$$\lim_{\varepsilon \to 0} \frac{F(x, \nabla u + \varepsilon \nabla \varphi) - F(x, \nabla u)}{\varepsilon} = \nabla_\xi F(x, \nabla u) \cdot \nabla \varphi$$

for a.e. $x \in \Omega$, (5.15) follows from the Lebesgue dominated convergence theorem provided that we find an L^1-majorant, independent of ε, for the integrand in (5.16).
To this end, write

$$\Delta F = F(x, \xi_0 + \varepsilon \xi_1) - F(x, \xi_0).$$

By the mean value theorem there is an $\varepsilon' \in (0, \varepsilon)$ with

$$\Delta F / \varepsilon = \xi_1 \cdot \nabla_\xi F(x, \xi_0 + \varepsilon' \xi_1).$$

Setting $\xi_0 = \nabla u$ and $\xi_1 = \nabla \varphi$ we deduce from Lemma 5.9 that for a.e. $x \in \Omega$

$$|\Delta F / \varepsilon| \le c(p, \delta) w(x) (|\nabla \varphi| |\nabla u|^{p-1} + |\nabla \varphi|^p) = g(x).$$

Since u and φ are in $L^{1,p}(\Omega; \mu)$, Hölder's inequality implies that $g \in L^1(\Omega)$ is the desired majorant. The theorem is proved. □

DEFINITION. A function $u \in H^{1,p}(\Omega; \mu)$ is called an F-extremal in Ω with boundary values $\vartheta \in H^{1,p}(\Omega; \mu)$ if $u - \vartheta \in H_0^{1,p}(\Omega; \mu)$ and

$$I_F(u, \Omega) \le I_F(v, \Omega)$$

whenever $v - \vartheta \in H_0^{1,p}(\Omega; \mu)$. A function $u \in H_{loc}^{1,p}(\Omega; \mu)$ is called a (free) F-extremal in Ω if u is an F-extremal with boundary values u in each open set $D \Subset \Omega$.

It is immediate that an F-extremal with boundary values is a free F-extremal. We also have:

5.17. Lemma. *Suppose that $u \in L^{1,p}(\Omega; \mu)$ is a free F-extremal in Ω. Then*

$$I_F(u, \Omega) \le I_F(v, \Omega)$$

whenever $v - u \in L_0^{1,p}(\Omega; \mu)$.

PROOF: Fix v such that $v - u \in L_0^{1,p}(\Omega; \mu)$ and let $\varphi_j \in C_0^\infty(\Omega)$ be a sequence with $\nabla \varphi_j$ converging to $\nabla(v - u)$ in $L^p(\Omega; \mu)$. The convexity inequality (5.7) implies that

$$F(x, \nabla u + \nabla \varphi_j) \le F(x, \nabla v) + \nabla_\xi F(x, \nabla u + \nabla \varphi_j) \cdot (\nabla u + \nabla \varphi_j - \nabla v).$$

Since u is a free F-extremal, we obtain by Lemma 5.9 that

$$I_F(u, \Omega) \le I_F(\nabla u + \nabla \varphi_j, \Omega) \le I_F(v, \Omega) +$$
$$c \left(\int_\Omega |\nabla u + \nabla \varphi_j|^p \, d\mu \right)^{(p-1)/p} \left(\int_\Omega |\nabla u + \nabla \varphi_j - \nabla v|^p \, d\mu \right)^{1/p},$$

where c is independent of j. Because $\nabla u + \nabla \varphi_j$ converges to ∇v in $L^p(\Omega; \mu)$, it follows that

$$I_F(u, \Omega) \le I_F(v, \Omega)$$

as desired. $\square$

5.18. Theorem. *A function $u \in H_{loc}^{1,p}(\Omega; \mu)$ is an F-extremal in Ω if and only if*

$$(5.19) \qquad\qquad\qquad - \operatorname{div} \nabla_\xi F(x, \nabla u) = 0$$

in Ω, that is

$$\int_\Omega \nabla_\xi F(x, \nabla u) \cdot \nabla \varphi \, dx = 0$$

for all $\varphi \in C_0^\infty(\Omega)$.

PROOF: We may assume that Ω is bounded and $u \in H^{1,p}(\Omega; \mu)$. Then the claim follows from Theorem 5.13 if we choose

$$\mathcal{K} = \{ v \in H^{1,p}(\Omega; \mu) \colon u - v \in H_0^{1,p}(\Omega; \mu) \}.$$

$\square$

Equation (5.19) is called the *Euler equation* of the variational integral I_F. In some cases it takes a simple appearance. For instance, as alluded to in Chapter 3, for the weighted p-Dirichlet integral

$$\int |\nabla u|^p \, d\mu$$

the Euler equation is the weighted p-Laplace equation

$$- \operatorname{div}(w(x)|\nabla u|^{p-2} \nabla u) = 0,$$

which in the unweighted case reduces to

$$- \operatorname{div}(|\nabla u|^{p-2} \nabla u) = 0.$$

5.20. F-superextremals with obstacles

We formulate the obstacle problem in terms of variational integrals; this makes the essence of the problem clearer.

DEFINITION. Suppose that Ω is bounded. Let $\psi \colon \Omega \to [-\infty, \infty]$ be an arbitrary function and call it an *obstacle*. Further, let $\vartheta \in H^{1,p}(\Omega; \mu)$ be a "boundary function", and as in Chapter 3 write

$$\mathcal{K}_{\psi,\vartheta}(\Omega) = \{ v \in H^{1,p}(\Omega; \mu) \colon v - \vartheta \in H_0^{1,p}(\Omega; \mu) \text{ and } v \geq \psi \text{ a.e. in } \Omega \}.$$

A function $u \in \mathcal{K}_{\psi,\vartheta}(\Omega)$ is called an *F-superextremal with obstacle ψ and boundary values ϑ* if

$$I_F(u, \Omega) \leq I_F(v, \Omega)$$

for all $v \in \mathcal{K}_{\psi,\vartheta}(\Omega)$. Thus u minimizes the variational integral $I_F(v, \Omega)$ among all functions v which, roughly speaking, coincide with ϑ on the boundary $\partial\Omega$ and lie above the obstacle ψ. Naturally, this problem makes sense only if $\mathcal{K}_{\psi,\vartheta}(\Omega)$ is nonempty. Moreover, as in Chapter 3 we assume that the notation $\mathcal{K}_{\psi,\vartheta}(\Omega)$ includes the assumption Ω is bounded.

A function $u \in H_{loc}^{1,p}(\Omega; \mu)$ is called a *(free) F-superextremal* in Ω if u is an F-superextremal with both obstacle and boundary values u in each open set $D \Subset \Omega$.

Then F-extremals can be interpreted as F-superextremals with ψ identically $-\infty$.

The following theorem shows that this obstacle problem is a special case of what we considered in Chapter 3.

5.21. Theorem.
Suppose that $\psi \colon \Omega \to [-\infty, \infty]$ and $\vartheta \in H^{1,p}(\Omega; \mu)$. Then a function $u \in \mathcal{K}_{\psi,\vartheta}(\Omega)$ is an F-superextremal with obstacle ψ and boundary values ϑ if and only if u is a solution to the obstacle problem in $\mathcal{K}_{\psi,\vartheta}(\Omega)$ with $\mathcal{A} = \nabla_\xi F$.

PROOF: This follows from Theorem 5.13 by choosing $\mathcal{K} = \mathcal{K}_{\psi,\vartheta}(\Omega)$. $\square$

Theorem 5.21 enables us to apply the results of Chapter 3 to F-superextremals. For example, while F-superextremals are not continuous in general (see Chapter 12), an F-superextremal with continuous obstacle is always continuous by Theorem 3.67.

5.22. Existence of F-superextremals

The preceding discussion shows that the existence of F-superextremals and F-extremals can be derived from the general theory of monotone operators; see Chapter 3 and Appendix I. However, the direct methods of the calculus of variations provide a conceptually easier approach. We explore this route and establish the existence of F-superextremals with obstacles. Our proof is not the most general but it is elementary; it relies on the homogeneity assumption (5.5) for the kernel F.

In the space $L^p(\Omega; \mu; \mathbf{R}^n)$ we employ not only the usual norm

$$||u||_p = (\int_\Omega |u|^p \, d\mu)^{1/p}$$

but also the F-*norm*, defined by

$$||u||_F = (\int_\Omega F(x, u(x)) \, dx)^{1/p} .$$

Recall that we usually write $L^p(\Omega; \mu)$ for $L^p(\Omega; \mu; \mathbf{R}^n)$.

5.23. Lemma.   *The F-norm $|| \cdot ||_F$ is a norm in $L^p(\Omega; \mu)$, equivalent to $|| \cdot ||_p$.*

PROOF: By (5.3) and (5.5) only the triangle inequality calls for a proof. To establish it, let $u, v \in L^p(\Omega; \mu)$ and write $||u||_F = s$, $||v||_F = t$, and $r = s + t$. We may assume that $s, t > 0$. Now

$$\frac{u + v}{r} = \frac{s}{r}(\frac{u}{s}) + \frac{t}{r}(\frac{v}{t})$$

and the convexity of F yields

$$F(x, \frac{u + v}{r}) \leq \frac{s}{r}F(x, \frac{u}{s}) + \frac{t}{r}F(x, \frac{v}{t})$$

for a.e. $x \in \Omega$. Integrating and using the homogeneity assumption we obtain

$$r^{-p} \int_\Omega F(x, u + v) \, dx \leq \frac{s}{r}s^{-p} \int_\Omega F(x, u) \, dx + \frac{t}{r}t^{-p} \int_\Omega F(x, v) \, dx$$

$$= \frac{s + t}{r} = 1 ,$$

which implies

$$||u + v||_F \leq ||u||_F + ||v||_F$$

as required. The equivalence of $|| \cdot ||_p$ and $|| \cdot ||_F$ follows from assumption (5.3).   □

Note that by using Lemma 5.23 we easily infer that

$$\lim_{j \to \infty} I_F(u_j, \Omega) = I_F(u, \Omega)$$

as soon as $\nabla u_j \to \nabla u$ in $L^p(\Omega; \mu)$; we already used this in the proof of Lemma 5.17.

The use of the F-norm in $L^p(\Omega; \mu)$ leads to a simple proof for the following fundamental lower semicontinuity theorem.

5.24. Lower semicontinuity theorem. *Suppose that $u, u_j \in L^{1,p}(\Omega; \mu)$ are such that ∇u_j converge to ∇u weakly in $L^p(\Omega; \mu)$. Then*

$$I_F(u, \Omega) \leq \liminf_{j \to \infty} I_F(u_j, \Omega).$$

PROOF: By Lemma 5.23, the sequence ∇u_j converges to ∇u weakly in $L^p(\Omega; \mu)$ equipped with the F-norm $\|\cdot\|_F$. By the weak lower semicontinuity of norms (see Remark 5.25 below)

$$\|\nabla u\|_F \leq \liminf_{j \to \infty} \|\nabla u_j\|_F,$$

which is the desired result. □

5.25. REMARK. In the proof of 5.24 we used the weak lower semicontinuity of norms: if $x_j \to x$ weakly in a normed space $(X, \|\cdot\|)$, then

$$\liminf_{j \to \infty} \|x_j\| \geq \|x\|.$$

This is a well-known consequence of the Hahn–Banach theorem (Dunford and Schwartz 1958, Lemma II.3.27; Yosida 1980, p. 120).

5.26. Theorem. *Suppose that $\mathcal{K} \subset L^{1,p}(\Omega; \mu)$ is a nonempty convex set such that*

$$\nabla \mathcal{K} = \{\nabla u : u \in \mathcal{K}\}$$

is a closed subset of $L^p(\Omega; \mu)$. Then there is $u \in \mathcal{K}$ such that

$$I_F(u, \Omega) = \min\{I_F(v, \Omega) : v \in \mathcal{K}\}.$$

PROOF: Let $u_j \in \mathcal{K}$ be a minimizing sequence, i.e.

$$I_F(u_j, \Omega) \to I_0 = \inf\{I_F(v, \Omega) : v \in \mathcal{K}\}.$$

Since $\mathcal{K} \neq \emptyset$, $0 \leq I_0 < \infty$, and we may assume that

$$I_F(u_j, \Omega) \leq I_0 + 1$$

for all j. Thus (5.3) implies that

$$\int_\Omega |\nabla u_j|^p \, d\mu \leq M < \infty,$$

where M is independent of j. Thus a subsequence which we still denote by ∇u_j converges weakly in $L^p(\Omega; \mu)$ to a function g. Next we invoke the Mazur lemma 1.29 and infer that a sequence of convex combinations of functions ∇u_j converges strongly to g in $L^p(\Omega; \mu)$. Since $\nabla \mathcal{K}$ is closed and convex, $g \in \nabla \mathcal{K}$ and hence there is $u \in \mathcal{K}$ with $\nabla u = g$. Now the lower semicontinuity theorem 5.24 implies that

$$I_0 \leq I_F(u, \Omega) \leq \liminf_{j \to \infty} I_F(u_j, \Omega) = I_0,$$

and so u is the desired minimizer. □

5.27. Theorem. *Suppose that Ω is bounded, that $\psi \colon \Omega \to [-\infty, \infty]$, and that $\vartheta \in H^{1,p}(\Omega; \mu)$. If $\mathcal{K}_{\psi,\vartheta}(\Omega) \neq \emptyset$, there exists a unique F-superextremal with obstacle ψ and boundary values ϑ.*

PROOF: Since the set $\mathcal{K}_{\psi,\vartheta} = \mathcal{K}_{\psi,\vartheta}(\Omega)$ is a nonempty convex subset of $L^{1,p}(\Omega; \mu)$, the existence follows from Theorem 5.26 if we show that

$$\nabla \mathcal{K} = \{\nabla v \colon v \in \mathcal{K}_{\psi,\vartheta}\}$$

is closed in $L^p(\Omega; \mu)$. To accomplish this, let $u_j \in \mathcal{K}_{\psi,\vartheta}$ be a sequence such that ∇u_j converges to a function g in $L^p(\Omega; \mu)$. Since $u_j - u_k \in H_0^{1,p}(\Omega; \mu)$, the Poincaré inequality (1.5) implies that

$$\int_\Omega |u_j - u_k|^p \, d\mu \leq c \int_\Omega |\nabla u_j - \nabla u_k|^p \, d\mu \,.$$

It follows that $u_j - \vartheta$ is a Cauchy sequence in $H_0^{1,p}(\Omega; \mu)$ and it converges to a function $v \in H_0^{1,p}(\Omega; \mu)$ with $\nabla v = g - \nabla\vartheta$. We show that $u = v + \vartheta \in \mathcal{K}_{\psi,\vartheta}$ which implies that $\nabla \mathcal{K}$ is closed because $\nabla u = g$. We already know that $u - \vartheta = v \in H_0^{1,p}(\Omega; \mu)$, whence it suffices to verify that $u \geq \psi$ a.e. in Ω. This follows by selecting a subsequence of $u_j - \vartheta$ that converges a.e. to v. Thus $u \in \mathcal{K}_{\psi,\vartheta}$ and the existence part is thereby established.

For the uniqueness, suppose that $u_1, u_2 \in \mathcal{K}_{\psi,\vartheta}$ are two distinct minimizers. Since $v = (u_1 + u_2)/2 \in \mathcal{K}_{\psi,\vartheta}$ and since the set $\{\nabla u_1 \neq \nabla u_2\}$ has positive measure, the strict convexity of F implies

$$I_F(v, \Omega) < \frac{1}{2}(I_F(u_1, \Omega) + I_F(u_2, \Omega)) = \min_{\mathcal{K}_{\psi,\vartheta}} I_F(u, \Omega) \,.$$

This contradiction completes the proof. $\square$

If we choose $\psi \equiv -\infty$ in Theorem 5.27, we obtain the existence of F-extremals.

5.28. Theorem. *Suppose that Ω is bounded and that $\vartheta \in H^{1,p}(\Omega; \mu)$. Then there exists a unique F-extremal u in Ω with $u - \vartheta \in H_0^{1,p}(\Omega; \mu)$.*

In Theorem 3.71 the solutions to obstacle problems with continuous obstacles are characterized in terms of solutions and supersolutions of the equation

$$-\operatorname{div} \mathcal{A}(x, \nabla u) = 0 \,.$$

Now, in light of Theorem 5.21 F-superextremals are supersolutions with $\mathcal{A} = \nabla_\xi F$. Thus we have:

5.29. Theorem. *Suppose that Ω is bounded, that $\psi: \Omega \to \mathbf{R}$ is continuous, that $\vartheta \in H^{1,p}(\Omega; \mu)$, and that $u \in \mathcal{K}_{\psi,\vartheta}(\Omega)$ is a free F-superextremal. Then u is an F-superextremal with obstacle ψ and boundary values ϑ if and only if there is a continuous function v in Ω such that $v = u$ a.e. and v is a solution to the equation*

$$- \operatorname{div} \nabla_\xi F(x, \nabla v) = 0$$

in the open set $\{v > \psi\}$.

5.30. Variational F-capacity

The variational (p, μ)-capacity was studied in Chapter 2. We close this chapter by introducing an extension of this concept, a capacity where the kernel $|\xi|^p w(x)$ is replaced by a kernel F satisfying the assumptions in Section 5.1.

DEFINITION. If K is a compact subset of Ω, the *F-capacity* of the *condenser* (K, Ω) is

$$\operatorname{cap}_F(K, \Omega) = \inf \int_\Omega F(x, \nabla u) \, dx,$$

where the infimum is taken over all u in

$$W(K, \Omega) = \{u \in C_0^\infty(\Omega): u \geq 1 \text{ in } K\}.$$

The F-capacity is extended for an arbitrary set $E \subset \Omega$ by using the Choquet device as in Chapter 2, namely

$$\operatorname{cap}_F(E, \Omega) = \inf_U \sup_K \operatorname{cap}_F(K, \Omega),$$

where U runs through all open subsets of Ω containing E and K runs through all compact subsets of U.

By (5.3) the F-capacity is equivalent to the (p, μ)-capacity:

$$\gamma \operatorname{cap}_{p,\mu}(E, \Omega) \leq \operatorname{cap}_F(E, \Omega) \leq \delta \operatorname{cap}_{p,\mu}(E, \Omega)$$

whenever $E \subset \Omega$; the (p, μ)-capacity is obtained by choosing $F(x, \xi) = w(x)|\xi|^p$. It is clear that the properties of the F-capacity are similar to those of the (p, μ)-capacity; in particular, they have the same sets of capacity zero. However, the F-capacity provides a precise tool for studying the variational integral I_F.

The capacity theory of Chapter 2 can be developed for F-capacities as well; we list here some of the results whose proofs are similar to the proofs of the corresponding theorems in Chapter 2, and are therefore omitted.

The first theorem shows that the F-capacity is a Choquet capacity.

5.31. Theorem. *The set function $E \mapsto \mathrm{cap}_F(E, \Omega)$, $E \subset \Omega$, enjoys the following properties:*

(i) *If $E_1 \subset E_2$, then $\mathrm{cap}_F(E_1, \Omega) \leq \mathrm{cap}_F(E_2, \Omega)$.*

(ii) *If $\Omega_1 \subset \Omega_2$ are open and $E \subset \Omega_1$, then $\mathrm{cap}_F(E, \Omega_2) \leq \mathrm{cap}_F(E, \Omega_1)$.*

(iii) *If K_1 and K_2 are compact, then*

$$\mathrm{cap}_F(K_1 \cup K_2, \Omega) + \mathrm{cap}_F(K_1 \cap K_2, \Omega) \leq \mathrm{cap}_F(K_1, \Omega) + \mathrm{cap}_F(K_2, \Omega).$$

(iv) *If $K_j \subset \Omega$ is a decreasing sequence of compact sets and $K = \bigcap_j K_j$, then*

$$\mathrm{cap}_F(K, \Omega) = \lim_{j \to \infty} \mathrm{cap}_F(K_j, \Omega).$$

(v) *If $E_1 \subset E_2 \subset \cdots \subset E = \bigcup_{j=1}^{\infty} E_j \subset \Omega$, then*

$$\mathrm{cap}_F(E, \Omega) = \lim_{j \to \infty} \mathrm{cap}_F(E_j, \Omega).$$

(vi) *If $E = \bigcup_j E_j \subset \Omega$, then*

$$\mathrm{cap}_F(E, \Omega) \leq \sum_j \mathrm{cap}_F(E_j, \Omega).$$

PROOF: The proof of Theorem 2.2 applies essentially word for word. $\square$

The F-capacity also satisfies the properties of Theorem 2.6.

5.32. Theorem. *If $E_1 \subset \Omega_1 \subset E_2 \subset \Omega_2 \subset \ldots$ and $\Omega = \cup_j \Omega_j$, then*

$$\mathrm{cap}_F(E_1, \Omega) \leq \left(\sum_j \mathrm{cap}_F(E_j, \Omega_j)^{1/(1-p)} \right)^{1-p}.$$

PROOF: Mimic the proof of Theorem 2.6. $\square$

DEFINITION. Suppose that $K \subset \Omega$ is compact. We call a continuous function $u \in L^{1,p}(\Omega; \mu)$ in Ω an *F-capacitary function for the condenser* (K, Ω) if

(i) $\mathrm{cap}_F(K, \Omega) = I_F(u, \Omega)$,

(ii) u is F-superextremal in Ω,

(iii) $u = 1$ in K, and

(iv) $\lim_{x \to y} u(x) = 0$ for all $y \in \partial\Omega$.

Note that the F-capacitary function is an F-extremal in $\Omega \setminus K$. The concept of F-capacitary function will be generalized in later chapters, where we also give conditions for its existence; see in particular Theorems 9.34, 9.35, and 9.38.

NOTES TO CHAPTER 5. The variational problems in this chapter belong to so-called regular problems, i.e. the kernel $F(x, \xi)$ is convex in the second variable; this is closely related to the monotonicity of the mapping $\mathcal{A}(x, \xi) = \nabla_\xi F(x, \xi)$. Roughly speaking, the regularity makes it possible to apply lower semicontinuity theorems in proving the existence of minimizers, and the strict convexity takes care of the uniqueness.

Lower semicontinuity results for general variational integrals are widely studied; they play a crucial role when the direct methods of the calculus of variations are applied. For modern treatments of the subject, see Morrey (1966) and Giaquinta (1983). Our elementary proof for the lower semicontinuity is taken from Kilpeläinen (1985) and it cannot be generalized to variational kernels which do not satisfy a homogeneity assumption like (5.5). In general, it is required that the variational kernel be convex with respect to ∇u (Dacorogna 1982; Morrey 1966; Reshetnyak 1967c).

In the unweighted case for $p \geq n$ (in fact, for $p > n - 1$) the existence proof for superextremals with continuous obstacles can be based on the use of monotone functions (Granlund et al. 1983; Martio 1975). This idea goes back to Lebesgue who used monotone functions in solving the Dirichlet problem in the plane, see Morrey (1966, 4.3).

6

$\mathcal{A}$-harmonic functions

It was established in Chapter 3 that each weak solution of the equation

(6.1) $$-\operatorname{div}\mathcal{A}(x,\nabla h)=0$$

can be redefined in a set of measure zero so that it becomes continuous (Theorem 3.70).

DEFINITION. A function $h:\Omega\to\mathbf{R}$ is said to be $\mathcal{A}$-harmonic in Ω if it is a continuous weak solution of (6.1) in Ω. We use the notation

$$\mathcal{H}(\Omega)=\mathcal{H}_{\mathcal{A}}(\Omega)=\{h:h\text{ is }\mathcal{A}\text{-harmonic in }\Omega\}.$$

We emphasize that throughout this book $\mathcal{A}$ denotes a mapping which satisfies assumptions (3.3)–(3.7).

$\mathcal{A}$-harmonic functions and their more general offspring, $\mathcal{A}$-superharmonic functions which will be introduced in Chapter 7, constitute the main object of study in this nonlinear potential theory. The prefix "nonlinear" is justified, as $\mathcal{A}$-harmonic functions generally do not form a linear space. However, we can compensate somewhat for this lack of linearity by using the immediate fact that λh and $h+\lambda$ are $\mathcal{A}$-harmonic whenever h is $\mathcal{A}$-harmonic and λ is a real constant; this holds because $\mathcal{A}$ depends only on x and ∇h, and $\mathcal{A}$ is homogeneous in ∇h.

This chapter covers the most fundamental properties of $\mathcal{A}$-harmonic functions. We start with an important inequality.

6.2. Harnack's inequality. *Let h be a nonnegative $\mathcal{A}$-harmonic function in Ω. Then there is a constant $c=c(n,p,\beta/\alpha,c_\mu)>0$ such that*

(6.3) $$\sup_B h\le c\inf_B h$$

whenever B is a ball in Ω such that $2B\subset\Omega$.

PROOF: We may assume that $h>0$ in Ω. If $4B\subset\Omega$, we deduce from the weak Harnack inequalities (Lemma 3.34 and Theorem 3.51) that

(6.4) $$\sup_B h\le c\Big(\fint_{2B}h^s\,d\mu\Big)^{1/s}\le c\inf_B h,$$

where $c=c(n,p,\beta/\alpha,c_\mu)>0$.

To prove the general case, observe that if $2B\subset\Omega$, then any two points on $\overline{B}$ can be joined by a chain of balls $B_1,B_2,\ldots,B_N$ such that $4B_i\subset\Omega$ and $B_i\cap B_{i+1}\neq\emptyset$, where N is an absolute constant. A repeated use of (6.4) in the balls B_i implies the desired result. $\qquad\square$

The requirement $2B \subset \Omega$ in Harnack's inequality 6.2 is chosen just for convenience; in fact, as the proof shows, we could assume $\lambda B \subset \Omega$ for any $\lambda > 1$, but then the constant c depends also on λ.

It is crucial to notice the uniformity in Harnack's inequality (6.3): *the constant c does not depend on the size of the ball*. This fact will be in constant use throughout this book. We also have that

$$\sup_K h \leq c \inf_K h$$

for all nonnegative $\mathcal{A}$-harmonic functions h in a domain Ω and compact $K \subset \Omega$; the constant c depends only on n, p, β/α, c_μ, K, and Ω.

Harnack's inequality quickly leads to the strong maximum principle.

6.5. Strong maximum principle.   *A nonconstant $\mathcal{A}$-harmonic function in a domain Ω cannot attain its supremum or infimum.*

PROOF: If $u(x_0) = \max_\Omega u = M$, then the function $v = M - u$ is nonnegative and $\mathcal{A}$-harmonic in Ω. Since $v(x_0) = 0$, it follows from Harnack's inequality that $v \equiv 0$ in Ω. The minimum is treated similarly.   □

The uniform Harnack inequality can be iterated to obtain the local Hölder continuity of $\mathcal{A}$-harmonic functions.

6.6. Theorem.   *Suppose that h is $\mathcal{A}$-harmonic in Ω. If $0 < r < R < \infty$ are such that $B(x_0, R) \subset \Omega$, then*

$$(6.7) \qquad \operatorname{osc}(h, B(x_0, r)) \leq 2^\kappa \left(\frac{r}{R}\right)^\kappa \operatorname{osc}(h, B(x_0, R)),$$

where $\kappa \in (0, 1]$ depends only on n, p, β/α, and c_μ.

PROOF: Let $0 < \rho \leq R$. We use the abbreviations $B(\rho) = B(x_0, \rho)$, $M(\rho) = \sup_{B(\rho)} h$, and $m(\rho) = \inf_{B(\rho)} h$. By applying the Harnack inequality to the $\mathcal{A}$-harmonic function $h - m(\rho)$ we obtain

$$(6.8) \qquad M(\frac{\rho}{2}) - m(\rho) \leq c_0 \left(m(\frac{\rho}{2}) - m(\rho)\right),$$

where $c_0 = c_0(n, p, \beta/\alpha, c_\mu) \geq 1$. Set $\Lambda = (c_0 - 1)/c_0$. We claim that

$$(6.9) \qquad \operatorname{osc}(h, B(\frac{\rho}{2})) \leq \Lambda \operatorname{osc}(h, B(\rho)).$$

To achieve (6.9) suppose first that

$$m(\frac{\rho}{2}) - m(\rho) \leq c_0^{-1}(M(\rho) - m(\rho)).$$

Then (6.8) yields

$$\mathrm{osc}(h, B(\tfrac{\rho}{2})) = M(\tfrac{\rho}{2}) - m(\rho) + m(\rho) - m(\tfrac{\rho}{2})$$
$$\leq (c_0 - 1)(m(\tfrac{\rho}{2}) - m(\rho)) \leq \Lambda\,\mathrm{osc}(h, B(\rho))\,.$$

Also, if

$$m(\tfrac{\rho}{2}) - m(\rho) > c_0^{-1}(M(\rho) - m(\rho))\,,$$

then

$$\mathrm{osc}(h, B(\tfrac{\rho}{2})) \leq M(\rho) - m(\rho) - (m(\tfrac{\rho}{2}) - m(\rho))$$
$$< \Lambda\,(M(\rho) - m(\rho))\,.$$

Thus (6.9) always holds.

To complete the proof, we iterate (6.9). Choose the integer $m \geq 1$ such that $2^{m-1} \leq R/r < 2^m$. Then (6.9) implies

$$\mathrm{osc}(h, B(r)) \leq \Lambda^{m-1}\,\mathrm{osc}(h, B(2^{m-1}r)) \leq \Lambda^{m-1}\,\mathrm{osc}(h, B(R))\,.$$

Set $\kappa = (-\log \Lambda)/(\log 2) \leq 1$ and obtain

$$\left(\frac{r}{R}\right)^{\kappa} \geq 2^{-\kappa}(2^{m-1})^{-\kappa} = 2^{-\kappa}\Lambda^{m-1}\,,$$

whence

$$\mathrm{osc}(h, B(r)) \leq 2^{\kappa}\left(\frac{r}{R}\right)^{\kappa}\mathrm{osc}(h, B(R))\,,$$

which is the desired estimate. $\square$

The Hölder continuity estimate (6.7) has two immediate consequences, *Liouville's theorem* and an equicontinuity property of $\mathcal{A}$-harmonic functions.

6.10. Liouville's theorem.   *If h is a bounded $\mathcal{A}$-harmonic function in $\mathbf{R}^n$, then h is constant.*

6.11. Corollary.   *If h is a nonnegative $\mathcal{A}$-harmonic function in $\mathbf{R}^n$, then h is constant.*

PROOF: If h is $\mathcal{A}$-harmonic and nonnegative in $\mathbf{R}^n$, it is bounded by Harnack's inequality, and hence constant by Liouville's theorem. $\square$

A stronger version of Liouville's theorem follows from Theorem 6.6. Namely, if h is $\mathcal{A}$-harmonic in $\mathbf{R}^n$ with

$$\lim_{x \to \infty} \frac{|h(x)|}{|x|^\kappa} = 0,$$

where κ is the Hölder continuity exponent in (6.7), then h is constant.

We recall that a family $\mathcal{F}$ of real-valued functions in Ω is *equicontinuous* in Ω if for each $x \in \Omega$ and $\varepsilon > 0$ there is a $\delta > 0$ such that $|x - y| < \delta$ implies $|u(x) - u(y)| < \varepsilon$ for all $u \in \mathcal{F}$.

6.12. Theorem. *Every locally uniformly bounded family of $\mathcal{A}$-harmonic functions in Ω is equicontinuous in Ω.*

We next record two important convergence theorems; the first is a reformulation of Theorem 3.78.

6.13. Theorem. *Let h_i, $i = 1, 2, ...$, be a sequence of $\mathcal{A}$-harmonic functions in Ω such that $h_i \to h$ locally uniformly in Ω. Then h is $\mathcal{A}$-harmonic.*

6.14. Harnack's convergence theorem. *Suppose that h_i, $i = 1, 2, ...$, is an increasing sequence of $\mathcal{A}$-harmonic functions in a domain Ω. Then the function $h = \lim_{i \to \infty} h_i$ is either $\mathcal{A}$-harmonic or identically $+\infty$ in Ω.*

PROOF: If $h(x) < \infty$ for some $x \in \Omega$, it follows from Harnack's inequality that h is locally bounded in Ω. Thus the sequence h_i is equicontinuous (Theorem 6.12), and by Ascoli's theorem it converges locally uniformly. Then h is $\mathcal{A}$-harmonic by Theorem 6.13. $\quad\square$

As another application of Harnack's inequality we next show that an $\mathcal{A}$-harmonic function that is bounded below has a limit at an isolated boundary point; the limit value need not be finite.

6.15. Lemma. *Let U be either a punctured ball $B(x_0, r_0) \setminus \{x_0\}$ or the complement of a closed ball. If $h \in \mathcal{H}(U)$ is nonnegative, the limit*

$$\lim_{x \to x_0} h(x) \in [0, \infty]$$

exists. Here $x_0 = \infty$ if U is the complement of a ball.

PROOF: We prove only the case $U = B(x_0, r_0) \setminus \{x_0\}$; the proof for $x_0 = \infty$ is completely analogous. Write

$$M = \limsup_{x \to x_0} h(x)$$

and

$$m = \liminf_{x \to x_0} h(x).$$

We may assume that $m < \infty$. Fix $\varepsilon > 0$. Then there is a radius $r_1 > 0$ such that $h > m - \varepsilon$ on $B(x_0, r_1) \setminus \{x_0\}$, and hence the $\mathcal{A}$-harmonic function $v = h - m + \varepsilon$ is positive in $B(x_0, r_1) \setminus \{x_0\}$. Pick a sequence x_i tending to x_0 such that

$$v(x_i) \geq \min(M - m, i)$$

and that

$$|x_{i+1} - x_0| < |x_i - x_0| < r_0/2.$$

Since each sphere of a ball of radius r can be covered by $N = N(n)$ balls of radii $r/2$, we deduce from the Harnack inequality that

$$v \geq c \min(M - m, i)$$

on $\partial B(x_0, |x_i - x_0|)$, where the constant $c > 0$ is independent of i. Now an application of the maximum principle in the annulus

$$\{x : |x_{i+1} - x_0| < |x - x_0| < |x_i - x_0|\}$$

shows that

$$v \geq c \min(M - m, i)$$

there. Consequently,

$$\varepsilon = \liminf_{x \to x_0} v(x) \geq c(M - m),$$

which yields $M = m$ as desired. $\square$

6.16. Boundary continuity of $\mathcal{A}$-harmonic functions

We demonstrated in Chapter 3 that if Ω is bounded and $\vartheta \in H^{1,p}(\Omega; \mu)$, then there exists a unique $\mathcal{A}$-harmonic function h in Ω such that $h - \vartheta \in H_0^{1,p}(\Omega; \mu)$. Suppose now that ϑ is continuous on $\overline{\Omega}$ and that x_0 is a boundary point of Ω. A fundamental problem of potential theory is: when does $\lim_{x \to x_0} h(x)$ exist and equal $\vartheta(x_0)$? For the Laplace equation this problem has a long and colorful history; it was finally settled by Wiener (1924b) and in his resolution surfaced what is now called the *Wiener criterion*. This device has become one of the central concepts of modern potential theory. We adopt the following form of the Wiener criterion: we say that a set E is (p, μ)-*thick* at x_0 if

$$(6.17) \qquad \int_0^1 \left(\frac{\mathrm{cap}_{p,\mu}(E \cap B(x_0, t), B(x_0, 2t))}{\mathrm{cap}_{p,\mu}(B(x_0, t), B(x_0, 2t))} \right)^{1/(p-1)} \frac{dt}{t} = \infty.$$

It is next shown that if $\complement\Omega$ is (p,μ)-thick at x_0, then the solution h to the Dirichlet problem described above has limit $\vartheta(x_0)$ at x_0. In fact, we give an estimate for the modulus of continuity of h at a boundary point in terms of Wiener type integrals. In Theorem 6.31 below we give a geometric criterion that implies the (p,μ)-thickness of the complement of a ball or the complement of a polyhedron.

For brevity, we write

$$\varphi_{p,\mu}(z,E,t) = \frac{\text{cap}_{p,\mu}(E \cap B(z,t), B(z,2t))}{\text{cap}_{p,\mu}(B(z,t), B(z,2t))}$$

for the capacity density function in (6.17).

6.18. Theorem. *Suppose that Ω is bounded. Let $\vartheta \in H^{1,p}(\Omega;\mu) \cap C(\overline{\Omega})$ and let h be the unique $\mathcal{A}$-harmonic function in Ω with $\vartheta - h \in H_0^{1,p}(\Omega;\mu)$. If $x_0 \in \partial\Omega$, then for $0 < r \le \rho$ it is true that*

$$\text{osc}(h, \Omega(r)) \le \text{osc}(\vartheta, \partial\Omega \cap \overline{B}(x_0, 2\rho))$$
$$+ \text{osc}(\vartheta, \partial\Omega) \exp(-c \int_r^\rho \varphi_{p,\mu}(x_0, \complement\Omega, t)^{1/(p-1)} \frac{dt}{t}),$$

where $\Omega(r) = \Omega \cap B(x_0, r)$ and $c = c(n, p, \beta/\alpha, c_\mu) > 0$.

Customarily, the boundary regularity is first proved for domains with a nice boundary. This fact is then used in proving the general Wiener criterion. In our approach we treat all open sets simultaneously.

Some preparatory work is in order. We first define the $\mathcal{A}$-potential of a compact set as follows. Let Ω be bounded and $K \subset \Omega$ a compact set. Furthermore, let $\psi \in C_0^\infty(\Omega)$ be such that $\psi = 1$ on K. The $\mathcal{A}$-harmonic function u in $\Omega \setminus K$ with $u - \psi \in H_0^{1,p}(\Omega \setminus K; \mu)$ is called the $\mathcal{A}$-*potential* of K in Ω and denoted by

$$\Re(K, \Omega) = \Re_{\mathcal{A}}(K, \Omega).$$

Because $u - \psi$ has a zero extension in $H^{1,p}(\mathbf{R}^n; \mu)$, we tacitly assume that $\Re(K, \Omega)$ is a quasicontinuous function in $H_0^{1,p}(\Omega;\mu)$ with $\Re(K, \Omega) = 1$ everywhere on K (see Theorem 4.5). In particular, if $\text{cap}_{p,\mu} K > 0$, then $\Re(K, \Omega)$ is not identically zero in $\Omega \setminus K$. The definition of $\Re(K, \Omega)$ is independent of the particular choice of ψ, for if $\tilde{\psi}$ is another such function and $\tilde{u} \in \mathcal{H}(\Omega \setminus K)$ is such that $\tilde{u} - \tilde{\psi} \in H_0^{1,p}(\Omega \setminus K; \mu)$, then $u - \tilde{u} \in H_0^{1,p}(\Omega \setminus K; \mu)$ and by the uniqueness we have $u = \tilde{u}$ in $\Omega \setminus K$.

Although the definition for the $\mathcal{A}$-potential here slightly differs from the one given later in Chapter 8, the concept is the same (Lemma 8.5).

The following sharp capacity estimate is crucial in many applications.

6.19. Lemma. *Suppose that Ω is bounded, $K \subset \Omega$ is compact, and $u = \Re(K, \Omega)$. If $0 < \gamma < 1$ and $K_\gamma = \{x \in \Omega : u(x) \geq \gamma\}$, then*

$$\frac{1}{c}\gamma^{p-1} \operatorname{cap}_{p,\mu}(K_\gamma, \Omega) \leq \operatorname{cap}_{p,\mu}(K, \Omega) \leq c\gamma^{p-1} \operatorname{cap}_{p,\mu}(K_\gamma, \Omega),$$

where $c = (\beta/\alpha)^{p+1}$.

PROOF: Before starting the proof, we remark that if we knew a priori that u is continuous on $\overline{\Omega}$, the proof would be much easier. At this point, however, there is no guarantee for such an assumption even in the case when Ω and K are smooth.

We begin by establishing the double inequality

$$(6.20) \quad (\frac{\beta}{\alpha})^{-p} \int_{\{u<\gamma\}} |\nabla u|^p \, d\mu \leq \gamma^p \operatorname{cap}_{p,\mu}(K_\gamma, \Omega) \leq \int_{\{u<\gamma\}} |\nabla u|^p \, d\mu.$$

Because we have defined u to be 1 on K and because u is continuous in $\Omega \setminus K$, the set K_γ is closed in Ω and $K \subset K_\gamma$. However, K_γ need not be compact in Ω.

Fix $\varepsilon > 0$ and exhaust K_γ by compact sets $C_1 \subset C_2 \subset \cdots \subset K_\gamma$ with $K \subset C_1$. Let $u_j \in C_0^\infty(\Omega)$ be a function such that $0 \leq u_j \leq 1$, $u_j = 1$ on C_j, and

$$\int_\Omega |\nabla u_j|^p \, d\mu \leq \operatorname{cap}_{p,\mu}(C_j, \Omega) + \varepsilon.$$

Now

$$\min(\gamma^{-1}u, 1) - u_j \in H_0^{1,p}(\Omega \setminus C_j; \mu)$$

(cf. Theorem 4.5) and hence the function $\min(\gamma^{-1}u, 1)$ can be approximated in $H^{1,p}(\Omega; \mu)$ by functions which are admissible for the condenser (C_j, Ω); therefore

$$\gamma^p \operatorname{cap}_{p,\mu}(C_j, \Omega) \leq \int_{\Omega \setminus C_j} |\nabla u|^p \, d\mu.$$

The second inequality of (6.20) follows from this because the capacities $\operatorname{cap}_{p,\mu}(C_j, \Omega)$ converge to $\operatorname{cap}_{p,\mu}(K_\gamma, \Omega)$ (Theorem 2.2). Moreover, this shows that $\operatorname{cap}_{p,\mu}(K_\gamma, \Omega)$ is finite.

To establish the first inequality of (6.20), note that by the above reasoning the sequence u_j is bounded in $H_0^{1,p}(\Omega; \mu)$. Hence using the weak sequential compactness of $H_0^{1,p}(\Omega; \mu)$ (Theorem 1.31) and the Mazur lemma

1.29, it is not difficult to construct a sequence v_k consisting of convex combinations of the functions u_j such that v_k is admissible for the condenser (C_k, Ω), that v_k converges in $H_0^{1,p}(\Omega; \mu)$ to a function v, and that

$$\int_\Omega |\nabla v_k|^p \, d\mu \leq \mathrm{cap}_{p,\mu}(C_k, \Omega) + \varepsilon$$

for all $k = 1, 2, \ldots$. This implies

$$\int_\Omega |\nabla v|^p \, d\mu \leq \mathrm{cap}_{p,\mu}(K_\gamma, \Omega) + \varepsilon.$$

Now using Theorem 4.3 we may assume that v is (locally) a (p, μ)-quasiuniform limit of a subsequence of v_j, so that v is quasicontinuous and equals 1 q.e. on K_γ. Therefore

$$u - \gamma v \in H_0^{1,p}(\Omega \setminus K_\gamma; \mu)$$

by Theorem 4.5, and the quasiminimizing property of $\mathcal{A}$-harmonic functions (3.15) implies

$$(\frac{\beta}{\alpha})^{-p} \int_{\Omega \setminus K_\gamma} |\nabla u|^p \, d\mu \leq \gamma^p \int_{\Omega \setminus K_\gamma} |\nabla v|^p \, d\mu \leq \gamma^p \mathrm{cap}_{p,\mu}(K_\gamma, \Omega) + \gamma^p \varepsilon.$$

Thus (6.20) follows.

Next, employing (6.20) and the fact that u is $\mathcal{A}$-harmonic in $\Omega \setminus K$, we have

$$\mathrm{cap}_{p,\mu}(K_\gamma, \Omega) \leq \gamma^{-p} \int_{\{u < \gamma\}} |\nabla u|^p \, d\mu \leq \alpha^{-1} \gamma^{-p} \int_{\{u < \gamma\}} \mathcal{A}(x, \nabla u) \cdot \nabla u \, dx$$

$$= \alpha^{-1} \gamma^{1-p} \int_\Omega \mathcal{A}(x, \nabla u) \cdot \nabla \min(u/\gamma, 1) \, dx$$

$$= \alpha^{-1} \gamma^{1-p} \int_\Omega \mathcal{A}(x, \nabla u) \cdot \nabla u \, dx$$

$$\leq \frac{\beta}{\alpha} \gamma^{1-p} \int_\Omega |\nabla u|^p \, d\mu$$

$$\leq (\frac{\beta}{\alpha})^{p+1} \gamma^{1-p} \mathrm{cap}_{p,\mu}(K, \Omega),$$

where we also utilized Theorem 4.5 which guarantees that

$$\min(u/\gamma, 1) - u \in H_0^{1,p}(\Omega \setminus K; \mu).$$

Similarly,

$$\operatorname{cap}_{p,\mu}(K,\Omega) \leq \int_\Omega |\nabla u|^p \, d\mu$$

$$\leq \frac{1}{\alpha} \int_\Omega \mathcal{A}(x,\nabla u) \cdot \nabla u \, dx = \frac{1}{\alpha\gamma} \int_{\{u<\gamma\}} \mathcal{A}(x,\nabla u) \cdot \nabla u \, dx$$

$$\leq \frac{\beta}{\alpha\gamma} \int_{\{u<\gamma\}} |\nabla u|^p \, d\mu \leq (\frac{\beta}{\alpha})^{p+1} \gamma^{p-1} \operatorname{cap}_{p,\mu}(K_\gamma,\Omega).$$

The lemma follows.   □

If $\mathcal{A}(x,\xi) = w(x)|\xi|^{p-2}\xi$, then $\alpha = \beta = 1$, and we have the equality

$$\operatorname{cap}_{p,\mu}(\{u \geq \gamma\},\Omega) = \gamma^{1-p} \operatorname{cap}_{p,\mu}(K,\Omega)$$

for the (p,μ)-potential u of K in Ω. Note also that using the F-capacity defined in Chapter 5 the double inequality of Lemma 6.19 becomes an equality

$$\operatorname{cap}_F(\{v \geq \gamma\},\Omega) = \gamma^{1-p} \operatorname{cap}_F(K,\Omega),$$

where v is the F-extremal in $\Omega \setminus K$ with boundary values 1 on K and 0 on $\partial\Omega$; this follows from Euler's theorem for homogeneous functions (see (5.11)).

The first important application of Lemma 6.19 is an estimate which relates the values of an $\mathcal{A}$-potential to a capacity density formula.

6.21. Lemma.   Let $K \subset B = B(x_0,r)$ be a compact set. If $u = \Re(K,2B)$, then

$$(6.22) \qquad\qquad u(x) \geq c \left(\frac{\operatorname{cap}_{p,\mu}(K,2B)}{\operatorname{cap}_{p,\mu}(B,2B)} \right)^{1/(p-1)}$$

for all x in B. The constant $c > 0$ depends only on n, p, β/α, and c_μ.

PROOF: Let M and m denote, respectively, the maximum and minimum of u on the sphere $\{x \in \mathbf{R}^n : |x - x_0| = \frac{3}{2}r\}$. We may assume that $M > 0$. By using a covering argument and the Harnack inequality we infer that $M \leq cm$ with $c = c(n,p,\beta/\alpha,c_\mu)$. Thus it follows from the previous lemma that

$$\operatorname{cap}_{p,\mu}(K,2B) \leq c\,M^{p-1} \operatorname{cap}_{p,\mu}(\{u \geq M\},2B)$$

$$\leq c\,m^{p-1} \operatorname{cap}_{p,\mu}(\{u \geq M\},2B).$$

Next, by the comparison lemma 3.18 and the maximum principle, the set $\{u \geq M\}$ is contained in $\frac{3}{2}\overline{B}$ so that

$$\mathrm{cap}_{p,\mu}(\{u \geq M\}, 2B) \leq \mathrm{cap}_{p,\mu}(\frac{3}{2}\overline{B}, 2B).$$

Finally, since

$$\mathrm{cap}_{p,\mu}(B, 2B) \approx r^{-p}\mu(B) \approx \mathrm{cap}_{p,\mu}(\frac{3}{2}\overline{B}, 2B)$$

(Lemma 2.14), estimate (6.22) follows from the comparison lemma. □

Using the method in the proof of Lemma 6.21 we obtain two other capacity estimates. Let K be a compact subset of $B = B(x_0, r)$, $\lambda > 1$ and $u = \Re(K, \lambda B)$. Then there is a constant $c = c(n, p, \beta/\alpha, \lambda, c_\mu) > 0$ such that

$$(6.23) \qquad u \geq c \left(\frac{\mathrm{cap}_{p,\mu}(K, \lambda B)}{\mathrm{cap}_{p,\mu}(\tilde{\lambda}B, \lambda B)} \right)^{1/(p-1)}$$

in B, where $\tilde{\lambda} = \frac{1}{2}(1 + \lambda)$. This follows from the proof of Lemma 6.21 by replacing 2 with λ and $\frac{3}{2}$ with $\tilde{\lambda}$. The constant c can be chosen to be independent of λ if $\lambda \geq 2$. Moreover, if $\lambda \leq 2$, one can replace $\tilde{\lambda}$ by 1 in (6.23).

Also, for $x \in \partial B(x_0, \tilde{\lambda}r)$ it holds that

$$(6.24) \qquad u(x) \leq c \left(\frac{\mathrm{cap}_{p,\mu}(K, \lambda B)}{\mathrm{cap}_{p,\mu}(B, \lambda B)} \right)^{1/(p-1)},$$

where the positive constant c depends on n, p, β/α, c_μ, and λ; the constant c is independent of λ if $\lambda \geq 2$. To prove (6.24), let $m = \min_{\partial B(x_0, \tilde{\lambda}r)} u$. We have by Lemma 6.19 that

$$\mathrm{cap}_{p,\mu}(K, \lambda B) \geq c\, m^{p-1}\, \mathrm{cap}_{p,\mu}(\{u \geq m\}, \lambda B) \geq c\, m^{p-1}\, \mathrm{cap}_{p,\mu}(B, \lambda B),$$

and hence (6.24) follows because by Harnack's inequality $m \geq c\, u(x)$.

Next we construct an $\mathcal{A}$-harmonic barrier at a thick boundary point. Recall the convention stated before Lemma 6.19: the $\mathcal{A}$-potential $\Re(K, \Omega)$ is defined pointwise in Ω and $\Re(K, \Omega) = 1$ on K.

6.25. Lemma. *Let* $x_0 \in \partial\Omega$, $\rho > 0$, *and*

$$u = 1 - \Re(\complement\Omega \cap \overline{B}(x_0, \rho), B(x_0, 2\rho)).$$

Then for all $r \leq \rho$

$$(6.26) \qquad u \leq \exp(-c \int_r^\rho \varphi_{p,\mu}(x_0, \complement\Omega, t)^{1/(p-1)} \frac{dt}{t})$$

in $B(x_0, r)$, *where the constant* $c > 0$ *depends only on* n, p, β/α, *and* c_μ.

PROOF: Abbreviate $B = B(x_0, 2\rho)$. Fix $r \leq \rho$ and let k be the integer with $2^{-k}\rho < r \leq 2^{1-k}\rho$. Then write for $i = 1, 2, \ldots$

$$v_i = \Re(\complement\Omega \cap 2^{-i}\overline{B}, 2^{1-i}B)$$

and

$$a_i = \varphi_{p,\mu}(x_0, \complement\Omega, 2^{1-i}\rho).$$

Since $e^t \geq 1 + t$, estimate (6.22) implies that

$$v_i \geq c\, a_i^{1/(p-1)} \geq 1 - \exp(-c\, a_i^{1/(p-1)}) \quad \text{in } 2^{-i}B,$$

where $c = c(n, p, \beta/\alpha, c_\mu) > 0$. Next, let $u_1 = u$. Then $1 - u_1 = v_1$ and therefore

$$u_1 \leq \exp(-c\, a_1^{1/(p-1)}) \quad \text{in } \frac{1}{2}B$$

or

$$\exp(c\, a_1^{1/(p-1)})u_1 \leq 1 \quad \text{in } \frac{1}{2}B.$$

Define u_i recursively by

$$u_i = \exp(c\, a_{i-1}^{1/(p-1)})u_{i-1},$$

$i = 2, 3, \ldots$, where c is the constant of (6.22). We show by induction that

$$u_i \leq \exp(-c\, a_i^{1/(p-1)}) \quad \text{in } 2^{-i}B.$$

If this is true for u_{i-1}, we have

$$u_i \leq 1 \quad \text{in } 2^{1-i}B.$$

On the other hand, since $u_i = 0$ q.e. in $B \setminus \Omega$, Theorem 4.5 implies that

$$\min(1 - u_i - v_i, 0) \in H_0^{1,p}(2^{1-i}B \cap \Omega; \mu).$$

Since both $1 - u_i$ and v_i are $\mathcal{A}$-harmonic in $2^{1-i}B \cap \Omega$, the comparison lemma 3.18 ensures

$$1 - u_i \geq v_i \quad \text{in } 2^{-i}B.$$

In conclusion,

$$u_i \leq 1 - v_i \leq \exp(-c\, a_i^{1/(p-1)}) \quad \text{in } 2^{-i}B$$

for all $i = 1, 2, \ldots$. We obtain

$$u \leq \exp\left(-c\sum_{i=1}^{k} a_i^{1/(p-1)}\right) \quad \text{in } 2^{-k}B.$$

Since it easily follows from the doubling property of the measure μ that

$$\int_r^\rho \varphi_{p,\mu}(x_0, \complement\Omega, t)^{1/(p-1)}\, \frac{dt}{t} \leq c\sum_{i=1}^{k} a_i^{1/(p-1)},$$

(see Lemma 12.10), we arrive at (6.26) as $B(x_0, r) \subset 2^{-k}B$. □

PROOF OF THEOREM 6.18: We may obviously assign the value of ϑ at x_0 to be 0. Fix $\rho > 0$ and let u be the function of Lemma 6.25. Write

$$s_1 = u \max_{\partial\Omega} \vartheta + \max_{\partial\Omega \cap \overline{B}(x_0, 2\rho)} \vartheta.$$

To show that $\eta = \min(s_1, h) - h$ is in $H_0^{1,p}(\Omega(2\rho); \mu)$, we apply Theorem 4.5. Indeed, as we may assume that u is quasicontinuous in $\mathbf{R}^n$ and $u = 1$ on $\complement B(x_0, 2\rho)$, we have $s_1 \geq \max_{\partial\Omega} \vartheta \geq h$ on $\partial B(x_0, 2\rho) \cap \Omega$. Moreover, since h has a quasicontinuous extension in $\mathbf{R}^n$ which coincides with ϑ q.e. on $\partial\Omega$, it follows that $s_1 \geq \vartheta = h$ q.e. on $\partial\Omega \cap \overline{B}(x_0, 2\rho)$. Hence $\eta = 0$ q.e. on $\partial(\Omega(2\rho))$ and, therefore, $\eta \in H_0^{1,p}(\Omega(2\rho); \mu)$.

Since both s_1 and h are $\mathcal{A}$-harmonic in $\Omega(2\rho)$, the comparison lemma 3.18 implies $s_1 \geq h$ in $\Omega(2\rho)$. Similarly, by setting

$$s_2 = \min_{\partial\Omega \cap \overline{B}(x_0, 2\rho)} \vartheta + u \min_{\partial\Omega} \vartheta,$$

we obtain $s_2 \leq h$ in $\Omega(2\rho)$. In conclusion,

$$\text{osc}(h, \Omega(r)) \leq \sup_{\Omega(r)} s_1 - \inf_{\Omega(r)} s_2$$

$$\leq \text{osc}(\vartheta, \partial\Omega \cap \overline{B}(x_0, 2\rho)) + \text{osc}(\vartheta, \partial\Omega) \sup_{\Omega(r)} u,$$

which, by virtue of Lemma 6.25, establishes Theorem 6.18. □

DEFINITION. A boundary point x_0 of a bounded open set Ω is said to be (*Sobolev $\mathcal{A}$-*)*regular* if, for each function $\vartheta \in H^{1,p}(\Omega; \mu) \cap C(\overline{\Omega})$, the $\mathcal{A}$-harmonic function h in Ω with $h - \vartheta \in H_0^{1,p}(\Omega; \mu)$ satisfies

$$\lim_{x \to x_0} h(x) = \vartheta(x_0).$$

Furthermore, we say that a bounded open set Ω is *regular* if each $x_0 \in \partial\Omega$ is regular.

Theorem 6.18 implies that thick boundary points are regular:

6.27. Theorem. Suppose that Ω is bounded and that $\complement\Omega$ is (p, μ)-thick at $x_0 \in \partial\Omega$. If $\vartheta \in H^{1,p}(\Omega; \mu) \cap C(\overline{\Omega})$ and if h is the $\mathcal{A}$-harmonic function in Ω with $h - \vartheta \in H_0^{1,p}(\Omega; \mu)$, then

$$\lim_{x \to x_0} h(x) = \vartheta(x_0).$$

In other words, x_0 is an $\mathcal{A}$-regular boundary point of Ω.

PROOF: This actually follows from the proof of Theorem 6.18. We showed there that under the assumption $\vartheta(x_0) = 0$,

$$s_2 \le h \le s_1$$

in $B(x_0, 2\rho) \cap \Omega$. Since both s_1 and $|s_2|$ can be made as small as we please near x_0, the result follows.   □

6.28. Corollary. If the complement of Ω is (p, μ)-thick at each point on $\partial\Omega$, then Ω is regular.

The following theorem generalizes Corollary 6.28 for solutions to obstacle problems.

6.29. Theorem. Let Ω be bounded, let $\psi \in C(\Omega)$, and let $\vartheta \in C(\overline{\Omega}) \cap H^{1,p}(\Omega; \mu)$ be such that $\vartheta \ge \psi$ in Ω. If u is the continuous solution to the obstacle problem in $\mathcal{K}_{\psi,\vartheta}(\Omega)$, then

$$\lim_{x \to x_0} u(x) = \vartheta(x_0)$$

whenever $\complement\Omega$ is (p, μ)-thick at $x_0 \in \partial\Omega$.

PROOF: Suppose that $\complement\Omega$ is (p, μ)-thick at $x_0 \in \partial\Omega$. We first show that

(6.30) $$\limsup_{x \to x_0} u(x) \le \vartheta(x_0).$$

Suppose that the open set $U = \{x \in \Omega : u(x) > \vartheta(x)\}$ is nonempty and that $x_0 \in \partial U$. Since $u = \vartheta$ on $\partial U \cap \Omega$ and since $u - \vartheta \in H_0^{1,p}(\Omega; \mu)$, it follows from Theorem 4.5 that $u - \vartheta \in H_0^{1,p}(U; \mu)$. Alternatively, the conclusion $u - \vartheta \in H_0^{1,p}(U; \mu)$ can be seen as follows: since $(u - \vartheta)^+ \in H_0^{1,p}(\Omega; \mu)$, there are nonnegative smooth functions φ_i in $C_0^\infty(\Omega)$ converging to $(u-\vartheta)^+$ in $H_0^{1,p}(\Omega; \mu)$; the functions $\min(\varphi_i, (u - \vartheta)^+)$ are compactly supported in Ω and they converge to $(u - \vartheta)^+ = u - \vartheta$ in $H_0^{1,p}(U; \mu)$ by Corollary 1.22.

The function u is the unique solution of (3.8) in U with $u - \vartheta \in H_0^{1,p}(U; \mu)$ (Theorem 3.67). Since $U \subset \Omega$, the set $\complement U$ is also (p, μ)-thick at $x_0 \in \partial U$, and hence Corollary 6.28 yields

$$\lim_{\substack{x \to x_0 \\ x \in U}} u(x) = \vartheta(x_0).$$

This proves (6.30).

To complete the proof let h be the unique continuous solution to (3.8) in Ω with $h - \vartheta \in H_0^{1,p}(\Omega; \mu)$. It follows from the comparison lemma 3.18 that $u \geq h$ in Ω. Thus

$$\liminf_{x \to x_0} u(x) \geq \lim_{x \to x_0} h(x) = \vartheta(x_0)$$

by Corollary 6.28. Combining this with (6.30) we conclude the proof. $\square$

We next give a practical condition which guarantees the regularity of Ω for all mappings $\mathcal{A}$. The problem of regular boundary points is pursued further in Chapter 9, where we discuss the question from the potential theoretic point of view.

We say that a set E has a *corkscrew* at $x \in E$ if there are constants $c \geq 1$ and $r_0 > 0$ such that the ball $B(x, r)$ contains a ball $B(y, r/c) \subset E$ whenever $0 < r < r_0$. The complement of a bounded open set Ω is said to satisfy a *corkscrew condition* if $\complement\Omega$ has a corkscrew at each boundary point $x \in \partial\Omega$.

It is easy to see that if a bounded open set Ω satisfies an *exterior cone condition*, i.e. if for each boundary point $x \in \partial\Omega$ there is a closed truncated cone in $\complement\Omega$ with vertex at x, then Ω satisfies a corkscrew condition.

6.31. Theorem.   *If the complement of Ω has a corkscrew at $x_0 \in \partial\Omega$, then $\complement\Omega$ is (p, μ)-thick at x_0. In particular, if the complement of Ω satisfies a corkscrew condition, then Ω is regular.*

PROOF: It suffices to show that the Wiener criterion (6.17) holds at $x_0 \in \partial\Omega$. For all r positive and small enough, $B = B(x_0, r)$ contains a ball

$B(y, r/c)$ which does not intersect Ω with $c \geq 1$ independent of r. Therefore, using the Poincaré inequality as in Lemma 2.14 and the doubling property of μ we obtain

$$\text{cap}_{p,\mu}(\complement\Omega \cap B(x_0, r), B(x_0, 2r)) \geq \text{cap}_{p,\mu}(B(y, r/c), B(x_0, 2r))$$
$$\geq \text{cap}_{p,\mu}(B(y, r/c), B(y, 3r)) \geq cr^{-p}\mu(B(y, 3r))$$
$$\geq cr^{-p}\mu(B(x_0, r)) \geq c \, \text{cap}_{p,\mu}(B(x_0, r), B(x_0, 2r)).$$

This means that

$$\frac{\text{cap}_{p,\mu}(\complement\Omega \cap B(x_0, r), B(x_0, 2r))}{\text{cap}_{p,\mu}(B(x_0, r), B(x_0, 2r))} \geq c > 0$$

for r small enough, so the assertion is proved. $\square$

It is an easy exercise to show that if there is a constant $c > 0$ and a sequence of radii r_j tending to 0 such that $B(x_0, r_j) \setminus \Omega$ contains a ball $B(y_j, r_j/c)$ for all j, then x_0 is regular.

DEFINITION. A bounded open set Ω is called a *polyhedron* if $\partial\Omega = \partial\overline{\Omega}$ and if $\partial\Omega$ is contained in a finite union of $(n-1)$-hyperplanes.

It is easily seen that a domain Ω can be exhausted by connected polyhedra $P_1 \Subset P_2 \Subset \cdots \Subset \Omega$, $\cup P_j = \Omega$. Indeed, first exhaust Ω by domains $D_1 \Subset D_2 \Subset \cdots \Subset \Omega$. Then cover $\overline{D}_j$ by a finite union $\mathcal{Q}$ of open cubes $Q_{j_i} \Subset D_{j+1}$ such that each Q_{j_i} intersects D_j. Letting

$$P_j = \text{int } \overline{\mathcal{Q}}$$

we obtain the desired exhaustion. Consequently, each open set can be exhausted by polyhedra.

Simple geometry shows that polyhedra and balls satisfy a cone condition; therefore their complements satisfy a corkscrew condition, and we arrive at the following corollary.

6.32. Corollary. *All polyhedra and all balls are regular. In particular, every open set can be exhausted by regular open sets.*

We next show that points of positive (p, μ)-capacity are regular. This fact is not surprising, but the proof is somewhat involved. However, if $w \equiv 1$ and $p > n$, then estimate (2.13) implies

$$\varphi(x_0, \complement\Omega, t) \geq c > 0$$

for all Ω and all $x_0 \in \partial\Omega$, where $0 < t \leq 1$ and $c = c(n, p) > 0$; so the following theorem in that case follows from Corollary 6.28.

6.33. Theorem. *Suppose that* $\text{cap}_{p,\mu}\{x_0\} > 0$. *Then* $\{x_0\}$ *is* (p,μ)-*thick at* x_0. *In particular, if* $x_0 \in \partial\Omega$, *then* x_0 *is a regular boundary point of* Ω.

PROOF: Suppose that $\{x_0\}$ is not (p,μ)-thick at x_0. Then we may choose a ball $B = B(x_0, r)$ such that

$$(6.34) \qquad c_0 \sum_{j=1}^{\infty} a_j^{1/(p-1)} \leq \frac{1}{2},$$

where

$$a_j = \varphi(x_0, \{x_0\}, 2^{-j}r)$$

(see Lemma 12.10) and the constant $c_0 = c_0(n, p, \beta/\alpha, c_\mu) > 0$ is the constant c of (6.24) with $\lambda = 2$.

Abbreviate $B_j = 2^{-j}B$ and let

$$u_j = \Re(\{x_0\}, 2B_j)$$

be the $\mathcal{A}$-potential of $\{x_0\}$ in $2B_j$. Because $\lim_{x \to x_0} u_j(x) = \delta$ exists by Lemma 6.15, because $\text{cap}_{p,\mu}\{x_0\} > 0$, and because u_j is quasicontinuous by definition, we see that u_j is continuous in $2B_j$ and $\delta = u_j(x_0) = 1$. Observe also that $\lim_{x \to y} u_j(x) = 0$ for all $y \in \partial 2B_j$, since each such y is a regular boundary point of $2B \setminus \{x_0\}$.

We have by (6.24) that

$$(6.35) \qquad u_j \leq c_0\, a_j^{1/(p-1)} \qquad \text{on } \partial B_j.$$

Now we iterate (6.35). Let

$$b_j = c_0\, a_j^{1/(p-1)}$$

and

$$v_j = \frac{v_{j-1} - b_j}{1 - b_j},$$

where $v_0 = u_1$. Then it follows, by induction, from Lemma 1.26 and (6.35) that $\min(u_{j+1} - v_j, 0) \in H_0^{1,p}(B_j \setminus \{x_0\}; \mu)$ and hence the comparison lemma 3.18 implies $v_j \leq u_{j+1}$ in B_j. Therefore,

$$u_1 \leq v_1 + b_1 \leq v_2 + b_2 + b_1 \leq v_j + \sum_{k=1}^{j} b_k \leq u_{j+1} + \sum_{k=1}^{j} b_k \qquad \text{in } B_j.$$

Now (6.35) and (6.34) imply

$$u_1 \leq b_{j+1} + \sum_{k=1}^{j} b_j \leq \frac{1}{2} \qquad \text{on } \partial B_{j+1}$$

for $j = 1, 2, \ldots$, and hence

$$\liminf_{x \to x_0} u_1(x) \leq \frac{1}{2} < 1 = u_1(x_0)$$

which contradicts the continuity of u_1. The theorem follows. $\qquad \square$

Theorem 6.18 also implies the following boundary Hölder continuity estimate.

6.36. Corollary. *Suppose that Ω is bounded, that $x_0 \in \partial\Omega$, and that*

$$\varphi(x_0, \complement\Omega, t) \geq c_0 > 0 \text{ for } 0 < t \leq \rho.$$

If $\vartheta \in C(\overline{\Omega}) \cap H^{1,p}(\Omega; \mu)$ and if h is the A-harmonic function in Ω with $h - \vartheta \in H_0^{1,p}(\Omega; \mu)$, then for $0 < r \leq \rho$

$$(6.37) \qquad \operatorname{osc}(h, \Omega(r)) \leq \operatorname{osc}(\vartheta, \partial\Omega \cap \overline{B}(x_0, 2\rho)) + \operatorname{osc}(\vartheta, \partial\Omega)\left(\frac{r}{\rho}\right)^\delta,$$

where $\delta = \delta(n, p, \beta/\alpha, c_0, c_\mu) > 0$ and $\Omega(r) = \Omega \cap B(x_0, r)$.

We need a version of Corollary 6.36 where it is not required that h belong to $H^{1,p}(\Omega; \mu)$.

6.38. Theorem. *Suppose that Ω is regular, that $x_0 \in \partial\Omega$, and that*

$$\varphi(x_0, \complement\Omega, t) \geq c_0 > 0 \text{ for } 0 < t \leq \rho.$$

If h is an A-harmonic function in Ω, continuous on $\overline{\Omega}$, then for $0 < r \leq \rho$

$$(6.39) \qquad \operatorname{osc}(h, \Omega(r)) \leq \operatorname{osc}(h, \partial\Omega \cap \overline{B}(x_0, 2\rho)) + \operatorname{osc}(h, \partial\Omega)\left(\frac{r}{\rho}\right)^\delta,$$

where $\delta = \delta(n, p, \beta/\alpha, c_0, c_\mu) > 0$.

PROOF: This follows from Corollary 6.36 by an approximation. Indeed, by the Tietze extension theorem we first extend the function h to a continuous function $\tilde{h} \colon \mathbf{R}^n \to \mathbf{R}$ with $\tilde{h} = h$ on $\overline{\Omega}$. Next we choose a mollifier $\eta \in C_0^\infty(B(0,1))$, $0 \leq \eta \leq 1$, such that $\int_{\mathbf{R}^n} \eta \, dx = 1$. Then the convolutions

$$\varphi_j(x) = \int_{\mathbf{R}^n} j^n \eta(jy)\tilde{h}(x - y) \, dy$$

are in $C^\infty(\mathbf{R}^n)$ and converge to h uniformly on $\overline{\Omega}$ (see Section 1.10). Now let h_j be the unique A-harmonic function in Ω with $h_j - \varphi_j \in H_0^{1,p}(\Omega; \mu)$. By Corollary 6.36 it suffices to show that the sequence h_j converges to h uniformly on $\overline{\Omega}$.

To this end, fix $\varepsilon > 0$. Since $h_j = \varphi_j \to h$ uniformly on $\partial\Omega$, there is an index j_ε such that

$$h - \varepsilon < h_j < h + \varepsilon$$

on $\partial\Omega$ for $j \geq j_\varepsilon$. Thus there is an open $D \Subset \Omega$ such that the same inequalities hold on $\Omega \setminus D$. Therefore, $\min(h + \varepsilon - h_j, 0)$ and $\min(h_j - (h - \varepsilon), 0)$ belong to $H_0^{1,p}(D; \mu)$, and it follows from the comparison lemma 3.18 that

$$h - \varepsilon \leq h_j \leq h + \varepsilon$$

on $\overline{\Omega}$ for $j \geq j_\varepsilon$. Hence $h = \lim h_j$ uniformly on $\overline{\Omega}$, and the theorem is proved. $\square$

It was left open in Chapter 2 whether a closed ball has the same capacity as its interior. It easily follows from the Wiener estimate (6.37) that

$$(6.40) \qquad \mathrm{cap}_{p,\mu}(\overline{B}, \Omega) = \mathrm{cap}_{p,\mu}(B, \Omega)$$

whenever $B \Subset \Omega$ is a ball.

6.41. Boundary Hölder continuity of $\mathcal{A}$-harmonic functions

We conclude this chapter with an investigation of global Hölder continuity of $\mathcal{A}$-harmonic functions. More precisely, we consider the following problem: suppose that $h \in C(\overline{\Omega})$ is $\mathcal{A}$-harmonic in a bounded open set Ω and that h satisfies the Hölder condition

$$(6.42) \qquad |h(x) - h(y)| \le M|x - y|^{\delta}$$

for all points x and y on the boundary of Ω. Is it then true that h satisfies a similar inequality for all points x and y on the closure $\overline{\Omega}$? In general, the answer is negative. An example given by Krol' and Maz'ya (1972) shows that a solution to the p-Laplace equation

$$- \mathrm{div}(|\nabla h|^{p-2} \nabla h) = 0$$

with boundary data in $C^{\infty}(\mathbf{R}^n)$ may fail to satisfy (6.42) for any $\delta > 0$ in a neighborhood of a boundary point. In their example the domain has a sharp inwardly directed spine and $p \in (1, n-1]$.

Therefore, to obtain a positive conclusion we need to place an additional constraint on Ω.

DEFINITION. We say that $\complement\Omega$ satisfies a (p, μ)-*capacity density condition* if for some $c_0 > 0$ and $r_0 > 0$,

$$(6.43) \qquad \frac{\mathrm{cap}_{p,\mu}(\complement\Omega \cap \overline{B}(x_0, r), B(x_0, 2r))}{\mathrm{cap}_{p,\mu}(\overline{B}(x_0, r), B(x_0, 2r))} \ge c_0$$

whenever $0 < r < r_0$ and $x_0 \in \partial\Omega$. We also say that $\complement\Omega$ is *uniformly* (p, μ)-*thick*.

As an example we observe that $\complement\Omega$ satisfies a (p, μ)-capacity density condition if $\complement\Omega$ has a corkscrew in a uniform sense at each point on the boundary. We use condition (6.43) again in Chapter 11 in connection with $\mathcal{A}$-harmonic measures.

6.44. Theorem. *Suppose that Ω is bounded and that $\complement\Omega$ satisfies a (p, μ)-capacity density condition with constants c_0 and $r_0 \leq 1$. Let $h \in C(\overline{\Omega})$ be A-harmonic in Ω. If there are constants $M \geq 0$ and $0 < \delta \leq 1$ such that*

$$(6.45) \qquad |h(x) - h(y)| \leq M|x - y|^\delta$$

for all $x, y \in \partial\Omega$, then

$$(6.46) \qquad |h(x) - h(y)| \leq M_1|x - y|^{\delta_1}$$

for all $x, y \in \overline{\Omega}$. Moreover, $\delta_1 = \delta_1(n, p, \beta/\alpha, c_\mu, \delta, c_0) > 0$ and one can choose $M_1 = 80\, M r_0^{-2} \max(1, (\operatorname{diam}\Omega)^2)$.

Theorem 6.44 follows easily from a Wiener estimate on the boundary and the following elementary lemma.

6.47. Lemma. *Suppose that Ω is bounded and that $h \in C(\overline{\Omega})$ is A-harmonic in Ω. If there are constants $L \geq 0$ and $0 < \gamma \leq 1$ such that*

$$(6.48) \qquad |h(x) - h(x_0)| \leq L|x - x_0|^\gamma$$

for all $x \in \Omega$ and $x_0 \in \partial\Omega$, then

$$(6.49) \qquad |h(x) - h(y)| \leq L_1|x - y|^{\gamma_1}$$

for all $x, y \in \overline{\Omega}$. One can choose $\gamma_1 = \min(\gamma, \kappa)$, where κ is the Hölder exponent in (6.7), and $L_1 = 20\, L \max(1, (\operatorname{diam}\Omega)^{\gamma-\gamma_1})$.

PROOF: It is clearly enough to prove (6.49) for $x, y \in \Omega$. We consider two separate cases.

    Case (i): $|x - y| < \frac{1}{2}\operatorname{dist}(x, \partial\Omega)$. Choose $x_0 \in \partial\Omega$ such that $|x_0 - x| = \operatorname{dist}(x, \partial\Omega) = r$. Then if $z \in \Omega$ is such that $|x - z| \leq \frac{r}{2}$, we have

$$\begin{aligned}
|h(x) - h(z)| &\leq |h(x) - h(x_0)| + |h(x_0) - h(z)| \\
&\leq L|x - x_0|^\gamma + L|x_0 - z|^\gamma \\
&\leq L(r^\gamma + (\frac{3r}{2})^\gamma) \leq \frac{5}{2}Lr^\gamma.
\end{aligned}$$

Therefore by the Hölder continuity estimate (6.7)

$$\begin{aligned}
|h(x) - h(y)| &\leq 2^\kappa \left(\frac{|x - y|}{r/2}\right)^\kappa \operatorname{osc}(h, B(x, r/2)) \\
&\leq 20\, L|x - y|^\kappa r^{-\kappa} r^\gamma.
\end{aligned}$$

If $\gamma < \kappa$, then $|x - y|^\kappa r^{-\kappa} \le |x - y|^\gamma r^{-\gamma}$, and hence

$$|h(x) - h(y)| \le 20\,L|x - y|^\gamma .$$

If $\gamma \ge \kappa$, then $r^{\gamma - \kappa} \le (\operatorname{diam}\Omega)^{\gamma - \kappa}$, and hence

$$|h(x) - h(y)| \le 20\,L(\operatorname{diam}\Omega)^{\gamma - \kappa}|x - y|^\kappa .$$

Consequently, (6.49) is proved in case (i).

Case (ii): $|x - y| \ge \frac{1}{2}\operatorname{dist}(x, \partial\Omega)$. Let again $x_0 \in \partial\Omega$ be such that $|x_0 - x| = \operatorname{dist}(x, \partial\Omega)$. We have

$$\begin{aligned}
|h(x) - h(y)| &\le |h(x) - h(x_0)| + |h(x_0) - h(y)| \\
&\le L|x - x_0|^\gamma + L|x_0 - y|^\gamma \le L(|x - x_0|^\gamma + (|x - x_0| + |x - y|)^\gamma) \\
&\le L(2^\gamma|x - y|^\gamma + 3^\gamma|x - y|^\gamma) \le 5L|x - y|^\gamma ,
\end{aligned}$$

and the lemma follows.   □

It is clear from the proof that Lemma 6.47 is valid (with obvious modifications) for any function u, continuous on $\overline{\Omega}$, that satisfies a Hölder continuity estimate similar to (6.7).

PROOF OF THEOREM 6.44: It is no loss of generality to assume that $\min h = 0$. Observe that the minimum and the maximum of h on $\overline{\Omega}$ are attained on the boundary $\partial\Omega$. If $x_0 \in \partial\Omega$ and $x \in \Omega$ with $r = |x - x_0| < r_0^2 \le 1$, the boundary Hölder estimate (6.39) with $\rho = r^{1/2}$ yields

$$\begin{aligned}
|h(x) - h(x_0)| &\le M\,4^{\delta/2}r^{\delta/2} + \max_{\overline{\Omega}} h\,r^{c_1} \\
&\le L\,r^{\min(\delta/2, c_1)} ,
\end{aligned}$$

where $c_1 = c_1(n, p, \beta/\alpha, c_\mu, c_0) > 0$ and

$$L = 4M \max(1, \operatorname{diam}\Omega) \ge 4\max(M, \max h) .$$

Thus, for $|x - x_0| < r_0^2$, we have

$$|h(x) - h(x_0)| \le L\,|x - x_0|^\gamma ,$$

where $\gamma = \min(\delta/2, c_1) = \gamma(n, p, \beta/\alpha, c_\mu, c_0) > 0$.

On the other hand, if $|x - x_0| \ge r_0^2$, then

$$\begin{aligned}
|h(x) - h(x_0)| &\le 2\max h\,r_0^{-2\gamma}|x - x_0|^\gamma \\
&\le L\,r_0^{-2}\,|x - x_0|^\gamma ,
\end{aligned}$$

where L and γ are as above.

The conditions of Lemma 6.47 are thus met, and the theorem is proved.

NOTES TO CHAPTER 6. Harnack's inequality for solutions to non-linear elliptic equations was first proved in Serrin (1963, 1964). Our proof that Harnack's inequality implies the Hölder continuity of solutions is the standard one, which first appeared in Moser (1961). It is simple but does not give good estimates for κ. Sharp Hölder estimates for p-harmonic functions in the plane were found by Iwaniec and Manfredi (1989). A different proof for the Hölder continuity can be given in the (unweighted) borderline case $p = n$ (Granlund et al. 1983). The convergence theorem 6.14 was proved in Heinonen and Kilpeläinen (1988a) but must have been known before that.

The equivalence of the Wiener test and the boundary continuity for linear elliptic equations with measurable coefficients was first established by Littman et al. (1963). They demonstrated that the regular points are the same as the regular points for the Laplacian, thus extending the fundamental work of Wiener (1924a). In the nonweighted case Theorem 6.18 was proved in Maz'ya (1976). For a general treatment, see Gariepy and Ziemer (1977). In the unweighted case Maz'ya's result admits the converse. This was established by Lindqvist and Martio (1985) when $p > n - 1$ and recently by Kilpeläinen and Malý (1992b) for all $p > 1$. For weighted nonlinear equations the converse is still open. For a good account of this and related problems, Adams (1992).

The work of Littman et al. (1963) was extended to linear weighted equations by Fabes et al. (1982b); see also Biroli and Marchi (1986). For weighted nonlinear equations Theorem 6.18 appears to be new; see, however, Stredulinsky (1984). Our proof is similar in spirit to Maz'ya's except that we treat all open sets simultaneously; that is, no a priori results in polyhedra are required.

The boundary Hölder continuity of $\mathcal{A}$-harmonic functions (in the unweighted case) belongs to folklore although there seems to be nothing in print. It would be interesting to know if one can choose $\delta_1 = \min\{\delta, \kappa\}$ in domains that are sufficiently smooth, say, balls. A similar problem for harmonic functions has been studied by Hinkkanen (1988). The uniform thickness condition (6.43) has been employed in many analysis papers; see e.g. Pommerenke (1984), Lewis (1988), and Jones and Wolff (1988).

Harnack's inequality does not imply the *unique continuation* property: "If u is $\mathcal{A}$-harmonic in a domain Ω and $u = 0$ in an open nonempty subset of Ω, then $u = 0$ throughout Ω." Planar $\mathcal{A}$-harmonic functions have the unique continuation property (Bers and Schechter 1964; Aronszajn et al. 1962). The unique continuation for $\mathcal{A}$-harmonic functions is not known even for the unweighted p-Laplacian when $n > 2$ and $p \neq 2$. It is known to fail for certain linear and nonlinear equations included in our studies (Pliš 1963; Miller 1974; Garofalo and Lin 1986; Martio 1988a).

7

$\mathcal{A}$-superharmonic functions

To develop a potential theory, it is necessary to define superharmonic functions.

DEFINITION. A function $u : \Omega \to \mathbf{R} \cup \{\infty\}$ is $\mathcal{A}$-*superharmonic* in Ω if

  (i) u is lower semicontinuous,

  (ii) $u \not\equiv \infty$ in each component of Ω, and

  (iii) for each open $D \Subset \Omega$ and each $h \in C(\overline{D}) \cap \mathcal{H}(D)$ the inequality $u \geq h$ on ∂D implies $u \geq h$ in D.

We write $\mathcal{S}(\Omega) = \mathcal{S}_{\mathcal{A}}(\Omega)$ for the class of all $\mathcal{A}$-superharmonic functions in Ω. A function v in Ω is called $\mathcal{A}$-*subharmonic* in Ω if $-v$ is $\mathcal{A}$-superharmonic in Ω; the class of all $\mathcal{A}$-subharmonic functions in Ω is denoted by $-\mathcal{S}(\Omega)$.

It is well known that in the case of the Laplacian, i.e. $\mathcal{A}(x, \xi) = \xi$, this definition is one of the equivalent characterizations of superharmonic functions, often defined via a super mean value property (Doob 1984, p. 22; Helms 1969; Radó 1949). The definition for $\mathcal{A}$-superharmonic functions does not have a local character. However, $\mathcal{A}$-superharmonicity is a local property as we shall see in Theorem 7.27.

In this chapter, we first study general properties of $\mathcal{A}$-superharmonic functions. Among other things, we establish the comparison principle and the minimum principle, and introduce the Poisson modification of an $\mathcal{A}$-superharmonic function. In the second half of the chapter we investigate the connection between $\mathcal{A}$-superharmonic functions and supersolutions of equation (3.8). It is remarkable, and vital to our theory, that locally bounded $\mathcal{A}$-superharmonic functions are supersolutions; in particular, they belong locally to $H^{1,p}(\Omega; \mu)$. Conversely, each supersolution has an $\mathcal{A}$-superharmonic representative. To summarize, $\mathcal{A}$-superharmonic functions could be characterized as the closure of (pointwise defined) supersolutions with respect to upper directed monotone convergence. It is important to notice that, in general, there are $\mathcal{A}$-superharmonic functions, necessarily unbounded, which do not belong locally to $H^{1,p}(\Omega; \mu)$ and thus are not supersolutions; singular solutions are constructed in Section 7.38. The aforementioned interplay between supersolutions and $\mathcal{A}$-superharmonic functions enables us to characterize removable sets for $\mathcal{A}$-superharmonic and $\mathcal{A}$-harmonic functions as well as to establish integrability properties of $\mathcal{A}$-superharmonic functions.

We begin with some elementary remarks. Although $\mathcal{S}(\Omega)$ is not, in general, closed under addition, we have

7.1. Lemma. *If u is $\mathcal{A}$-superharmonic, then $\lambda u + \tau$ is $\mathcal{A}$-superharmonic whenever λ and τ are real numbers and $\lambda \geq 0$.*

The following two lemmas are easy to prove. The first describes the lattice property of $\mathcal{S}(\Omega)$ and the second reveals that $\mathcal{S}(\Omega)$ is closed under both upper directed and locally uniform convergence.

7.2. Lemma. *If u and v are $\mathcal{A}$-superharmonic in Ω, then the function $\min(u, v)$ is $\mathcal{A}$-superharmonic.*

7.3. Lemma. *Suppose that u_i, $i = 1, 2, \ldots$, are $\mathcal{A}$-superharmonic in Ω. If the sequence u_i either is increasing or converges uniformly on compact subsets of Ω, then in each component of Ω the limit function $u = \lim_{i \to \infty} u_i$ is $\mathcal{A}$-superharmonic unless $u \equiv \infty$.*

Next, we give a version of the fundamental convergence theorem. It will be refined further in Chapter 8.

7.4. Theorem. *Suppose that $\mathcal{F}$ is a family of $\mathcal{A}$-superharmonic functions in Ω, locally uniformly bounded below. Then the lower semicontinuous regularization s of $\inf \mathcal{F}$,*

$$s(x) = \lim_{r \to 0} \inf_{B(x,r)} (\inf \mathcal{F}),$$

is $\mathcal{A}$-superharmonic in Ω.

PROOF: Since $\mathcal{F}$ is locally uniformly bounded below, s is lower semicontinuous. Fix an open set $D \Subset \Omega$ and let $h \in C(\overline{D}) \cap \mathcal{H}(D)$ be such that $h \leq s$ on ∂D. Then $h \leq u$ in D whenever $u \in \mathcal{F}$. It follows from the continuity of h that $h \leq s$ in D, and the lemma is proved.   □

Theorem 7.4 above can be used to give a quick proof for the following convexity theorem.

7.5. Theorem. *Let $u : \Omega \to (a, b)$, $-\infty \leq a < b \leq \infty$, be $\mathcal{A}$-superharmonic. If $f : (a, b) \to \mathbf{R}$ is concave and increasing, then $f \circ u$ is $\mathcal{A}$-superharmonic.*

If u is $\mathcal{A}$-harmonic, then $f \circ u$ is $\mathcal{A}$-superharmonic if f is merely concave.

PROOF: If $f : (a, b) \to \mathbf{R}$ is concave and increasing, then for all $t \in (a, b)$,

$$f(t) = \inf\{\lambda t + \sigma : \lambda, \sigma \in \mathbf{R}, \, \lambda \geq 0, \lambda s + \sigma \geq f(s) \text{ for all } s \in (a, b)\};$$

this is the usual definition for concave functions in terms of the supporting lines. Now if u is $\mathcal{A}$-superharmonic, then $\lambda u + \sigma$ is $\mathcal{A}$-superharmonic and hence $f \circ u$ is A-superharmonic because it is a lower semicontinuous infimum of a family of A-superharmonic functions.

If f is only concave and u is A-harmonic, then in the definition of f we have $\lambda \in \mathbf{R}$, and the same reasoning applies. □

The open interval (a, b) in Theorem 7.5 can be replaced by the half open interval $(a, b]$. Indeed, the case where u is A-harmonic is immediate by the maximum principle. If $f : (a, b] \to \mathbf{R}$ is concave and increasing, then f is continuous. Then, if we choose a sequence $b_k \in (a, b)$ that increases to b, we see that $f \circ u$ is A-superharmonic as it is an increasing limit of A-superharmonic functions $f \circ \min(u, b_k)$.

In the subharmonic counterpart of Theorem 7.5 it should be assumed that f is convex and increasing or that f is convex and u is A-harmonic.

The nonlinear potential theory of A-superharmonic functions rests on the validity of the following *comparison principle*.

7.6. Comparison principle. *Suppose that u is A-superharmonic and that v is A-subharmonic in Ω. If*

$$(7.7) \qquad \limsup_{y \to x} v(y) \leq \liminf_{y \to x} u(y)$$

for all $x \in \partial\Omega$, and also for $x = \infty$ if Ω is unbounded, and if both sides of (7.7) are not simultaneously ∞ or $-\infty$, then $v \leq u$ in Ω.

PROOF: Fix $x \in \Omega$ and $\varepsilon > 0$. Choose a regular open set $D \Subset \Omega$ such that $x \in D$ and that $v < u + \varepsilon$ on ∂D (Corollary 6.32); because of (7.7) a sufficiently large D will do. Then let $\varphi_i \in C^\infty(\Omega)$ be a decreasing sequence converging to v on $\overline{D}$. Since ∂D is compact and $u + \varepsilon$ is lower semicontinuous, $\varphi_i \leq u + \varepsilon$ on ∂D for some i. Consequently, if we let h be the A-harmonic function in D such that h is continuous up to the boundary of D and coincides with φ_i on ∂D, then $v \leq h \leq u + \varepsilon$ on ∂D and therefore $v \leq h \leq u + \varepsilon$ in D. By letting $\varepsilon \to 0$ we obtain $v(x) \leq u(x)$, as desired. □

7.8. Lemma. *A function h is A-harmonic if and only if it is both A-superharmonic and A-subharmonic. That is,*

$$\mathcal{H}(\Omega) = \mathcal{S}(\Omega) \cap -\mathcal{S}(\Omega).$$

PROOF: By the proof of the comparison principle an A-harmonic function is both A-superharmonic and A-subharmonic. To establish the converse, let $u \in \mathcal{S}(\Omega) \cap -\mathcal{S}(\Omega)$. Then u is continuous and it suffices to show that

$u \in \mathcal{H}(B)$ whenever $B \Subset \Omega$ is a ball. Thus fix a ball $B \Subset \Omega$ and $\varepsilon > 0$. Choose $\varphi \in C^\infty(\mathbf{R}^n)$ such that

$$|u - \varphi| < \varepsilon$$

on $\overline{B}$. Let h_ε be the $\mathcal{A}$-harmonic function in B, continuous on $\overline{B}$, with $h_\varepsilon = \varphi$ on ∂B. Then
$$h_\varepsilon - \varepsilon \le u \le h_\varepsilon + \varepsilon$$

on ∂B, and since u is $\mathcal{A}$-superharmonic and $\mathcal{A}$-subharmonic, the same inequalities hold in B. Thus h_ε converges to u uniformly on $\overline{B}$ as $\varepsilon \to 0$, whence $u \in \mathcal{H}(B)$ (Theorem 6.13), as desired. □

As in classical linear theories, $\mathcal{A}$-superharmonic functions are much more flexible to deal with than $\mathcal{A}$-harmonic functions. The following pasting lemma, which is used many times in this book, demonstrates this flexibility. See also Section 7.29.

7.9. Pasting lemma. *Suppose that $D \subset \Omega$ is open, that u is $\mathcal{A}$-super-harmonic in Ω, and that v is $\mathcal{A}$-superharmonic in D. If the function*

$$s = \begin{cases} \min(u, v) & \text{in } D \\ u & \text{in } \Omega \setminus D \end{cases}$$

is lower semicontinuous, then it is $\mathcal{A}$-superharmonic in Ω.

PROOF: Let $G \Subset \Omega$ be open and $h \in C(\overline{G}) \cap \mathcal{H}(G)$ be such that $h \le s$ on ∂G. Then $h \le u$ in $\overline{G}$. In particular, since s is lower semicontinuous,

$$\lim_{\substack{y \to x \\ y \in D \cap G}} h(y) \le u(x) = s(x) \le \liminf_{\substack{y \to x \\ y \in D \cap G}} v(y)$$

for all $x \in \partial D \cap G$. Thus

$$\lim_{\substack{y \to x \\ y \in D \cap G}} h(y) \le s(x) \le \liminf_{\substack{y \to x \\ y \in D \cap G}} s(y)$$

for all $x \in \partial(D \cap G)$, and the comparison principle implies $h \le s$ in $D \cap G$. Therefore $h \le s$ in G and the lemma is proved. □

In the definition we require that an $\mathcal{A}$-superharmonic function u not be identically ∞ in any component of Ω. We shall see in Chapter 10 that the set where u takes the value ∞ cannot have positive (p, μ)-capacity. Before that, the following weaker result is needed.

7.10. Lemma.   *If u is $\mathcal{A}$-superharmonic in Ω, then the set $\{x \in \Omega : u(x) < \infty\}$ is dense in Ω.*

PROOF: We may assume that Ω is a domain. Suppose that there is a ball $B \Subset \Omega$ such that $u = \infty$ in $\overline{B}$. Pick $y \in \Omega$ with $u(y) < \infty$ and choose a connected polyhedron $D \Subset \Omega$ containing $\overline{B}$ and y. Then $\complement(D \setminus \overline{B})$ is (p, μ)-thick at each point of its boundary and therefore $D \setminus \overline{B}$ is regular (Corollaries 6.28 and 6.32). Consequently, we may choose a function h in $C(\overline{D} \setminus B) \cap \mathcal{H}(D \setminus \overline{B})$ taking the value 0 on ∂D and 1 on ∂B. If $m = \inf_{\overline{D}} u$, then $m > -\infty$ because u is lower semicontinuous. By the comparison principle $ih \leq u - m$ in $D \setminus \overline{B}$ for each integer i. Since $h(y) > 0$ by the minimum principle, we have

$$\lim_{i \to \infty} ih(y) = \infty \,.$$

This contradicts the inequality

$$ih(y) \leq u(y) - m < \infty \,,$$

and the lemma follows.   □

7.11. Corollary.   *If $u \in \mathcal{S}(\Omega)$, then $u \in \mathcal{S}(D)$ whenever $D \subset \Omega$ is open.*

Lemma 7.10 leads to the strict minimum principle for $\mathcal{A}$-superharmonic functions.

7.12. Theorem.   *A nonconstant $\mathcal{A}$-superharmonic function u cannot attain its infimum in a domain Ω.*

PROOF: Suppose that $u(x) = m = \inf_\Omega u$ and that $u(y) > m$. Then $u > m$ in an open set D. Let $v_i = i(u - m)$. Since $v_i(x) = 0$, the function $v = \lim_{i \to \infty} v_i$ is $\mathcal{A}$-superharmonic in Ω (Lemma 7.3). On the other hand, $v = \infty$ in D which contradicts Lemma 7.10.   □

7.13. Poisson modification
The Poisson modification carries the idea of local smoothing of an $\mathcal{A}$-superharmonic function. Suppose that u is $\mathcal{A}$-superharmonic in Ω and that $D \Subset \Omega$ is a regular open set. Let

$$u_D = \inf\{v : v \in \mathcal{S}(D), \liminf_{y \to x} v(y) \geq u(x) \text{ for each } x \in \partial D\}$$

and define the *Poisson modification* $P(u, D)$ of u in D to be the function

$$P(u, D) = \begin{cases} u & \text{in } \Omega \setminus D \\ u_D & \text{in } D \,. \end{cases}$$

7.14. Lemma. *The Poisson modification $P(u, D)$ is $\mathcal{A}$-superharmonic in Ω, $\mathcal{A}$-harmonic in D, and $P(u, D) \leq u$ in Ω.*

PROOF: Clearly $P(u, D) \leq u$ in Ω. Next, choose an increasing sequence $\varphi_i \in C^\infty(\mathbf{R}^n)$ which converges to u in $\overline{D}$. Let h_i be the unique function in $C(\overline{D}) \cap \mathcal{H}(D)$ such that $h_i = \varphi_i$ on ∂D. The sequence h_i is increasing by the comparison principle, and hence Harnack's convergence theorem implies that the function

$$h = \lim_{i \to \infty} h_i$$

is $\mathcal{A}$-harmonic in D; note that $h \leq u$ so that Lemma 7.10 ensures that h is finite somewhere. Since

$$\liminf_{x \to y} h(x) \geq \lim_{i \to \infty} \varphi_i(y) = u(y)$$

for $y \in \partial D$, it follows that $h \geq P(u, D)$ in D. On the other hand, the comparison principle implies that $h_i \leq P(u, D)$ in D for all i and therefore $P(u, D)|_D = h$ is $\mathcal{A}$-harmonic in D. This reasoning also shows that $P(u, D)$ is lower semicontinuous, and hence $\mathcal{A}$-superharmonic by the pasting lemma 7.9.   □

In fact, in the proof above we gave an equivalent way to define the Poisson modification of u in D. Namely, in D $P(u, D)$ is the limit of any increasing sequence of functions $h_i \in \mathcal{H}(D) \cap C(\overline{D})$ such that the boundary values of h_i increase to u.

7.15. Supersolutions and $\mathcal{A}$-superharmonic functions

Next we study the relationship between $\mathcal{A}$-superharmonic functions and supersolutions of equation (3.8). We begin by showing that a supersolution is always $\mathcal{A}$-superharmonic.

7.16. Theorem. *Suppose that u is a supersolution of (3.8) in Ω with*

$$(7.17) \qquad\qquad u(x) = \operatorname{ess\,lim\,inf}_{y \to x} u(y)$$

for each $x \in \Omega$. Then u is $\mathcal{A}$-superharmonic.

PROOF: Since u is locally essentially bounded below (Theorem 3.41), $u > -\infty$, and the lower semicontinuity of u follows from (7.17). Moreover, since $u \in H^{1,p}_{loc}(\Omega; \mu)$, it cannot be identically ∞ in any component of Ω. Finally, let $D \Subset \Omega$ be open and let $h \in C(\overline{D}) \cap \mathcal{H}(D)$ be such that $h \leq u$ in ∂D. Fix $\varepsilon > 0$ and choose an open set $G \Subset D$ such that $u + \varepsilon > h$ in $D \setminus G$. Since the function $\min(u + \varepsilon - h, 0)$ has compact support, it belongs to $H^{1,p}_0(G; \mu)$. Hence the comparison lemma 3.18 implies $u + \varepsilon \geq h$ a.e. in G, hence a.e. in D. By (7.17) this inequality holds at each point in D. The theorem is proved since ε was arbitrary.   □

We proved in Theorem 3.63 that each supersolution u can be redefined in a set of measure zero such that the "ess lim inf" property (7.17) holds. Thus we obtain:

7.18. Corollary. *If u is a supersolution of (3.8) in Ω, then there is an $\mathcal{A}$-superharmonic function v in Ω such that $v = u$ a.e.*

It follows from Theorem 7.22 below that an $\mathcal{A}$-superharmonic representative of a supersolution is unique.

Next we show that $\mathcal{A}$-superharmonic functions can be approximated locally from below by continuous supersolutions (which are $\mathcal{A}$-superharmonic by the previous discussion). A global approximation is established later in Chapter 8.

7.19. Theorem. *Suppose that u is $\mathcal{A}$-superharmonic in Ω and that $D \Subset \Omega$ is an open set. Then there is an increasing sequence of supersolutions $u_i \in C(\overline{D}) \cap H^{1,p}(D; \mu)$ such that $u = \lim_{i\to\infty} u_i$ in D. Moreover, the functions u_i are $\mathcal{A}$-superharmonic in D.*

PROOF: We assume without loss of generality that D is a polyhedron. Choose an increasing sequence of functions $\varphi_i \in C^\infty(\mathbf{R}^n)$ converging to u in $\overline{D}$ and let u_i be the solution to the obstacle problem in $\mathcal{K}_{\varphi_i}(D)$. Then $u_i \in C(\overline{D}) \cap H^{1,p}(D; \mu)$ and $u_i = \varphi_i$ on ∂D (Theorems 3.67 and 6.29). We show that u_i is the desired sequence.

To this end, note that the minimal property of the solution to the obstacle problems (Lemma 3.22) implies that the sequence u_i is increasing. Moreover, in light of Theorems 3.67 and 7.16, each u_i is $\mathcal{A}$-superharmonic in D and $\mathcal{A}$-harmonic in the open set $U_i = \{x \in D : u_i(x) > \varphi_i(x)\}$. Since

$$\liminf_{x\to y} u(x) \geq u(y) \geq \varphi_i(y) = \lim_{x\to y} u_i(x)$$

for all $y \in \partial U_i$, the comparison principle implies that $u \geq u_i$ in U_i. Hence $u \geq u_i$ in D. Thus

$$u = \lim_{i\to\infty} \varphi_i \leq \lim_{i\to\infty} u_i \leq u$$

in D, and the proof is complete.   □

By Theorem 3.75 the limit of a locally bounded and increasing sequence of supersolutions is again a supersolution. Thus we obtain:

7.20. Corollary. *If u is $\mathcal{A}$-superharmonic in Ω and locally bounded above, then $u \in H^{1,p}_{loc}(\Omega; \mu)$ and u is a supersolution of (3.8) in Ω.*

Applying Corollary 7.20 to the $\mathcal{A}$-superharmonic functions $\min(u, i)$, $i = 1, 2, \ldots$, we have by virtue of the remark after Theorem 3.77:

7.21. Corollary. *If an $\mathcal{A}$-superharmonic function belongs to $H^{1,p}_{loc}(\Omega; \mu)$, it is a supersolution of (3.8) in Ω.*

To make the picture complete, we next show that $\mathcal{A}$-superharmonic functions share the important "ess lim inf" property of supersolutions.

7.22. Theorem. *If u is $\mathcal{A}$-superharmonic in Ω, then*

$$u(x) = \operatorname*{ess\,lim\,inf}_{y \to x} u(y)$$

for each $x \in \Omega$.

7.23. Corollary. *Suppose that both u and v are $\mathcal{A}$-superharmonic in Ω. If $u = v$ a.e. in Ω, then $u(x) = v(x)$ for all $x \in \Omega$.*

For the proof of Theorem 7.22 we require the following lemma.

7.24. Lemma. *Suppose that u is $\mathcal{A}$-superharmonic in Ω and $u = 0$ a.e. in Ω. Then $u(x) = 0$ for each $x \in \Omega$.*

PROOF: Because u is lower semicontinuous it is nonpositive. Thus, by localizing the problem, we may assume by Corollary 7.20 that $u \in H^{1,p}(\Omega; \mu)$. It suffices to show that $u = 0$ in a given ball $B \Subset \Omega$. Let $v = P(u, B)$ be the Poisson modification of u in B. Since $u - v \in H^{1,p}_0(\Omega; \mu)$ is nonnegative and since v is a supersolution in Ω (Corollary 7.20), we have

$$\alpha \int_\Omega |\nabla v|^p \, d\mu \leq \int_\Omega \mathcal{A}(x, \nabla v) \cdot \nabla v \, dx$$
$$\leq \int_\Omega \mathcal{A}(x, \nabla v) \cdot \nabla u \, dx = 0 \,;$$

note that $\nabla u = 0$. Thus $\nabla v = 0$ in Ω and hence $v = 0$ a.e. in Ω. By continuity $v(x) = 0$ for all $x \in B$. This establishes the desired result since $v \leq u$ in Ω. $\qquad \square$

PROOF OF THEOREM 7.22: Fix $x \in \Omega$. If

$$\lambda = \operatorname*{ess\,lim\,inf}_{y \to x} u(y) \,,$$

then

$$\lambda \geq \liminf_{y \to x} u(y) \geq u(x)$$

by the lower semicontinuity of u. To establish the reverse inequality, pick $\gamma < \lambda$. Then there is a radius $r > 0$ such that $B = B(x, r) \subset \Omega$ and $u \geq \gamma$ a.e. in B. According to Lemma 7.24, the $\mathcal{A}$-superharmonic function

$$v = \min(u, \gamma) - \gamma$$

is identically 0 in B. In particular, $u(x) \geq \gamma$ and the lemma follows since $\gamma < \lambda$ was arbitrary. $\qquad \square$

It is convenient to collect the results of this section as follows.

7.25. Theorem.

(i) If u is a function in $H^{1,p}_{loc}(\Omega; \mu)$, then u is $\mathcal{A}$-superharmonic if and only if u is a supersolution of (3.8) with

(7.26)
$$u(x) = \operatorname*{ess\,lim\,inf}_{y \to x} u(y)$$

for each $x \in \Omega$.

(ii) If u is $\mathcal{A}$-superharmonic and locally bounded, then u is in $H^{1,p}_{loc}(\Omega; \mu)$.

(iii) If u is $\mathcal{A}$-superharmonic in Ω, then (7.26) holds everywhere in Ω.

(iv) If u is a supersolution of (3.8), then (7.26) holds a.e. in Ω, and there is an $\mathcal{A}$-superharmonic function v in Ω such that $v = u$ a.e.

For an $\mathcal{A}$-superharmonic function u the truncations $\min(u, i)$ are supersolutions of (3.8). Thus, roughly speaking, the class $S(\Omega)$ of all $\mathcal{A}$-superharmonic functions in Ω can be described as the closure of (pointwise defined) supersolutions of (3.8) with respect to upper directed monotone convergence. We show in Theorem 7.39 that if $\mathrm{cap}_{p,\mu}\{x\} = 0$, then there is an $\mathcal{A}$-superharmonic function u, defined in a neighborhood U of x, such that u does not belong to $H^{1,p}_{loc}(U; \mu)$. Therefore, in general, the class of $\mathcal{A}$-superharmonic functions is larger than the class of supersolutions. Observe that an $\mathcal{A}$-superharmonic function u in $H^{1,p}_{loc}(\Omega; \mu)$ need not be locally bounded (see Theorem 7.48 below).

The definition of an $\mathcal{A}$-superharmonic function is not local; it requires testing in all open sets $D \Subset \Omega$. However, Theorem 7.25 reveals the local nature of $\mathcal{A}$-superharmonic functions. This is difficult to establish directly.

7.27. Theorem.
A function u is $\mathcal{A}$-superharmonic in Ω if and only if each $x \in \Omega$ has a neighborhood $D \subset \Omega$ such that $u|_D$ is $\mathcal{A}$-superharmonic in D.

PROOF: Since we may assume that u is bounded, the statement reduces to the corresponding property of supersolutions. This, in turn, follows easily by using a partition of unity. $\qquad \square$

We alert the reader that Theorem 7.27, which is by no means trivial, will be used often in this book and most of the time without an explicit reference. As the first application, we present another version of the pasting lemma.

7.28. Lemma.
Suppose that $\Omega = D \cup G$, where D and G are open sets. Let $u \in S(D)$ and $v \in S(G)$ be such that $u \leq v$ in $D \cap G$. If the function

$$s = \begin{cases} v & \text{on } \Omega \setminus D \\ u & \text{in } D \end{cases}$$

is lower semicontinuous, it is $\mathcal{A}$-superharmonic in Ω.

PROOF: Because $\min(u,v) = u$ in $D \cap G$, it follows from the pasting lemma 7.9 that s is $\mathcal{A}$-superharmonic in G. Therefore s is $\mathcal{A}$-superharmonic both in D and in G, and hence in $G \cup D = \Omega$ by Theorem 7.27.   □

7.29. Extension of $\mathcal{A}$-superharmonic functions

Now we proceed to show that each $\mathcal{A}$-superharmonic function $u \in \mathcal{S}(\Omega)$ is locally a restriction of an entire $\mathcal{A}$-superharmonic function $\tilde{u} \in \mathcal{S}(\mathbf{R}^n)$. In general, there does not exist a similar extension property for $\mathcal{A}$-harmonic functions; for example, an ordinary harmonic function is real analytic and hence its values in a domain Ω are already determined by its values in any ball $B \subset \Omega$.

7.30. Theorem.

Let $E \subset \mathbf{R}^n$ be a compact set such that $\mathbf{R}^n \setminus E$ is connected and that E is (p, μ)-thick at each point on E. If u is $\mathcal{A}$-superharmonic in a connected neighborhood Ω of E, then there is an $\mathcal{A}$-superharmonic function s in $\mathbf{R}^n$ such that $s = u$ on E.

We first prove an auxiliary result.

7.31. Lemma.

Let $E \subset \mathbf{R}^n$ be a compact set such that $\mathbf{R}^n \setminus E$ is connected and that E is (p, μ)-thick at each point on E. Then there is a function

$$v \in C(\mathbf{R}^n) \cap \mathcal{S}(\mathbf{R}^n) \cap \mathcal{H}(\mathbf{R}^n \setminus E)$$

such that $v = 0$ in E and that $v < 0$ in $\mathbf{R}^n \setminus E$.

PROOF: Let B_i be a sequence of concentric open balls containing E such that $B_i \Subset B_{i+1}$ and $\cup_i B_i = \mathbf{R}^n$. Fix a point $y_0 \in \partial B_1$. Let u_i be the $\mathcal{A}$-harmonic function in the open set $B_i \setminus E$, continuous up to the boundary, such that $u_i = 0$ on ∂E and $u_i = 1$ on ∂B_i; because $B_i \setminus E$ is regular, u_i can be found. Define

$$v_i = \frac{-u_i}{u_i(y_0)} .$$

Then v_i is a negative $\mathcal{A}$-harmonic function in $B_i \setminus E$ with $v_i(y_0) = -1$. It follows from Harnack's inequality that the sequence v_i is locally bounded, and hence equicontinuous by Corollary 6.12. Thus Ascoli's theorem together with Theorem 6.13 implies that a subsequence v_k converges locally uniformly to a nonpositive $\mathcal{A}$-harmonic function v in $\mathbf{R}^n \setminus E$ with $v(y_0) = -1$. The maximum principle then implies that $v < 0$.

Next we show that there is a constant $c > 0$ such that

$$cv_1 \le v < 0$$

in $B_1 \setminus E$. To achieve this, first use Harnack's inequality to infer that $v_k \geq -c$ on the sphere ∂B_1, where $c > 0$ is independent of k. Then the comparison principle yields $v_k \geq cv_1$ in $B_1 \setminus E$ and so $cv_1 \leq v < 0$. Consequently,

$$\lim_{x \to y} v(x) = 0$$

for $y \in \partial E$. Finally, define v to be 0 on E, take a look at the pasting lemma 7.9, and conclude that v is the desired $\mathcal{A}$-superharmonic function in $\mathbf{R}^n$.

$\square$

PROOF OF THEOREM 7.30: Let v be the $\mathcal{A}$-superharmonic function provided by Lemma 7.31. Since v has a negative limit value at infinity (Lemma 6.15), the comparison principle guarantees that there is $c < 0$ such that

$$V = \{x : v(x) > c\} \Subset \Omega.$$

We may assume that $u \geq 0$ on $\overline{V}$. Choose open sets W and U such that

$$E \subset W \Subset U \Subset V$$

and that $U \setminus E$ is regular. If $h = v - c$, then there is a constant $\lambda > 0$ such that $\lambda h \geq P(u, U \setminus E)$ on ∂W. Thus the pasting lemma 7.9 implies that

$$\tilde{s} = \begin{cases} \min(\lambda h, P(u, U \setminus E)) & \text{in } V \setminus \overline{W} \\ P(u, U \setminus E) & \text{in } \overline{W} \end{cases}$$

is $\mathcal{A}$-superharmonic in V. Moreover, since $\min(\lambda h, \tilde{s}) = \tilde{s}$ in $V \setminus \overline{W}$ and

$$\lim_{x \to y} \tilde{s}(x) = \lim_{x \to y} \lambda h(x) = 0$$

for $y \in \partial V$, the second version of the pasting lemma (Lemma 7.28) implies that the function

$$s = \begin{cases} \lambda h & \text{in } \mathbf{R}^n \setminus V \\ \tilde{s} & \text{in } V \end{cases}$$

is $\mathcal{A}$-superharmonic in $\mathbf{R}^n$. Moreover, since $s = u$ on E it is a desired $\mathcal{A}$-superharmonic extension of u.

$\square$

If E is a closed ball, then the assumptions of Theorem 7.30 are clearly satisfied. Lemma 7.31 is later employed in a situation where E is a point of positive (p, μ)-capacity; then the function v can be regarded as a singular solution in the complement of a point (cf. Theorem 6.33).

It is easily shown that the connectedness assumption of $\mathbf{R}^n \setminus E$ cannot be ignored. For example, let B be an open ball and let $\Omega = 2B \setminus \overline{B}$ be an annulus; then Ω is regular and hence there is a function $h \in \mathcal{H}(\Omega) \cap C(\overline{\Omega})$ such that $h = 0$ on ∂B and $h = 1$ on $\partial 2B$. Choose numbers $1 < \lambda_1 < \lambda_2 < 2$ such that $h \geq 3/4$ on $\partial \lambda_2 B$ and $h \leq 1/4$ on $\partial \lambda_1 B$. If we let E be the closed annulus $\lambda_2 \overline{B} \setminus \lambda_1 B$, then there does not exist any $\mathcal{A}$-superharmonic function u in $\mathbf{R}^n$ with $u = h$ on E, because $u \geq \min_{\partial \lambda_2 B} u \geq 3/4$ in $\lambda_2 B$ by the comparison principle.

7.32. Removable sets

We next show that sets of (p, μ)-capacity zero are removable for both positive $\mathcal{A}$-superharmonic functions and bounded $\mathcal{A}$-harmonic functions. These results are sharp.

7.33. Lemma. Let E be a relatively closed set in Ω of zero (p, μ)-capacity. If u is a supersolution in $\Omega \setminus E$ and $u \in H^{1,p}_{loc}(\Omega; \mu)$, then u is a supersolution in Ω. In particular, if $u \in H^{1,p}_{loc}(\Omega; \mu)$ is $\mathcal{A}$-harmonic in $\Omega \setminus E$, then u has a continuous representative which is $\mathcal{A}$-harmonic in Ω.

PROOF: Since the problem is local, we may assume that $u \in H^{1,p}(\Omega; \mu)$. Let $\varphi \in C_0^\infty(\Omega)$ be nonnegative. Since E has zero (p, μ)-capacity, we have $H_0^{1,p}(\Omega; \mu) = H_0^{1,p}(\Omega \setminus E; \mu)$ (Theorem 2.43) and so $\varphi \in H_0^{1,p}(\Omega \setminus E; \mu)$. Thus, since $|E| = 0$ (Lemma 2.10), it holds that

$$0 \leq \int_{\Omega \setminus E} \mathcal{A}(x, \nabla u) \cdot \nabla \varphi \, dx = \int_\Omega \mathcal{A}(x, \nabla u) \cdot \nabla \varphi \, dx,$$

as desired. □

7.34. Lemma. Let E be a relatively closed set in Ω of zero (p, μ)-capacity and let u be a supersolution in $\Omega \setminus E$. If each point $x \in E \cap \Omega$ has a neighborhood $U \subset \Omega$ such that u is bounded in $U \setminus E$, then u is a supersolution in Ω.

PROOF: If we show that $u \in H^{1,p}_{loc}(\Omega; \mu)$, the previous lemma implies the claim. Let B be a ball such that $2B \Subset \Omega$ and u is bounded in $2B$. It suffices to show that $u \in H^{1,p}(B; \mu)$. We may assume that $u \leq 0$ in $2B$. Since $C_{p,\mu}(E) = 0$, we may choose a sequence of functions $\varphi_i \in C_0^\infty(\mathbf{R}^n)$ such that $0 \leq \varphi_i \leq 1$, that $\varphi_i = 1$ in a neighborhood of $E \cap 2\overline{B}$, and that $\varphi_i \to 0$ in $H^{1,p}(\mathbf{R}^n; \mu)$ (see Corollary 2.39 and Lemma 2.36). Then let $\eta \in C_0^\infty(2B)$ be a nonnegative function such that $\eta = 1$ in B. Now it follows from the Caccioppoli estimate (3.29) that

$$\int_{2B \setminus E} |\nabla u|^p |\eta(1 - \varphi_i)|^p \, d\mu \leq c \sup_{2B \setminus E} |u|^p \int_{2B} |\nabla(\eta(1 - \varphi_i))|^p \, d\mu.$$

Consequently, since $\eta(1 - \varphi_i)$ converges to η in $H^{1,p}(\Omega; \mu)$, we conclude

$$\int_{B \setminus E} |\nabla u|^p \, d\mu \leq c < \infty,$$

and hence $u \in H^{1,p}(B \setminus E; \mu)$. On the other hand, since E has zero (p, μ)-capacity, we have $H^{1,p}(B \setminus E; \mu) = H^{1,p}(B; \mu)$ (Theorem 2.44), and the lemma follows. □

The next theorem extends a well-known result of classical potential theory.

7.35. Theorem. *Let E be a relatively closed set in Ω of zero (p, μ)-capacity. If u is A-superharmonic in $\Omega \setminus E$ with $\liminf_{y \to x} u(y) > -\infty$ for all $x \in E \cap \Omega$, then*

$$u(x) = \operatorname{ess\,lim\,inf}_{y \to x} u(y), \quad x \in \Omega,$$

defines an A-superharmonic function in Ω.

PROOF: If $u_k = \min(u, k)$, $k = 1, 2, \ldots$, then u_k is a supersolution of (3.8) in Ω by the previous lemma and Corollary 7.20. According to Theorem 7.16 the representative of u_k defined by

$$u_k(x) = \operatorname{ess\,lim\,inf}_{y \to x} u_k(y)$$

is A-superharmonic in Ω. The theorem follows by letting $k \to \infty$. □

The following theorem is an obvious corollary to Theorem 7.35.

7.36. Theorem. *Let E be a relatively closed set in Ω of zero (p, μ)-capacity. Then each bounded A-harmonic function h in $\Omega \setminus E$ can be extended to E so that the extension is A-harmonic in Ω.*

The removability theorems 7.35 and 7.36 are sharp. In fact, let B be a ball and $C \Subset B$ a compact set such that $\operatorname{cap}_{p,\mu}(C, B) > 0$. If $1 - h$ is the restriction to $B \setminus C$ of the A-potential of C in B, then by the minimum principle every A-superharmonic extension $\tilde{h}$ of h in B satisfies $\tilde{h} \geq 1$ in B. This, however, is untenable because by Lemma 6.19 $h(x) < 1$ for some x in $B \setminus C$.

In linear potential theories one usually allows an exceptional set in the comparison inequality (7.7) and still retains the same conclusion. Easy examples show that a singleton of positive capacity cannot be ignored. It is not known if a point of (p, μ)-capacity zero can be neglected in general.

However, the connection between $\mathcal{A}$-superharmonic functions and supersolutions enables us to prove the following weaker result, which is often useful in practical situations. Of course, it would be desirable to dispense with the imposed integrability assumption.

DEFINITION. We say that a set $E \subset \overline{\Omega}$ has *zero (inner)* (p,μ)-*capacity relative to* Ω if for each compact subset K of E we have

$$\inf \int_\Omega (|\varphi|^p + |\nabla\varphi|^p)\, d\mu = 0\,,$$

where the infimum is taken over all $\varphi \in C(\Omega \cup K) \cap H^{1,p}(\Omega; \mu)$ such that $\varphi = 1$ on K.

By using the Sobolev capacity we easily infer that if $\mathrm{cap}_{p,\mu} E = 0$, then E has zero (p, μ)-capacity relative to Ω. The converse is not true, as easy examples quickly confirm.

7.37. Lemma. *Suppose that Ω is bounded and that $E \subset \partial\Omega$ has zero* (p, μ)-*capacity relative to Ω. Let $u \in \mathcal{S}(\Omega)$ and $v \in -\mathcal{S}(\Omega)$ be bounded such that*

$$\limsup_{y \to x} v(y) \le \liminf_{y \to x} u(y)$$

for all $x \in \partial\Omega \setminus E$. If either $u \in L^{1,p}(\Omega; \mu)$ or $v \in L^{1,p}(\Omega; \mu)$, then $v \le u$ in Ω.

PROOF: Since for each $\varepsilon > 0$ the set

$$\{x \in \partial\Omega\colon \liminf_{y \to x} u(y) + \varepsilon > \limsup_{y \to x} v(y)\}$$

is open on $\partial\Omega$, we may assume that E is compact. By symmetry, we may also assume that $u \in H^{1,p}(\Omega; \mu)$.

Now we may choose a decreasing sequence of functions $\varphi_i \in C(\Omega \cup E) \cap H^{1,p}(\Omega; \mu)$ such that

$$0 \le \varphi_i \le M = \sup|u| + \sup|v|\,,$$

that $\varphi_i = M$ on E, and that

$$\|\varphi_i\|_{H^{1,p}(\Omega;\mu)} \to 0\,.$$

Indeed, we may choose a sequence of nonnegative functions $\tilde\varphi_j \in C(\Omega \cup E) \cap H^{1,p}(\Omega; \mu)$ such that $\tilde\varphi_j = M$ on E and that $\tilde\varphi_j \to 0$ in $H^{1,p}(\Omega; \mu)$.

Then let $\varphi_1 = \min(M, \tilde{\varphi}_1)$ and let, for $i \geq 1$, $\varphi_{i+1} = \min(\varphi_i, \tilde{\varphi}_j)$, where the index j is chosen so large that

$$\| \min(\varphi_i, \tilde{\varphi}_j) \|_{H^{1,p}(\Omega; \mu)} \leq \frac{1}{2} \|\varphi_i\|_{H^{1,p}(\Omega; \mu)} ;$$

this is possible because $\min(\varphi_i, \tilde{\varphi}_j) \to 0$ in $H^{1,p}(\Omega; \mu)$ as $j \to \infty$ by Lemma 1.22.

Then write

$$\psi_i = u + \varphi_i$$

and let u_i be the $\mathcal{A}$-superharmonic solution to the obstacle problem in $\mathcal{K}_{\psi_i}$. By the "ess lim inf" property (7.22) we have that $u_i \geq \psi_i$ in Ω. Hence

$$\limsup_{y \to x} v(y) \leq \liminf_{y \to x} u_i(y)$$

for all $x \in \partial\Omega$ and for each i, and by the comparison principle $u_i \geq v$. By virtue of Theorem 3.79, the sequence u_i converges to u a.e. Therefore, by appealing again to the "ess lim inf" property, we infer that $u \geq v$ in Ω, and the lemma follows. $\qquad\square$

7.38. Singular solutions

Next we examine the behavior of $\mathcal{A}$-harmonic functions near an isolated singularity.

7.39. Theorem. *Suppose that Ω is bounded and that $x_0 \in \Omega$. If $\mathrm{cap}_{p,\mu}\{x_0\} = 0$, then there is a function $u \in \mathcal{S}(\Omega) \cap \mathcal{H}(\Omega \setminus \{x_0\})$ such that*

$$\lim_{x \to x_0} u(x) = \infty = u(x_0)$$

and that

$$\lim_{x \to y} u(x) = 0$$

at each regular boundary point $y \in \partial\Omega$. Moreover, $u \notin H^{1,p}_{loc}(\Omega; \mu)$ and so it is not a supersolution of (3.8) in Ω.

PROOF: Suppose first that $u \in \mathcal{S}(\Omega) \cap \mathcal{H}(\Omega \setminus \{x_0\})$ with

$$\lim_{x \to x_0} u(x) = \infty .$$

If $u \in H^{1,p}_{loc}(\Omega; \mu)$, it would follow from Lemma 7.33 that u can be extended to be $\mathcal{A}$-harmonic in Ω. This is impossible because u is unbounded near x_0.

Now we construct the desired function u. We may assume that Ω is a domain. Let $B = B(x_0, r) \Subset \Omega$ be a ball and $B_i = i^{-1}B$, $i = 1, 2, \ldots$. Choose a function $\varphi \in C_0^\infty(\Omega)$ such that $\varphi = 1$ in $\overline{B}$ and let h_i be the $\mathcal{A}$-harmonic function in $\Omega \setminus \overline{B}_i$ such that $h_i - \varphi \in H_0^{1,p}(\Omega \setminus \overline{B}_i)$. By setting $h_i = 1$ in $\overline{B}_i$ we have $h_i \in \mathcal{S}(\Omega) \cap \mathcal{H}(\Omega \setminus \overline{B}_i)$. Define

$$u_i = \frac{h_i}{\max_{\partial B} h_i}.$$

Then $u_i \in \mathcal{S}(\Omega) \cap \mathcal{H}(\Omega \setminus \overline{B}_i)$ and it follows from the Harnack inequality that the sequence u_i is locally bounded in $\Omega \setminus \{x_0\}$; hence $\{u_i : i \geq j\}$ is equicontinuous in $\Omega \setminus \overline{B}_j$ (Corollary 6.12). Now we easily find a subsequence of u_i that converges locally uniformly in $\Omega \setminus \{x_0\}$ to a function $u \in \mathcal{H}(\Omega \setminus \{x_0\})$ (Theorem 6.13). Moreover, $u \in \mathcal{S}(\Omega)$ by the removability theorem 7.35.

It follows from the comparison lemma 3.18 that $0 \leq u_i \leq u_1$ in $\Omega \setminus \overline{B}$. Thus $0 \leq u \leq u_1$ in $\Omega \setminus \overline{B}$ and

$$\lim_{x \to y} u(x) = 0$$

whenever y is a regular point on $\partial\Omega$.

To complete the proof, we show that

$$\lim_{x \to x_0} u(x) = \infty.$$

First note that the limit $\lim_{x \to x_0} u(x)$ exists (Lemma 6.15). Suppose that this limit is finite so that u is bounded near x_0. Then Theorem 7.36 implies that u is $\mathcal{A}$-harmonic in Ω and hence $u \leq u_1$ in Ω by the comparison principle. We show that this leads to a contradiction. Let $D_j \Subset \Omega$ be an increasing sequence of regular domains invading Ω and let

$$v_j = P(u_1, D_j).$$

Then $v_j \in H_0^{1,p}(\Omega; \mu)$ and $u \leq v_j \leq u_1$ in Ω. Moreover, v_j decreases to an $\mathcal{A}$-harmonic function v in Ω. Since $u_1 - v_j \in H_0^{1,p}(\Omega; \mu)$ is nonnegative, the quasiminimizing property of $\mathcal{A}$-superharmonic functions 3.13 implies

$$\int_\Omega |\nabla v_j|^p \, d\mu \leq c \int_\Omega |\nabla u_1|^p \, d\mu < \infty.$$

Then it follows from the weak completeness of $H_0^{1,p}(\Omega; \mu)$ (Theorem 1.32) that $v \in H_0^{1,p}(\Omega; \mu)$ and necessarily $v \equiv 0$ in Ω. Because $0 \leq u \leq v$, u vanishes identically in Ω. This contradicts the inequality $u \geq c > 0$ on ∂B, which is a consequence of the Harnack inequality.

In conclusion, we have

$$\lim_{x \to x_0} u(x) = \infty,$$

and the proof is complete. $\qquad\qquad\qquad\qquad\qquad\qquad\qquad\qquad\qquad\qquad$ $\square$

The assumption that $\mathrm{cap}_{p,\mu}\{x_0\} = 0$ in Theorem 7.39 is necessary because the existence of a singular solution requires that $\mathrm{cap}_{p,\mu}\{x_0\} = 0$. This phenomenon will be investigated in more detail in Chapter 10, where we show that if $\mathrm{cap}_{p,\mu}\{x_0\} > 0$, then each $\mathcal{A}$-superharmonic function is necessarily finite at x_0. However, in that case we could call the $\mathcal{A}$-potential of $\{x_0\}$ a singular solution since x_0 is not a removable singularity.

The next theorem describes the growth of singular solutions.

7.40. Theorem. *Let u be an $\mathcal{A}$-subharmonic function in a punctured ball $B \setminus \{x_0\}$, $B = B(x_0, R)$, such that $\limsup_{x \to y} u(x) \leq 0$ for all $y \in \partial B$. Then either $u \leq 0$ everywhere in $B \setminus \{x_0\}$ or*

$$\liminf_{r \to 0} M_u(r) \left(\mathrm{cap}_{p,\mu}(B(x_0, r), B) \right)^{1/(p-1)} > 0 \,,$$

where $M_u(r) = \max_{\partial B(x_0, r)} u$ for $r < R$.

PROOF: If $u(x') > 0$ for some $x' \in B$, then the maximum principle implies $M_u(r) > 0$ for each $r < |x'| = r'$, and we may thus assume that $r' < R/2$. The comparison principle yields

$$u(x') \leq M_u(r)\omega_r(x') \,,$$

where $\omega_r = \Re(\overline{B}(x_0, r), B)$ is the $\mathcal{A}$-potential of $\overline{B}(x_0, r)$ in B. Let

$$m' = \min_{\partial B(x_0, r')} \omega_r \,.$$

Then the set $K' = \{x \in B : \omega_r(x) \geq m'\}$ contains $B(x_0, r')$ by the maximum principle; in particular, we obtain from Lemma 6.19 that

$$m' \leq c \left(\frac{\mathrm{cap}_{p,\mu}(B(x_0, r), B)}{\mathrm{cap}_{p,\mu}(K', B)} \right)^{1/(p-1)}$$

$$\leq \left(\frac{\mathrm{cap}_{p,\mu}(B(x_0, r), B)}{\mathrm{cap}_{p,\mu}(B(x_0, r'), B)} \right)^{1/(p-1)}$$

$$\leq c \left(\mathrm{cap}_{p,\mu}(B(x_0, r), B) \right)^{1/(p-1)} \,,$$

where c is independent of r. Finally, since $\omega_r(x') \leq cm'$ for $r < r'/2$ by Harnack's inequality, the theorem follows by combining the above estimates. $\square$

For solutions we have an upper estimate as well.

7.41. Theorem. *Suppose that h is $\mathcal{A}$-harmonic in a punctured ball $B \setminus \{x_0\}$, $B = B(x_0, R)$, such that $\lim_{x \to y} h(x) = 0$ for all $y \in \partial B$. If $h(x_1) = 1$ for some point x_1 on the sphere $\partial B(x_0, R/2)$, then*

$$\frac{1}{c} \, \mathrm{cap}_{p,\mu}(B(x_0, r), B)^{1/(1-p)} \le h(x) \le c \, \mathrm{cap}_{p,\mu}(B(x_0, r), B)^{1/(1-p)}$$

for $|x - x_0| = r < R/4$, where $c = c(n, p, \beta/\alpha, c_\mu) > 0$.

PROOF: Let $\omega_r = \Re(\overline{B}(x_0, r), B)$. If M_r and m_r are, respectively, the maximum and the minimum values of h on the sphere $\partial B(x_0, r)$, then it follows from the comparison principle and the potential estimates (6.24) and (6.22) that

$$\frac{1}{M_r} = \frac{h(x_1)}{M_r} \le \omega_r(x_1) \le c \left(\frac{\mathrm{cap}_{p,\mu}\left(B(x_0, r), B\right)}{\mathrm{cap}_{p,\mu}\left(\frac{1}{2}B, B\right)} \right)^{1/(p-1)}$$
$$\le c \, \mathrm{cap}_{p,\mu}\left(B(x_0, r), B\right)^{1/(p-1)}$$

and

$$\frac{1}{m_r} = \frac{h(x_1)}{m_r} \ge \omega_r(x_1) \ge c \, \mathrm{cap}_{p,\mu}\left(B(x_0, r), B\right)^{1/(p-1)}.$$

The theorem follows because $M_r \le c m_r$ by Harnack's inequality. $\square$

More explicit estimates for the growth of singular solutions can be obtained from the capacity estimates in Chapter 2. For example, let $w = 1$ and let h be as in Theorem 7.41. Then

$$h(x) \approx |x|^{(p-n)/(p-1)}, \quad 1 < p < n,$$

and

$$h(x) \approx \log \frac{1}{|x|}, \quad p = n,$$

for $|x| < R/2$.

7.42. Integrability of $\mathcal{A}$-superharmonic functions

An $\mathcal{A}$-superharmonic function is by definition lower semicontinuous, and hence locally bounded below. However, as shown in Theorem 7.39, an $\mathcal{A}$-superharmonic function in Ω need not be locally bounded above nor is it necessarily in $H^{1,p}_{loc}(\Omega; \mu)$. This section is devoted to integrability questions for $\mathcal{A}$-superharmonic functions.

In what follows $\varkappa > 1$ is the number appearing in the Sobolev inequality III. We first prove a lemma about general Sobolev functions.

7.43. Lemma. *Let Ω be bounded and let u be a nonnegative, a.e. finite function on Ω. Suppose that for all $k = 1, 2, \dots$*

$$\min(u, k) \in H_0^{1,p}(\Omega; \mu)$$

and

(7.44) $$\int_\Omega |\nabla \min(u, k)|^p \, d\mu \le Mk$$

for some constant M, independent of k.

(i) *If $0 < q < \varkappa p / (\varkappa(p-1) + 1)$, then*

$$\int_\Omega |\nabla \min(u, k)|^{q(p-1)} \, d\mu \le c \,,$$

where $c = c(p, c_\mu, q, M, \mu(\Omega), \operatorname{diam} \Omega)$.

(ii) *If $0 < s < \varkappa(p-1)$, then*

$$\int_\Omega u^s \, d\mu < \infty \,.$$

PROOF: We first prove (i). It follows from the Sobolev embedding theorem (**III**) and (7.44) that

$$k^{\varkappa p} \mu(\{k \le u < 2k\}) \le \int_{\{k \le u < 2k\}} \min(u, 2k)^{\varkappa p} \, d\mu$$

$$\le \int_\Omega \min(u, 2k)^{\varkappa p} \, d\mu \le c \Big(\int_\Omega |\nabla \min(u, 2k)|^p \, d\mu \Big)^\varkappa$$

$$\le c \, (Mk)^\varkappa \,.$$

Using the Hölder inequality we obtain

$$\int_{\{k \le u < 2k\}} |\nabla \min(u, 2k)|^{q(p-1)} \, d\mu$$

$$\le \mu\big(\{k \le u < 2k\}\big)^{1 - q(p-1)/p} \Big(\int_{\{k \le u < 2k\}} |\nabla \min(u, 2k)|^p \, d\mu \Big)^{q(p-1)/p}$$

$$\le c \, k^{p_1} \,,$$

where $c = c(p, q, c_\mu, M, \mu(\Omega), \operatorname{diam} \Omega) > 0$ and

$$p_1 = \varkappa(1-p)\Big(1 - \frac{q(p-1)}{p}\Big) + \frac{q(p-1)}{p} = (p-1)\Big(q \frac{\varkappa(p-1) + 1}{p} - \varkappa\Big) < 0 \,.$$

Hence for $k \leq 2^{\ell}$

$$\int_{\Omega} |\nabla \min(u, k)|^{q(p-1)} \, d\mu$$

$$\leq \int_{\{u<1\}} |\nabla \min(u, 1)|^{q(p-1)} \, d\mu + \sum_{j=1}^{\ell} \int_{\{2^{j-1} \leq u < 2^j\}} |\nabla \min(u, 2^j)|^{q(p-1)} \, d\mu$$

$$\leq M + c \sum_{j=1}^{\infty} 2^{p_1 j} < c,$$

where $c = c\big(p, q, c_{\mu}, M, \mu(\Omega), \operatorname{diam}(\Omega)\big)$, and (i) follows.

To prove (ii) we proceed similarly. As in (i) it holds that

$$\mu(\{k \leq u \leq 2k\}) \leq c \, k^{\varkappa(1-p)},$$

and using the Hölder inequality and the Sobolev inequality (**III**) we obtain

$$\int_{\{k \leq u \leq 2k\}} u^s \, d\mu \leq \mu(\{k \leq u \leq 2k\})^{1-s/\varkappa p} \Big(\int_{\Omega} \min(u, 2k)^{\varkappa p} \, d\mu \Big)^{s/\varkappa p}$$

$$\leq c \, k^{s(p-1)/p - \varkappa(p-1)} \Big(\int_{\Omega} |\nabla \min(u, 2k)|^p \, d\mu \Big)^{s/p}$$

$$\leq c \, k^{s(p-1)/p - \varkappa(p-1) + s/p}.$$

Since $p_2 = s(p-1)/p - \varkappa(p-1) + s/p = s - \varkappa(p-1)$ is negative, we have

$$\int_{\Omega} u^s \, d\mu \leq \mu(\Omega) + \sum_{j=1}^{\infty} \int_{\{2^{j-1} \leq u < 2^j\}} u^s \, d\mu$$

$$\leq \mu(\Omega) + c \sum_{j=1}^{\infty} 2^{p_2 j} < \infty,$$

and the proof is complete.   □

DEFINITION. Suppose that u is a function in Ω such that $\min(u, k) \in H^{1,p}_{loc}(\Omega; \mu)$ for all nonnegative integers k. Then we define the *weak gradient* of u to be the function

$$Du = \lim_{k \to \infty} \nabla \min(u, k).$$

Since for $k \geq j$ we have $\nabla \min(u, k) = \nabla \min(u, j)$ a.e. in the set $\{u \leq j\}$, the weak gradient Du is an a.e. defined function. Moreover, Du is defined for all $\mathcal{A}$-superharmonic functions (Corollary 7.20). If $u \in H_{loc}^{1,p}(\Omega; \mu)$, then $\nabla u = Du$. .In particular, if $w^{1/(1-p)}$ is locally integrable and Du is locally integrable (with respect to the Lebesgue measure), then Du is the distributional gradient of u (see Section 1.9).

The main result of this section is that for an $\mathcal{A}$-superharmonic u the weak gradient Du is in $L_{loc}^{q(p-1)}(\Omega; \mu)$ for some $q > 1$. We start with a global result, where u vanishes on the boundary.

7.45. Theorem. *Let u be a nonnegative $\mathcal{A}$-superharmonic function in a bounded open set Ω such that*

$$\min(u, k) \in H_0^{1,p}(\Omega; \mu), \quad k = 1, 2, \ldots .$$

Then $Du \in L^{q(p-1)}(\Omega; \mu)$ for each q with

$$0 < q < \frac{\varkappa p}{\varkappa(p-1) + 1}.$$

Moreover, if $0 < s < \varkappa(p-1)$, then u is in $L^s(\Omega; \mu)$.

PROOF: We show that u verifies the hypotheses of Lemma 7.43, from which the assertion follows. Let

$$a_k = \int_{\{k-1 \leq u \leq k\}} \mathcal{A}(x, Du) \cdot Du \, dx.$$

Then

$$\int_{\{k-1 \leq u \leq k\}} |\nabla \min(u, k)|^p \, d\mu \leq \frac{a_k}{\alpha},$$

and if we show that the sequence a_k is decreasing, we have that

$$\int_{\Omega} |\nabla \min(u, k)|^p \, d\mu \leq k \frac{a_1}{\alpha},$$

which is the desired estimate (7.44).

To this end, let

$$v_k = (1 - |u - k|)^+.$$

Then $v_k \in H_0^{1,p}(\Omega; \mu)$ is a nonnegative function, and since $\min(u, k+1)$ is a supersolution in $H^{1,p}(\Omega; \mu)$, we have

$$0 \leq \int_{\Omega} \mathcal{A}\big(x, \nabla \min(u, k+1)\big) \cdot \nabla v_k \, dx$$

$$= \int_{\{k-1 \leq u < k\}} \mathcal{A}\big(x, Du\big) \cdot Du \, dx - \int_{\{k \leq u < k+1\}} \mathcal{A}\big(x, Du\big) \cdot Du \, dx$$

$$= a_k - a_{k+1}.$$

In conclusion, $a_{k+1} \leq a_k$ and the theorem is proved. $\qquad \square$

For general $\mathcal{A}$-superharmonic functions we have a local version.

7.46. Theorem. *If u is $\mathcal{A}$-superharmonic in Ω, then $u \in L_{loc}^s(\Omega; \mu)$ and $Du \in L_{loc}^{q(p-1)}(\Omega; \mu)$ whenever $0 < s < \varkappa(p-1)$ and*

$$0 < q < \frac{\varkappa p}{\varkappa(p-1)+1}.$$

PROOF: Let B be a ball with $2B \Subset \Omega$. It suffices to show that both $|u|^s$ and $|Du|^{q(p-1)}$ are μ-integrable in B. Since $m = \inf_{2B} u > -\infty$, we may replace u with $u-m+1$ and thus assume that $u \geq 1$ in $2B$. Now the Poisson modification $P(u, 2B \setminus \overline{B})$ is $\mathcal{A}$-harmonic in the annulus $2B \setminus \overline{B}$. Hence there is an $\mathcal{A}$-harmonic function h in $2B \setminus \frac{3}{2}B$ such that h is continuous up to the boundary, $h \in H^{1,p}(2B \setminus \frac{3}{2}B; \mu)$, $h = 0$ on $\partial 2B$, and $h = P(u, 2B \setminus \overline{B})$ on $\partial \frac{3}{2}B$. It follows from the comparison principle and the pasting lemma 7.9 that the function

$$v = \begin{cases} P(u, 2B \setminus \overline{B}) & \text{on } \overline{\frac{3}{2}B} \\ h & \text{in } 2B \setminus \overline{\frac{3}{2}B} \end{cases}$$

is $\mathcal{A}$-superharmonic in $2B$. Moreover, since $v = u$ in B and since the truncations $\min(v, k)$ belong to $H_0^{1,p}(2B; \mu)$, the assertion follows from Theorem 7.45.   □

AN ALTERNATIVE PROOF FOR THEOREM 7.46: Let B be a ball such that $4B \subset \Omega$. We may assume that $u > 0$ in $4B$. For $k = 1, 2, \ldots$ write $u_k = \min(u, k)$. Then u_k is a supersolution in Ω and hence the weak Harnack inequality (Corollary 3.59) implies for $0 < s < \varkappa(p-1)$ that

$$\int_B u_k^s \, d\mu \leq c \, (\text{ess} \inf_B u)^s < \infty,$$

where the constant c is independent of k. Letting $k \to \infty$ we obtain $u^s \in L^1(B; \mu)$, as desired.

The integrability of the weak gradient Du follows from this result combined with the estimate in Lemma 3.57. Indeed, if

$$1 \leq q < \frac{\varkappa p}{\varkappa(p-1)+1}$$

and if $\varepsilon > 0$, we have by Lemma 3.57 that

$$\int_B |\nabla u_k|^{q(p-1)} d\mu$$

$$= \int_B |\nabla u_k|^{q(p-1)} u_k^{-(1+\varepsilon)(p-1)q/p} u_k^{(1+\varepsilon)(p-1)q/p} d\mu$$

$$\leq \Big(\int_B |\nabla u_k|^p u_k^{-1-\varepsilon} d\mu \Big)^{q(p-1)/p} \Big(\int_B u_k^{(1+\varepsilon)q(p-1)/(p-q(p-1))} d\mu \Big)^{(p-q(p-1))/p}$$

$$\leq c \Big(\int_{2B} u^{p-1-\varepsilon} d\mu \Big)^{q(p-1)/p} \Big(\int_B u^{(1+\varepsilon)q(p-1)/(p-q(p-1))} d\mu \Big)^{(p-q(p-1))/p}.$$

Now the last two integrals are finite if we choose $0 < \varepsilon < p - 1$ such that

$$\frac{(1+\varepsilon)q(p-1)}{p-q(p-1)} < \varkappa(p-1).$$

An $L^q(B; \mu)$-estimate for $|Du|^{p-1}$ is obtained by letting $k \to \infty$. $\qquad\square$

7.47. EXAMPLE. We display an example which shows that the integrability results are sharp. Let $\gamma \geq 0$ and $w(x) = |x|^\gamma$. Then w is p-admissible for all $p > 1$ and

$$p_0 = \inf\{q \geq 1 \colon w \in A_q\} = 1 + \frac{\gamma}{n};$$

see Section 1.6. Let $p \leq n + \gamma$ and $\mathcal{A}(x, \xi) = w(x)|\xi|^{p-2}\xi$. If

$$u(x) = \int_{|x|}^1 t^\delta \, dt,$$

where $\delta = (1 - n - \gamma)/(p - 1)$, then it is an easy task to check that u is $\mathcal{A}$-harmonic in the punctured space $\mathbf{R}^n \setminus \{0\}$, and $\mathcal{A}$-superharmonic in $\mathbf{R}^n$ if we set $u(0) = \infty$. Moreover, if $B = B(0, 1)$ and $\delta < -1$, then

$$\int_B u^s \, d\mu = c \int_0^1 t^{n+s(\delta+1)+\gamma} \frac{dt}{t} < \infty$$

if and only if

$$s < \frac{n+\gamma}{\gamma + n - p}(p - 1) = s_0.$$

If $\delta = -1$, then let $s_0 = \infty$, and it follows that (for all $p \leq n + \gamma$)

$$\int_B u^s \, d\mu < \infty$$

if and only if $s < s_0$. Similarly, we have that

$$\int_B |Du|^{(p-1)q}\, d\mu = c \int_0^1 t^{n+\gamma+q(1-\gamma-n)}\, \frac{dt}{t} < \infty$$

if and only if

$$q < \frac{n+\gamma}{n+\gamma-1} = q_0 .$$

Now let us figure out what we obtain from Theorem 7.46. Theorem 15.23 implies that in order to apply Theorem 7.46 we may choose any

$$\varkappa < \frac{n}{n - p/p_0} = \frac{n+\gamma}{n+\gamma-p} = \kappa .$$

Thus

$$\int_B u^s\, d\mu < \infty$$

if

$$s < \frac{n+\gamma}{n+\gamma-p}(p-1) = s_0 ,$$

and

$$\int_B |Du|^{(p-1)q}\, d\mu < \infty$$

if

$$q < \begin{cases} \dfrac{\kappa p}{\kappa(p-1)+1} = q_0 & \text{if } p < n+\gamma \\[2ex] \dfrac{p}{p-1} = q_0 & \text{if } p = n+\gamma. \end{cases}$$

Consequently, these examples show that Theorem 7.46 is sharp.

In the unweighted case Theorem 7.46 claims that for an $\mathcal{A}$-superharmonic function u, both $|u|^s$ and $|Du|^q$ are locally integrable whenever

$$0 < s < \begin{cases} \frac{n(p-1)}{n-p} & \text{if } p < n \\[1ex] \infty & \text{if } p \geq n \end{cases}$$

and

$$0 < q < \begin{cases} \frac{n(p-1)}{n-1} & \text{if } p < n \\[1ex] p & \text{if } p \geq n. \end{cases}$$

Note that Du is locally integrable if $p > 2 - 1/n$ and hence Du is the distributional gradient of u. These results are sharp for $p \leq n$, because the function u defined in Example 7.47 is, essentially,

$$v(x) = \begin{cases} |x|^{(p-n)/(p-1)} & \text{if } p < n \\ -\log|x| & \text{if } p = n. \end{cases}$$

Then v is p-superharmonic in $\mathbf{R}^n$, i.e. $\mathcal{A}$-superharmonic when (3.8) is the p-Laplace equation

$$- \operatorname{div}(|\nabla u|^{p-2} \nabla u) = 0 .$$

A direct computation with v shows that if $p \le 2 - 1/n$, the distributional derivatives of v do not exist as functions in any neighborhood of the origin.

However, for $p > n$, the above integrability results are not sharp in the unweighted case because then it follows from the Sobolev embedding theorem that $\mathcal{A}$-superharmonic functions are continuous and hence locally bounded. Consequently, they belong to $H^{1,p}_{loc}(\Omega; dx)$ by Theorem 7.25.

We close this chapter by establishing a useful result asserting that $\log u$ is in $H^{1,p}_{loc}(\Omega; \mu)$ whenever u is a positive $\mathcal{A}$-superharmonic, although u need not belong to $H^{1,p}_{loc}(\Omega; \mu)$.

7.48. Theorem. *Suppose that u is a positive $\mathcal{A}$-superharmonic function in Ω. Then $\log u$ is a supersolution of (3.8). In particular, $\log u \in H^{1,p}_{loc}(\Omega; \mu)$ and the estimate*

$$\int_E |\nabla \log u|^p \, d\mu \le c \operatorname{cap}_{p,\mu}(E, \Omega)$$

holds whenever $E \subset \Omega$ is measurable; here $c = (p\beta/\alpha(p-1))^p$.

PROOF: Because $t \mapsto \log t$ is concave and increasing for $t > 0$, the function $\log u$ is $\mathcal{A}$-superharmonic (Theorem 7.5). So it suffices to show that $\log u \in H^{1,p}_{loc}(\Omega; \mu)$ (Theorem 7.25 and Theorem 3.53). To this end, let B be a ball with $2B \subset \Omega$ and let $u_k = \min(u, k)$, $k = 1, 2, \dots$. Then $\log u$ is bounded below in B, and for $0 < s < \varkappa(p-1)$, we have that

$$\int_B |\log u|^p \, d\mu \le c \int_B u^s \, d\mu < \infty$$

(Theorem 7.46). By Lemma 3.47, $\log u_k$ belongs to $H^{1,p}_{loc}(\Omega; \mu)$ and

$$\int_B |\nabla \log u_k|^p \, d\mu \le c ,$$

where the constant $c < \infty$ is independent of k. Hence the sequence $\log u_k$ is bounded in $H^{1,p}(B; \mu)$, and we infer from the weak compactness of $H^{1,p}(B; \mu)$ that $\log u \in H^{1,p}(B; \mu)$ (Theorem 1.32). The theorem follows. $\qquad\square$

NOTES TO CHAPTER 7. Quoting Doob (1984, p. 794): "F. Riesz inaugurated the systematic study of superharmonic and subharmonic functions". See Riesz (1926, 1930) and Radó (1949). The definition of superharmonic functions in terms of harmonic functions is the standard approach of axiomatic potential theory. In connection with nonlinear partial differential equations, this definition was first given by Granlund et al. (1983). Papers by Granlund et al. (1982, 1985, 1986), Lindqvist (1986), and Lindqvist and Martio (1985, 1988) display the early results and applications of $\mathcal{A}$-superharmonic functions. The proof of Theorem 7.5 is from Gardiner and Klimek (1986). The general (unweighted) form of Theorem 7.25 was first proved in Heinonen and Kilpeläinen (1988a), where also removability results for $\mathcal{A}$-superharmonic functions were established. The removability theorem 7.36 and the behavior of singular solutions can be found in Serrin (1964, 1965). The extension of $\mathcal{A}$-superharmonic functions was considered in Kilpeläinen (1989). The integrability of $\mathcal{A}$-superharmonic functions was first studied in Lindqvist (1986) and our second proof for Theorem 7.46 is from there; the other approach is that of Kilpeläinen and Malý (1990).

8
Balayage

The theory of balayage (sweeping) fits naturally into the nonlinear framework; it can be thought of as a basic form of the obstacle problem. In this chapter we first introduce the balayage of a function, and the rest is devoted to the proof of the fundamental convergence theorem.

We first recall that the *lower semicontinuous regularization* $\hat{u}$ of any function $u \colon E \to [-\infty, \infty]$ is defined by

$$\hat{u}(x) = \lim_{r \to 0} \inf_{E \cap B(x,r)} u .$$

Then $\hat{u} \leq u$ on E. If u is locally bounded below, then $\hat{u}$ is lower semicontinuous; indeed, $\hat{u}$ is the greatest lower semicontinuous minorant of u.

DEFINITION. Let $\psi : \Omega \to (-\infty, \infty]$ be a function that is locally bounded below, and let

$$\Phi^\psi = \Phi^\psi(\Omega) = \Phi^\psi(\Omega; \mathcal{A})$$
$$= \{u : u \text{ is } \mathcal{A}\text{-superharmonic in } \Omega \text{ and } u \geq \psi \text{ in } \Omega \ \}.$$

Then the function

$$R^\psi = R^\psi(\Omega) = R^\psi(\Omega; \mathcal{A}) = \inf \Phi^\psi$$

is called the *réduite* and its lower semicontinuous regularization

$$\hat{R}^\psi = \hat{R}^\psi(\Omega) = \hat{R}^\psi(\Omega; \mathcal{A})$$

the *balayage* of ψ in Ω.

If Φ^ψ is empty, then $\hat{R}^\psi \equiv \infty$. However, to exclude this trivial case, we assume hereafter that Φ^ψ *is nonempty*.

We proved in Theorem 7.4 that the lower semicontinuous regularization of the infimum of a family of $\mathcal{A}$-superharmonic functions such as Φ^ψ is $\mathcal{A}$-superharmonic. Hence we have:

8.1. Theorem. *The balayage $\hat{R}^\psi$ is $\mathcal{A}$-superharmonic in Ω.*

DEFINITION. If u is a nonnegative function on a set $E \subset \Omega$, we write

$$\Phi_E^u = \Phi^\psi, \quad R_E^u = R^\psi, \quad \text{and} \quad \hat{R}_E^u = \hat{R}^\psi ,$$

157

where

$$\psi = \begin{cases} u & \text{on } E \\ 0 & \text{on } \Omega \setminus E. \end{cases}$$

The function $\hat{R}_E^u$ is called the *balayage of u relative to E*. Furthermore, if $u \equiv c$, a constant, we write $\hat{R}_E^c = \hat{R}_E^u$, etc. Especially, we call the function $\hat{R}_E^1$ the *$\mathcal{A}$-potential* of E in Ω.

$\mathcal{A}$-potentials will play an important role in the further development of the theory and their properties will be examined thoroughly in this and subsequent chapters. This definition extends the one used in Chapter 6 (see Lemma 8.5 and Theorem 9.33).

The fundamental convergence theorem in nonlinear potential theory reads as follows.

8.2. Fundamental convergence theorem. *Suppose that $\mathcal{F}$ is a nonempty family of $\mathcal{A}$-superharmonic functions in Ω. Let $\hat{s}$ be the lower semicontinuous regularization of the function $s = \inf \mathcal{F}$. If $\mathcal{F}$ is locally uniformly bounded below, then $\hat{s}$ is $\mathcal{A}$-superharmonic in Ω and $\hat{s} = s$ except on a set of (p, μ)-capacity zero.*

We have already seen in Lemma 7.4 that the function $\hat{s}$ is $\mathcal{A}$-superharmonic. To show that $\hat{s} = s$ quasieverywhere, we require several lemmas; some of them are used in later chapters as well. In particular, the first topological lemma is often employed.

Recall that a family $\mathcal{U}$ of functions defined on the same set is *downward directed* if for each $u, v \in \mathcal{U}$ there is $s \in \mathcal{U}$ with $s \leq \min(u, v)$.

8.3. Choquet's topological lemma. *Suppose that $E \subset \mathbf{R}^n$ and that $\mathcal{U} = \{u_\gamma : \gamma \in I\}$ is a family of functions $u_\gamma : E \to [-\infty, \infty]$. Let $u = \inf \mathcal{U}$. If $\mathcal{U}$ is downward directed, then there is a decreasing sequence of functions $v_j \in \mathcal{U}$ with limit v such that the lower semicontinuous regularizations $\hat{u}$ and $\hat{v}$ coincide.*

PROOF: We first show that there is a countable subset J of I such that the lower semicontinuous regularization of

$$\inf\{u_\gamma : \gamma \in J\}$$

coincides with $\hat{u}$. For this, let $B_1, B_2, \ldots$ be an enumeration of all balls in $\mathbf{R}^n$ with rational center and radius such that each ball appears infinitely many times in the sequence $B_1, B_2, \ldots$. For $k = 1, 2, \ldots$, let $E_k = B_k \cap E$; we assume that $E_k \neq \emptyset$. Choose $x_k \in E_k$ such that

$$u(x_k) \leq \inf_{E_k} u + \frac{1}{k}.$$

provided the infimum is not $-\infty$; otherwise choose $x_k \in E_k$ with $u(x_k) < -k$. Next let $\gamma_k \in I$ be an index with

$$u_{\gamma_k}(x_k) \le u(x_k) + \frac{1}{k}$$

if $u(x_k) > -\infty$; if $u(x_k) = -\infty$, make an obvious modification. Now the set $J = \{\gamma_k \in I : k = 1, 2, \ldots\}$ is countable and for $s = \inf\{u_\gamma : \gamma \in J\}$ we have

$$\inf_{E_k} s \le u_{\gamma_k}(x_k) \le \inf_{E_k} u + \frac{2}{k}$$

for all $k = 1, 2, \ldots$ if $u(x_k) > -\infty$; if $u(x_k) = -\infty$, then $\inf_{E_k} s_k < -k$. Because each set $E \cap B(x, r)$ with $x \in E$ contains infinitely many of the sets E_k, it follows that

$$\hat{s} \le \hat{u}$$

in E. The reverse inequality is trivial and hence $\hat{s} = \hat{u}$ as desired.

To complete the proof, let $v_1 = v_{\gamma_1}$ and for $j = 2, 3, \ldots$ choose a function $v_j \in \mathcal{U}$ such that

$$v_j \le \min(v_{j-1}, u_{\gamma_j}).$$

Then v_j is the desired decreasing sequence because

$$u \le \lim_{j \to \infty} v_j \le s.$$

$\square$

8.4. Lemma. *The balayage $\hat{R}_E^u$ is $\mathcal{A}$-harmonic in $\Omega \setminus \overline{E}$ and coincides with R_E^u there. If, in addition, $u \in \mathcal{S}(\Omega)$, then $\hat{R}_E^u = u$ in the interior of E.*

PROOF: Observe first that if v_1 and v_2 are in $\boldsymbol{\Phi}_E^u$, then so is $\min(v_1, v_2)$. Hence the family $\boldsymbol{\Phi}_E^u$ is downward directed and we may invoke Choquet's topological lemma 8.3: there is a decreasing sequence of functions $v_i \in \boldsymbol{\Phi}_E^u$ with the limit v such that

$$\hat{v}(x) = \hat{R}_E^u(x)$$

for all $x \in \Omega$.

Next, pick a ball $B \subset\subset \Omega \setminus \overline{E}$ and let s_i be the Poisson modification of v_i in B, where v_i is the sequence obtained above. Then s_i is $\mathcal{A}$-harmonic in B, $s_i \in \boldsymbol{\Phi}_E^u$, and $s_{i+1} \le s_i \le v_i$. Hence

$$R_E^u \le s = \lim s_i \le \lim v_i = v$$

and we conclude that

$$\hat{R}_E^u = \hat{s}$$

in B. This establishes the first assertion because $\hat{s} = s$ is $\mathcal{A}$-harmonic in B by Harnack's convergence theorem. The second part of the lemma is immediate since u is lower semicontinuous and $u \in \boldsymbol{\Phi}_E^u$. $\square$

Next we prove that for compact E and bounded Ω the $\mathcal{A}$-potential $\hat{R}^1_E$ coincides with the $\mathcal{A}$-potential of Chapter 6. The case where Ω is not necessarily bounded is treated in Theorem 9.33.

8.5. Lemma. *Suppose that K is a compact subset of a bounded open set Ω and that $u = \hat{R}^1_K(\Omega)$ is the $\mathcal{A}$-potential of K in Ω. Let $\varphi \in C_0^\infty(\Omega)$ be such that $\varphi = 1$ on K. Then $u|_{\Omega \setminus K}$ is the unique $\mathcal{A}$-harmonic function in $\Omega \setminus K$ with $u - \varphi \in H_0^{1,p}(\Omega \setminus K; \mu)$.*

PROOF: First exhaust $\complement K$ by polyhedra $D_1 \Subset D_2 \Subset \cdots \Subset \complement K$ such that $\partial\Omega \subset D_1$. Then let Ω_i be an exhaustion of Ω such that Ω_i are polyhedra, $\Omega_1 \Subset \Omega_2 \Subset \cdots \Subset \Omega$, and $\Omega \setminus D_1 \subset \Omega_1$. We may assume that $\varphi = 1$ on $\Omega \setminus D_1$ and that $\varphi \in C_0^\infty(\Omega_1)$. Let then $h_{i,j}$ be the $\mathcal{A}$-harmonic function in $\Omega_i \cap D_j$ with $h_{i,j} - \varphi \in H_0^{1,p}(\Omega_i \cap D_j; \mu)$. Now $h_{i,j}$ can be extended continuously to $\mathbf{R}^n$ by setting $h_{i,j} = 1$ on $\Omega_i \setminus D_j$ and $h_{i,j} = 0$ in $\complement\Omega_i$. Hence we have by the pasting lemma 7.9 that $h_{i,j}$ is $\mathcal{A}$-superharmonic in Ω_i. By the comparison principle $h_{i,j} \leq h_{i+1,j} \leq 1$ so that Harnack's convergence theorem and Lemma 7.3 imply that

$$h_j = \lim_{i \to \infty} h_{i,j}$$

is $\mathcal{A}$-harmonic in $\Omega \cap D_j$ and $\mathcal{A}$-superharmonic in Ω. Moreover, $h_j = 1$ on $\Omega \setminus D_j$ and hence

$$h_j \geq \hat{R}^1_K(\Omega) = u.$$

Since each of the functions $h_{i,j}$ quasiminimizes the (p,μ)-energy, we obtain

$$\int_{\Omega_i \cap D_j} |\nabla h_{i,j}|^p d\mu \leq c \int_{\Omega \setminus K} |\nabla\varphi|^p d\mu.$$

Therefore $h_{i,j} - \varphi$ is a bounded sequence in $H_0^{1,p}(\Omega \cap D_j; \mu)$ and the limit function $h_j - \varphi$ is in $H_0^{1,p}(\Omega \cap D_j; \mu)$ (Theorem 1.32). Since h_j is $\mathcal{A}$-harmonic, we similarly conclude that the sequence h_j decreases to an $\mathcal{A}$-harmonic function h in $\Omega \setminus K$ and that $h - \varphi \in H_0^{1,p}(\Omega \setminus K; \mu)$.

To complete the proof, we show that $h = u$ in $\Omega \setminus K$. Since $h_j \geq u$ on $\Omega \cap D_j$, we have $h \geq u$. On the other hand, if we pick a function $v \in \Phi^1_K$ and fix $\varepsilon > 0$, then

$$(1 + \varepsilon)v \geq 1$$

on $\partial D_j \cap \Omega$ for some j. Therefore

$$(1 + \varepsilon)v \geq h_{i,j}$$

in $\Omega_i \cap D_j$ and hence

$$(1 + \varepsilon)v \geq h_j$$

in $\Omega \cap D_j$. We conclude that

$$(1 + \varepsilon)v \geq h$$

in $\Omega \setminus K$. Letting $\varepsilon \to 0$ and taking the infimum over all $v \in \Phi_E^1$ we obtain

$$R_K^1 \geq h$$

in $\Omega \setminus K$. Since $u = R_K^1$ in $\Omega \setminus K$ (Lemma 8.4), we have $u \geq h$. Consequently, $u = h$ and the lemma is proved. □

Lemma 8.5 can be used to establish the existence of the *capacitary function* for a condenser (K, Ω), where K is a compact subset of Ω.

8.6. Theorem. *Suppose that K is a compact subset of a bounded open set Ω and that $u = \hat{R}_K^1(\Omega)$ is the $\mathcal{A}$-potential of K in Ω. Then*

$$\text{cap}_{p,\mu}(K, \Omega) \leq \int_\Omega |\nabla u|^p \, d\mu \leq (\frac{\beta}{\alpha})^p \, \text{cap}_{p,\mu}(K, \Omega) \, .$$

In particular, if $\mathcal{A}(x, \xi) = w(x)|\xi|^{p-2}\xi$, then

$$\text{cap}_{p,\mu}(K, \Omega) = \int_\Omega |\nabla u|^p \, d\mu \, .$$

PROOF: Let φ be in $C_0^\infty(\Omega)$ such that $\varphi = 1$ on K. Since $u - \varphi \in H_0^{1,p}(\Omega; \mu)$ by the previous lemma, u can be approximated in $H^{1,p}(\Omega; \mu)$ by functions that are admissible for the condenser (K, Ω), so that the first inequality readily follows.

The second inequality holds because u is $\mathcal{A}$-harmonic in $\Omega \setminus K$ and therefore quasiminimizes the weighted p-energy with the constant $(\beta/\alpha)^p$; see 3.13. □

We extend Theorem 8.6 for more general condensers in Theorems 9.35 and 9.38.

8.7. Lemma. *Let K be a compact subset of a bounded open set Ω. Then $\hat{R}_K^1 = 1$ (p, μ)-quasieverywhere on K.*

PROOF: If $K \subset \Omega' \subset \Omega$, then

$$\hat{R}_K^1(\Omega') \leq \hat{R}_K^1(\Omega) \leq 1$$

in Ω' and hence it is no loss of generality to assume that Ω is a polyhedron. Fix $\gamma \in (0, 1)$. If

$$K_\gamma = \{x \in K : \hat{R}_K^1(x) \leq \gamma\},$$

then K_γ is compact and

$$v = \hat{R}^1_{K_\gamma} \leq \hat{R}^1_K \leq \gamma$$

on K_γ. Since Ω is a polyhedron, $\complement\Omega$ is (p,μ)-thick at each point of $\partial\Omega$; hence

$$\lim_{y\to x} v(y)/\gamma = 0$$

for all $x \in \partial\Omega$ (see Theorem 6.27). Note that v/γ is $\mathcal{A}$-harmonic in $\Omega \setminus K_\gamma$ and $\varphi - v \in H^{1,p}_0(\Omega \setminus K_\gamma; \mu)$ for any $\varphi \in C^\infty_0(\Omega)$, $\varphi = 1$ on K_γ (Lemma 8.5). Moreover,

$$v/\gamma \leq 1,$$

and it follows from the comparison principle that

$$v/\gamma \leq s$$

in Ω whenever $s \in \Phi^1_{K_\gamma}$. Consequently,

$$v/\gamma \leq v$$

which is possible only if $v \equiv 0$. It follows from Theorem 8.6 that

$$\mathrm{cap}_{p,\mu}(K_\gamma, \Omega) = 0.$$

Because the capacity is countably subadditive, we find

$$\mathrm{cap}_{p,\mu}\{x \in K \colon \hat{R}^1_K(x) < 1\} = 0,$$

and the lemma follows.   □

8.8. Lemma. *Suppose that u is a nonnegative $\mathcal{A}$-superharmonic function in a ball B. If K is a compact subset of B, then*

$$\lim_{x\to y} \hat{R}^u_K(B)(x) = 0$$

for all $y \in \partial B$.

PROOF: Let $v \in \Phi^u_K$, let $B_0 \Subset B$ be an open ball containing K, and let D be the annulus

$$D = B \setminus \overline{B}_0.$$

Replacing v by its Poisson modification in a narrow annulus containing ∂B_0, we may assume that

$$v < M < \infty$$

on ∂B_0. Let h be the unique $\mathcal{A}$-harmonic function in D with boundary values 0 on ∂B and M on ∂B_0. Then the $\mathcal{A}$-superharmonic function

$$s = \begin{cases} v & \text{in } \overline{B}_0 \\ \min(v, h) & \text{in } D \end{cases}$$

belongs to Φ^u_K by the pasting lemma 7.9. Hence $0 \leq \hat{R}^u_K \leq s$ in B and the lemma follows.   □

Recall that a boundary point x_0 of a bounded open set Ω is (Sobolev) regular if

$$\lim_{x \to x_0} h(x) = \vartheta(x_0)$$

whenever $\vartheta \in C(\overline{\Omega}) \cap H^{1,p}(\Omega; \mu)$ and h is the $\mathcal{A}$-harmonic function in Ω with $h - \vartheta \in H_0^{1,p}(\Omega; \mu)$; moreover, Ω is regular if all its boundary points are regular (Chapter 6).

8.9. Lemma. *Suppose that x_0 is a boundary point of a bounded open set Ω. If*

$$\hat{R}_{\overline{B} \setminus \Omega}^1 (2B)(x_0) = 1$$

whenever B is a ball with rational center and radius containing x_0, then x_0 is a regular boundary point.

PROOF: Let ϑ be in $C(\overline{\Omega}) \cap H^{1,p}(\Omega; \mu)$ and let h be the $\mathcal{A}$-harmonic function in Ω with $h - \vartheta \in H_0^{1,p}(\Omega; \mu)$. There is no loss of generality in assuming that $\vartheta(x_0) = 0$ and that $\max |\vartheta| \leq 1$ in $\overline{\Omega}$.

Fix $\varepsilon > 0$. Choose a ball B with rational center and radius such that $x_0 \in B$, $\partial(2B) \cap \Omega \neq \emptyset$, and $|\vartheta| < \varepsilon$ in $2\overline{B} \cap \overline{\Omega}$. Then define

$$u = \begin{cases} 1 - \hat{R}_{\overline{B} \setminus \Omega}^1 (2B) + \varepsilon & \text{in } 2B \cap \Omega \\ 1 + \varepsilon & \text{elsewhere in } \Omega. \end{cases}$$

It is clear from the pasting lemma 7.9 and Lemma 8.8 that u is $\mathcal{A}$-superharmonic in Ω. Since $u \geq \vartheta$ in Ω, an appeal to Lemma 1.25 shows that the function

$$\min(h - \vartheta, u - \vartheta) = \min(h, u) - \vartheta$$

belongs to $H_0^{1,p}(\Omega; \mu)$. Because u is a supersolution, we conclude from the comparison lemma 3.22 that $u \geq h$ a.e. in Ω, and because u and h are continuous the inequality actually holds everywhere in Ω. In conclusion,

$$\limsup_{x \to x_0} h(x) \leq \lim_{x \to x_0} u(x) = \varepsilon.$$

Similar reasoning reveals that $-u \leq h$ in Ω, whence

$$\liminf_{x \to x_0} h(x) \geq - \lim_{x \to x_0} u(x) = -\varepsilon.$$

Since ε was arbitrary, we obtain

$$\lim_{x \to x_0} h(x) = 0 = \vartheta(x_0),$$

as desired. $\square$

The fact that quasievery boundary point is regular is classically called the *Kellogg property*. We next establish it for bounded sets. The general case is treated later in Theorem 9.11, after a discussion of regular boundary points in arbitrary open sets.

We say that a boundary point of Ω is *irregular* if it is not regular.

8.10. Theorem. *The set of irregular boundary points of a bounded open set has (p, μ)-capacity zero.*

PROOF: Let Ω be bounded and let $E \subset \partial\Omega$ be the set of all irregular boundary points. In view of Lemma 8.9 we can choose a countable collection of balls B_i such that each x in E belongs to some B_i with

$$\hat{R}^1_{\overline{B}_i \setminus \Omega}(2B_i)(x) < 1.$$

Thus E is a subset of the countable union

$$\bigcup_i \{x \in \overline{B}_i \setminus \Omega : \hat{R}^1_{\overline{B}_i \setminus \Omega}(2B_i)(x) < 1\},$$

and hence of (p, μ)-capacity zero by Lemma 8.7. The theorem follows. $\square$

We require one more lemma before we turn to the proof of the fundamental convergence theorem.

8.11. Lemma. *Suppose that u is nonnegative and $\mathcal{A}$-superharmonic in Ω and that $E \subset \Omega$ is compact. Then the set*

$$S = \{x \in \Omega : \hat{R}^u_E(x) < R^u_E(x)\}$$

has (p, μ)-capacity zero.

PROOF: As in the proof of Theorem 8.10 the claim follows from Lemma 8.7 if we show that each $x \in S$ is contained in a ball $B = B_x$ with rational center and radius such that

$$\hat{R}^1_{\overline{B} \cap E}(2B)(x) < 1;$$

note that $S \subset E$ by Lemma 8.4. To this end, pick $x \in S$ and choose λ with $\hat{R}^u_E(x) < \lambda < u(x)$. By the lower semicontinuity of u, x is contained in a ball B with rational center and radius such that $u \geq \lambda$ in $2B \subset \Omega$. This means that

$$\lambda \hat{R}^1_{\overline{B} \cap E}(2B)(x) = \hat{R}^\lambda_{\overline{B} \cap E}(2B)(x) \leq \hat{R}^u_E(x) < \lambda,$$

as required. $\square$

PROOF OF THEOREM 8.2: In light of Lemma 7.4 it suffices to show that $\hat{s} = s$ quasieverywhere. We may assume that $\mathcal{F}$ is downward directed; indeed we may replace $\mathcal{F}$ with

$$\bigcup_{j=0}^{\infty} \mathcal{F}_j,$$

where $\mathcal{F}_0 = \mathcal{F}$ and

$$\mathcal{F}_j = \mathcal{F}_{j-1} \cup \{\min(u,v): u,v \in \mathcal{F}_{j-1}\}.$$

Next, by applying Choquet's topological lemma 8.3 we may assume that $\mathcal{F}$ consists of a decreasing sequence of $\mathcal{A}$-superharmonic functions s_i with $s = \lim s_i$. Since the capacity is subadditive, it suffices to demonstrate that the set

$$S_j = \left\{x \in \Omega : \hat{s}(x) + \frac{1}{j} < s(x)\right\}$$

has zero (p,μ)-capacity for each positive integer j. To reduce the problem further, fix j and let $K \subset S_j$ be compact; since S_j is a Borel set and hence capacitable (Theorem 2.5), it suffices to show that $\mathrm{cap}_{p,\mu} K = 0$.

To this end, let $D \Subset \Omega$ be an open neighborhood of K. We may assume that $\hat{s}$ is nonnegative in D. Writing $v = \hat{s} + 1/j$, we observe that each s_i belongs to $\boldsymbol{\Phi}_K^v(D)$, and hence that $\hat{R}_K^v(D) \leq \hat{s}$ in D. This implies

$$\hat{R}_K^v(D) < \hat{s} + \frac{1}{j} = v = R_K^v(D)$$

on K, and hence it follows from Lemma 8.11 that K is of (p,μ)-capacity zero. This completes the proof of Theorem 8.2. $\square$

The fundamental convergence theorem implies the following extension of Lemma 8.7.

8.12. Corollary. *Suppose that E is a subset of Ω. Then $\hat{R}_E^1(\Omega) = 1$ (p,μ)-quasieverywhere on E.*

We close this chapter by establishing a regularity result for the balayage; it leads to a corollary which slightly improves Theorem 7.19.

First we require a lemma which generalizes Lemma 8.4.

8.13. Lemma. *Suppose that there is an $\mathcal{A}$-subharmonic function v in an open set $D \subset \Omega$ such that $\psi \leq v \leq R^\psi(\Omega)$ in D. Then the balayage $\hat{R}^\psi(\Omega)$ is $\mathcal{A}$-harmonic in D and coincides with $R^\psi(\Omega)$ there.*

PROOF: Let $G \Subset D$ be regular. Choose a decreasing sequence $u_j \in \boldsymbol{\Phi}^\psi(\Omega)$ which converges to a function u with $\hat{u} = \hat{R}^\psi$ (see Choquet's topological lemma 8.3). Then the Poisson modifications $P(u_j, G)$ belong to $\mathbf{\Phi}^\psi(\Omega)$ since

$$P(u_j, G) \geq v \geq \psi$$

in G. Now $P(u_j, G)$ decreases to an $\mathcal{A}$-harmonic function h in G with $\hat{u} \leq h \leq u$ in G. By continuity, $h = \hat{u}$ as required.   □

8.14. Theorem.   *If ψ is continuous in Ω, then $\hat{R}^\psi$ is continuous and $\hat{R}^\psi \geq \psi$ in Ω. Moreover, $\hat{R}^\psi$ is $\mathcal{A}$-harmonic in the open set $\{\hat{R}^\psi > \psi\}$.*

PROOF: It is obvious that $\hat{R}^\psi \geq \psi$ in Ω. To prove the continuity, fix $x_0 \in \Omega$ and $\varepsilon > 0$. Since $\hat{R}^\psi(x_0) \geq \psi(x_0)$, we may choose a ball $B = B(x_0, r_0) \Subset \Omega$ such that

$$\hat{R}^\psi + \varepsilon \geq \psi(x_0) + \frac{\varepsilon}{2} \geq \psi$$

on $\overline{B}$. Let $v = P(\hat{R}^\psi + \varepsilon, B)$. Since v is $\mathcal{A}$-harmonic in B, the minimum principle yields $v \geq \psi(x_0) + \varepsilon/2 \geq \psi$ in B. On the other hand, since $\hat{R}^\psi \geq \psi$ in Ω, we have $v \geq \psi$ in Ω and hence $v \in \mathbf{\Phi}^\psi$. Thus $v \geq \hat{R}^\psi$ in Ω and we obtain

$$\limsup_{x \to x_0} \hat{R}^\psi(x) \leq \lim_{x \to x_0} v(x) = v(x_0) \leq \hat{R}^\psi(x_0) + \varepsilon.$$

Since ε was arbitrary and since the balayage $\hat{R}^\psi$ is lower semicontinuous, $\hat{R}^\psi$ is continuous at x_0.

If $\hat{R}^\psi(x_0) > \psi(x_0)$, then there is $\lambda \in \mathbf{R}$ with $\hat{R}^\psi > \lambda > \psi$ in a neighborhood of x_0. Thus $\hat{R}^\psi$ is $\mathcal{A}$-harmonic there by Lemma 8.13 and the proof is complete.   □

8.15. Theorem.   *Suppose that u is $\mathcal{A}$-superharmonic in Ω. Then there is an increasing sequence of continuous $\mathcal{A}$-superharmonic functions u_i in Ω such that $u = \lim u_i$. Moreover, each u_i is a supersolution of (3.8).*

PROOF: Since u is lower semicontinuous we may choose an increasing sequence of continuous functions $f_i \in C(\Omega)$ with $u = \lim f_i$. Then we easily verify that $u_i = \hat{R}^{f_i}(\Omega)$ is the desired sequence (see Theorem 8.14 and Corollary 7.20).   □

NOTES TO CHAPTER 8.   The modern theory of balayage is based on Brelot (1945) but the idea goes back to Gauss. Poincaré coined the term "balayage". The fundamental convergence theorem in its present form for ordinary superharmonic functions is due to Brelot (1938) and Cartan (1945). See Doob (1984) for a historical account.

In the nonlinear potential theory the method of balayage was first studied in Lehtola (1986) in an axiomatic setting. Heinonen and Kilpeläinen (1988b) introduced the balayage as a tool to study $\mathcal{A}$-harmonic functions; the fundamental convergence theorem 8.2 in the unweighted case was proved there. The definition for the balayage is the same as in linear theories but the proofs are different for the most part. This chapter is taken from Heinonen and Kilpeläinen (1988b, c) and Kilpeläinen (1989).

For the ordinary superharmonic functions it is well known that if $\complement\Omega$ is nonpolar, then

$$\hat{R}^{\psi}(\Omega) = \hat{Q}^{\psi}(\Omega),$$

where $\hat{Q}^{\psi}$ is defined similarly to $\hat{R}^{\psi}$ but the infimum is taken over the family of all superharmonic functions that lie above ψ quasieverywhere. In the (unweighted) nonlinear theory this is only known to be true if Ω is bounded and $\hat{Q}^{\psi} \in H^{1,p}(\Omega; dx)$; see Heinonen and Kilpeläinen (1988c).

9

Perron's method, barriers, and resolutivity

Perron's method is a method to solve the Dirichlet problem in a given open set with arbitrary boundary data. Perhaps surprisingly, there are no difficulties in extending this important method to the nonlinear theory. In this chapter we study general Perron solutions, boundary regularity in terms of barriers, and resolutivity.

As usual, we assume throughout this chapter that Ω is an open subset of $\mathbf{R}^n$. However, we make the convention that if Ω is unbounded, then *the point at infinity belongs to the boundary of* Ω. All topological notions are therefore understood with respect to the compactified space $\overline{\mathbf{R}}^n = \mathbf{R}^n \cup \{\infty\}$. This convention allows a unified treatment of the Dirichlet problem for all open sets Ω. We also recall that the functions in $C(E)$, $E \subset \overline{\mathbf{R}}^n$, are assumed to be continuous and real valued. Thus each function in $C(\partial\Omega)$ is bounded, even if Ω is unbounded.

9.1. Perron solutions

DEFINITION. Let $f : \partial\Omega \to [-\infty, \infty]$ be a function. The *upper class* $\mathcal{U}_f$ of f consists of all functions u such that

   (i) u is $\mathcal{A}$-superharmonic in Ω,
   (ii) u is bounded below, and
   (iii) $\liminf_{x \to y} u(x) \geq f(y)$ for all $y \in \partial\Omega$.

The *lower class* $\mathcal{L}_f$ is defined analogously via $\mathcal{A}$-subharmonic functions: $v \in \mathcal{L}_f$ if

   (i) v is $\mathcal{A}$-subharmonic in Ω,
   (ii) v is bounded above, and
   (iii) $\limsup_{x \to y} v(x) \leq f(y)$ for all $y \in \partial\Omega$.

Obviously, $v \in \mathcal{L}_f$ if and only if $-v \in \mathcal{U}_{-f}$.

DEFINITION. The function

$$\overline{H}_f = \overline{H}_f(\Omega) = \inf\{u : u \in \mathcal{U}_f\}$$

is the *upper Perron solution* and

$$\underline{H}_f = \underline{H}_f(\Omega) = \sup\{u : u \in \mathcal{L}_f\}$$

the *lower Perron solution* of f in Ω. If $\mathcal{U}_f = \emptyset$ (or $\mathcal{L}_f = \emptyset$), then we naturally set $\overline{H}_f = \infty$ (and $\underline{H}_f = -\infty$, respectively).

Then $\underline{H}_f = -\overline{H}_{-f}$ and the comparison principle yields $\underline{H}_f \leq \overline{H}_f$; requirement (ii) is imposed to guarantee this. Moreover, it is immediate that $\overline{H}_f \leq \overline{H}_g$ if $f \leq g$, and for $\lambda \in \mathbf{R}$ it holds that

$$\overline{H}_\lambda = \lambda = \underline{H}_\lambda,$$
$$\overline{H}_{f+\lambda} = \overline{H}_f + \lambda, \quad \text{and} \quad \underline{H}_{f+\lambda} = \underline{H}_f + \lambda.$$

Furthermore, if $\lambda > 0$ or if $\lambda \geq 0$ and f is finite valued, we have

$$\overline{H}_{\lambda f} = \lambda \overline{H}_f, \quad \underline{H}_{\lambda f} = \lambda \underline{H}_f,$$
$$\text{and} \quad \overline{H}_{-\lambda f} = -\lambda \underline{H}_f.$$

We also obtain from the comparison principle that if h is a bounded $\mathcal{A}$-harmonic function in Ω such that the limit

$$f(y) = \lim_{x \to y} h(x)$$

exists for each $y \in \partial\Omega$, then

$$\overline{H}_f = h = \underline{H}_f;$$

thus the Perron solution agrees with the classical solution to the Dirichlet problem, provided the latter exists. It should be observed that if h is not bounded, then the existence of the limit

$$f(y) = \lim_{x \to y} h(x), \quad y \in \partial\Omega,$$

is not enough to guarantee that $\overline{H}_f = h$; for a counterexample consider a singular $\mathcal{A}$-harmonic function in the punctured ball $B(0,1) \setminus \{0\}$ if $\text{cap}_{p,\mu}\{0\} = 0$.

The following result is fundamental in potential theory. Its proof is classical.

9.2. Theorem. *Let D be a component of Ω. Then one of the following alternatives is true:*

   (i) $\overline{H}_f$ *is $\mathcal{A}$-harmonic in D,*
   (ii) $\overline{H}_f \equiv \infty$ *in D,*
   (iii) $\overline{H}_f \equiv -\infty$ *in D.*

A similar statement is true if $\overline{H}_f$ is replaced by $\underline{H}_f$.

PROOF: If the upper class is not empty, it has the property that $\min(u,v) \in \mathcal{U}_f$ and $P(u,G) \in \mathcal{U}_f$ whenever u, $v \in \mathcal{U}_f$ and $G \Subset \Omega$ is regular. Hence the conclusion follows from Choquet's topological lemma as in the proof of Lemma 8.13. Indeed, let $u_j \in \mathcal{U}_f$ be a decreasing sequence converging to a function u such that the lower semicontinuous regularizations of u and $\overline{H}_f$ are equal in D. Then pick a regular domain $G \Subset D$ and consider the Poisson modifications $P(u_j, G)$. Now $P(u_j, G) \in \mathcal{U}_f$ and by Harnack's convergence theorem the limit function

$$\lim_{j \to \infty} P(u_j, G)$$

is either $\mathcal{A}$-harmonic or identically $-\infty$ in G. It is easy to see that this leads to the desired result. $\square$

For the next theorem recall that a family $\mathcal{F}$ of functions from a set E into $[-\infty, \infty]$ is downward directed if for each pair f_1, $f_2 \in \mathcal{F}$, there is an $f \in \mathcal{F}$ with $f \leq \min(f_1, f_2)$.

9.3. Theorem. *Suppose that $\mathcal{F}$ is a downward directed family of upper semicontinuous functions $f \colon \partial\Omega \to [-\infty, \infty)$. If $g = \inf \mathcal{F}$, then*

$$\overline{H}_g = \inf\{\overline{H}_f \colon f \in \mathcal{F}\}.$$

PROOF: Write $h = \inf_{\mathcal{F}} \overline{H}_f$. Since $g \leq f$ for $f \in \mathcal{F}$, we have $\overline{H}_g \leq h$ in Ω. To establish the reverse inequality, fix $\varepsilon > 0$ and let $u \in \mathcal{U}_g$. Since the functions in $\mathcal{F}$ are upper semicontinuous, the sets

$$\{y \in \partial\Omega \colon \liminf_{x \to y} u(x) + \varepsilon > f(y)\}, \quad f \in \mathcal{F},$$

form an open covering of $\partial\Omega$. Since $\partial\Omega$ is a compact set in $\overline{\mathbf{R}}^n$, and since $\mathcal{F}$ is downward directed, we find a function $f \in \mathcal{F}$ such that

$$\liminf_{x \to y} u(x) + \varepsilon > f(y)$$

for all $y \in \partial\Omega$. In other words, $u + \varepsilon \in \mathcal{U}_f$ and hence $u + \varepsilon \geq \overline{H}_f \geq h$. Consequently,

$$\overline{H}_g + \varepsilon \geq h,$$

and since $\varepsilon > 0$ was arbitrary,

$$\overline{H}_g \geq h,$$

as desired. $\square$

9.4. Corollary.   *Let* $f_j \colon \partial\Omega \to [-\infty, \infty)$ *be a decreasing sequence of upper semicontinuous functions and* $f = \lim f_j$. *Then*

$$\overline{H}_f = \lim_{j \to \infty} \overline{H}_{f_j}.$$

9.5. Regular boundary points

In Chapter 6 we termed a boundary point x_0 of a bounded open set Ω Sobolev regular if

$$\lim_{x \to x_0} h(x) = \vartheta(x_0)$$

whenever $\vartheta \in H^{1,p}(\Omega; \mu) \cap C(\overline{\Omega})$ and h is the $\mathcal{A}$-harmonic function in Ω with $h - \vartheta \in H_0^{1,p}(\Omega; \mu)$. Traditionally, and more correctly, regularity is defined in connection with Perron solutions.

DEFINITION. A boundary point x_0 of an open set Ω is called $\mathcal{A}$-*regular*, or simply *regular*, if

$$\lim_{x \to x_0} \overline{H}_f(x) = f(x_0)$$

for each continuous $f \colon \partial\Omega \to \mathbf{R}$. A point is *irregular* if it is not regular. If all boundary points of Ω are regular, then Ω is a *regular open set*.

If Ω is bounded, then these two definitions for regular boundary points coincide as we shall see in Theorem 9.20. Since $\overline{H}_f = -\underline{H}_{-f}$, we arrive at the same concept of regularity if we replace the upper Perron solution by the lower Perron solution. We emphasize that it is not required that Ω be bounded, nor that x_0 be a finite boundary point.

It easily follows from the removability theorem 7.36 and the maximum principle that Ω has no finite regular boundary points if $\mathrm{cap}_{p,\mu} \, \complement\Omega = 0$; in fact, all Perron solutions are constants in this case (see Theorem 10.13).

The next lemma is the first step in showing that regularity is a local property.

9.6. Lemma.   *A boundary point* x_0 *of* Ω *is regular if and only if*

$$\lim_{x \to x_0} \overline{H}_f(x) = f(x_0)$$

for each bounded $f \colon \partial\Omega \to \mathbf{R}$, continuous at x_0.

PROOF: Only the necessity calls for a proof. Let $x_0 \in \partial\Omega$ be regular and fix $\varepsilon > 0$. Let U be an open neighborhood of x_0 such that $|f - f(x_0)| < \varepsilon$ on $U \cap \partial\Omega$. Then choose a continuous function $g \colon \partial\Omega \to [f(x_0) + \varepsilon, \sup|f| + \varepsilon]$ such that $g(x_0) = f(x_0) + \varepsilon$ and $g = \sup|f| + \varepsilon$ on $\partial\Omega \setminus U$. Now $g \geq f$ on $\partial\Omega$ and hence we have

$$\limsup_{x \to x_0} \overline{H}_f(x) \leq \lim_{x \to x_0} \overline{H}_g(x) = g(x_0) = f(x_0) + \varepsilon.$$

Similarly, we see that

$$\liminf_{x \to x_0} \overline{H}_f(x) \geq f(x_0) - \varepsilon.$$

Thus we conclude

$$\lim_{x \to x_0} \overline{H}_f(x) = f(x_0),$$

and the lemma is proved.   □

9.7. Barriers

One of our goals in this chapter is to characterize regular boundary points in terms of barriers. This is a well-known device in linear theories and the definition here is similar.

DEFINITION. A function u is said to be a *barrier (relative to Ω) at x_0* if

(i) u is $\mathcal{A}$-superharmonic in Ω,
(ii) $\liminf_{x \to y} u(x) > 0$ for each $y \in \partial\Omega \setminus \{x_0\}$, and
(iii) $\lim_{x \to x_0} u(x) = 0$.

By the minimum principle a barrier is always nonnegative and only $\mathbf{R}^n$ admits a barrier that is not strictly positive: the zero function is a barrier at ∞ relative to $\mathbf{R}^n$. Moreover, it is clear that if u is a strictly positive barrier relative to Ω at x_0 and Ω' is an open subset of Ω with $x_0 \in \partial\Omega'$, then u is a barrier relative to Ω'.

If $\mathrm{cap}_{p,\mu}(\complement\Omega) = 0$, then no finite boundary point has a barrier; in fact, in view of the removability theorem 7.35, the existence of a barrier would violate the minimum principle. If $\mathrm{cap}_{p,\mu}(B, \mathbf{R}^n) > 0$ for some ball B, there always is a barrier at ∞. If $\mathrm{cap}_{p,\mu}(B, \mathbf{R}^n) = 0$ and $\mathrm{cap}_{p,\mu}(\complement\Omega) = 0$, then ∞ does not have a barrier relative to Ω unless $\Omega = \mathbf{R}^n$, cf. Theorem 9.22 below.

9.8. Theorem. *A boundary point x_0 of an open set Ω is regular if and only if there is a barrier relative to Ω at x_0.*

If $\Omega = \mathbf{R}^n$, Theorem 9.8 is immediate as the only boundary point $x_0 = \infty$ of $\mathbf{R}^n$ has the zero function as a barrier; moreover, ∞ is a regular boundary point of $\mathbf{R}^n$ because $\overline{H}_\lambda = \lambda$ for all $\lambda \in \mathbf{R}$.

PROOF OF THE SUFFICIENCY PART OF THEOREM 9.8: We may assume that Ω is a proper subset of $\mathbf{R}^n$. Let u be a barrier at x_0. Then $u > 0$ by the minimum principle. Let $f: \partial\Omega \to \mathbf{R}$ be continuous and assume without loss of generality that $f(x_0) = 0$. Let $\varepsilon > 0$ and choose an open neighborhood U of x_0 such that $\partial U \cap \Omega \neq \emptyset$ and that $|f| < \varepsilon$ in $\overline{U} \cap \partial\Omega$. Fix $\lambda > \max|f|$ and write

$$v = \begin{cases} \lambda & \text{in } \mathbf{R}^n \setminus U \\ \frac{\lambda}{m}\min(u, m) & \text{in } U \cap \Omega, \end{cases}$$

where

$$m = \inf\{u(x) : x \in \partial U \cap \Omega\} > 0\,.$$

Then v is $\mathcal{A}$-superharmonic in $\complement\overline{U} \cup \Omega$ by the pasting lemma 7.9. Moreover, $v + \varepsilon$ is in the upper class $\mathcal{U}_f$, and hence

$$\limsup_{x \to x_0} \overline{H}_f(x) \leq \limsup_{x \to x_0} v(x) + \varepsilon = \varepsilon\,.$$

Similarly, $-(v + \varepsilon)$ is in the lower class $\mathcal{L}_f$, and we obtain

$$\liminf_{x \to x_0} \overline{H}_f(x) \geq \liminf_{x \to x_0} \underline{H}_f(x) \geq \liminf_{x \to x_0}(-v(x) - \varepsilon) = -\varepsilon\,.$$

Since ε was arbitrary, we have

$$\lim_{x \to x_0} \overline{H}_f(x) = 0 = f(x_0)\,,$$

as desired.   □

A look at the above proof reveals that the function v is a barrier relative to $\complement\overline{U} \cup \Omega$ at x_0. Thus the existence of a barrier is a local property:

9.9. Proposition. *Let $G \subset \Omega \neq \mathbf{R}^n$ be an open set and let $x_0 \in \partial\Omega \cap \partial G$ be such that $U \cap G = U \cap \Omega$ for some open neighborhood U of x_0. Then there is a barrier relative to Ω at x_0 if and only if there is a barrier relative to G at x_0.*

For the proof of the necessity in Theorem 9.8 we need to extend some of the lemmas from Chapter 8. Keep in mind that regularity is now defined in terms of Perron solutions.

9.10. Lemma. *A boundary point $x_0 \in \partial\Omega \setminus \{\infty\}$ is regular if*

$$\hat{R}^1_{\overline{B}\setminus\Omega}(2B)(x_0) = 1$$

whenever B is a ball with rational center and radius containing x_0. In particular, if $\complement\Omega$ is (p,μ)-thick at x_0, then x_0 is regular.

PROOF: The proof of Lemma 8.9 applies almost verbatim. Just replace ϑ by a function $f \in C(\partial\Omega)$ and h by $\overline{H}_f$, and note that the function u belongs to $\mathcal{U}_f$ and the function $-u$ to $\mathcal{L}_f$. The second assertion follows from Lemma 8.5 and Theorem 6.27.   □

It is shown in Theorem 9.17 below that the condition in Lemma 9.10 actually characterizes regular boundary points.

Now we have the general form of the *Kellogg property*.

9.11. Theorem.   *The set of finite irregular boundary points of an open set Ω has (p, μ)-capacity zero.*

PROOF: The assertion is derived from Lemma 9.10 as Theorem 8.10 is derived from Lemma 8.9.   □

9.12. Corollary.   *A point $x_0 \in \mathbf{R}^n$ with $\operatorname{cap}_{p,\mu}\{x_0\} > 0$ is always regular.*

It is important to know that each component of the complement of a closed set of positive (p, μ)-capacity has a finite regular boundary point.

9.13. Theorem.   *Suppose that $\complement\Omega$ has positive (p, μ)-capacity and that G is a component of Ω. Then there is a point $x_0 \in \partial G \setminus \{\infty\}$ which is regular relative to G.*

PROOF: The boundary of G is of positive (p, μ)-capacity, since $\complement G$ is of positive (p, μ)-capacity. For if $\operatorname{cap}_{p,\mu} \partial G = 0$, then $\complement G$, being of positive (p, μ)-capacity, must have an interior point and hence $\complement \partial G$ has at least two components. However, this is impossible since a set of (p, μ)-capacity zero does not separate $\mathbf{R}^n$ (Lemma 2.46). Therefore $\operatorname{cap}_{p,\mu} \partial G > 0$ and the assertion follows from Theorem 9.11.   □

Before completing the proof of Theorem 9.8 we present another version of the extension lemma 7.30.

9.14. Lemma.   *Suppose that u is $\mathcal{A}$-superharmonic in a neighborhood of a closed ball $\overline{B} \subset \Omega$. If $\complement\Omega$ has positive (p, μ)-capacity, then there is an $\mathcal{A}$-superharmonic function $\tilde{u}$ in Ω such that $\tilde{u} = u$ in B and $\tilde{u}$ is bounded below.*

PROOF: Since it suffices to show that u has a desired extension to the component of Ω which contains B, we may assume that Ω is connected. Choose a ball $B_0 \Subset \Omega$ such that $u \in \mathcal{S}(B_0)$ and that $B \Subset B_0$. We may assume that $u > 0$ in B_0 and replace u by its balayage $v = \hat{R}^u_B(B_0)$. Then $v = u$ in B and $\lim_{x \to y} v(x) = 0$ for all $y \in \partial B_0$ (Lemma 8.8). Let $m = \min_{\overline{B}} v$. Then the set

$$K = \{x \in B_0 : v(x) \geq m\}$$

is compact and contains $\overline{B}$. Now put

$$s = \begin{cases} \overline{H}_f & \text{in } \Omega \setminus K \\ m & \text{in } K, \end{cases}$$

where $f = m$ in K, $f = 0$ on $\partial\Omega$, and the Perron solution $\overline{H}_f$ is taken in $\Omega \setminus K$. To show that $\lim_{x \to y} s(x) = m$ for all $y \in K$, pick $\tilde{s} \in \mathcal{U}_f$. Then $\tilde{s}$

is nonnegative and $\tilde{s} \geq v$ in $B_0 \setminus K$ by the comparison principle; note that v is $\mathcal{A}$-harmonic in $B_0 \setminus \overline{B}$. Thus

$$\liminf_{x \to y} s(x) \geq m$$

for all $y \in \partial K$ and, consequently,

$$\lim_{x \to y} s(x) = m$$

for all $y \in K$. Now the pasting lemma 7.9 yields that $s \in \mathcal{S}(\Omega) \cap \mathcal{H}(\Omega \setminus K)$. By Theorem 9.13, there is a point y on $\partial\Omega$ such that $\lim_{x \to y} s(x) = 0$. Hence it follows from the maximum principle that $s < m$ on ∂B_0. Thus for $M = \max_{\partial B_0} s$ and $\delta = m/(m - M) > 0$ we have

$$\delta(s - m) \leq -m = v - m$$

on ∂B_0. By the comparison principle the same inequality holds in $B_0 \setminus K$. Therefore, invoking the second version of the pasting lemma (Lemma 7.28), we observe that the function

$$\tilde{u} = \begin{cases} v & \text{in } K \\ \delta(s - m) + m & \text{in } \Omega \setminus K \end{cases}$$

is the desired $\mathcal{A}$-superharmonic extension of u.   □

PROOF OF THE NECESSITY OF THEOREM 9.8: Suppose first that $x_0 \neq \infty$ is a regular boundary point. It follows from the removability theorem 7.35 and the comparison principle that $\text{cap}_{p,\mu} \complement\Omega > 0$. Indeed, if $\text{cap}_{p,\mu} \complement\Omega = 0$, pick $f \in C(\overline{\mathbf{R}}^n)$ such that $0 \leq f \leq 1$, $f(x_0) = 0$, and $f(\infty) = 1$. Because each $u \in \mathcal{U}_f(\Omega)$ actually belongs to $\mathcal{U}_f(\mathbf{R}^n)$ (Theorem 7.35), it follows from the comparison principle that $\overline{H}_f \geq 1$, an obvious contradiction.

If $\text{cap}_{p,\mu}\{x_0\} > 0$, then $\{x_0\}$ is (p, μ)-thick at x_0 (Theorem 6.33) and the existence of a barrier is provided by Lemma 7.31. Thus we may assume that $\text{cap}_{p,\mu}(\complement\Omega \setminus \{x_0\}) > 0$. Then choose a ball $B = B(x_0, r_0)$ so small that the complement of $\Omega \cup 2B$ has positive (p, μ)-capacity. Define

$$f(x) = \frac{(r_0 - |x - x_0|)^+}{r_0}$$

and write

$$v = \hat{R}^f(2B).$$

Then $0 \leq v \leq 1$, v is continuous in $2B$, and $\lim_{x \to y} v(x) = 0$ for each $y \in \partial 2B$ (see Theorem 8.14 and Lemma 8.8). Moreover, since v is $\mathcal{A}$-harmonic in the set $\{f < v\}$, it follows from the maximum principle that $v = 1$ only at x_0.

Next pick a compact set $K \subset \complement(\Omega \cup 2B)$ of positive (p, μ)-capacity and let $D = \complement K$. We use Lemma 9.14 and infer from the pasting lemma 7.9 that there is a bounded $\mathcal{A}$-superharmonic function $\tilde{v}$ in D such that v and $\tilde{v}$ coincide in B and that $\tilde{v} \leq 1 - \varepsilon < 1$ in $D \setminus \overline{B}$ for some $\varepsilon > 0$. If we define

$$\tilde{v}(y) = \liminf_{x \to y} \tilde{v}(x)$$

for $y \in K$, then the function $w = 1 - \overline{H}_{\tilde{v}}$ is $\mathcal{A}$-harmonic in Ω and $\lim_{x \to x_0} w(x) = 0$ by Lemma 9.6. Moreover, since $\tilde{v} \in \mathcal{U}_{\tilde{v}}$, we have $w \geq 1 - \tilde{v}$; hence

$$\liminf_{x \to y} w(x) > 0$$

for $y \in \partial \Omega \setminus \{x_0\}$. Thus w is a barrier relative to Ω at x_0.

Suppose then that $x_0 = \infty$. If $\mathrm{cap}_{p,\mu} \complement\Omega = 0$, choose a continuous boundary function f such that $0 \leq f \leq 1$, $f(\infty) = 0$, and $f = 1$ somewhere on $\partial\Omega$; note that the case when $\Omega = \mathbf{R}^n$ is trivial. Because the point at infinity is regular, we may choose a nonconstant function u from the upper class $\mathcal{U}_f$ with $0 \leq u \leq 1$. By the removability theorem u is indeed $\mathcal{A}$-superharmonic in all of $\mathbf{R}^n$; hence $u > 0$ in $\mathbf{R}^n$. If B is a ball, then the balayage $\hat{R}_B^v(\mathbf{R}^n)$ of $v = u - \inf u$ is the desired barrier because it has limit 0 at infinity (Lemma 6.15).

Now we are free to assume that $\mathrm{cap}_{p,\mu} \complement\Omega > 0$. In that case the proof is similar to the case where x_0 was finite, and we leave the details to the reader.

Theorem 9.8 follows. $\square$

A particularly useful consequence of Theorem 9.8 and Proposition 9.9 is the fact that regularity is a local property.

9.15. Corollary. *Let Ω and G be proper open subsets of $\mathbf{R}^n$ and let $x_0 \in \partial\Omega \cap \partial G$. If there is an open neighborhood U of x_0 such that*

$$U \cap \Omega = U \cap G,$$

then x_0 is regular relative to Ω if and only if it is regular relative to G.

We also have:

9.16. Corollary. *Suppose that $D \subset \Omega \neq \mathbf{R}^n$ is open and that $x_0 \in \partial\Omega \cap \partial D$. If x_0 is regular relative to Ω, then x_0 is regular relative to D.*

Now we combine some characterizations for the regularity. Keep in mind that a point of positive (p, μ)-capacity is always regular.

9.17. Theorem. *Suppose that $x_0 \in \partial\Omega \setminus \{\infty\}$ is such that $\mathrm{cap}_{p,\mu}\{x_0\} = 0$. Then the following are equivalent:*

(i) *x_0 is regular.*

(ii) *There is a barrier at x_0 relative to Ω.*

(iii) *If $U \Subset V$ are bounded open sets with $x_0 \in U$, then*

$$\hat{R}^u_{\overline{U}\setminus\Omega}(V)(x_0) = u(x_0)$$

whenever $u \in \mathcal{S}(V)$ is nonnegative.

(iv) *For all balls B containing x_0 it holds that*

$$\hat{R}^1_{\overline{B}\setminus\Omega}(2B)(x_0) = 1.$$

PROOF: The equivalence of (i) and (ii) is the content of Theorem 9.8.

To prove that (i) implies (iii), let $f_j \in C(\mathbf{R}^n)$ be an increasing sequence such that $\lim f_j = u$ in $\overline{U}$ and $f_j = 0$ in ∂V. Define

$$g_j = \begin{cases} \overline{H}_{f_j}(V \setminus (\overline{U} \setminus \Omega)) & \text{in } V \setminus (\overline{U} \setminus \Omega) \\ f_j & \text{in } \overline{U} \setminus \Omega. \end{cases}$$

It follows from Corollary 9.15 that

$$f_j(x_0) = \lim_{x \to x_0} g_j(x) \le \liminf_{x \to x_0} R^u_{\overline{U}\setminus\Omega}(V)(x) = \hat{R}^u_{\overline{U}\setminus\Omega}(V)(x_0).$$

Hence

$$u(x_0) \le \hat{R}^u_{\overline{U}\setminus\Omega}(V)(x_0).$$

The opposite inequality is trivial and (iii) follows.

That (iii) implies (iv) is trivial; Lemma 9.10 shows that (iv) implies (i), and the loop is complete. $\qquad\square$

Next we study the behavior of an $\mathcal{A}$-superharmonic function in a neighborhood of a finite regular boundary point x_0 of Ω with $\mathrm{cap}_{p,\mu}\{x_0\} = 0$.

9.18. Theorem. *Let $x_0 \neq \infty$ be a regular boundary point of Ω such that $\mathrm{cap}_{p,\mu}\{x_0\} = 0$. If u is $\mathcal{A}$-superharmonic in a neighborhood of x_0, then*

$$\liminf_{\substack{x \to x_0 \\ x \in \complement\Omega}} u(x) = u(x_0).$$

PROOF: Suppose, on the contrary, that there is an $\mathcal{A}$-superharmonic function u in a neighborhood of x_0 such that

$$(9.19) \qquad \liminf_{\substack{x \to x_0 \\ x \in \complement\Omega}} u(x) > u(x_0).$$

By adding a constant and truncating we may assume that $u \in H^{1,p}(3B; \mu)$ for some ball B centered at x_0 and that $u = 1$ in $\complement\Omega \cap \overline{B} \setminus \{x_0\}$. Then choose a function $f \in C^\infty(\mathbf{R}^n)$ such that $f = 1$ in $\overline{B}$ and $f \leq u$ on the sphere $\partial 2B$. It follows from the generalized comparison lemma 7.37 that $\overline{H}_f \leq u$ in $2B \setminus (\overline{B} \cap \complement\Omega)$, where $\overline{H}_f$ is the Perron solution in $2B \setminus (\overline{B} \cap \complement\Omega)$. We thus have

$$1 = \lim_{\substack{x \to x_0 \\ x \in \Omega}} \overline{H}_f(x) \leq \liminf_{\substack{x \to x_0 \\ x \in \Omega}} u(x) = u(x_0) < 1\,,$$

where the second equality follows from the "ess lim inf" property (Theorem 7.22). This is a contradiction, and the theorem follows. $\qquad\square$

The above theorem may be called Cartan's lemma. It is well known that property (9.19) characterizes regular boundary points in classical potential theory; in general, this question is open in weighted nonlinear potential theory (see Notes to this chapter).

9.20. Theorem. *Suppose that Ω is bounded and that $x_0 \in \partial\Omega$. Then x_0 is regular if and only if x_0 is Sobolev regular.*

PROOF: By Theorem 9.17 it suffices to show that x_0 is Sobolev regular if and only if

$$\hat{R}^1_{\overline{B}\setminus\Omega}(2B)(x_0) = 1$$

whenever $B = B(x_0, r)$ is a ball. That this condition implies Sobolev regularity was proved in Lemma 8.9.

Let then x_0 be Sobolev regular and let B be a ball that contains x_0. By Lemma 8.5 we have

$$\hat{R}^1_{\overline{B}\setminus\Omega}(2B)(x_0) = \lim_{\substack{x \to x_0 \\ x \in \Omega}} \hat{R}^1_{\overline{B}\setminus\Omega}(2B)(x) = 1\,,$$

and the theorem follows. $\qquad\square$

9.21. Bounded, nonconstant $\mathcal{A}$-superharmonic functions

It is an old issue to try to classify the Riemann surfaces or Riemannian manifolds which carry nonconstant bounded superharmonic functions; the reader who is familiar with that theory recognizes the following results, when restricted to the classical Laplacian case.

It is clear that there exists a positive, nonconstant $\mathcal{A}$-superharmonic function in $\mathbf{R}^n$ exactly when there is a bounded, nonconstant $\mathcal{A}$-superharmonic function in $\mathbf{R}^n$. It turns out that the existence of such functions depends on whether or not the point at infinity is regular for the Dirichlet problem (relative to all unbounded open sets).

9.22. Theorem. *The following are equivalent:*

(i) *The point at infinity is a regular boundary point for each unbounded* Ω.

(ii) *The point at infinity is a regular boundary point for the complement of a closed ball.*

(iii) *There exists a nonconstant bounded $\mathcal{A}$-superharmonic function in* $\mathbf{R}^n$.

(iv) $\mathrm{cap}_{p,\mu}(B, \mathbf{R}^n)$ *is positive for each ball B.*

(v) $\mathrm{cap}_{p,\mu}(B, \mathbf{R}^n)$ *is positive for some ball B.*

PROOF: It is trivial that (i) implies (ii). To prove that (ii) implies (iii), let $\Omega = \complement \overline{B}$ for a ball B such that the point at infinity is regular for Ω. Then Ω is regular by the fact that $\overline{B}$ is (p, μ)-thick at each of its points. Thus there is an $\mathcal{A}$-harmonic function h in Ω, continuous on $\overline{\Omega}$ with boundary values 1 on ∂B and 0 at ∞. Extend h to B by setting $h = 1$ there. This gives a nonconstant $\mathcal{A}$-superharmonic function h in $\mathbf{R}^n$ with $0 \le h \le 1$.

For the implication (iii)$\Rightarrow$(iv), let u be a bounded, nonconstant $\mathcal{A}$-superharmonic function in $\mathbf{R}^n$. We may suppose that $\inf u = 0$. Pick an open ball $B \subset \mathbf{R}^n$ and observe that $m = \min_{\overline{B}} u > 0$. If $\mathrm{cap}_{p,\mu}(\overline{B}, \mathbf{R}^n) = 0$, we can exhaust $\mathbf{R}^n$ by concentric balls $B \Subset B_1 \Subset B_2 \Subset \ldots$ such that

$$\mathrm{cap}_{p,\mu}(\overline{B}, B_j) \to 0.$$

Let now u_j be the $\mathcal{A}$-potential $\hat{R}^1_{\overline{B}}(B_j)$. Then u_j is $\mathcal{A}$-harmonic in $B_j \setminus \overline{B}$ and an appeal to the Harnack convergence theorem shows that u_j increases to an $\mathcal{A}$-harmonic function h in $\mathbf{R}^n \setminus \overline{B}$. Since $h \le u/m$, we deduce that h is not constant; note that $h(x) \to 1$ as $x \to y$ for all $y \in \partial B$.

On the other hand, we have by the quasiminimizing property 3.13 that

$$\int_{B_j} |\nabla u_j|^p \, d\mu \le \left(\frac{\beta}{\alpha}\right)^p \mathrm{cap}_{p,\mu}(\overline{B}, B_j),$$

and hence the sequence ∇u_j is bounded in $L^p(\mathbf{R}^n;\mu)$. Since u_j converges to h, we infer from Lemma 1.33 that $h \in L^{1,p}(\complement\overline{B};\mu)$ and that $\nabla u_j \to \nabla h$ weakly in $L^p(\complement\overline{B};\mu)$. Thus the weak lower semicontinuity of norms (Remark 5.25) yields

$$\int_{\complement\overline{B}} |\nabla h|^p \, d\mu \leq \liminf_{j\to\infty} \int_{\complement\overline{B}} |\nabla u_j|^p \, d\mu$$
$$\leq (\frac{\beta}{\alpha})^p \lim_{j\to\infty} \mathrm{cap}_{p,\mu}(\overline{B}, B_j) = 0 \,.$$

Consequently, h is constant in $\complement\overline{B}$, which is a contradiction. Therefore

$$\mathrm{cap}_{p,\mu}(\overline{B}, \mathbf{R}^n) > 0 \,,$$

and (iv) follows.

Assertion (iv) clearly implies (v).

To establish the last link, we show that (i) follows from (v). For this, suppose that

$$\mathrm{cap}_{p,\mu}(\overline{B}, \mathbf{R}^n) = \delta > 0$$

for a ball B. We show that ∞ has a strictly positive barrier v relative to $\mathbf{R}^n$. Then v is a barrier at ∞ relative to each unbounded open set and assertion (i) follows from Theorem 9.8.

To this end, for $j = 1, 2, \ldots$ let $B_j = (j+1)B$ and let u_j be the $\mathcal{A}$-potential $\hat{R}^1_{\overline{B}}(B_j)$. Then u_j increases to an $\mathcal{A}$-harmonic function u in $\complement\overline{B}$. We first show that u is not constant. Indeed, we see as in the proof of (iii) $\Rightarrow$ (iv) above that $\nabla u_j \to \nabla u$ weakly in $L^p(\complement\overline{B};\mu)$. Thus by the Mazur lemma 1.29 there is a sequence $\tilde{u}_j$ of convex combinations of u_j's,

$$\tilde{u}_j = \sum_{k=1}^{j} \lambda_{k,j} u_k \,, \quad \sum_{k=1}^{j} \lambda_{k,j} = 1 \,, \quad \lambda_{k,j} \geq 0 \,,$$

such that $\nabla \tilde{u}_j \to \nabla u$ in $L^p(\complement\overline{B};\mu)$. Thus if u is constant, there is an index j_δ such that

$$\int_{\complement\overline{B}} |\nabla \tilde{u}_{j_\delta}|^p \, d\mu < \delta \,.$$

But the function $\tilde{u}_{j_\delta}$ is admissible for the condenser $(\overline{B}, B_{j_\delta})$, whence

$$\delta > \mathrm{cap}_{p,\mu}(\overline{B}, B_{j_\delta}) \geq \mathrm{cap}_{p,\mu}(\overline{B}, \mathbf{R}^n) \,,$$

contrary to our assumptions. We conclude that u is not constant.

To complete the proof we invoke Lemma 6.15 and the minimum principle and infer that

$$\lim_{x \to \infty} u(x) = \inf u < 1 .$$

Because

$$\lim_{x \to y} u(x) = 1$$

for all $y \in \partial B$, it follows that if we set $u = 1$ on $\overline{B}$, then u is $\mathcal{A}$-super-harmonic in $\mathbf{R}^n$. Hence $v = u - \inf u$ is the desired positive barrier relative to $\mathbf{R}^n$ at ∞. Thus (i) follows and the theorem is proved.   □

If Ω is fixed, Theorem 9.22 takes the following form:

9.23. Theorem. *Suppose that ∞ is an irregular boundary point of an unbounded open set G. If $\Omega \subset \mathbf{R}^n$ is open, then the following are equivalent:*

(i) $\mathrm{cap}_{p,\mu} \complement\Omega > 0$.
(ii) *There exists a bounded $\mathcal{A}$-superharmonic function u in Ω such that u is nonconstant in each component of Ω.*
(iii) $\mathrm{cap}_{p,\mu}(B, \Omega)$ *is positive for each ball $B \Subset \Omega$.*
(iv) $\mathrm{cap}_{p,\mu}(B, \Omega)$ *is positive for some ball $B \Subset \Omega$.*

PROOF: The proof is quite similar to that of Theorem 9.22. Note that since sets of zero (p, μ)-capacity do not separate $\mathbf{R}^n$ (Lemma 2.46), we may assume that Ω is connected. The proofs for (i) $\Rightarrow$ (ii) $\Rightarrow$ (iii) go essentially as in Theorem 9.22; we only need to note that there is at least one regular point of $\Omega \setminus \overline{B}$ on $\partial\Omega$, whenever $B \Subset \Omega$ is a ball, and that polyhedra can replace the balls B_j in the exhaustion argument. The implication (iii) $\Rightarrow$ (iv) is trivial.

To prove the remaining implication (iv) $\Rightarrow$ (i), observe first that necessarily

$$\mathrm{cap}_{p,\mu}(\overline{B}, \mathbf{R}^n) = 0$$

because ∞ is an irregular boundary point of G (Theorem 9.22). Then fix $\varepsilon > 0$ and choose a function $\varphi \in C_0^\infty(\mathbf{R}^n)$, $0 \le \varphi \le 1$, such that $\varphi = 1$ in $\overline{B}$ and that

$$\int_{\mathbf{R}^n} |\nabla\varphi|^p \, d\mu < \varepsilon .$$

If $\mathrm{cap}_{p,\mu} \complement\Omega = 0$, then $\mathrm{cap}_{p,\mu} K = 0$, where $K = \mathrm{spt}\,\varphi \cap \complement\Omega$. Now we can choose a function $\eta \in C_0^\infty(\mathbf{R}^n)$ such that $0 \le \eta \le 1$, $\eta = 1$ in a neighborhood of K, $\eta = 0$ on $\overline{B}$, and

$$\int_{\mathbf{R}^n} |\nabla\eta|^p \, d\mu < \varepsilon .$$

Then $\psi = (1 - \eta)\varphi$ is admissible for the condenser $(\overline{B}, \Omega)$ and hence

$$
\begin{aligned}
\mathrm{cap}_{p,\mu}(\overline{B}, \Omega) &\leq \int_\Omega |\nabla\psi|^p \, d\mu \\
&\leq 2^{p-1}\Big(\int_{\mathbf{R}^n} |\nabla\varphi|^p \, d\mu + \int_{\mathbf{R}^n} |\nabla\eta|^p \, d\mu\Big) \\
&\leq 2^p \varepsilon .
\end{aligned}
$$

Since ε was arbitrary, this violates the assumption that $\mathrm{cap}_{p,\mu}(\overline{B}, \Omega) > 0$; consequently $\complement\Omega$ cannot be of zero (p, μ)-capacity.

The theorem is proved. $\square$

In the unweighted case ($w = 1$) Theorem 9.22 reveals that there exists a nonconstant bounded $\mathcal{A}$-superharmonic function in $\mathbf{R}^n$, or that the point at infinity is regular for each unbounded Ω, if and only if $1 < p < n$. This is due to the fact that $\mathrm{cap}_p(B, \mathbf{R}^n) > 0$ for some ball B if and only if $1 < p < n$ (see Example 2.12).

9.24. Resolutivity

It is natural to ask which one of the two Perron solutions $\overline{H}_f$ and $\underline{H}_f$ is the "correct" solution to the Dirichlet problem. In classical potential theory it is known that $\overline{H}_f = \underline{H}_f$ for all bounded Borel measurable functions on $\partial\Omega$. In the nonlinear case, this problem of *resolutivity* is largely open. We next show that the Perron method gives a satisfactory solution to the Dirichlet problem when the boundary function is continuous.

DEFINITION. We say that a function $f : \partial\Omega \to [-\infty, \infty]$ is $(\mathcal{A}\text{-})resolutive$ if the upper and the lower Perron solutions $\overline{H}_f$ and $\underline{H}_f$ coincide and are $\mathcal{A}$-harmonic in Ω.

The resolutivity of f does not imply that

$$
\lim_{x \to y} \overline{H}_f(x) = f(y)
$$

for $y \in \partial\Omega$. However, the converse is trivial; more precisely, if there exists a bounded $\mathcal{A}$-harmonic function h in Ω such that

$$
\lim_{x \to y} h(x) = f(y)
$$

for all $y \in \partial\Omega$, then it follows from the comparison principle that f is resolutive. In particular, if Ω is regular, then continuous functions on $\partial\Omega$ are resolutive. If Ω is not regular, the resolutivity of continuous functions is not immediate.

9.25. Theorem. *If the complement of Ω has positive (p, μ)-capacity, then every continuous $f \colon \partial\Omega \to \mathbf{R}$ is resolutive.*

If $\mathrm{cap}_{p,\mu}\, \complement\Omega = 0$ and if there exists a nonconstant bounded $\mathcal{A}$-superharmonic function in $\mathbf{R}^n$, then each boundary function, finite at infinity, is resolutive (Theorem 10.13(ii)). On the other hand, if Ω is a proper subset of $\mathbf{R}^n$ with $\mathrm{cap}_{p,\mu}\, \complement\Omega = 0$ and there exist no nonconstant bounded $\mathcal{A}$-superharmonic functions in $\mathbf{R}^n$, then a resolutive function on $\partial\Omega$ is necessarily constant (Theorem 10.13(i)); in particular, not all continuous functions are resolutive. Thus, in light of the Kellogg property (see Theorem 9.13) and Theorem 9.22 we may reformulate Theorem 9.25: *if Ω has a regular boundary point, then real-valued continuous functions on $\partial\Omega$ are resolutive.* As noted above, if Ω does not have any regular boundary points, then no nonconstant function of $\partial\Omega$ is resolutive.

In preparation for the proof of Theorem 9.25, we establish some results of independent interest. We begin with the following theorem which connects the balayage to the obstacle problem.

9.26. Theorem. *Let f be a continuous real-valued function in $\overline{\mathbf{R}}^n$ and let $u = \hat{R}^f(\Omega)$. Then u is continuous, $u \geq f$, and*

$$\lim_{x \to x_0} u(x) = f(x_0)$$

whenever $x_0 \in \partial\Omega \setminus \{\infty\}$ is a regular boundary point.

If, in addition, $f \in L^{1,p}(\Omega; \mu)$, then $u \in L^{1,p}(\Omega; \mu)$, $u - f \in L_0^{1,p}(\Omega; \mu)$, and

$$\int_\Omega \mathcal{A}(x, \nabla u) \cdot \nabla\varphi \, dx \geq 0$$

whenever $\varphi \in H_0^{1,p}(\Omega; \mu)$ with $\varphi \geq f - u$ a.e.

PROOF: Since f is continuous and hence bounded, $u \in C(\Omega)$ is bounded and $u \geq f$ (Theorem 8.14). To prove the boundary limit assertion, let s be a barrier at x_0. Fix $\varepsilon > 0$ and let B be a ball centered at x_0 such that $\partial B \cap \Omega \neq \emptyset$ and that $|f(x) - f(x_0)| < \varepsilon$ for all $x \in B$. Since

$$\inf\{s(x) : x \in \partial B \cap \Omega\} > 0,$$

there is $\lambda > 0$ such that

$$v(x) = \begin{cases} \min(\lambda s(x) + \varepsilon + f(x_0), \sup|f|) & x \in B \cap \Omega \\ \sup|f| & x \in \Omega \setminus B \end{cases}$$

is lower semicontinuous and hence $\mathcal{A}$-superharmonic in Ω with $v \geq f$ in Ω (the pasting lemma 7.9). Then

$$\limsup_{x \to x_0} u(x) \leq \lim_{x \to x_0} v(x) \leq f(x_0) + \varepsilon.$$

and since $u \geq f$, we obtain

$$\lim_{x \to x_0} u(x) = f(x_0),$$

as desired.

Next, suppose that $f \in L^{1,p}(\Omega; \mu)$ and let $D_1 \Subset D_2 \Subset \ldots \Subset \Omega$ be regular open sets (for example, polyhedra) such that $\cup_i D_i = \Omega$ and let u_i be the $\mathcal{A}$-superharmonic solution to the obstacle problem in D_i with both obstacle and boundary values f. Since $u \in H^{1,p}_{loc}(\Omega; \mu)$ and $u \geq f$, we obtain from Lemma 3.22 that $u_1 \leq u_2 \leq \ldots \leq u$. Thus $v = \lim u_i$ is $\mathcal{A}$-superharmonic in Ω and $u \geq v \geq f$. On the other hand, the definition for balayage implies $u \leq v$ so that $u = v$. Since $u_i - f \in H^{1,p}_0(\Omega; \mu)$ (by setting $u_i = f$ on $\Omega \setminus D_i$) and since

$$\int_{D_i} |\nabla u_i|^p \, d\mu \leq c \int_{D_i} |\nabla f|^p \, d\mu \leq c \int_{\Omega} |\nabla f|^p \, d\mu < \infty$$

by the quasiminimizing property, we deduce that $\nabla u_i - \nabla f$ converges weakly to $\nabla u - \nabla f$ in $L^p(\Omega; \mu)$ and, therefore, that $u - f \in L^{1,p}_0(\Omega; \mu)$ (see Lemma 1.33).

To complete the proof, fix $\varphi \in H^{1,p}_0(\Omega; \mu)$ with $\varphi \geq f - u$ a.e. in Ω. If $\varphi_j \in C^\infty_0(\Omega)$ is a sequence that converges to φ in $H^{1,p}(\Omega; \mu)$, then the functions $\eta_j = \max(\varphi_j, f - u)$ have compact supports and thus belong to $H^{1,p}_0(\Omega; \mu)$. Moreover, $\eta_j \to \varphi$ in $H^{1,p}(\Omega; \mu)$ (see Lemma 1.22). Then fix j and choose an index i_j such that $\operatorname{spt} \eta_j \subset D_{i_j}$. Since u is the solution to the obstacle problem in $\mathcal{K}_{f,u}(D_{i_0})$ by Lemma 3.81, we have

$$\int_\Omega \mathcal{A}(x, \nabla u) \cdot \nabla \eta_j \, dx = \int_{D_{i_j}} \mathcal{A}(x, \nabla u) \cdot \nabla \eta_j \, dx \geq 0.$$

Because $u \in L^{1,p}(\Omega; \mu)$ and $\eta_j \to \varphi$ in $H^{1,p}(\Omega; \mu)$, it follows that

$$\int_\Omega \mathcal{A}(x, \nabla u) \cdot \nabla \varphi \, dx = \lim_{j \to \infty} \int_\Omega \mathcal{A}(x, \nabla u) \cdot \nabla \eta_j \, dx \geq 0,$$

as desired. Theorem 9.26 is proved.   □

9.27. REMARKS. (a) Let f and $u = \hat{R}^f(\Omega)$ be as in the previous theorem. If, in addition, Ω is bounded and $f \in H^{1,p}(\Omega; \mu)$, then $u \in H^{1,p}(\Omega; \mu)$ and $u - f \in H^{1,p}_0(\Omega; \mu)$. To see this, let u_i be as in the proof of Theorem 9.26 and note that $u_i - f \in H^{1,p}_0(\Omega; \mu)$ converges weakly to $u - f$ as the $L^p(\Omega; \mu)$-norms are uniformly bounded.

(b) The continuity assumption in Theorem 9.26 can be relaxed to an extent. Namely, assume that f is bounded above and that $x_0 \in \partial\Omega \setminus \{\infty\}$ is regular. If f is continuous at x_0 and if $\hat{R}^f \geq f$ in $B(x_0, r) \cap \Omega$ for some $r > 0$, then

$$\lim_{x \to x_0} \hat{R}^f(x) = f(x_0).$$

Moreover, the second part of the claim remains true under the assumptions: $f \in L^{1,p}(\Omega; \mu)$, f bounded above, and for all open $D \Subset \Omega$ the $\mathcal{A}$-superharmonic solution v to the obstacle problem in $\mathcal{K}_f(D)$ lies above f everywhere in D. Observe that, for example, each lower semicontinuous f verifies the last condition.

These are easily established by repeating the argument used to prove Theorem 9.26.

(c) Suppose that u and f are as in Theorem 9.26. If Ω is unbounded and ∞ is a regular boundary point of each unbounded open subset of Ω, then

$$\lim_{x \to \infty} u(x) = f(\infty).$$

The proof is a slight modification of the proof of Theorem 9.26.

The next theorem gives a solution to the Dirichlet problem with Sobolev boundary values in unbounded open sets.

9.28. Theorem.   *Suppose that the complement of Ω has positive (p, μ)-capacity and that f is in $C(\overline{\mathbf{R}}^n) \cap L^{1,p}(\Omega; \mu)$. Then f is resolutive and $\overline{H}_f - f \in L_0^{1,p}(\Omega; \mu)$.*

PROOF: As in the proof of Theorem 9.26 let $u = \hat{R}^f(\Omega)$ and let D_i be an exhaustion of Ω by regular open sets. Write u_i for the Poisson modification $P(u, D_i)$ of u in D_i. Then $u \geq u_1 \geq u_2 \geq \ldots \geq \overline{H}_f$, and hence the limit function $h^* = \lim u_i$ is $\mathcal{A}$-harmonic in Ω with $h^* \geq \overline{H}_f$. Moreover, a minor modification to the proof of Theorem 9.26 shows that ∇u_i converges weakly to ∇h^* in $L^p(\Omega; \mu)$ and hence that $f - h^*$ belongs to $L_0^{1,p}(\Omega; \mu)$.

A similar construction applied to the $\mathcal{A}$-subharmonic function $v = -\hat{R}^{-f}$ provides an $\mathcal{A}$-harmonic function h_* satisfying

$$v \leq h_* \leq \underline{H}_f \leq \overline{H}_f \leq h^*$$

in Ω and $f - h_* \in L_0^{1,p}(\Omega; \mu)$.

Because $h^* - h_* \in L_0^{1,p}(\Omega; \mu)$, we have

$$\int_\Omega (\mathcal{A}(x, \nabla h^*) - \mathcal{A}(x, \nabla h_*)) \cdot (\nabla h^* - \nabla h_*)\, dx = 0$$

by (3.12), and it follows from the strict monotonicity of $\mathcal{A}$ that, in each component Ω_0 of Ω, $h^* = h_* + c$ for some constant c. To complete the proof, we show that $c = 0$. Since $\text{cap}_{p,\mu} \mathsf{C}\Omega > 0$, Ω_0 has a finite regular boundary point x_0 (Theorem 9.13) and therefore by Theorem 9.26

$$\limsup_{\substack{x \to x_0 \\ x \in \Omega_0}} h_*(x) \leq \limsup_{\substack{x \to x_0 \\ x \in \Omega_0}} h^*(x) \leq \lim_{\substack{x \to x_0 \\ x \in \Omega_0}} u(x)$$

$$= f(x_0) = \lim_{\substack{x \to x_0 \\ x \in \Omega_0}} v(x) \leq \liminf_{\substack{x \to x_0 \\ x \in \Omega_0}} h_*(x) \leq \liminf_{\substack{x \to x_0 \\ x \in \Omega_0}} h^*(x).$$

This shows that $c = 0$, and the proof is complete.   □

9.29. Corollary. If $f \in C(\overline{\Omega}) \cap H^{1,p}(\Omega; \mu)$ and if Ω is bounded, then
$$\overline{H}_f - f \in H_0^{1,p}(\Omega; \mu).$$
In particular, $\overline{H}_f$ is the unique $\mathcal{A}$-harmonic function with Sobolev boundary values f.

PROOF: By the Tietze extension theorem we may assume that $f \in C(\overline{\mathbf{R}}^n)$. Proceeding as in the proof of Theorem 9.28, we obtain that $u_i - f$ converges weakly to $\overline{H}_f - f$ in $H_0^{1,p}(\Omega; \mu)$ as $\hat{R}^f - f \in H_0^{1,p}(\Omega; \mu)$ by Remark 9.27.

An alternative proof can be based on the generalized comparison principle (Lemma 7.37). Namely, if h is the $\mathcal{A}$-harmonic function in Ω with $h - f \in H_0^{1,p}(\Omega; \mu)$, we have that
$$\lim_{x \to y} h(x) = f(y)$$
for q.e. $x \in \partial\Omega$ by the Kellogg property. Lemma 7.37 then guarantees that $h \leq u$ for each u in the upper class $\mathcal{U}_f$. Similarly we infer that $h \geq v$ for each v in the lower class $\mathcal{L}_f$. This means that
$$h \leq \overline{H}_f = \underline{H}_f \leq h,$$
as required.   □

Before the proof of Theorem 9.25 we require one more lemma, which shows that the mapping $f \mapsto \overline{H}_f$ is continuous in the topology of uniform convergence.

9.30. Lemma. If f_i are real-valued resolutive functions on $\partial\Omega$ such that $f_i \to f$ uniformly, then f is resolutive and $\lim \overline{H}_{f_i} = \overline{H}_f$.

PROOF: For given positive ε we have that $|f_i - f| < \varepsilon$ for i large enough, and hence
$$\overline{H}_f - \varepsilon \leq \overline{H}_{f_i} = \underline{H}_{f_i} \leq \underline{H}_f + \varepsilon.$$
This implies
$$\lim_{i \to \infty} \overline{H}_{f_i} = \underline{H}_f = \overline{H}_f$$
and since $\overline{H}_f$ is finite, it is also $\mathcal{A}$-harmonic. The lemma follows.   □

PROOF OF THEOREM 9.25: By the Tietze extension theorem we may assume that f is continuous on $\overline{\mathbf{R}}^n$. Furthermore, we may assume that $f(\infty) = 0$. Since f can be approximated uniformly by smooth functions φ_j with compact support in $\mathbf{R}^n$ and since φ_j's are resolutive (Theorem 9.28), the claim follows from Lemma 9.30. □

We close this section with the following simple observation.

9.31. Proposition. *Let Ω be regular. If f is bounded and lower semi-continuous on $\partial\Omega$, then f is resolutive in Ω.*

PROOF: It suffices to show that $\underline{H}_f \geq \overline{H}_f$. Let f_j be an increasing sequence of continuous functions that converges to f on $\partial\Omega$. Since Ω is regular, we have that

$$\liminf_{x \to y} \underline{H}_f(x) \geq \lim_{x \to y} \underline{H}_{f_j}(x) = f_j(y)$$

for all $y \in \partial\Omega$. Because $f_j \to f$ it follows that $\underline{H}_f$ belongs to the upper class $\mathcal{U}_f$. Hence $\underline{H}_f \geq \overline{H}_f$, as desired. □

9.32. Perron solutions and $\mathcal{A}$-potentials

In this last section of the present chapter we apply the general Perron method and provide generalizations to results in Chapter 8. First we connect $\mathcal{A}$-potentials to Perron solutions.

9.33. Theorem. *Suppose that E is a relatively closed subset of Ω and that $u \in \mathcal{S}(\Omega)$ is nonnegative. Let f be a function such that*

$$f = \begin{cases} u & \text{on } \partial E \cap \Omega \\ 0 & \text{on } \partial\Omega. \end{cases}$$

Then

$$\hat{R}_E^u(\Omega) = \overline{H}_f(\Omega \setminus E)$$

in $\Omega \setminus E$.

If, in addition, $f \in C(\overline{\Omega}) \cap L^{1,p}(\Omega; \mu)$, then $\overline{H}_f - f \in L_0^{1,p}(\Omega \setminus E; \mu)$. In particular, if Ω is bounded, then $\overline{H}_f - f \in H_0^{1,p}(\Omega \setminus E; \mu)$.

PROOF: That $\overline{H}_f \leq \hat{R}_E^u$ in $\Omega \setminus E$ is immediate. To prove the reverse inequality, pick $v \in \mathcal{U}_f$ and define

$$s = \begin{cases} \min(u, v) & \text{in } \Omega \setminus E \\ u & \text{on } E. \end{cases}$$

By the pasting lemma $s \in \mathcal{S}(\Omega)$. Thus $s \geq \hat{R}_E^u$ and hence $v \geq \hat{R}_E^u$ in $\Omega \setminus E$. This establishes the first assertion.

The remaining assertions follow from Theorem 9.26 and Corollary 9.29. □

9.34. Theorem. *Suppose that $u \in \mathcal{S}(\Omega)$ is nonnegative and $E \Subset \Omega$. Then*

$$\lim_{x \to x_0} \hat{R}^u_E(x) = 0$$

whenever x_0 is a regular boundary point of Ω.

In particular, $\lim_{x \to x_0} \hat{R}^u_E(x) = 0$ (p, μ)-q.e. on $\partial\Omega$.

PROOF: Choose a polyhedron $D \Subset \Omega$ with $E \Subset D$. Employing the Poisson modification in a neighborhood of ∂D, we find a function $v \in \Phi^u_E(\Omega)$ such that v is bounded on ∂D. Then

$$0 \le \hat{R}^u_E \le \hat{R}^v_{\overline{D}}$$

and the assertion follows because by Theorem 9.33

$$\lim_{x \to y} \hat{R}^v_{\overline{D}}(x) = 0$$

whenever $y \in \partial\Omega$ is regular, and because (p, μ)-q.e. point on $\partial\Omega$ is regular by the Kellogg property. $\square$

We close this chapter with a study of capacitary functions. We establish the existence of the capacitary function for an arbitrary condenser, provided its capacity is finite. This extends Theorem 8.6.

9.35. Theorem. *Suppose that $E \subset \Omega$ is such that $\mathrm{cap}_{p,\mu}(E, \Omega) < \infty$. Let u be the (p, μ)-potential of E in Ω, that is*

$$u = \hat{R}^1_E(\Omega; \mathcal{A}) \quad \text{with} \quad \mathcal{A}(x, \xi) = w(x)|\xi|^{p-2}\xi.$$

Then $u \in L^{1,p}_0(\Omega; \mu)$ and

$$\int_\Omega |\nabla u|^p \, d\mu = \mathrm{cap}_{p,\mu}(E, \Omega).$$

In particular, if Ω is bounded, then $u \in H^{1,p}_0(\Omega; \mu)$.

PROOF: Because the only restriction placed on the sets E and Ω is that the capacity $\mathrm{cap}_{p,\mu}(E, \Omega)$ be finite, the proof is somewhat awkward with several approximations.

Let U be an open set in Ω with $\mathrm{cap}_{p,\mu}(U, \Omega) < \infty$. Exhaust U by polyhedra $U_1 \Subset U_2 \Subset \cdots \Subset U$. Fix j and let v_j be the (p, μ)-potential $v_j = \hat{R}^1_{\overline{U}_j}(\Omega)$. We first show that

$$(9.36) \qquad \int_\Omega |\nabla v_j|^p \, d\mu = \mathrm{cap}_{p,\mu}(\overline{U}_j, \Omega).$$

By Theorem 9.33, $v_j - \varphi \in L_0^{1,p}(\Omega \setminus \overline{U}_j; \mu)$ whenever φ is admissible for the condenser $(\overline{U}_j, \Omega)$. Since v_j is (p, μ)-harmonic in $\Omega \setminus \overline{U}_j$, we have

$$\text{cap}_{p,\mu}(\overline{U}_j, \Omega) \leq \int_\Omega |\nabla v_j|^p \, d\mu \leq \int_\Omega |\nabla \varphi|^p \, d\mu \,,$$

and (9.36) follows by taking the infimum over all such φ.

Noting that

$$\int_\Omega |\nabla v_j|^p \, d\mu = \text{cap}_{p,\mu}(\overline{U}_j, \Omega) \leq \text{cap}_{p,\mu}(U, \Omega) < \infty \,,$$

we see that $\nabla v_j \to \nabla \hat{R}_U^\tau$ weakly in $L^p(\Omega; \mu)$ (Lemma 1.33). Mazur's lemma implies that for each j, $\nabla \hat{R}_U^1$ can be approximated in $L^p(\Omega; \mu)$ by gradients of functions that are admissible for the condenser $(\overline{U}_j, \Omega)$. Hence by the lower semicontinuity of norms (see 5.25)

$$\text{cap}_{p,\mu}(U, \Omega) = \lim_{j \to \infty} \text{cap}_{p,\mu}(\overline{U}_j, \Omega) \leq \int_\Omega |\nabla \hat{R}_U^1|^p \, d\mu$$

$$\leq \liminf_{j \to \infty} \int_\Omega |\nabla v_j|^p \, d\mu = \lim_{j \to \infty} \text{cap}_{p,\mu}(\overline{U}_j, \Omega)$$

$$= \text{cap}_{p,\mu}(U, \Omega) \,.$$

Thus the assertion of the theorem is proved if $E = U$ is open.

To treat the general case, suppose that E is an arbitrary subset of Ω with

$$\text{cap}_{p,\mu}(E, \Omega) < \infty \,.$$

There is no loss of generality in assuming that Ω is a domain. Let $\varphi_j \in C_0^\infty(\Omega)$ be an increasing sequence of nonnegative functions that converges pointwise to $\hat{R}_E^1$ in Ω and let $U \subset \Omega$ be a nonempty open set containing E such that $\text{cap}_{p,\mu}(U, \Omega) < \infty$. Replacing, if necessary, φ_j with $(1 - \varepsilon_j)\varphi_j$, where ε_j decreases to zero, we are free to assume that $\varphi_j < \hat{R}_U^1$ in Ω.

Fix an integer j and exhaust Ω by polyhedra $D_1 \Subset D_2 \Subset \cdots \Subset \Omega$. Because the $\mathcal{A}$-potentials $\hat{R}_{U \cap D_i}^1(\Omega)$ increase to $\hat{R}_U^1(\Omega)$ and $\mathcal{A}$-potentials are lower semicontinuous, we find an index i_j such that

$$\hat{R}_{U \cap D_{i_j}}^1(\Omega) > \varphi_j \,;$$

note that $\hat{R}_{U \cap D_i}^1(\Omega) > 0$ for large i and that the support of φ_j is compact. Now let $U_j = U \cap D_{i_j}$. Similarly, since $\hat{R}_{U_j}^1(D_k)$ increase to $\hat{R}_{U_j}^1(\Omega)$ (as $k \to \infty$), there is k_j such that

$$\hat{R}_{U_j}^1(D_k) \geq \varphi_j$$

in D_k for $k \geq k_j$. Write

$$v_{j,k} = \hat{R}^1_{U_j}(D_k),$$
$$u_j = \hat{R}^{\varphi_j}(\Omega),$$

and

$$u_{j,k} = \hat{R}^{\varphi_j}(D_k).$$

Then $v_{j,k}$ increases to $\hat{R}^1_{U_j}(\Omega)$ as $k \to \infty$, and extending $v_{j,k}$ as zero to $\Omega \setminus D_k$ we have

$$\int_\Omega |\nabla v_{j,k}|^p \, d\mu = \mathrm{cap}_{p,\mu}(U_j, D_k) \to \mathrm{cap}_{p,\mu}(U_j, \Omega) < \infty.$$

Hence we obtain from Lemma 1.33 that $\nabla v_{j,k} \to \nabla \hat{R}^1_{U_j}(\Omega)$ weakly in $L^p(\Omega; \mu)$. Moreover,

$$\lim_{k \to \infty} u_{j,k} = u_j.$$

Since $v_{j,k} - u_{j,k}$, $k \geq k_j$, is nonnegative and belongs to $H^{1,p}_0(D_k; \mu)$, the (quasi)minimizing property implies that

$$\int_\Omega |\nabla u_{j,k}|^p \, d\mu \leq \int_\Omega |\nabla v_{j,k}|^p \, d\mu.$$

Hence $\nabla v_{j,k} - \nabla u_{j,k}$ converges weakly in $L^p(\Omega; \mu)$ to $\nabla \hat{R}^1_{U_j}(\Omega) - \nabla u_j$, and consequently

$$0 \leq \int_\Omega \mathcal{A}(x, \nabla u_j) \cdot (\nabla v_{j,k} - \nabla u_{j,k}) \, dx \to \int_\Omega \mathcal{A}(x, \nabla u_j) \cdot (\nabla \hat{R}^1_{U_j} - \nabla u_j) \, dx$$

as $k \to \infty$. Thus

$$\int_\Omega |\nabla u_j|^p \, d\mu \leq \int_\Omega |\nabla \hat{R}^1_{U_j}|^p \, d\mu = \mathrm{cap}_{p,\mu}(U_j, \Omega) \leq \mathrm{cap}_{p,\mu}(U, \Omega) < \infty;$$

recall that $\mathcal{A}(x, \xi) = w(x)|\xi|^{p-2}\xi$ here. Since u_j increases to $u = \hat{R}^1_E$, it follows from Lemma 1.33 that ∇u_j converges to ∇u weakly in $L^p(\Omega; \mu)$ and that $u \in L^{1,p}_0(\Omega; \mu)$ (if Ω is bounded we have that $u \in H^{1,p}_0(\Omega; \mu)$ by Theorem 1.32). Moreover, we infer from the weak lower semicontinuity of norms that

$$\int_\Omega |\nabla u|^p \, d\mu \leq \liminf_{j \to \infty} \int_\Omega |\nabla u_j|^p \, d\mu \leq \lim_{j \to \infty} \mathrm{cap}_{p,\mu}(U_j, \Omega) = \mathrm{cap}_{p,\mu}(U, \Omega).$$

Finally, taking the infimum over all open neighborhoods U of E in Ω we obtain

$$\int_\Omega |\nabla u|^p \, d\mu \leq \mathrm{cap}_{p,\mu}(E, \Omega).$$

To prove the reverse inequality, let $0 < \varepsilon < 1$. Then $U_\varepsilon = \{u > 1 - \varepsilon\}$ is open and contains the set

$$\tilde E = \{x \in E : u(x) = 1\}.$$

Then $\mathrm{cap}_{p,\mu}(\tilde E, \Omega) = \mathrm{cap}_{p,\mu}(E, \Omega)$ by the fundamental convergence theorem. If $U \Subset U_\varepsilon$ is a polyhedron, we have by (9.36)

$$\int_\Omega |\nabla \hat R^1_{\overline U}|^p \, d\mu = \mathrm{cap}_{p,\mu}(\overline U, \Omega).$$

But since we already know that the (p, μ)-potentials u and $\hat R^1_{\overline U}$ are in $L_0^{1,p}(\Omega; \mu)$, it easily follows that

$$\min(1, \frac{u}{1 - \varepsilon}) - \hat R^1_{\overline U} \in L_0^{1,p}(\Omega \setminus \overline U).$$

Since $\hat R^1_{\overline U}$ is (p, μ)-harmonic in $\Omega \setminus \overline U$, we obtain

$$(9.37) \qquad \mathrm{cap}_{p,\mu}(\overline U, \Omega) = \int_\Omega |\nabla \hat R^1_{\overline U}|^p \, d\mu \leq (1 - \varepsilon)^{-p} \int_\Omega |\nabla u|^p \, d\mu.$$

Thus by taking the supremum over all polyhedra U in U_ε we have

$$\mathrm{cap}_{p,\mu}(E, \Omega) \leq \mathrm{cap}_{p,\mu}(U_\varepsilon, \Omega) \leq (1 - \varepsilon)^{-p} \int_\Omega |\nabla u|^p \, d\mu.$$

We now let $\varepsilon \to 0$, thus concluding the proof. $\qquad \square$

A corresponding result holds for general $\mathcal{A}$-potentials.

9.38. Theorem. *Suppose that $E \subset \Omega$ is such that $\mathrm{cap}_{p,\mu}(E, \Omega) < \infty$. If $u = \hat R^1_E(\Omega; \mathcal{A})$ is the $\mathcal{A}$-potential of E in Ω, then*

$$\mathrm{cap}_{p,\mu}(E, \Omega) \leq \int_\Omega |\nabla u|^p \, d\mu \leq (\frac{\beta}{\alpha})^p \, \mathrm{cap}_{p,\mu}(E, \Omega).$$

PROOF: We leave the proof as an exercise for the patient reader. The proof closely follows that of Theorem 9.35. When proving the estimate (9.37), let $\hat R^1_{\overline U}$ be the (p, μ)-potential (not the $\mathcal{A}$-potential) of $\overline U$ and obtain (9.37) as above. $\qquad \square$

NOTES TO CHAPTER 9. Much has been said and written about the long and interesting history of the Dirichlet problem for harmonic functions. We do not attempt to repeat it here but refer to Doob (1984) and Monna (1975). Brelot (1939) removed the early ambiguities in Perron's method (also known as the PWB method after Perron, Wiener, and Brelot) and the modern treatments more or less follow his work.

After some preliminary work, Perron's device readily applies to nonlinear problems. The first work on this seems to be by Granlund et al. (1986) although the approach there is different from the one presented here. For an earlier work, see Beckenbach and Jackson (1953). Most of this chapter is an adaptation of Kilpeläinen (1989) to the weighted case. For the unweighted nonlinear equations the Kellogg property can be derived from the work by Hedberg (1972) and Hedberg and Wolff (1983). In the situation of Theorem 8.10 a simpler proof is available, as observed by Kilpeläinen (1989). The Kellogg property, as well as other topics, in homogeneous groups has been considered by Vodop'yanov (1989). For a further discussion of the Dirichlet problem, see Kilpeläinen and Malý (1989).

For a long time it was an open problem whether the Wiener criterion characterizes regular boundary points. In the unweighted case this was proved by Lindqvist and Martio (1985) for $p > n - 1$, and recently a complete solution was given by Kilpeläinen and Malý (1992b). It is not known whether the same is true with weights; the methods in the aforementioned papers do not seem to extend to the general situation. For the weighted linear equations the answer is affirmative as shown by Fabes et al. (1982b); this also follows by using the methods in Chapter 12.

Barriers in the nonlinear potential theory have been studied by Lehtola (1986), Granlund et al. (1986), and Kilpeläinen (1989). The resolutivity of continuous functions in the borderline case $p = n$ was proved by Lindqvist and Martio (1985) and later in all (unweighted) cases by Kilpeläinen (1989).

The existence of bounded nonconstant A-superharmonic functions on manifolds has been studied by Holopainen (1990, 1992) and Holopainen and Rickman (1991). The classic reference here is Ahlfors and Sario (1960).

10

Polar sets

We have seen in the previous chapters that a set of (p, μ)-capacity zero often plays the role of an exceptional set. Such a set is removable for bounded $\mathcal{A}$-harmonic and $\mathcal{A}$-superharmonic functions. The set of irregular boundary points is of (p, μ)-capacity zero. Moreover, $H_0^{1,p}(\Omega; \mu) = H_0^{1,p}(\Omega \setminus E; \mu)$ exactly when E is a relatively closed subset of Ω with (p, μ)-capacity zero. In this chapter we give a potential theoretic characterization for those sets as polar sets. This is a well-known topic in classical potential theory.

DEFINITION. A set E in $\mathbf{R}^n$ is called $\mathcal{A}$-polar, or simply polar, if there is an open neighborhood Ω of E and an $\mathcal{A}$-superharmonic function u in Ω such that $u = \infty$ on E.

It follows from Lemma 7.10 that an $\mathcal{A}$-polar set cannot have interior points but we can say much more: the main result in this chapter asserts that E is $\mathcal{A}$-polar if and only if E has (p, μ)-capacity zero. In particular, the $\mathcal{A}$-polarity depends only on p and μ.

The second theorem implies that the above definition, a priori a local one, could be given globally. Namely, for each $\mathcal{A}$-polar set E there is an entire $\mathcal{A}$-superharmonic function u in $\mathbf{R}^n$ such that $u = \infty$ in E; besides, things can be arranged in such a way that u is finite at any prescribed point in the complement of E.

We begin by stating the principal theorems of this chapter.

10.1. Theorem. *Let E be a set in $\mathbf{R}^n$. Then the following are equivalent:*

(i) *E is $\mathcal{A}$-polar.*

(ii) *There is an open neighborhood Ω of E such that if u is nonnegative and $\mathcal{A}$-superharmonic in Ω, the balayage $\hat{R}_E^u$ vanishes identically in Ω.*

(iii) *E is of (p, μ)-capacity zero.*

(iv) *There is a nonnegative lower semicontinuous function f such that $f \in H^{1,p}(\mathbf{R}^n; \mu)$ and $f = \infty$ on E.*

10.2. Theorem. *Suppose that $E \subset \mathbf{R}^n$ is $\mathcal{A}$-polar and that $x_0 \notin E$. Then there is an $\mathcal{A}$-superharmonic function u in $\mathbf{R}^n$ such that $u = \infty$ in E and $u(x_0) < \infty$.*

10.3. Corollary. *$\mathcal{A}$-polarity depends only on p and μ, not on $\mathcal{A}$.*

10.4. Corollary. *A countable union of $\mathcal{A}$-polar sets is $\mathcal{A}$-polar.*

Our proof for 10.1 will go

$$(\mathrm{i}) \; \Rightarrow \; (\mathrm{ii}) \; \Rightarrow \; (\mathrm{iii}) \; \Rightarrow \; (\mathrm{iv}) \; \Rightarrow \; (\mathrm{i}),$$

and in the proof of the last implication we show, in effect, that Theorem 10.1(iv) implies the conclusion in Theorem 10.2. Each of the implications will be stated as an individual lemma.

10.5. Lemma. *Suppose that u is nonnegative and $\mathcal{A}$-superharmonic in Ω and that $E \subset \Omega$.*

(i) *If E is $\mathcal{A}$-polar, there is an open neighborhood V of E, $V \subset \Omega$, such that $R^u_E(V)$ has a zero in each component of V.*

(ii) *If $R^u_E(\Omega)(x) = 0$ for some $x \in \Omega$, then $\hat{R}^u_E(\Omega) \equiv 0$ in the x-component of Ω.*

PROOF: Since $0 \le \hat{R}^u_E \le R^u_E$, the minimum principle implies (ii). To prove (i), let U be an open neighborhood of E and let v be an $\mathcal{A}$-superharmonic function in U such that $v = \infty$ on E. We can choose V to be the open set $\Omega \cap \{x \in U : v(x) > 0\}$. Then $\lambda v \ge R^u_E(V)$ for each positive λ. Because v is not identically ∞ in any component of V (Lemma 7.10), the lemma follows.   □

10.6. Lemma. *Suppose that u is a positive $\mathcal{A}$-superharmonic function in Ω and that $E \subset \Omega$. If $\hat{R}^u_E(\Omega) \equiv 0$, then $\mathrm{cap}_{p,\mu} E = 0$.*

PROOF: This follows immediately from the fundamental convergence theorem 8.2.   □

10.7. Lemma. *If $\mathrm{cap}_{p,\mu} E = 0$, then there exists a nonnegative lower semicontinuous function f in $H^{1,p}(\mathbf{R}^n; \mu)$ such that $f = \infty$ in E.*

PROOF: Fix a positive integer k and write $B = B(0, k)$, $E' = E \cap \frac{1}{2}B$. Choose for each $i = 1, 2, \ldots$ an open neighborhood U_i of E' such that $U_i \subset \frac{1}{2}B$ and that

$$\mathrm{cap}_{p,\mu}(U_i, B) < i^{-2p}.$$

Let $\varphi_i = \hat{R}^1_{U_i}(B)$ be the (p, μ)-potential of U_i in B; then $\varphi_i \in H^{1,p}_0(B; \mu)$ and

$$\int_B |\nabla \varphi_i|^p \, d\mu = \mathrm{cap}_{p,\mu}(U_i, B) < i^{-2p}$$

(see Theorem 9.35). Moreover, if we set $\varphi_i = 0$ on $\complement B$, then φ_i is lower semicontinuous and by the Poincaré inequality (1.5) we have

$$\|\varphi_i\|_{1,p} \le c\, i^{-2},$$

where $c = c(n, p, c_\mu, k)$. This means that the sum $\psi_k = \sum_i \varphi_i$ has finite norm in $H^{1,p}(\mathbf{R}^n; \mu)$; moreover, ψ_k is nonnegative, lower semicontinuous, and $\psi_k = \infty$ on $E' = E \cap \frac{1}{2}B$. Finally, choose positive numbers $\lambda_1, \lambda_2, \ldots$ such that

$$f = \sum_{k=1}^{\infty} \lambda_k \psi_k$$

is the desired function. The lemma is proved.   □

We are now in the position where the implications (i) $\Rightarrow$ (ii) $\Rightarrow$ (iii) $\Rightarrow$ (iv) in Theorem 10.1 are established. Next we are going to complete the loop by proving that (iv) implies the conclusion in Theorem 10.2. The following lemma is the main step.

10.8. Lemma.   *Suppose that $u \in H^{1,p}(\Omega; \mu)$ is $\mathcal{A}$-superharmonic and that Ω is bounded. Let $M \in \mathbf{R}$, $E \subset \mathbf{R}^n$, and*

$$v = \inf\{s \in H^{1,p}(\Omega; \mu) \cap \mathcal{S}(\Omega) : s \geq u \text{ in } \Omega, \, s \geq M \text{ on } E \cap \Omega\}.$$

If there exists a nonnegative lower semicontinuous function f in $H^{1,p}(\Omega; \mu)$ such that $f = \infty$ on E, then the lower semicontinuous regularization $\hat{v}$ of v coincides with u in Ω.

PROOF: Clearly $u \leq \hat{v}$ in Ω. Write $\psi_j = u + j^{-1}f$, $j = 1, 2, \ldots$, and let v_j be the $\mathcal{A}$-superharmonic solution to the obstacle problem in Ω with both obstacle and boundary values ψ_j. Then the lower semicontinuity of ψ_j and the "ess lim inf" property of $\mathcal{A}$-superharmonic functions (Theorem 7.22) imply that $v_j \geq \psi_j$ in Ω. In particular $v_j = \infty$ on $E \cap \Omega$ so that $v_j \geq v$.

We invoke Theorem 3.79 and deduce from it that $v_0 = \lim v_j$ coincides a.e. with u, whence

$$u(x) = \operatorname*{ess\,lim\,inf}_{y \to x} u(y) = \operatorname*{ess\,lim\,inf}_{y \to x} v_0(y) \geq \liminf_{y \to x} v(y) \geq \hat{v}(x)$$

for each $x \in \Omega$. It follows that $\hat{v} = u$ as desired.   □

PROOF OF THEOREM 10.2: Suppose that the set E satisfies (iv) of Theorem 10.1 and let $x_0 \notin E$. We construct an entire $\mathcal{A}$-superharmonic function u in $\mathbf{R}^n$, finite at x_0 and ∞ on E. This proves Theorem 10.2 because we have just established that each $\mathcal{A}$-polar set E has property (iv).

Write $D_j = B(x_0, 2^j)$, $B_j = B(x_0, 2^{-j})$, and $E_j = (E \cap D_j) \setminus B_j$, $j = 1, 2, \ldots$. Then $E = \cup_j E_j$. We construct a sequence of $\mathcal{A}$-superharmonic functions u_j inductively as follows. Let $u_1 = 0$ and write

$$v_2 = \inf\{s \in H^{1,p}(D_3; \mu) \cap \mathcal{S}(D_3) : u_1 \leq s \leq 2 \text{ in } D_3, \, s = 2 \text{ on } E_2\}.$$

Then $\hat{v}_2 = u_1$ in D_3 by the previous lemma; moreover, since u_1 is $\mathcal{A}$-harmonic in B_2, also $v_2 = \hat{v}_2$ is $\mathcal{A}$-harmonic there (see the proof of Lemma 8.4). Thus we may choose an $\mathcal{A}$-superharmonic function u_2 in D_3 such that $u_1 \leq u_2 \leq 2$ in D_2, $u_2 = 2$ on E_2, and $u_2(x_0) < \frac{1}{4}$. Moreover, replacing u_2 by its Poisson modification in B_3 we may select u_2 so that it is $\mathcal{A}$-harmonic in B_3. By invoking the extension theorem 7.30 and truncating if needed, we may assume that $u_2 \leq 2$ is $\mathcal{A}$-superharmonic in all of $\mathbf{R}^n$. Then it follows that $u_2 \in H^{1,p}_{loc}(\mathbf{R}^n; \mu)$ (Corollary 7.20).

Now suppose that the functions u_1, u_2, ..., u_{j-1} have been selected for $j \geq 3$. Let

$$v_j = \inf\{s \in H^{1,p}(D_{j+1}; \mu) \cap \mathcal{S}(D_{j+1}) : u_{j-1} \leq s \leq j \text{ in } D_{j+1}, \, s = j \text{ on } E_j\}.$$

Because $\hat{v}_j = u_{j-1}$ and because u_{j-1} is $\mathcal{A}$-harmonic in B_j, we find a function u_j such that u_j is $\mathcal{A}$-superharmonic in D_{j+1}, $u_{j-1} \leq u_j \leq j$ in D_j, $u_j = j$ on E_j, and $u_j(x_0) < u_{j-1}(x_0) + 2^{-j}$. Again by replacing u_j by its Poisson modification in B_{j+1} and extending we may assume that u_j is $\mathcal{A}$-harmonic in B_{j+1} and $\mathcal{A}$-superharmonic in $\mathbf{R}^n$. Moreover, we may assume that $u_j \leq j$ and therefore that $u_j \in H^{1,p}_{loc}(\mathbf{R}^n; \mu)$.

To complete the proof, we observe that $u_{j+k} \leq u_{j+k+1}$ in D_j, $k = 1, 2, ...$, whence $u = \lim u_j$ is $\mathcal{A}$-superharmonic in $\mathbf{R}^n$. Further, $u = \infty$ on E and

$$u(x_0) \leq \sum_{j=1}^{\infty} 2^{-j} < \infty$$

as required.

The proofs for Theorems 10.1 and 10.2 are now complete.   □

As an application of Theorem 10.1 we prove the following result.

10.9. Theorem. *If u is $\mathcal{A}$-superharmonic, then u is (p, μ)-quasicontinuous.*

PROOF: First note that u is finite (p, μ)-quasieverywhere by Theorem 10.1. We may assume that u is nonnegative. Thus replacing u with $\arctan u$, we may in light of Theorem 7.5 (see also the remark after it) assume that u is bounded; therefore it belongs to $H^{1,p}_{loc}(\Omega; \mu)$. Choose an increasing sequence of continuous $\mathcal{A}$-superharmonic functions u_j in Ω which converges to u (Theorem 8.15). Let $D \Subset \Omega$ be an open set. Because the functions u_j are bounded supersolutions, it follows from the Caccioppoli estimate (3.29) that the sequence u_j is bounded in $H^{1,p}(D; \mu)$; hence it converges to u weakly in $H^{1,p}(D; \mu)$ (Theorem 1.32). Appealing to the Mazur lemma 1.29, we find for each k a sequence of convex combinations $v_{j,k}$ of the functions u_j, $j \geq k$, such that $v_{j,k} \to u$ in $H^{1,p}(D;\mu)$ as $j \to \infty$. For fixed k we can now choose a function $\psi_k \in C(D) \cap H^{1,p}(D;\mu)$ such that $u_k \leq \psi_k \leq u$ and

$$\|\psi_k - u\|_{H^{1,p}(D;\mu)} < 1/k \, .$$

Because a subsequence of $\psi_k \in C(D)$ converges to u pointwise and locally quasiuniformly by Theorem 4.3, it follows that u is quasicontinuous. □

We next show that if Ω has regular boundary points, the construction in the proof of Theorem 10.2 can be refined so as to lead to a nonnegative $\mathcal{A}$-superharmonic function in Ω, infinite in a given $\mathcal{A}$-polar set contained in Ω and finite in a prescribed point. Recall that we include ∞ in $\partial\Omega$ if Ω is unbounded. We begin with the following lemma similar to Lemma 10.8.

10.10. Lemma. *Suppose either that* $\mathrm{cap}_{p,\mu} \complement \Omega > 0$ *or that* ∞ *is a regular boundary point of each unbounded open set. Let* $f \in L^{1,p}(\Omega;\mu)$ *be bounded and lower semicontinuous and* g_j *a decreasing sequence of bounded lower semicontinuous functions in* $H^{1,p}(\mathbf{R}^n;\mu)$ *such that* $g_j \to 0$ *in* $H^{1,p}(\mathbf{R}^n;\mu)$. *If there is a compact set* $K \subset \Omega$ *such that* $\mathrm{spt}\, g_j \subset K$ *and* f *is continuous in* $\overline{\Omega} \setminus K$, *then the lower semicontinuous regularization* $\hat{v}$ *of the limit function* $v = \lim \hat{R}^{f+g_j}(\Omega)$ *coincides with the balayage* $u = \hat{R}^f(\Omega)$.

PROOF: Let $f_j = f + g_j$ and

$$v_j = \hat{R}^{f_j} \, .$$

Then $v_j \geq f_j$ since f_j is lower semicontinuous, and hence referring to Theorem 9.26 and to Remark 9.27 we infer that $v_j \in L^{1,p}(\Omega;\mu)$ and $v_j - f_j \in L_0^{1,p}(\Omega;\mu)$. Now the same argument that was used in the proof of Theorem 3.79 shows

$$(10.11) \qquad \int_\Omega \mathcal{A}(x, \nabla v) \cdot \nabla\varphi \, dx \geq 0$$

whenever $\varphi \in H_0^{1,p}(\Omega;\mu)$ is such that $\varphi \geq f - v$.

We next show that (10.11) holds with $\varphi = u - v$. For this exhaust Ω by regular open sets $D_1 \Subset D_2 \Subset \cdots \Subset \Omega$, $\cup_k D_k = \Omega$, such that $K \subset D_1$. Since f and f_1 are lower semicontinuous, it is easy to see that $\hat{R}^f(D_k)$ increases to u and $\hat{R}^{f_1}(D_k)$ increases to v_1 as $k \to \infty$. If we define both $\hat{R}^f(D_k)$ and $\hat{R}^{f_1}(D_k)$ to be f in $\Omega \setminus D_k$, then

$$\hat{R}^f(D_k) - f \in H_0^{1,p}(D_{k+i};\mu)$$

and

$$\hat{R}^{f_1}(D_k) - f \in H_0^{1,p}(D_{k+i};\mu)$$

for all $i = 0, 1, 2, \ldots$ by Theorem 9.26. Consequently, it follows from the quasiminimizing property that both sequences $\nabla \hat{R}^f(D_k)$ and $\nabla \hat{R}^{f_1}(D_k)$ are bounded in $L^p(\Omega; \mu)$. Denoting $\varphi_k = \min(v, \hat{R}^{f_1}(D_k))$ we have that $\varphi_k \to v$ in Ω, and hence $\nabla \hat{R}^f(D_k) - \nabla \varphi_k$ converges weakly in $L^p(\Omega; \mu)$ to $\nabla u - \nabla v$ as $k \to \infty$ (Lemma 1.33). Because $\hat{R}^f(D_k) - \varphi_k \geq f - v$ in Ω we infer from (10.11) via the weak convergence that

$$0 \leq \lim_{k \to \infty} \int_\Omega \mathcal{A}(x, \nabla v) \cdot \nabla\big(\hat{R}^f(D_k) - \varphi_k\big) \, dx = \int_\Omega \mathcal{A}(x, \nabla v) \cdot \nabla(u - v) \, dx \,.$$

Similarly, since $\varphi_k - \hat{R}^f(D_k)$ is nonnegative and belongs to $H_0^{1,p}(\Omega; \mu)$ and since u is a supersolution in $L^{1,p}(\Omega; \mu)$ we obtain

$$\int_\Omega \mathcal{A}(x, \nabla u) \cdot \nabla(v - u) \, dx \geq 0 \,.$$

Hence

$$0 \geq \int_\Omega \big(\mathcal{A}(x, \nabla v) - \mathcal{A}(x, \nabla u)\big) \cdot \big(\nabla v - \nabla u\big) \, dx \,.$$

Since the integrand is nonnegative a.e., it follows that

$$\int_\Omega \big(\mathcal{A}(x, \nabla v) - \mathcal{A}(x, \nabla u)\big) \cdot \big(\nabla v - \nabla u\big) \, dx = 0 \,,$$

and so $\nabla u = \nabla v$ a.e. in Ω by (3.6). Consequently, if G is a component of Ω, then $v - u = c =$ constant a.e. in G. Because there is a regular point $x_0 \in \partial G$ (Theorem 9.13), we have by Remark 9.27 that

$$0 \leq c \leq \limsup_{x \to x_0} v(x) - \lim_{x \to x_0} u(x) \leq \lim_{x \to x_0} v_j(x) - f(x_0) = 0 \,.$$

Thus $u = v$ a.e. in Ω. Because $\hat{v} = v$ a.e. by the fundamental convergence theorem, $\hat{v} = u$ a.e. and hence everywhere in Ω (Corollary 7.23). The proof is complete. $\qquad \square$

It is easy to see that Lemma 10.10 still holds if instead of assuming that f is continuous in $\overline{\Omega} \setminus K$ we assume that f is continuous in $(\Omega \cup R) \setminus K$, where R is the set of the regular boundary points of Ω.

Now we can prove:

10.12. Theorem. *Suppose either that* $\mathrm{cap}_{p,\mu} \complement\Omega > 0$ *or that* ∞ *is a regular boundary point of each unbounded open set. If* $E \subset \Omega$, *then the following are equivalent:*

(i) E *is a polar set.*

(ii) *If* $x_0 \in \Omega \setminus E$, *there exists a nonnegative* $\mathcal{A}$-*superharmonic function* u *in* Ω *with* $u = \infty$ *in* E *and* $u(x_0) < \infty$.

(iii) *If* f *is a nonnegative function in* Ω, *then* $\hat{R}_E^f = 0$.

PROOF: The implications (ii) $\Rightarrow$ (iii) and (iii) $\Rightarrow$ (i) are trivial (note that one may choose $f = \infty$ in (iii)). Thus it suffices to show that (i) implies (ii). The proof of this is similar to the corresponding assertion in Theorem 10.2. Fix a ball $B = B(x_0, r) \Subset \Omega$ and exhaust Ω by open sets $D_1 \Subset D_2 \Subset \ldots$ such that $B \Subset D_1$. Write $B_j = j^{-1}B$ and

$$E_j = E \cap (D_j \setminus B_j).$$

We construct a sequence of $\mathcal{A}$-superharmonic functions u_j inductively. Let $u_0 \equiv 0$ and suppose that $u_{j-1} \in \mathcal{S}(\Omega)$ is chosen such that $0 \le u_{j-1} \le j-1$, $u_{j-1} \in L^{1,p}(\Omega; \mu)$ is continuous in $\Omega \cup R$ and in B_j where R is the set of the regular boundary points of Ω, $u_{j-1} = j - 1$ in E_{j-1}, and

$$u_{j-1}(x_0) \le \sum_{k=0}^{j-1} 2^{-k}.$$

Now we select u_j as follows. Since $E_j \subset D_j \setminus B_j$ is $\mathcal{A}$-polar, we may choose a decreasing sequence of lower semicontinuous functions $g_{k,j} \in H^{1,p}(\mathbf{R}^n; \mu)$ such that $g_{k,j} \to 0$ in $H^{1,p}(\mathbf{R}^n; \mu)$, $0 \le g_{k,j} \le j$, $g_{k,j} = j$ in E_j, and spt $g_{k,j} \subset D_{j+1} \setminus \overline{B}_{j+1}$. If

$$v_{k,j} = \hat{R}^{u_{j-1} + g_{k,j}}(\Omega),$$

then $u_{j-1} \le v_{k,j} \le j$, $v_{k,j} - u_{j-1} \in L_0^{1,p}(\Omega; \mu)$, $v_{k,j}$ is continuous in

$$(\Omega \cup R) \setminus (D_{j+1} \setminus B_{j+1}),$$

and $v_{k,j} = j$ in E_j. Moreover, by Lemma 10.10 the lower semicontinuous regularization of $\lim_{k\to\infty} v_{k,j}$ equals u_{j-1}. Since u_{j-1} is continuous in B_{j+1}, it follows that $v_{k,j} \to u_{j-1}$ locally uniformly in B_{j+1}. Thus we may choose an index k_j such that for $u_j = v_{k_j,j}$ we have

$$u_j(x_0) \le u_{j-1}(x_0) + 2^{-j} \le \sum_{k=0}^{j} 2^{-k}.$$

Since the constructed sequence is nonnegative and increasing and since the limit $\lim_{j\to\infty} u_j = \infty$ on E, the function $u = \lim_{j\to\infty} u_j$ is the desired $\mathcal{A}$-superharmonic function as

$$u(x_0) \le \sum_{k=0}^{\infty} 2^{-k} < \infty.$$

The proof is complete. $\square$

We close this chapter by observing that the Perron method is interesting only if the complement of Ω is nonpolar. Recall that ∞ is a regular boundary point of every unbounded open set exactly when there exist nonconstant bounded $\mathcal{A}$-superharmonic functions in $\mathbf{R}^n$ (Theorem 9.22).

10.13. Theorem. *Suppose that* $\text{cap}_{p,\mu} \complement\Omega = 0$ *and let* $f : \partial\Omega \to [-\infty, \infty]$ *be a function.*

   (i) *If there are no nonconstant bounded $\mathcal{A}$-superharmonic functions in* $\mathbf{R}^n$*, then* $\overline{H}_f \equiv \sup f$ *and* $\underline{H}_f \equiv \inf f$.

   (ii) *If there exist nonconstant bounded $\mathcal{A}$-superharmonic functions in* $\mathbf{R}^n$*, then* $\overline{H}_f \equiv \underline{H}_f \equiv f(\infty)$.

PROOF: First note that each function in $\mathcal{U}_f$ can be extended uniquely to $\mathbf{R}^n$ as an $\mathcal{A}$-superharmonic function.

If all bounded $\mathcal{A}$-superharmonic functions in $\mathbf{R}^n$ are constants, the assertion follows directly from the removability theorem 7.35.

Consider then the alternative that there are nonconstant bounded $\mathcal{A}$-superharmonic functions in $\mathbf{R}^n$. We show that $\overline{H}_f = f(\infty)$, for then also

$$\underline{H}_f = -\overline{H}_{-f} = -(-f(\infty)) = f(\infty).$$

By the minimum principle $\overline{H}_f \geq f(\infty)$, so that there is no loss of generality in assuming that $f(\infty) < \infty$. Let $\lambda > f(\infty)$ and fix $x \in \Omega$. By Theorem 10.12 there exists a nonnegative $\mathcal{A}$-superharmonic function u in $\mathbf{R}^n$ such that $u = \infty$ in $\complement\Omega$ and $u(x) < \infty$. Then $\overline{H}_{f-\lambda}(x) \leq \varepsilon u(x)$ for any positive ε, and we conclude that $\overline{H}_f(x) \leq \lambda$ in Ω. Since $\lambda > f(\infty)$ was arbitrary and since $\overline{H}_f \geq f(\infty)$, assertion (ii) follows. $\square$

NOTES TO CHAPTER 10. Polar sets are the small sets of potential theory. They were introduced by Brelot (1941). Later Cartan (1945) characterized polar sets as sets of zero capacity.

In the nonlinear (unweighted) theory polar sets were first considered by Lindqvist and Martio (1988); they proved the essential parts of Theorem 10.1 in the $p = n$ case. For all $p > 1$ Theorem 10.1 appeared in Heinonen and Kilpeläinen (1988b). Theorem 10.2 was proved by Kilpeläinen (1989), while Theorem 10.12 is new even without weights. Classical results here are due to Evans, Deny, Choquet, and others. See Doob (1984, p. 795) and Helms (1969, pp. 269–271) for historical remarks.

It is not known in the nonlinear theory whether there is, for a given G_δ-set $E \subset \mathbf{R}^n$, an $\mathcal{A}$-superharmonic function u in $\mathbf{R}^n$ such that $E = \{x \in \mathbf{R}^n : u(x) = \infty\}$.

The quasicontinuity of classical superharmonic functions (Theorem 10.9) follows from Fuglede (1971a).

11
$\mathcal{A}$-harmonic measure

In classical potential theory harmonic measure provides an important tool for many applications. In this chapter we study a similar concept, termed $\mathcal{A}$-harmonic measure, in the $\mathcal{A}$-potential theory. Although the nonlinearity ruins its measure character, the $\mathcal{A}$-harmonic measure can still be used to estimate $\mathcal{A}$-harmonic functions. It is also a substitute for the ordinary harmonic measure in the theory of quasiregular mappings, as we shall see in Chapter 14.

We begin with the definition. Let E be a set on the boundary $\partial\Omega$ of Ω and let χ_E be the characteristic function of E. The upper class $\mathcal{U}_E$ of E is the upper class $\mathcal{U}_{\chi_E}$, which consists of all functions u such that

  (i) u is $\mathcal{A}$-superharmonic in Ω,
  (ii) $u \geq 0$, and
  (iii) $\liminf_{x \to y} u(x) \geq \chi_E(y)$ for all $y \in \partial\Omega$.

We recall that if Ω is unbounded, then the point at infinity is included in the boundary of Ω and (iii) is assumed to hold also for $y = \infty$.

DEFINITION. The function

$$\omega = \omega(E, \Omega; \mathcal{A}) = \overline{H}_{\chi_E} = \inf \mathcal{U}_E$$

is the $\mathcal{A}$-harmonic measure of E with respect to Ω.

Theorem 9.2 together with the facts that $1 \in \mathcal{U}_E$ and $u \geq 0$ for each $u \in \mathcal{U}_E$ implies that ω is $\mathcal{A}$-harmonic in Ω and $0 \leq \omega \leq 1$.

We also use the following convention throughout the remainder of this book: if $E \subset \overline{\Omega}$ and $E \cap \Omega$ is a relatively closed subset of Ω, we denote

$$\omega(E, \Omega; \mathcal{A}) = \omega(E \cap \partial(\Omega \setminus E), \Omega \setminus E; \mathcal{A})$$

unless a specific reference to the open set $\Omega \setminus E$ is required.

If Ω is regular and E is closed, we have alternative characterizations for the $\mathcal{A}$-harmonic measure $\omega(E, \Omega; \mathcal{A})$. Let $H(E, \Omega) = H_{\mathcal{A}}(E, \Omega)$ denote the class of all nonnegative functions $u \in C(\overline{\Omega})$ such that

  (a) u is $\mathcal{A}$-harmonic in Ω, and
  (b) $u \geq 1$ on E.

Let $S(E, \Omega) = S_{\mathcal{A}}(E, \Omega) \subset C(\overline{\Omega})$ be a similar class of $\mathcal{A}$-superharmonic functions, i.e. condition (a) is replaced by

  (a') u is $\mathcal{A}$-superharmonic in Ω.

11.1. Theorem. *Suppose that Ω is a regular open set. If E is a closed subset of $\partial\Omega$, then the A-harmonic measure $\omega = \omega(E, \Omega; A)$ can be characterized by the following equivalent ways:*

(a) $\omega = \inf\{u : u \in H(E, \Omega)\}$

(b) $\omega = \inf\{u : u \in S(E, \Omega)\}$

(c) $\omega = \lim \overline{H}_{f_i}$ *uniformly on compact subsets of Ω, where f_i is any decreasing sequence of continuous functions converging pointwise to χ_E on $\partial\Omega$.*

PROOF: Let $v_1 = \inf H(E, \Omega)$, $v_2 = \inf S(E, \Omega)$, and $v_3 = \lim \overline{H}_{f_i}$. First observe that

$$H(E, \Omega) \subset S(E, \Omega) \subset \mathcal{U}_E,$$

and so

$$\omega \leq v_2 \leq v_1.$$

Further, since Ω is regular, each of the functions $\overline{H}_{f_i}$ is actually continuous on $\overline{\Omega}$ and greater than or equal to 1 on E. Thus

$$\omega \leq v_2 \leq v_1 \leq v_3.$$

Now by Corollary 9.4, $v_3 = \omega$, and the convergence is locally uniform in Ω because the sequence $\overline{H}_{f_i}$ is decreasing and the limit function ω is continuous. This completes the proof. $\square$

In light of Corollary 9.4 the assumption that Ω is regular is superfluous in Theorem 11.1(c); for if E is closed, there is a decreasing sequence of continuous functions converging pointwise to χ_E.

In classical potential theory, i.e. in the case $A(x, \xi) = \xi$, $p = 2$, and $d\mu = dx$, harmonic measure is defined as a probability measure arising in the Dirichlet problem. Suppose first that Ω is a regular open set. Then for each $x \in \Omega$ the mapping $f \mapsto \overline{H}_f(x)$ is a positive linear functional on $C(\partial\Omega)$, and by the Riesz representation theorem there is a measure ω^x defined on Borel subsets of $\partial\Omega$ such that

$$(11.2) \qquad\qquad \overline{H}_f(x) = \int_{\partial\Omega} f \, d\omega^x.$$

It follows from Theorem 11.1 that $\omega^x(E) = \omega(E, \Omega; A)(x)$ for each closed, hence Borel, subset E of Ω. Thus the harmonic measure can be used to solve the Dirichlet problem via the representation in (11.2). Even if Ω is not regular, the mapping $f \mapsto \overline{H}_f(x)$ is a positive linear functional and representation (11.2) is valid. Indeed, the only assumption required is that

Ω be Greenian, that is the complement of Ω is nonpolar; see Doob (1984, pp. 114–115). It is not hard to see that in this case also we obtain the $\mathcal{A}$-harmonic measure $\omega(\cdot, \Omega; \mathcal{A})$ for $\mathcal{A}(x, \xi) = \xi$.

It is clear that the Riesz representation theorem can be used to define the $\mathcal{A}$-harmonic measure whenever equation (3.8) is linear. This is the case, for instance, when

$$\mathcal{A}(x, \xi) = \theta(x)\xi,$$

where $\theta(x)\xi \cdot \xi \approx w(x)|\xi|^2$ and $p = 2$. These measures are often called *elliptic harmonic measures*. In the $\mathcal{A}$-potential theory the mapping $f \mapsto \overline{H}_f(x)$ is, in general, nonlinear and no representation like (11.2) is possible.

We next establish some basic properties of $\mathcal{A}$-harmonic measures.

11.3. Theorem.

(a) (Monotonicity) If $E_1 \subset E_2 \subset \partial\Omega$, then

$$\omega(E_1, \Omega; \mathcal{A}) \le \omega(E_2, \Omega; \mathcal{A}).$$

(b) (Carleman's principle) If $E \subset \partial\Omega_1 \cap \partial\Omega_2$ and if $\Omega_1 \subset \Omega_2$, then

$$\omega(E, \Omega_1; \mathcal{A}) \le \omega(E, \Omega_2; \mathcal{A})$$

in Ω_1.

(c) If $C_1 \supset C_2 \supset \ldots$ are closed subsets of $\partial\Omega$ and $C = \bigcap_i C_i$, then

$$\lim_{i \to \infty} \omega(C_i, \Omega; \mathcal{A}) = \omega(C, \Omega; \mathcal{A})$$

uniformly on compact subsets of Ω.

PROOF: Property (a) follows immediately from the definition. Property (c) follows from Theorem 11.1 and the remark after its proof.

To prove (b), let u belong to the upper class $\mathcal{U}_E(\Omega_2)$ with respect to Ω_2. Then $u|_{\Omega_1}$ belongs to the upper class $\mathcal{U}_E(\Omega_1)$ with respect to Ω_1, and hence for every $x \in \Omega_1$

$$\omega(E, \Omega_1; \mathcal{A})(x) \le \inf\{u(x) \colon u \in \mathcal{U}_E(\Omega_2)\} = \omega(E, \Omega_2; \mathcal{A})(x),$$

and (b) follows. $\square$

Although the $\mathcal{A}$-harmonic measure need not be subadditive, it behaves well with respect to complements, at least for nice sets.

11.4. Theorem. *If* $E \subset \partial\Omega$, *then*

$$\omega(E, \Omega; A) \geq 1 - \omega(\partial\Omega \setminus E, \Omega; A).$$

Moreover,

$$\omega(E, \Omega; A) = 1 - \omega(\partial\Omega \setminus E, \Omega; A)$$

if and only if the characteristic function of E *is resolutive in* Ω.

PROOF: Let $f = \chi_E$ be the characteristic function of E. Then

$$\omega(\partial\Omega \setminus E, \Omega; A) = \overline{H}_{1-f} = 1 - \underline{H}_f$$
$$\geq 1 - \overline{H}_f = 1 - \omega(E, \Omega; A),$$

and the equality holds if and only if $\overline{H}_f = \underline{H}_f$.   □

Because bounded semicontinuous functions are resolutive in regular open sets by Proposition 9.31, we obtain

11.5. Corollary. *If* Ω *is regular and* $K \subset \partial\Omega$ *is compact, then*

$$\omega(K, \Omega; A) = 1 - \omega(\partial\Omega \setminus K, \Omega; A).$$

Often the $\mathcal{A}$-harmonic measure $\omega(E, \Omega; A)$ is vaguely described as an $\mathcal{A}$-harmonic function with boundary values 1 on E and 0 on $\partial\Omega \setminus E$. Although this is not generally true, it is useful to observe the following boundary behavior of $\omega(E, \Omega; A)$. The theorem is a direct consequence of Lemma 9.6.

11.6. Theorem. *Let* x_0 *be a regular boundary point of* Ω. *If* x_0 *has a neighborhood* V *such that* $V \cap \partial\Omega \subset E$, *then*

$$\lim_{x \to x_0} \omega(E, \Omega; A)(x) = 1.$$

Similarly, if x_0 *has a neighborhood* V *such that* $V \cap \partial\Omega \cap E = \emptyset$, *then*

$$\lim_{x \to x_0} \omega(E, \Omega; A)(x) = 0.$$

We recall some simple criteria which will be sufficient to conclude that x_0 is a regular boundary point of Ω. For instance, if Ω is piecewise smooth, say a polyhedron, then Ω is always regular (Corollary 6.32). Further, if $d\mu = dx$, then for $p > n - 1$ it suffices to have a nondegenerate continuum $C \subset \complement\Omega$ with $x_0 \in C$ (Maz'ya 1985, p. 392), and for $p > n$ every point $x_0 \in \partial\Omega$ is of positive p-capacity and, therefore, regular. Lemma 11.21

below provides a quantitative estimate for the boundary behavior in the situation of Theorem 11.6.

Next we relate the $\mathcal{A}$-harmonic measure to the $\mathcal{A}$-potential and to the capacitary function of a condenser. Suppose that E is a relatively closed subset of Ω and denote

$$\omega = \omega(E, \Omega; \mathcal{A}) = \omega(E \cap \partial(\Omega \setminus E), \Omega \setminus E, \mathcal{A}).$$

Set $\omega(x) = 1$ for $x \in E$ and let $\hat{\omega}$ be the lower semicontinuous regularization of ω in Ω. If u belongs to the upper class of E in $\Omega \setminus E$, then after putting $u = 1$ on E it follows from the pasting lemma that $\min(u, 1)$ is $\mathcal{A}$-superharmonic in Ω, and we obtain from Theorem 9.33 the following result:

11.7. Theorem. *The function $\hat{\omega}$ equals $\hat{R}_E^1$, the $\mathcal{A}$-potential of E in Ω. In particular, $\omega = \hat{R}_E^1$ in $\Omega \setminus E$.*

Now we have by Theorem 9.35 and by Theorem 11.7 the following:

11.8. Corollary. *Suppose that $\mathcal{A}(x, \xi) = |\xi|^{p-2}\xi w(x)$ and that E is a compact subset of Ω. Then the (p, μ)-capacity of the condenser (E, Ω) is obtained by*

$$\operatorname{cap}_{p,\mu}(E, \Omega) = \int_{\Omega \setminus E} |\nabla \omega|^p \, d\mu,$$

where $\omega = \omega(E, \Omega; \mathcal{A})$ is the $\mathcal{A}$-harmonic measure of E in Ω.

The $\mathcal{A}$-harmonic measure is the minimal nonnegative $\mathcal{A}$-harmonic function with "boundary values" 1 on the E. This extremal property is useful in estimating other $\mathcal{A}$-harmonic or $\mathcal{A}$-superharmonic functions. A typical example is the following estimate which yields, for instance, the two-constant theorem for quasiregular mappings (see Chapter 14).

11.9. Theorem. *Suppose that $E \subset \partial\Omega$ and let $\omega = \omega(E, \Omega; \mathcal{A})$. If v is an $\mathcal{A}$-subharmonic function in Ω with*

$$\limsup_{x \to y} v(x) \leq \begin{cases} M & \text{if } y \in E \\ m & \text{if } y \in \partial\Omega \setminus E, \end{cases}$$

where $M \geq m$, then

$$v(x) \leq (M - m)\omega(x) + m$$

for all $x \in \Omega$.

PROOF: If $M = m$, then the claim $v \leq M$ follows directly from the comparison principle. If $M > m$, then the function $v_1 = (v - m)(M - m)^{-1}$ is $\mathcal{A}$-subharmonic in Ω with

$$\limsup_{x \to y} v_1(x) \leq \begin{cases} 1 & \text{if } y \in E \\ 0 & \text{if } y \in \partial\Omega \setminus E. \end{cases}$$

Therefore, if u is any function from the upper class $\mathcal{U}_E$, then

$$\limsup_{x \to y} v_1(x) \le \liminf_{x \to y} u(x)$$

for each $y \in \partial\Omega$. Thus by the comparison principle $v_1 \le u$ in Ω for all u in $\mathcal{U}_E$, and we obtain the required inequality $v_1 \le \omega$. $\square$

Theorem 11.9 implies:

11.10. Corollary. *Suppose that v is A-subharmonic and bounded above in Ω. If $\omega(E, \Omega; A) = 0$ and if*

$$\limsup_{x \to y} v(x) \le m$$

for all $y \in \partial\Omega \setminus E$, then $v \le m$ in Ω.

For linear equations Corollary 11.10 provides an improved version of the comparison principle: sets of zero harmonic measure can be neglected when comparing solutions on the boundary. It is not known whether this is true in general.

We illustrate the use of Theorem 11.9 by establishing the Phragmén–Lindelöf principle for A-subharmonic functions. This principle lies behind various geometric and analytic estimates which together constitute Phragmén–Lindelöf type theorems. (Observe that a similar reasoning was used in Theorem 7.40.)

Fix $x_0 \in \Omega$, and for $R > 0$ let $\Omega_R = \Omega \cap B(x_0, R)$, $E_R = \partial B(x_0, R) \cap \partial\Omega_R$. Then Ω_R is a bounded open set and E_R a compact subset of $\partial\Omega_R$. Write further $\omega_R = \omega(E_R, \Omega_R; A)$.

11.11. The Phragmén–Lindelöf principle. *Suppose that v is A-subharmonic in an unbounded domain Ω with $\limsup_{x \to y} v(x) \le 0$ for each $y \in \partial\Omega \cap \mathbf{R}^n$. Then either $v \le 0$ in Ω or*

$$M(R) = \sup\{v(x) : x \in \partial B(x_0, R) \cap \Omega\}$$

grows so fast that

$$(11.12) \qquad\qquad \liminf_{R \to \infty} M(R) \, \omega_R(x) > 0$$

for each $x \in \Omega$.

PROOF: Suppose that $v(y_0) > 0$ for some $y_0 \in \Omega$. We deduce from the maximum principle and from the condition $\limsup_{x \to y} v(x) \le 0$ for $y \in \partial\Omega \cap \mathbf{R}^n$ that

$$M(R) = \sup\{v(x) : x \in \Omega_R\}.$$

Hence Theorem 11.9 yields $v \leq M(R)\,\omega_R$ for R so large that $y_0 \in B(x_0, R)$. Thus (11.12) follows for $x = y_0$.

Next if $x \in \Omega$, then for $R \geq R_x$ Harnack's inequality gives a constant c which is independent of R such that $\omega_R(y_0) \leq c\omega_R(x)$. Hence (11.12) follows for each x in Ω.   □

In certain situations some quantitative estimates for $M(R)$ in Theorem 11.11 are available; see Granlund *et al.* (1985), Lindqvist (1985), and Heinonen *et al.* (1989b). In the classical harmonic case there is an extensive literature of growth estimates of subharmonic functions in unbounded domains; see Hayman and Kennedy (1976) and Hayman (1990).

11.13. Sets of $\mathcal{A}$-harmonic measure zero

DEFINITION. A set $E \subset \partial\Omega$ is of *$\mathcal{A}$-harmonic measure zero* (*with respect to* Ω) if $\omega = \omega(E, \Omega; \mathcal{A}) \equiv 0$.

Note that by the minimum principle $\omega = 0$ if and only if each component of Ω contains a point x with $\omega(x) = 0$.

For general $\mathcal{A}$ it is a difficult and largely open problem to determine which sets are of $\mathcal{A}$-harmonic measure zero. The following criterion is far from being necessary.

11.14. Theorem. *Suppose that* $\mathrm{cap}_{p,\mu}\,\complement\Omega > 0$ *or that* ∞ *is a regular boundary point of each unbounded open set. If* $E \subset \partial\Omega$ *is $\mathcal{A}$-polar, then* $\omega(E, \Omega; \mathcal{A}) = 0$.

PROOF: Fix $x \in \Omega$. Since E is $\mathcal{A}$-polar, it is contained in a Borel $\mathcal{A}$-polar set, and by the monotonicity of $\mathcal{A}$-harmonic measure we may assume that E itself is a Borel set. Thus by Choquet's capacitability theorem (Theorem 2.5) there is an open set Ω_1 containing $\Omega \cup E$ such that $\mathrm{cap}_{p,\mu}\,\complement\Omega_1 > 0$ if $\mathrm{cap}_{p,\mu}\,\complement\Omega > 0$. In the other case we may choose $\Omega_1 = \mathbf{R}^n$. By Theorem 10.12 there is a nonnegative $\mathcal{A}$-superharmonic function u in Ω_1 such that $u = \infty$ on E and $u(x) < \infty$. Now the restrictions $\lambda u|_\Omega$ belong to the upper class $\mathcal{U}_E$ whenever $\lambda > 0$. Consequently, $0 \leq \omega(x) \leq \lambda u(x)$, and the desired result follows by letting $\lambda \to 0$.   □

In classical potential theory a set E on the boundary of the unit ball of $\mathbf{R}^n$ is of harmonic measure zero if and only if the $(n-1)$-measure of E is zero. $\mathcal{A}$-polar sets are, in general, much thinner. For example, if E is polar in the classical theory, then the Hausdorff dimension of E is at most $n-2$.

It is important to notice, however, that in Theorem 11.14 the conclusion $\omega = 0$ is essentially independent of Ω, and in this respect Theorem 11.14 is the best possible. To be more precise, consider a bounded open set Ω and a compact subset E of Ω. By Theorem 11.7 the $\mathcal{A}$-harmonic measure

$\omega = \omega(E, \Omega; \mathcal{A})$ coincides with the $\mathcal{A}$-potential $\hat{R}_E^1(\Omega; \mathcal{A})$ in $\Omega \setminus E$. It then follows from Theorem 10.1 and Theorem 11.14 that $\omega = 0$ if and only if E is $\mathcal{A}$-polar or, equivalently, E is of (p, μ)-capacity zero. Hence Theorem 11.14 cannot be improved in this situation. In the next theorem, after an appropriate definition, we rephrase this more succinctly.

DEFINITION. A set E in $\mathbf{R}^n$ is of *absolute $\mathcal{A}$-harmonic measure zero* if

$$\omega(E \cap \partial\Omega, \Omega; \mathcal{A}) = 0$$

for all bounded open sets Ω.

11.15. Theorem.   *A Borel set E is of absolute $\mathcal{A}$-harmonic measure zero if and only if it is $\mathcal{A}$-polar.*

PROOF: Suppose first that E is $\mathcal{A}$-polar, and let Ω be a bounded open set. As a subset of an $\mathcal{A}$-polar set, the set $E \cap \partial\Omega$ is $\mathcal{A}$-polar and hence by Theorem 11.14

$$\omega(E \cap \partial\Omega, \mathcal{A}; \Omega) = 0$$

as required.

To prove the converse, let E be of absolute $\mathcal{A}$-harmonic measure zero and let K be a compact subset of E. Fix a ball B such that K is contained in B. Then $\Omega = B \setminus K$ is open and $K_1 = K \cap \partial\Omega \subset E \cap \partial\Omega = E_1$. Since E is of absolute $\mathcal{A}$-harmonic measure zero, we obtain from the monotonicity of $\mathcal{A}$-harmonic measures that

$$\omega(K_1, \Omega; \mathcal{A}) \leq \omega(E_1, \Omega; \mathcal{A}) = 0 \,.$$

But now we are back in the situation of Theorem 11.7 which shows that the $\mathcal{A}$-potential of K in B vanishes. It then follows that $\text{cap}_{p,\mu} K = 0$ (Theorem 10.1), and since K was an arbitrary compact subset of E, E has the inner (p, μ)-capacity zero. Because E is a Borel set, and hence capacitable by the theorem of Choquet, we have that $\text{cap}_{p,\mu} E = 0$. Thus E is $\mathcal{A}$-polar (Theorem 10.1). $\qquad\square$

In particular, Theorem 11.15 together with Theorem 10.1 implies that Borel sets of absolute $\mathcal{A}$-harmonic measure zero are exactly the sets of (p, μ)-capacity zero. Thus to be of absolute $\mathcal{A}$-harmonic measure zero is *independent of $\mathcal{A}$*; this concept depends *only on p and the weight w*. For sets of $\mathcal{A}$-harmonic measure zero the situation is not so simple in general; it is possible to construct sets E on the boundary of the unit ball of $\mathbf{R}^n$ and linear operators div $\mathcal{A}$ with $p = 2$ and $d\mu = dx$ such that E has zero $(n-1)$-measure, and hence E is of zero harmonic measure in the classical case, while at the same time $\omega(E, B; \mathcal{A}) > 0$; see Caffarelli *et al.* (1981).

An interesting problem is to characterize those linear operators div $\mathcal{A}$ for which the associated $\mathcal{A}$-harmonic measure is absolutely continuous with respect to the surface measure on the boundary of a ball; for the latest developments here, see Fefferman *et al.* (1991) and the references therein.

Even though it is possible for a set to be of zero $\mathcal{A}$-harmonic measure for one $\mathcal{A}$ and positive for another $\mathcal{A}^*$ of the same type, it is still reasonable to seek conditions which guarantee that $\omega(E, \Omega; \mathcal{A}) = 0$ for all $\mathcal{A}$. We turn our attention to sets of zero $\mathcal{A}$-harmonic measure on the boundary of a regular open set Ω. We recall again that if Ω is unbounded, then $E \subset \partial\Omega$ may contain the point at infinity. A simple but useful lemma is needed.

11.16. Lemma. *Suppose that Ω is regular and that E is a closed subset of $\partial\Omega$. Then $\omega = \omega(E, \Omega; \mathcal{A}) = 0$ if and only if*

$$\sup_{x \in \Omega} \omega(x) < 1.$$

PROOF: Write $\lambda = \sup_{x \in \Omega} \omega(x)$. Since Ω is regular at each boundary point $y \in \partial\Omega$ and since E is closed, Theorem 11.6 yields

$$\lim_{x \to y} \omega(x) = 0$$

for every $y \in \partial\Omega \setminus E$. Now we can apply Theorem 11.9 with $v = \omega$, $M = \lambda$, and $m = 0$, and conclude that $\omega \le \lambda\omega$. If $\lambda < 1$, this is possible only if $\omega = 0$. $\qquad\square$

If the mapping

$$E \mapsto \omega^x(E) = \omega(E, \Omega; \mathcal{A})(x)$$

is a measure for fixed $x \in \Omega$ and if $\omega^x(E_i) = 0$ for $i = 1, 2, \ldots$, then also $\omega^x(\cup E_i) = 0$. For nonlinear equations very little is known in this respect. However, the following result can be established.

11.17. Theorem. *Suppose that Ω is regular and that E_1, $E_2 \subset \partial\Omega$ are closed sets of $\mathcal{A}$-harmonic measure zero. If $E_1 \cap E_2 = \emptyset$, then $E_1 \cup E_2$ is of $\mathcal{A}$-harmonic measure zero.*

PROOF: Recall that topological notions on $\partial\Omega$ are taken with respect to the extended space $\overline{\mathbf{R}}^n$.

By considering the components of Ω separately, we may assume that Ω is connected. Write $\omega = \omega(E_1 \cup E_2, \Omega; \mathcal{A})$ and suppose that $\omega > 0$. We first show that $E_1 \cup E_2 \ne \partial\Omega$. Noting that the theorem is trivially true if $E_1 = \emptyset$, we pick $x_0 \in E_1$. Suppose that $E_1 \cup E_2 = \partial\Omega$. Since $E_1 \cap E_2 = \emptyset$, there is a neighborhood U of x_0 such that $U \cap \partial\Omega \subset E_1$, and an appeal to Theorem 11.6 shows that $\lim_{x \to x_0} \omega(E_1, \Omega; \mathcal{A}) = 1$. This contradicts the assumption $\omega(E_1, \Omega; \mathcal{A}) = 0$, and thus $E_1 \cup E_2 \neq \partial\Omega$.

For $0 < t < 1$ consider the open set $A_t = \{x \in \Omega : \omega(x) > t\}$. By Lemma 11.16 we have that $\sup \omega = 1$ and hence $A_t \neq \emptyset$. Because ω satisfies the maximum principle and because $\lim_{x \to x_0} \omega(x) = 0$ for every $x_0 \in \partial\Omega \setminus (E_1 \cup E_2)$ by Theorem 11.6, it is easy to see that for t close enough to 1 there is a component A of A_t such that either $\overline{A} \subset \Omega \cup E_1$ or $\overline{A} \subset \Omega \cup E_2$. Assume, for instance, that $\overline{A} \subset \Omega \cup E_1$. Write

$$v(x) = \begin{cases} \omega(x) - t & \text{if } x \in A \\ 0 & \text{if } x \in \Omega \setminus A. \end{cases}$$

Now v is $\mathcal{A}$-subharmonic in Ω by the pasting lemma 7.9, and it satisfies

$$\limsup_{x \to y} v(x) \leq \begin{cases} 0 & \text{if } y \in \partial\Omega \setminus E_1 \\ 1 & \text{if } y \in E_1. \end{cases}$$

To complete the argument we apply Theorem 11.9 with $M = 1$, $m = 0$, and conclude that $v \leq \omega(E_1, \Omega; \mathcal{A}) = 0$. This means that $A = \emptyset$, a contradiction, and the theorem follows. □

It is not known whether the conclusion of Theorem 11.17 holds if the sets E_1 and E_2 intersect. However, Avilés and Manfredi (1992) have proved a result in this direction for the unweighted p-Laplacian.

Next we seek metric conditions for sets of $\mathcal{A}$-harmonic measure zero. We begin with the following lemma which is a slight generalization of Lemma 11.16.

11.18. Lemma. *Suppose that Ω is regular and that $E \subset \partial\Omega$ is closed. Then $\omega = \omega(E, \Omega; \mathcal{A}) = 0$ if and only if there is a sequence of neighborhoods U_i of E and a constant $\lambda < 1$ such that*

(i) *$\bigcap_i U_i \cap \Omega = \emptyset$, and*

(ii) *for each i and $x \in \Omega \cap \partial U_i$, $\omega(x) \leq \lambda$.*

PROOF: Since the condition is clearly necessary, only the converse calls for a proof. By Lemma 11.16 it suffices to show that

$$(11.19) \qquad\qquad \omega(x) \leq \lambda$$

for each $x \in \Omega$. Fix $x \in \Omega$ and choose an index i such that $x \notin U_i$. If $x \in \partial U_i$, then (11.19) is part of the hypothesis, and hence we may assume that $x \in \Omega \setminus \overline{U}_i$. Next, let V be the x-component of $\Omega \setminus \overline{U}_i$ and let $y \in \partial V$. If $y \in \Omega$, then $y \in \partial U_i$ and hence $\omega(y) \leq \lambda$; if $y \notin \Omega$, then $y \in \partial\Omega \setminus E$ and by Theorem 11.6 we have $\lim_{z \to y} \omega(z) = 0$. Consequently, $\lim_{x \to y} \omega(x) \leq \lambda$ for each $y \in \partial V$ and thus the comparison principle yields $\omega \leq \lambda$ in V. In particular $\omega(x) \leq \lambda$ and (11.19) follows. □

Recall that $\complement\Omega$ satisfies a (p, μ)-capacity density condition, or is uniformly (p, μ)-thick, if for some $c_0 > 0$ and $r_0 > 0$ it holds that

$$(11.20) \quad \mathrm{cap}_{p,\mu}(\overline{B}(x_0, r) \cap \complement\Omega, B(x_0, 2r)) \geq c_0 \, \mathrm{cap}_{p,\mu}(\overline{B}(x_0, r), B(x_0, 2r))$$

for all $x_0 \in \partial\Omega$, $x_0 \neq \infty$, and $r \leq r_0$. This condition was already employed in Chapter 6. If $\complement\Omega$ is uniformly (p, μ)-thick, then every finite boundary point of Ω is regular. It is good to keep in mind that $\complement\Omega$ satisfies (11.20), for instance, if it has a corkscrew in a uniform sense at each finite boundary point (see Chapter 6). In particular, complements of polyhedra are always uniformly (p, μ)-thick. In the special case $d\mu = dx$ and $p > n - 1$ there is a particularly amenable sufficient criterion for the uniform thickness: $\complement\Omega$ is uniformly thick if each boundary component is a continuum with diameter greater than some fixed positive constant (see Maz'ya 1985, p. 392).

The (p, μ)-capacity density condition can be used to estimate ω both from above and from below in a neighborhood of a boundary point x_0.

11.21. Lemma. *Suppose that Ω satisfies the (p, μ)-capacity density condition with constants c_0 and r_0. Let $E \subset \partial\Omega$ and $x_0 \in \partial\Omega$, $x_0 \neq \infty$. If for some r, $0 < r \leq r_0$, $B(x_0, 2r)$ does not intersect E, then*

$$\omega(x) = \omega(E, \Omega; \mathcal{A})(x) \leq \lambda < 1$$

for each $x \in B(x_0, r) \cap \Omega$. Furthermore, if $B(x_0, 2r) \cap \partial\Omega \subset E$, then

$$\omega(x) \geq \delta > 0$$

for each $x \in B(x_0, r) \cap \Omega$. The constants λ and δ depend only on n, p, β/α, c_μ, and c_0.

PROOF: We prove the first inequality only; the proof for the second assertion is analogous. Set $v = 1 - \omega$. Then v is $\mathcal{A}$-harmonic in Ω and $0 \leq v \leq 1$. Now all finite boundary points of Ω are regular, and hence if we set $v(x) = 1$ for x in $\complement\Omega \cap B(x_0, 2r)$, then v is continuous and $\mathcal{A}$-superharmonic in $B(x_0, 2r)$ by Theorem 11.6 and by the pasting lemma.

Next let $C = \complement\Omega \cap \overline{B}(x_0, r)$ and let $u = \hat{R}_C^1(B(x_0, 2r); \mathcal{A})$ be the $\mathcal{A}$-potential of C in $B(x_0, 2r)$. Then $v \geq u$ in $B(x_0, 2r)$, and thus Lemma 6.21 implies

$$(11.22) \qquad v(x) \geq u(x) \geq c_1 \left(\frac{\mathrm{cap}_{p,\mu}(C, 2B)}{\mathrm{cap}_{p,\mu}(B, 2B)} \right)^{1/(p-1)}$$

for a.e. $x \in B = B(x_0, r)$, where the constant c_1 depends only on n, p, β/α, and c_μ. Since v is continuous in Ω and since $\complement\Omega$ is uniformly (p, μ)-thick, we obtain

$$1 - \omega(x) = v(x) \geq c_1 c_0^{1/(p-1)} > 0$$

for each $x \in B \cap \Omega$, as desired. $\qquad\square$

We introduce the quasihyperbolic metric in a domain and use it to give metric conditions for sets of $\mathcal{A}$-harmonic measure zero. The quasihyperbolic metric is closely related to the hyperbolic (Poincaré) metric in planar domains; in fact, it easily follows from the Koebe one-quarter theorem that these two metrics are comparable in any simply connected proper subdomain in the plane.

DEFINITION. Let D be an open, proper subset of $\mathbf{R}^n$. The *quasihyperbolic distance* of x and y in D is the number

$$k_D(x,y) = \inf \int_\gamma d(z, \partial D)^{-1} ds,$$

where the infimum is taken over all rectifiable curves γ joining x to y in D; if no such curve exists, then x and y belong to different components of D, and we set $k_D(x,y) = \infty$. Here $d(z,A) = \mathrm{dist}(x,A)$ is the distance of the point z from the set A.

It is easy to see that k_D defines a complete metric in each component of D and that this metric generates the usual Euclidean topology.

11.23. Theorem. *Let Ω be a bounded domain whose complement is uniformly (p,μ)-thick and let E be a closed subset of $\partial\Omega$. Suppose that there exist $M < \infty$ and a sequence of neighborhoods U_i, $i = 1, 2, ...$, of E such that*

(a) $\bigcap_i U_i \cap \Omega = \emptyset$, *and*

(b) *for each i and $x \in \partial U_i \cap \Omega$ there is $y \in \partial\Omega$ with $k_{\complement E}(x,y) \leq M$.*

Then $\omega = \omega(E, \Omega; \mathcal{A}) = 0$.

PROOF: The ingredients of the proof are: Harnack's inequality, the potential estimate (6.22), and Lemma 11.18. Since Ω is bounded, we may assume that the (p,μ)-capacity density condition (11.20) holds for all $r < \mathrm{diam}(\Omega)$. We show that

(11.24) $\omega(x) \leq \lambda < 1$

for each $x \in \partial U_i \cap \Omega$, $i = 1, 2, ...$ Then Lemma 11.18 yields the conclusion $\omega = 0$.

To this end, fix i and let $x \in \partial U_i \cap \Omega$. Write $D = \complement E$ and choose $y \in \partial\Omega \setminus E$ with $k_D(x,y) \leq M$. Let γ be a rectifiable curve in D joining x to y with

(11.25) $\int_\gamma d(z, \partial D)^{-1} ds \leq M + 1.$

Next choose points z_1, z_2, ..., z_k on γ and radii r_1, r_2, ..., r_k inductively as follows. Set $z_1 = x$ and $r_1 = d(z_1, E)/4$. Assume that z_1, z_2, ..., z_j and r_1, r_2, ..., r_j have been chosen and let γ_j denote the part of γ from z_j to y. If $\partial\Omega \cap \overline{B}(z_j, r_j/2) \neq \emptyset$, we set $k = j$ and stop. If $\partial\Omega \cap \overline{B}(z_j, r_j/2) = \emptyset$, then we choose z_{j+1} to be the last point where γ_j meets the sphere $\partial B(z_j, r_j/4)$ and let $r_{j+1} = d(z_{j+1}, E)/4$. Since $y \in \partial\Omega \setminus E$, this process ends after a finite number, say k, of steps.

We obtain an upper bound for k in terms of M. Assume that $k > 2$. Fix $j = 1, 2, ..., k-1$ and let γ_j be the part of γ from z_j to z_{j+1}. Pick $z' \in E$ such that

$$4r_j = d(z_j, E) = |z_j - z'|.$$

Then for $z \in \gamma_j \cap B(z_j, r_j)$ we have

$$d(z, E) \leq |z - z'| \leq |z - z_j| + |z_j - z'| \leq r_j + 4r_j = 5r_j$$

and thus

$$\int_{\gamma_j} d(z, E)^{-1}\, ds \geq \int_{\gamma_j \cap B(z_j, r_j/4)} d(z, E)^{-1}\, ds \geq \frac{r_j}{20r_j} = \frac{1}{20}.$$

Therefore

$$\int_{\gamma} d(z, E)^{-1}\, ds \geq \sum_{j=1}^{k-1} \int_{\gamma_j} d(z, E)^{-1}\, ds \geq \frac{k-1}{20}$$

and we obtain from (11.25) that

$$k \leq 20M + 21.$$

By the construction there is $x_0 \in \partial\Omega \cap \overline{B}(z_k, r_k/2)$. Moreover,

$$B(x_0, 3r_k) \cap E = \emptyset$$

because $4r_k = d(z_k, E)$. It then follows from Lemma 11.21 that

$$(11.26) \qquad v(z) = 1 - \omega(z) \geq c_1 > 0$$

for each $z \in B(x_0, 3r_k/2) \cap \Omega$. Consequently, (11.26) holds for each $z \in B(z_k, r_k) \cap \Omega$.

To complete the proof, set $B_j = B(z_j, r_j/4)$ for $j = 1, 2, ..., k$. Then for $j \leq k - 1$, $2B_j \subset \Omega$, whence Harnack's inequality together with (11.26) yields

$$c_1 \leq \inf_{B_k \cap \Omega} v \leq \sup_{B_{k-1}} v \leq c_2 \inf_{B_{k-1}} v \leq \cdots \leq c_2^{k-1} \inf_{B_1} v,$$

where $c_2 = c_2(n, p, \beta/\alpha, c_\mu)$. Therefore, since $k \leq 20M + 21$, we obtain

$$\omega(x) = 1 - v(x) \leq 1 - \inf_{B_1} v \leq 1 - c_1 c_2^{1-k} \leq 1 - c_1 c_2^{-20(M+1)} < 1.$$

This is the required estimate (11.24) and the proof is complete. $\square$

Theorem 11.23 can be applied to several practical situations. To give a specific example, we consider porous sets on the boundary of a ball.

A set E on the boundary ∂B of the unit ball $B = B(0,1)$ is said to be δ-porous with respect to ∂B, $\delta > 0$, if there is a sequence r_i of positive numbers converging to 0 such that for each $x \in E$ and $i = 1, 2, \ldots$ the ball $B(x, r_i)$ contains a ball $B(y, \delta r_i)$ with $y \in \partial B$ and $B(y, \delta r_i) \cap E = \emptyset$.

11.27. Theorem. *Suppose that E is δ-porous with respect to the boundary of the unit ball B. Then $\omega = \omega(E, B; \mathcal{A}) = 0$.*

PROOF: First note that the complement of B satisfies a (p, μ)-capacity density condition. Because E is δ-porous, $\overline{E}$ is δ-porous as well, and we may assume that E is compact.

In order to use Theorem 11.23, let $U_i = E + B(0, r_i)$, $i = 1, 2, \ldots$; that is,

$$U_i = \bigcup_{x \in E} B(x, r_i),$$

where the sequence r_i is as in the porosity condition. We may clearly assume that $r_i \leq \frac{1}{4}$ for each i. Then U_i is a neighborhood of E and $\cap U_i = E \subset \partial B$. Next fix i and let $x \in \partial U_i \cap B$. Since E is compact, this means that $x \in \partial B(x_0, r_i)$ for some $x_0 \in E$ and that $d(x, E) = r_i$. The porosity condition gives a point $y \in \partial B$ such that $B(y, \delta r_i) \cap E = \emptyset$ and $B(y, \delta r_i) \subset B(x_0, r_i)$. Write $x_1 = (1 - r_i)x_0$ and let the curve $\gamma = \gamma_1 \cup \gamma_2$ consist of two curves γ_1 and γ_2 such that γ_1 is the shorter circular arc joining x to x_1 on $\partial B(x_0, r_i)$ and on the plane determined by x, x_1, and 0, whereas γ_2 is the straight line segment from x_1 to y. Then γ joins x to y in $\complement E$. It easily follows from elementary geometric considerations that $d(z, E) \geq r_i$ for each $z \in \gamma_1$ and that $d(z, E) \geq \delta/2 r_i$ for each $z \in \gamma_2$. The lengths of γ_1 and γ_2 satisfy $l(\gamma_1) \leq \pi r_i$ and $l(\gamma_2) \leq 2r_i$. Thus

$$k_{\complement E}(x, y) \leq \int_\gamma d(z, E)^{-1} ds = \int_{\gamma_1} d(z, E)^{-1} ds + \int_{\gamma_2} d(z, E)^{-1} ds$$

$$\leq \pi + \frac{4}{\delta} = M.$$

Hence the conditions of Theorem 11.23 are met, and therefore

$$\omega = \omega(E, B; \mathcal{A}) = 0.$$

The theorem follows.   □

A quintessential porous set is the Cantor ternary set on the boundary of the unit disk in the plane. In general, many fractal type constructions in $\mathbf{R}^n$ result in porous Cantor type sets.

If E is a set on the boundary of the unit ball B, we say that E is of *total harmonic measure zero* if $\omega(E, B; \mathcal{A}) = 0$ for all mappings $\mathcal{A}$. In particular, no restriction on p or w is imposed. Then Theorem 11.27 says that porous sets are sets of total harmonic measure zero. Although one can show that the Hausdorff dimension of a δ-porous set does not exceed a constant $\gamma(n, \delta) < n - 1$, there are, for any $\varepsilon > 0$, porous sets on ∂B with Hausdorff dimension greater than $n - 1 - \varepsilon$. On the other hand, if E is of total harmonic measure zero, then the $(n-1)$-measure of E is zero because classical potential theory is included in our presentation. Wu (1992) has recently improved Theorem 11.27 for p-harmonic measures in dimension two.

Sets of total harmonic measure zero cannot be characterized in terms of Hausdorff dimension. This was first discovered in connection with quasiconformal mappings. In particular, on the boundary of the unit disk B in the plane there are, for each $\gamma > 0$, compact sets E whose Hausdorff dimension is less than γ but $\omega(E, B; \mathcal{A}) > 0$ for certain linear operators div $\mathcal{A}$ in the unweighted case. See Tukia (1989), Martio (1989b). For $n > 2$ or $p \neq 2$ these questions are open. It seems that in order to find characterizations for sets of total harmonic measure zero, some other means should be found.

NOTES TO CHAPTER 11. Harmonic measure was used in function theory long before it was given a name. The monograph by Nevanlinna (1953) inspired much of the future research, and the discoveries of Lévy, Kakutani, Doob, and others linked the harmonic measure to probability. See Doob (1984).

The harmonic measures for operators $-$ div $\mathcal{A}$ first found their applications in the theory of quasiconformal mappings in the plane; important works include Beurling and Ahlfors (1956), Pfluger (1955), Caffarelli *et al.* (1981), Øksendal (1988). Granlund *et al.* (1982, 1985) were the first to consider the nonlinear situation with applications to quasiregular mappings; they also introduced the nonlinear Phragmén–Lindelöf principle. Further applications were given by Heinonen *et al.* (1989b) and Lindqvist (1985); see also Miklyukov (1980). Sets of absolute harmonic measure zero in the classical case and their function theoretic applications can be found in Nevanlinna (1953). Sets of $\mathcal{A}$-harmonic measure zero in the nonlinear situation were first studied by Granlund *et al.* (1982), and Theorem 11.17 was proved by them in the unweighted $p = n$ case; for recent progress related to Theorem 11.17, see Avilés and Manfredi (1992). Maz'ya (1976), Heinonen and Martio (1987), Martio (1987, 1989a), Heinonen (1988b), and Wu (1992) have given analytic and metric estimates. Gehring and Palka (1976) introduced the quasihyperbolic distance; its main use has been elsewhere in geometric function theory, see Vuorinen (1988). Contrary to the classical harmonic case $-\Delta u = 0$, where deep information is available (Carleson 1985, Makarov 1985, Bourgain 1987, Jones and Wolff 1988), the structure of the sets of $\mathcal{A}$-harmonic measure zero is largely unknown for nonlinear equations $-\operatorname{div}\mathcal{A} = 0$.

12

Fine topology

The fine topology was introduced to potential theory by Henri Cartan in 1940 as the weakest topology on $\mathbf{R}^n$ making all superharmonic functions continuous. Classical fine topology has found many applications; its connections to the theory of analytic functions and probability are well known. In this chapter we study a similar concept in the framework of $\mathcal{A}$-superharmonic functions. We establish that the fine topology associated with $\mathcal{A}$-superharmonic functions can be characterized by means of a Wiener criterion.

DEFINITION. The $\mathcal{A}$-*fine topology* $\tau_{\mathcal{A}}$ is the coarsest topology on $\mathbf{R}^n$ making all $\mathcal{A}$-superharmonic functions in $\mathbf{R}^n$ continuous.

An $\mathcal{A}$-superharmonic function u satisfies $-\infty < u \leq \infty$, and here the interval $(-\infty, \infty]$ is endowed with the usual topology. Topological concepts with respect to the $\mathcal{A}$-fine topology are equipped with the phrase "$\mathcal{A}$-fine" or "$\mathcal{A}$-finely", e.g. $\mathcal{A}$-finely open, $\mathcal{A}$-fine neighborhood.

Since $\mathcal{A}$-superharmonic functions are lower semicontinuous and since $\mathcal{S}(\mathbf{R}^n)$ is closed under truncations, the $\mathcal{A}$-fine topology $\tau_{\mathcal{A}}$ is the coarsest topology on $\mathbf{R}^n$ making all locally bounded $\mathcal{A}$-superharmonic functions continuous. Thus the $\mathcal{A}$-fine topology could be defined as the coarsest topology making all supersolutions of equation (3.8),

$$-\operatorname{div}\mathcal{A}(x, \nabla u) = 0\,,$$

continuous; then, of course, one should consider representatives which satisfy the "ess lim inf" property (see Theorem 7.25).

12.1. Lemma. *The $\mathcal{A}$-fine topology is finer than the Euclidean topology.*

PROOF: We fix a ball $B = B(x_0, r)$ in $\mathbf{R}^n$ and show that B contains a nonempty $\mathcal{A}$-finely open set U such that $x_0 \in U$. Because $\frac{1}{2}\overline{B}$ is (p, μ)-thick at each of its points, there is a function $v \in \mathcal{S}(\mathbf{R}^n) \cap \mathcal{H}(\mathbf{R}^n \setminus \frac{1}{2}\overline{B})$ such that $v = 0$ in $\frac{1}{2}\overline{B}$ and $v < 0$ in $\complement\frac{1}{2}\overline{B}$ (Lemma 7.31). Since v has a limit at infinity by Lemma 6.15, it follows from the maximum principle that the $\mathcal{A}$-finely open neighborhood $\{x \colon v(x) > \max_{\partial B} v\}$ of x_0 is contained in B.

$\square$

In general, the $\mathcal{A}$-fine topology is strictly finer than the Euclidean topology. For example, if there exists a sequence $x_j \to x$ such that each singleton $\{x_j\}$ is $\mathcal{A}$-polar or, equivalently, of (p,μ)-capacity zero, then there exists a function $u \in \mathcal{S}(\mathbf{R}^n)$ such that $u(x_j) = 1$ for all j but $u(x) = 0$ (Theorem 10.2). Hence u is discontinuous at x and $\tau_{\mathcal{A}}$ is strictly finer than the Euclidean topology. Note that $\{x\}$ may be nonpolar.

In the case of the p-Laplacian,

$$-\operatorname{div}(|\nabla u|^{p-2}\nabla u) = 0\,,$$

the fine topology coincides with the Euclidean topology if and only if $p > n$. This is due to the example above and to the fact that bounded p-superharmonic functions are locally in the Sobolev space $H^{1,p}(\,\cdot\,;dx)$, and hence continuous for $p > n$ by the "ess lim inf" property and by the usual Sobolev embedding theorem.

Since $\tau_{\mathcal{A}}$ is finer than the Euclidean topology, it is a Hausdorff topology. A natural subbase of $\tau_{\mathcal{A}}$ consists of the sets of the form $\{u > \lambda\}$ or $\{u < \lambda\}$ where $u \in \mathcal{S}(\mathbf{R}^n)$ and $\lambda \in \mathbf{R}$. So a base of $\tau_{\mathcal{A}}$ is formed by finite intersections of the sets of this type. It is often convenient to use a neighborhood base of a point x_0 consisting of the sets

$$(12.2) \qquad \bigcap_{i=1}^{k} \{x \in \overline{B} : u_i(x) \le \lambda\}\,,$$

where k is an integer, $\lambda > 0$, B is a ball centered at x_0, and each u_i is a locally bounded $\mathcal{A}$-superharmonic function in $\mathbf{R}^n$ with $u_i(x_0) = 0$. We show that (12.2) gives a neighborhood base of x_0. Indeed, because the sets

$$\{x \in B : u_i(x) < \lambda\} = \{u_i < \lambda\} \cap B$$

are $\mathcal{A}$-finely open, formula (12.2) defines an $\mathcal{A}$-fine neighborhood of x_0. Moreover, if U is an $\mathcal{A}$-fine neighborhood of x_0, we find locally bounded functions $v_j \in \mathcal{S}(\mathbf{R}^n)$ and constants $\lambda_j > 0$, $j = 1, 2, \ldots, m, m+1, \ldots \ell$, such that $v_j(x_0) = 0$ and

$$x_0 \in \bigcap_{j=1}^{m} \{v_j < \lambda_j\} \cap \bigcap_{j=m+1}^{\ell} \{v_j > \lambda_j\} \subset U\,.$$

By lower semicontinuity we may pick a ball $B = B(x_0, r)$ such that

$$\overline{B} \subset \bigcap_{j=m+1}^{\ell} \{v_j > \lambda_j\}\,.$$

Now by letting $u_j = \lambda_j^{-1} v_j$ we have $u_j(x_0) = 0$ and

$$x_0 \in \bigcap_{j=1}^{m} \{x \in \overline{B} : u_j \leq \frac{1}{2}\} \subset \bigcap_{j=1}^{m} \{v_j < \lambda_j\} \cap \bigcap_{j=m+1}^{\ell} \{v_j > \lambda_j\} \subset U$$

as desired.

The sets in (12.2) are Euclidean compact and $\mathcal{A}$-finely closed. This fact can be used for example to show that $\tau_{\mathcal{A}}$ is *completely regular* and *Baire*. To prove the first property is an easy exercise and for the latter the reader is asked to mimic the classical argument as in Doob (1984, p. 167). We recall that a topological space X is completely regular if for every point x in X and for every open neighborhood U of x there is a continuous function f on X, $0 \leq f \leq 1$, such that $f(x) = 0$ and $f = 1$ on $X \setminus U$. Further, X is Baire if the intersection of any sequence of open dense sets is dense. We do not need these concepts in this book.

Let $\mathcal{C} = \mathcal{C}_{\mathcal{A}}$ be the smallest convex cone that is closed under min-operation and contains all $\mathcal{A}$-superharmonic functions in $\mathbf{R}^n$. That is, $\mathcal{C}$ is the intersection of all sets $\mathcal{C}'$ with the properties

(i) $\mathcal{S}(\mathbf{R}^n) \subset \mathcal{C}'$, and

(ii) $u, v \in \mathcal{C}'$ and $\lambda \geq 0$ imply that $u + v$, λu, and $\min(u, v)$ belong to $\mathcal{C}'$.

The family of all $\mathcal{A}$-finely continuous functions in $\mathbf{R}^n$ is one of the sets $\mathcal{C}'$ that satisfy (i) and (ii) above, and therefore $\tau_{\mathcal{A}}$ is the coarsest topology on $\mathbf{R}^n$ making all functions in $\mathcal{C}$ continuous.

12.3. Lemma. *Suppose that $\Omega \subset \mathbf{R}^n$ is open. Then the topology induced by $\tau_{\mathcal{A}}$ on Ω is the coarsest topology in Ω making all functions in $\mathcal{S}(\Omega)$ continuous.*

PROOF: It is enough to show that for $u \in \mathcal{S}(\Omega)$ the sets

$$\{u < \lambda\},$$

$\lambda \in \mathbf{R}$, are $\mathcal{A}$-finely open. Fix $x \in \{u < \lambda\}$ and choose a ball $B \Subset \Omega$ centered at x. By the extension theorem 7.30 there is $v \in \mathcal{S}(\mathbf{R}^n)$ such that $v = u$ in B. Thus

$$x \in \{v < \lambda\} \cap B \subset \{u < \lambda\}$$

which establishes the claim.   □

12.4. $\mathcal{A}$-thinness

Classical fine topology can be used to characterize the regular boundary points for the Dirichlet problem. Namely, $x \in \partial\Omega$ is irregular in the potential theory of the Laplacian if and only if x is an isolated point of $\complement\Omega$

in the pertinent fine topology. We next pursue this problem in the general
$\mathcal{A}$-potential theory and prove that irregular boundary points are always
$\mathcal{A}$-finely isolated; as to the converse statement, a point of positive (p, μ)-
capacity can be $\mathcal{A}$-finely isolated on the boundary but it is never irregular
(Corollary 9.12). If an $\mathcal{A}$-finely isolated boundary point has (p, μ)-capacity
zero, then we know in some cases that it is irregular but a complete solution
to this problem is open in the weighted case (see Notes to this chapter).

In Chapter 9 we proved that a boundary point x_0 of Ω is irregular if $\{x_0\}$
is $\mathcal{A}$-polar and there is an $\mathcal{A}$-superharmonic function u in a neighborhood
of x_0 such that

$$\liminf_{\substack{x \to x_0 \\ x \in \mathcal{C}\Omega \setminus \{x_0\}}} u(x) > u(x_0).$$

See Theorem 9.18. It is natural to adopt the following definition:

DEFINITION. A set E is $\mathcal{A}$-*thin* at $x_0 \notin E$ if there exists an $\mathcal{A}$-super-
harmonic function u in a neighborhood of x_0 such that

$$\liminf_{\substack{x \to x_0 \\ x \in E}} u(x) > u(x_0).$$

We use the convention that if $x_0 \notin \overline{E}$, then the limit inferior is ∞. In light
of the extension theorem 7.30 we may always assume that $u \in \mathcal{S}(\mathbf{R}^n)$.

12.5. Theorem. *If a set E is $\mathcal{A}$-thin at $x_0 \notin E$, then $\mathcal{C}E$ is an $\mathcal{A}$-fine
neighborhood of x_0.*

PROOF: Choose a function $u \in \mathcal{S}(\mathbf{R}^n)$ with

$$\liminf_{\substack{x \to x_0 \\ x \in E}} u(x) > \gamma > u(x_0).$$

Then there is a ball $B = B(x_0, r)$ such that

$$u(x) \geq \gamma$$

whenever $x \in E \cap B$. Because the $\mathcal{A}$-fine neighborhood $\{x \in B : u(x) < \gamma\}$
of x_0 is contained in $\mathcal{C}E$, the claim follows.   □

12.6. Theorem. *A finite irregular boundary point of Ω is an $\mathcal{A}$-finely
isolated point of $\mathcal{C}\Omega$.*

PROOF: Let $x_0 \in \partial\Omega \setminus \{\infty\}$ be irregular. By Lemma 9.10 we find a ball B
such that $x_0 \in B$ and

$$u = \hat{R}^1_{\overline{B} \setminus \Omega}(2B)(x_0) = 1 - \delta < 1.$$

Moreover, by the fundamental convergence theorem 8.2 the set

$$E = \{x \in \overline{B} \setminus \Omega \colon u(x) < 1\} \setminus \{x_0\}$$

has zero (p, μ)-capacity. Hence there is an $\mathcal{A}$-superharmonic function v in $\mathbf{R}^n$, positive in B, such that $v(x_0) < \delta$ and $v|_E = \infty$ (Theorem 10.2). It follows that the set

$$\{x \in B \colon u(x) + v(x) < 1\}$$

is an $\mathcal{A}$-fine neighborhood of x_0 that does not intersect $\complement\Omega \setminus \{x_0\}$. □

It is clear that a boundary point $x \in \partial\Omega$ of positive (p, μ)-capacity may be $\mathcal{A}$-finely isolated on $\complement\Omega$, though it is regular. In general it is not known whether a regular boundary point x of Ω is an $\mathcal{A}$-fine accumulation point of $\complement\Omega$ if $\mathrm{cap}_{p,\mu}\{x\} = 0$. The answer is known to be affirmative in the unweighted case or if equation (3.8) is linear. Further, in Theorem 12.21 below we show that if $\{x_0\}$ has positive (p, μ)-capacity, then a set E is $\mathcal{A}$-thin at x_0 if and only if $\complement E$ is an $\mathcal{A}$-fine neighborhood of x_0. See Notes to this chapter.

12.7. (p, μ)-thinness
The Wiener test provides another way to generalize the classical concept of thinness.

DEFINITION. A set E is (p, μ)-*thin* at x_0 if

$$W_{p,\mu}(E, x_0) = \int\limits_0^1 \Big(\frac{\mathrm{cap}_{p,\mu}(E \cap B(x_0, t), B(x_0, 2t))}{\mathrm{cap}_{p,\mu}(B(x_0, t), B(x_0, 2t))} \Big)^{1/(p-1)} \frac{dt}{t} < \infty.$$

As in Chapter 6, we say that E is (p, μ)-*thick* at x_0 if E is not (p, μ)-thin at x_0.

It was proved in Theorem 6.33 that if $x_0 \in E$ and $\mathrm{cap}_{p,\mu}\{x_0\} > 0$, then E is (p, μ)-thick at x_0. On the other hand, if $x_0 \in \overline{E} \setminus E$, then E can be (p, μ)-thin at x_0 even if $\mathrm{cap}_{p,\mu}\{x_0\} > 0$. For example, if $p \leq n$ and $-n < \gamma < p - n$, then $w(x) = |x|^\gamma$ is a p-admissible weight and each point except the origin has (p, μ)-capacity zero while $\mathrm{cap}_{p,\mu}\{0\} > 0$ (see Example 2.22). Therefore, if E consists of a sequence $x_j \neq 0$ converging to 0, then E is (p, μ)-thin at x_0.

In Chapters 6 and 9 we demonstrated that if the complement of Ω is (p, μ)-thick at $x_0 \in \partial\Omega$, then x_0 is regular. In other words, $\complement\Omega$ is (p, μ)-thin at each finite irregular boundary point $x_0 \in \partial\Omega$; this is an analogue of Theorem 12.6.

The main goal of this section is to establish that (p, μ)-thinness characterizes the $\mathcal{A}$-fine topology:

12.8. Theorem. *A set U is an $\mathcal{A}$-fine neighborhood of a point x_0 if and only if x_0 is in U and the complement of U is (p, μ)-thin at x_0.*

Theorem 12.8 has interesting corollaries, which we discuss in the next section.

For the proof some auxiliary results are required. In many cases the Wiener integral $W_{p,\mu}(E, x_0)$ is conveniently replaced by a handier concept, the *Wiener sum*:

$$(12.9) \quad W_{p,\mu}^{\Sigma}(E, x_0) = \sum_{j=0}^{\infty} \left(\frac{\text{cap}_{p,\mu}(E \cap B(x_0, 2^{-j}), B(x_0, 2^{1-j}))}{\text{cap}_{p,\mu}(B(x_0, 2^{-j}), B(x_0, 2^{1-j}))} \right)^{1/(p-1)}.$$

12.10. Lemma. *There is a constant $c = c(n, p, c_\mu) \geq 1$ such that*

$$c^{-1} W_{p,\mu}(E, x_0) \leq W_{p,\mu}^{\Sigma}(E, x_0) \leq c(a_0^{1/(p-1)} + W_{p,\mu}(E, x_0))$$

for all $E \subset \mathbf{R}^n$ and $x_0 \notin E$, where

$$a_0 = \frac{\text{cap}_{p,\mu}(E \cap B(x_0, 1), B(x_0, 2))}{\text{cap}_{p,\mu}(B(x_0, 1), B(x_0, 2))}.$$

In particular, $W_{p,\mu}(E, x_0)$ is finite if and only if $W_{p,\mu}^{\Sigma}(E, x_0)$ is finite.

PROOF: If $t \leq s \leq 2t$, then we have by Lemmas 2.16 and 2.14

$$\text{cap}_{p,\mu}(E \cap B(x_0, t), B(x_0, 2t)) \approx \text{cap}_{p,\mu}(E \cap B(x_0, t), B(x_0, 2s))$$

and

$$\text{cap}_{p,\mu}(B(x_0, t), B(x_0, 2t)) \approx \text{cap}_{p,\mu}(B(x_0, s), B(x_0, 2s)),$$

where the constants in $\approx$ depend only on n, p, and c_μ. Thus for

$$2^{-1-j} \leq t \leq 2^{-j}$$

it holds that

$$\frac{\text{cap}_{p,\mu}(E \cap B(x_0, t), B(x_0, 2t))}{\text{cap}_{p,\mu}(B(x_0, t), B(x_0, 2t))}$$

$$\leq c \frac{\text{cap}_{p,\mu}(E \cap B(x_0, 2^{-j}), B(x_0, 2^{1-j}))}{\text{cap}_{p,\mu}(B(x_0, 2^{-j}), B(x_0, 2^{1-j}))}$$

$$\leq c \frac{\text{cap}_{p,\mu}(E \cap B(x_0, 2t), B(x_0, 4t))}{\text{cap}_{p,\mu}(B(x_0, 2t), B(x_0, 4t))}.$$

This implies

$$
\begin{aligned}
W_{p,\mu}(E, x_0) &= \int_0^1 \Big(\frac{\mathrm{cap}_{p,\mu}(E \cap B(x_0,t), B(x_0,2t))}{\mathrm{cap}_{p,\mu}(B(x_0,t), B(x_0,2t))} \Big)^{1/(p-1)} \frac{dt}{t} \\
&= \sum_{j=0}^{\infty} \int_{2^{-1-j}}^{2^{-j}} \Big(\frac{\mathrm{cap}_{p,\mu}(E \cap B(x_0,t), B(x_0,2t))}{\mathrm{cap}_{p,\mu}(B(x_0,t), B(x_0,2t))} \Big)^{1/(p-1)} \frac{dt}{t} \\
&\leq c \sum_{j=0}^{\infty} \Big(\frac{\mathrm{cap}_{p,\mu}(E \cap B(x_0,2^{-j}), B(x_0,2^{1-j}))}{\mathrm{cap}_{p,\mu}(B(x_0,2^{-j}), B(x_0,2^{1-j}))} \Big)^{1/(p-1)} \\
&= c W_{p,\mu}^{\Sigma}(E, x_0).
\end{aligned}
$$

Similarly

$$
W_{p,\mu}^{\Sigma}(E, x_0) \leq c \Big(\frac{\mathrm{cap}_{p,\mu}(E \cap B(x_0,1), B(x_0,2))}{\mathrm{cap}_{p,\mu}(B(x_0,1), B(x_0,2))} \Big)^{1/(p-1)} + c W_{p,\mu}(E, x_0),
$$

and the proof is complete. $\square$

12.11. Lemma. *Suppose that $E \subset \mathbf{R}^n$ and that $x_0 \notin E$.*

(i) *If E is (p,μ)-thin at x_0, there is an open neighborhood U of E such that U is (p,μ)-thin at x_0.*

(ii) *If E is a Borel set and (p,μ)-thick at x_0, there is a compact set $K \subset E \cup \{x_0\}$ such that K is (p,μ)-thick at x_0.*

PROOF: (i) Let $B_j = B(x_0, 2^{1-j})$. Since $V_1 \cup V_2$ is (p,μ)-thin at x_0 if both V_1 and V_2 are, we may assume that $E \cap \partial B_j = \emptyset$. Then let $U_0 = \mathbf{R}^n$ and for each $j = 1, 2, \ldots$ choose an open $U_j \subset B_j \cap U_{j-1}$ such that $E_j = E \cap B_j \subset U_j$ and that

$$
\Big(\frac{\mathrm{cap}_{p,\mu}(U_j, B_{j-1})}{\mathrm{cap}_{p,\mu}(B_j, B_{j-1})} \Big)^{1/(p-1)} \leq \Big(\frac{\mathrm{cap}_{p,\mu}(E_j, B_{j-1})}{\mathrm{cap}_{p,\mu}(B_j, B_{j-1})} \Big)^{1/(p-1)} + 2^{-j-1}.
$$

Then letting

$$
U = \bigcup_{j=0}^{\infty} (U_j \setminus \overline{B}_{j+1})
$$

we have that $E \subset U$, U is open, and

$$
\begin{aligned}
W_{p,\mu}^{\Sigma}(U, x_0) &\leq \sum_{j=0}^{\infty} \Big(\frac{\mathrm{cap}_{p,\mu}(U_j, B_{j-1})}{\mathrm{cap}_{p,\mu}(B_j, B_{j-1})} \Big)^{1/(p-1)} \\
&\leq W_{p,\mu}^{\Sigma}(E, x_0) + 1 < \infty.
\end{aligned}
$$

Thus U is the desired neighborhood of E.

(ii) Again let $B_j = B(x_0, 2^{1-j})$. Since the sets $E \cap B_j$ are Borel,

$$\operatorname{cap}_{p,\mu}(E \cap B_j, B_{j-1}) = \sup \operatorname{cap}_{p,\mu}(K, B_{j-1}),$$

where the supremum is taken over all compact $K \subset E \cap B_j$ (Theorem 2.5). Now for each j choose a compact $K_j \subset E \cap B_j$ such that

$$\left(\frac{\operatorname{cap}_{p,\mu}(E_j, B_{j-1})}{\operatorname{cap}_{p,\mu}(B_j, B_{j-1})}\right)^{1/(p-1)} \le \left(\frac{\operatorname{cap}_{p,\mu}(K_j, B_{j-1})}{\operatorname{cap}_{p,\mu}(B_j, B_{j-1})}\right)^{1/(p-1)} + 2^{-j}.$$

Then $K = \bigcup_j K_j \cup \{x_0\}$ is the desired compact set. $\square$

12.12. Theorem.   *Suppose that $x_0 \notin E$. If E is $\mathcal{A}$-thin at x_0, then E is (p, μ)-thin at x_0.*

PROOF: The following proof could be simplified if $\operatorname{cap}_{p,\mu}\{x_0\} = 0$ (cf. Theorem 9.18 and Lemma 12.11). To present a proof that covers all possibilities, we need to modify the argument used in the proof of Lemma 6.25.

We can assume that $x_0 \in \overline{E}$. Since E is $\mathcal{A}$-thin at x_0, there is $u \in \mathcal{S}(\mathbf{R}^n)$ such that

$$\liminf_{\substack{x \to x_0 \\ x \in E}} u(x) > 1 > u(x_0) > 0.$$

Let $B = B(x_0, r)$ be a ball such that $u > 0$ in B and $u > 1$ in $E \cap B$. Now if U is the open set

$$U = \{x \in B : u(x) > 1\},$$

we have for the $\mathcal{A}$-potential $v = \hat{R}^1_{U \cap \frac{1}{2}B}(B; \mathcal{A})$ that

$$0 < v(x_0) < 1$$

and

$$v|_{U \cap \frac{1}{2}B} = 1.$$

Write $B_j = 2^{-j}B$, $j = 0, 1, 2, \ldots$. If

$$s_j = \hat{R}^1_{U \cap B_j}(B_{j-1}; \mathcal{A}),$$

it follows from the capacity density estimate Lemma 6.21 by approximating that

$$(12.13) \qquad s_j \ge c a_j{}^{1/(p-1)} \ge 1 - \exp(-c a_j{}^{1/(p-1)})$$

in B_j, $j = 1, 2, \ldots$; here $c = c(n, p, \beta/\alpha, c_\mu) > 0$,

$$a_j = \frac{\mathrm{cap}_{p,\mu}(U \cap B_j, B_{j-1})}{\mathrm{cap}_{p,\mu}(B_j, B_{j-1})},$$

and the approximation involves an exhaustion of $U \cap B_j$ by compact sets.

Now we proceed as in the proof of Lemma 6.25. Letting $v_1 = 1 - v = 1 - s_1$ we have by (12.13) that

$$v_1 \leq \exp(-c a_1^{1/(p-1)})$$

in B_1. Then write for $j = 2, 3, \ldots$

$$v_j = \exp(c a_{j-1}^{1/(p-1)}) v_{j-1},$$

where c is the constant in (12.13). Then $1 - v_j \in \mathcal{S}(B_{j-1})$ is nonnegative in B_{j-1} and $1 - v_j = 1$ in $B_j \cap U$. Hence

$$1 - v_j \geq s_j \geq 1 - \exp(-c a_j^{1/(p-1)})$$

or

$$v_j \leq \exp(-c a_j^{1/(p-1)})$$

in B_j. It follows that

$$1 - v = v_1 \leq \exp\left(-c \sum_{k=1}^{j} a_k^{1/(p-1)}\right)$$

in B_j. Since $1 - v(x_0) = \delta > 0$, we have

$$\sum_{k=1}^{j} a_k^{1/(p-1)} \leq -c^{-1} \log \delta < \infty$$

for each j. By letting $j \to \infty$ we see that U, and hence E, is (p, μ)-thin at x_0. The theorem is proved. $\qquad \square$

Before we take up the proof of Theorem 12.8, we record a consequence of the basic capacity estimate in Lemma 6.19.

12.14. Lemma. *Suppose that $B = B(x_0, r)$ is a ball and that $E \subset \frac{1}{2}B$ is an open set. If $u = \hat{R}_E^1(B; \mathcal{A})$ is the $\mathcal{A}$-potential of E in B, then*

$$\min_{\partial B(x_0, \rho)} u \le c \Big(\frac{\mathrm{cap}_{p,\mu}(E, B)}{\mathrm{cap}_{p,\mu}(\frac{1}{2}B, B)} \Big)^{1/(p-1)}$$

whenever $\rho \in (\frac{r}{4}, \frac{r}{2})$, where $c = c(n, p, \beta/\alpha, c_\mu) > 0$.

PROOF: Fix $\rho \in (\frac{r}{4}, \frac{r}{2})$ and let $\gamma = \min_{\partial B(x_0, \rho)} u$. Then it follows from the minimum principle and Lemma 6.19 by approximating that

$$\mathrm{cap}_{p,\mu}(B(x_0, \rho), B) \le \mathrm{cap}_{p,\mu}(\{u \ge \gamma\}, B) \le (\alpha/\beta)^{p+1} \gamma^{1-p} \, \mathrm{cap}_{p,\mu}(E, B).$$

Since the doubling property of μ implies that

$$\mathrm{cap}_{p,\mu}(B(x_0, \rho), B) \approx \mathrm{cap}_{p,\mu}(\frac{1}{2}B, B),$$

where constants in $\approx$ depend only on n, p, and c_μ (Lemmas 2.14 and 2.16), the assertion follows.   □

PROOF OF THEOREM 12.8: Suppose first that U is an $\mathcal{A}$-fine neighborhood of x_0. The base construction at the beginning of this chapter gives $\mathcal{A}$-superharmonic functions $u_1, \ldots, u_k \in \mathcal{S}(\mathbf{R}^n)$ and a ball $B = B(x_0, r)$ such that

$$x_0 \in \bigcap_{j=1}^k \{x \in \overline{B} : u_j(x) < 1\} \subset U$$

and $u_j(x_0) = 0$. Then $\complement U \cap B$ is contained in

$$\bigcup_{j=1}^k \{x \in B : u_j(x) \ge 1\}$$

which is a finite union of sets that are $\mathcal{A}$-thin at x_0. By Theorem 12.12 each set

$$\{x \in B : u_j(x) \ge 1\}$$

is (p, μ)-thin at x_0; hence the union is (p, μ)-thin at x_0, and we deduce that $\complement U$ is (p, μ)-thin at x_0. This proves the first part of the theorem.

The converse is more involved. Suppose that $E = \complement U$ is (p, μ)-thin at $x_0 \in U$. We may assume that $E \subset B(x_0, \frac{1}{2})$ and, by Lemma 12.11, that E is open. Let

$$D = \bigcup_{j=1}^\infty \Big((E \cap B(x_0, 2^{-j})) \setminus \overline{B}(x_0, \frac{4}{3} 2^{-j-1}) \Big)$$

and

$$D' = \bigcup_{j=1}^{\infty} \left((E \cap B(x_0, \tfrac{15}{11}2^{-j})) \setminus \overline{B}(x_0, \tfrac{10}{11}2^{-j}) \right).$$

Then D and D' are open sets that are (p,μ)-thin at x_0. Moreover, $E \subset D \cup D'$. We construct an $\mathcal{A}$-superharmonic function v_0 in a neighborhood B_1 of x_0 such that $v_0 = 1$ in $D \cap B_1$ and that $v_0(x_0) < \frac{1}{2}$; then Theorem 12.5 implies that $\complement D$ is an $\mathcal{A}$-fine neighborhood of x_0. A similar construction shows that also $\complement D'$ is an $\mathcal{A}$-fine neighborhood of x_0, and this is sufficient because $\complement D \cap \complement D' \subset \complement E = U$.

To start the construction, write $B_j = B(x_0, 2^{-j})$, $D_j = D \cap B_j$ and let $u_j = \hat{R}^1_{D_j}(B_{j-1}; \mathcal{A})$ be the $\mathcal{A}$-potential of D_j in B_{j-1}, $j = 1, 2, \ldots$. If S_j is the sphere $\partial \frac{7}{6} B_{j+1}$, then $S_j \subset B_j \setminus \overline{B}_{j+1}$ and

$$\frac{\mathrm{dist}(S_j, \overline{D})}{\mathrm{dist}(S_j, x_0)} \geq \frac{1}{7} > 0.$$

Thus S_j can be covered by N balls B' such that $2B'$ lies in $B_j \setminus \overline{D}$ and the number N depends only on n. Therefore, since u_j is $\mathcal{A}$-harmonic in $B_j \setminus \overline{D}$, it follows from the Harnack inequality and Lemma 12.14 that

$$(12.15) \qquad u_j \leq c \left(\frac{\mathrm{cap}_{p,\mu}(E \cap B_j, B_{j-1})}{\mathrm{cap}_{p,\mu}(B_j, B_{j-1})} \right)^{1/(p-1)} = b_j$$

on S_j with constant $c = c(n, p, \beta/\alpha, c_\mu)$. Now we choose j_0 in such a way that

$$\sum_{j=j_0}^{\infty} b_j < \frac{1}{2}.$$

For notational simplicity we assume that $j_0 = 1$. We show that

$$v_0 = u_1 = \hat{R}^1_{D_1}(B_0; \mathcal{A})$$

is the desired function. Since $D_1 = D \cap B_1$ is open, $u_1 = 1$ in D_1, and it suffices to show that $u_1(x_0) < 1/2$. To this end, let

$$v_1 = \frac{u_1 - b_1}{1 - b_1}$$

and

$$s_1 = \begin{cases} \min(v_1, u_2) & \text{in } \frac{7}{6}B_2 \\ v_1 & \text{in } B_0 \setminus \frac{7}{6}B_2. \end{cases}$$

Since $v_1 \leq 0$ on S_1 by (12.15), the function s_1 is lower semicontinuous and it follows from the pasting lemma 7.9 that $s_1 \in \mathcal{S}(B_0)$. Moreover, v_1 is the minimal function in $\mathcal{S}(B_0)$ lying above

$$\psi_1 = \frac{\psi_0 - b_1}{1 - b_1},$$

where ψ_0 is the characteristic function of D_1. Therefore $s_1 \geq v_1$ in B_0. In particular, $u_2 \geq v_1$ in $\frac{7}{6}B_2$ and hence

$$v_0 - b_1 = (1 - b_1)v_1 \leq u_2 \leq b_2$$

on S_2. We continue recursively. Write

$$v_k = \frac{v_{k-1} - b_k}{1 - b_k}$$

$k = 1, 2, \ldots$. Then v_k is the minimal function in $\mathcal{S}(B_0)$ lying above

$$\psi_k = \frac{\psi_{k-1} - b_k}{1 - b_k}.$$

Since also $v_k \leq 0$ on S_k, the function

$$s_k = \begin{cases} \min(v_k, u_{k+1}) & \text{in } \frac{7}{6}B_{k+1} \\ v_k & \text{in } B_0 \setminus \frac{7}{6}B_{k+1} \end{cases}$$

is $\mathcal{A}$-superharmonic and greater than v_k in B_0. Thus

$$v_k \leq u_{k+1} \leq b_{k+1}$$

on S_{k+1} and hence

$$v_{k-1} - b_k = (1 - b_k)v_k \leq b_{k+1}$$

on S_{k+1}. By recursion we obtain

$$v_0 \leq \sum_{j=1}^{k+1} b_j$$

on S_{k+1}. Since v_0 is lower semicontinuous, we have that

$$v_0(x_0) \leq \liminf_{k \to \infty} \inf_{S_k} v_0 \leq \sum_{j=1}^{\infty} b_j < \frac{1}{2}$$

as desired. Theorem 12.8 is thereby proved. □

12.16. (p, μ)-fine topology

We define the (p, μ)-*fine topology* $\tau_{p,\mu}$ to be the collection of all sets U such that $\complement U$ is (p, μ)-thin at each point of U. Then it is an easy task to prove that $\tau_{p,\mu}$ is a topology, and Theorem 12.8 implies:

12.17. Theorem. *The $\mathcal{A}$-fine topology $\tau_{\mathcal{A}}$ coincides with the (p, μ)-fine topology $\tau_{p,\mu}$. In particular, $\tau_{\mathcal{A}}$ depends only on the structure of $\mathcal{A}$, not on $\mathcal{A}$ itself.*

We sometimes write $\tau_{p,\mu}$ instead of $\tau_{\mathcal{A}}$ and use the expressions "(p, μ)-fine" and "(p, μ)-finely".

12.18. Corollary. *A point x_0 is an $\mathcal{A}$-fine accumulation point of a set E if and only if $E \setminus \{x_0\}$ is (p, μ)-thick at x_0.*

12.19. Corollary. *An $\mathcal{A}$-polar set is $\mathcal{A}$-finely isolated.*

12.20. Corollary. *Suppose that E is an $\mathcal{A}$-finely compact set. Then only finitely many points of E are $\mathcal{A}$-polar.*

Note that Corollary 12.19 also follows from Theorem 10.2.

If a limit point x_0 of a set E is nonpolar, then we have the following characterization which is not known in general if x_0 is polar, i.e. if x_0 has zero (p, μ)-capacity; see Notes to this chapter.

12.21. Theorem. *Suppose that $\mathrm{cap}_{p,\mu}\{x_0\} > 0$ and that $E \subset \mathbf{R}^n$ is a set with $x_0 \notin E$. Then the following are equivalent:*

  (i) *E is $\mathcal{A}$-thin at x_0.*
  (ii) *$\complement E$ is an $\mathcal{A}$-fine neighborhood of x_0.*
  (iii) *E is (p, μ)-thin at x_0.*

PROOF: We already know that (i) implies (ii) (Theorem 12.5) and that (ii) implies (iii) (Theorem 12.8). To complete the loop we show that (i) follows from (iii). We may assume that $x_0 \in \overline{E}$. Choose an open set U that contains E and is (p, μ)-thin at x_0 (Lemma 12.11). Let u be the $\mathcal{A}$-potential

$$u = \hat{R}^1_{U \cap B_k}(B_{k-1}),$$

where $B_k = B(x_0, 2^{-k})$ and k is a positive integer, to be fixed later. We show that

$$1 = \liminf_{\substack{x \to x_0 \\ x \in U}} u(x) > u(x_0).$$

If not, then

$$u(x_0) = 1,$$

whence u is continuous at x_0. In particular, u can be approximated in $H_0^{1,p}(B_{k-1}; \mu)$ by functions that are admissible for testing the (p, μ)-capacity of the condenser $(\{x_0\}, B_{k-1})$ (cf. Chapter 4). We deduce

$$\text{cap}_{p,\mu}(\{x_0\}, B_{k-1}) \le \int_{B_{k-1}} |\nabla u|^p \, d\mu \le (\frac{\beta}{\alpha})^p \, \text{cap}_{p,\mu}(U \cap B_k, B_{k-1}),$$

where the last inequality follows from Theorem 9.38. On the other hand, since U is (p, μ)-thin and since $\{x_0\}$ is (p, μ)-thick at x_0 (Theorem 6.33), we find an integer k such that

$$\text{cap}_{p,\mu}(\{x_0\}, B_{k-1}) > (\frac{\beta}{\alpha})^p \, \text{cap}_{p,\mu}(U \cap B_k, B_{k-1}).$$

This contradicts the previous inequality, and the theorem follows. $\square$

If U is an $\mathcal{A}$-fine neighborhood of a point x_0, then U cannot have zero (outer) n-measure. To see this, choose an open neighborhood V of $\complement U \setminus \{x_0\}$ such that V is (p, μ)-thin at x_0. Approximating $V \cap B(x_0, r)$ from inside by compact sets we obtain from the Poincaré inequality (1.5) and the capacity estimate (2.15) that

$$(12.22) \qquad \frac{\mu(V \cap B(x_0, r))}{\mu(B(x_0, r))} \le c \, \frac{\text{cap}_{p,\mu}(V \cap B(x_0, r), B(x_0, 2r))}{\text{cap}_{p,\mu}(B(x_0, r), B(x_0, 2r))}.$$

By Lemma 12.10 we have that

$$\lim_{j \to \infty} \frac{\text{cap}_{p,\mu}(V \cap B(x_0, 2^{-j}), B(x_0, 2^{-j+1}))}{\text{cap}_{p,\mu}(B(x_0, 2^{-j}), B(x_0, 2^{-j+1}))} = 0$$

and so the capacity estimates (2.15) and (2.17) imply that the right hand side of (12.22) converges to zero as $r \to 0$. Hence $\complement V$ has μ-measure density 1 at x_0. Thus U cannot be of n-measure zero because $\complement V \subset U$. In fact, the argument shows that the μ-density of a (p, μ)-finely open set is 1 at each of its points. Hence $\tau_{\mathcal{A}}$ is smaller than the μ-density topology, where a set U is defined to be open if the complement of U has zero μ-density at each $x \in U$.

If μ is the Lebesgue n-measure and $p = 2$, then $\tau_{\mathcal{A}}$ is the fine topology of the Laplacian. Observe that the equation

$$- \text{div} \, \mathcal{A}(x, \nabla u) = 0$$

may still be nonlinear.

If all singletons are $\mathcal{A}$-polar, then each countable set is $\mathcal{A}$-finely closed. Since $\mathcal{A}$-fine neighborhoods cannot have zero μ-measure, no countable nonempty set is $\mathcal{A}$-finely open. Thus it is easily seen that if singletons are $\mathcal{A}$-polar, then

(a) $(\mathbf{R}^n, \tau_{\mathcal{A}})$ is not separable,
(b) a sequence x_j converges to x in $\tau_{\mathcal{A}}$ if and only if $x_j = x$ for all but finitely many j,
(c) no point has a countable $\mathcal{A}$-fine neighborhood base; consequently, $(\mathbf{R}^n, \tau_{\mathcal{A}})$ is not metrizable.

As noted earlier, in the unweighted case it follows from the Sobolev embedding theorem that for $p > n$ the p-fine topology $\tau_{p,dx}$ coincides with the Euclidean topology. Hence statements (a)–(c) do not hold in general.

12.23. $\mathcal{A}$-fine limits
We conclude this chapter by studying $\mathcal{A}$-fine limits. Two lemmas are needed.

12.24. Lemma. *Suppose that a set E is (p, μ)-thin at x_0. If B is a ball containing x_0, then*

$$\lim_{r \to 0} \operatorname{cap}_{p,\mu}(E \cap B(x_0, r), B) = 0.$$

PROOF: Since

$$\operatorname{cap}_{p,\mu}(E \cap B(x_0, r), B) \leq \operatorname{cap}_{p,\mu}(\overline{B}(x_0, r), B) \to \operatorname{cap}_{p,\mu}(\{x_0\}, B)$$

as $r \to 0$ (Theorem 2.2), the claim is trivial if $\operatorname{cap}_{p,\mu}\{x_0\} = 0$. Thus we assume that $\operatorname{cap}_{p,\mu}\{x_0\} > 0$. Moreover, invoking Lemma 2.16 we see that it suffices to find *some* ball B with

$$\operatorname{cap}_{p,\mu}(E \cap B(x_0, r), B) \to 0.$$

We are going to use the proof of Theorem 12.8. We reduce the situation a little more by assuming that E is open and write for $r > 0$

$$D(r) = \bigcup_{j=1}^{\infty} \left((E \cap B(x_0, r) \cap B(x_0, 2^{-j})) \setminus \overline{B}(x_0, \frac{4}{3} 2^{-j-1}) \right).$$

Observe that $D(\frac{1}{2})$ is the set D in the proof of Theorem 12.8; similarly define sets $D'(r)$ corresponding to D' there. Then by using the subadditivity of the (p, μ)-capacity we deduce that it is sufficient to verify that

$$\operatorname{cap}_{p,\mu}(D(r), B) \to 0$$

for some ball B containing x_0.

To show this, let B be the ball B_0 in the proof of 12.8 and write

$$u_i = \hat{R}^1_{D(2^{-i})}(B; \mathcal{A}).$$

It follows from the proof of 12.8 that

$$u_i(x_0) \le \frac{1}{2}$$

for all $i = 1, 2, \ldots$. Furthermore, functions u_i decrease to a function u that is $\mathcal{A}$-harmonic in $B \setminus \{x_0\}$. Because $u_i \le 1/2$ uniformly near ∂B (Theorem 6.18 and Lemma 8.5), the maximum principle implies that $u \le 1/2$ in B. Since $\mathrm{cap}_{p,\mu}\{x_0\} > 0$, we obtain from the fundamental convergence theorem that u is $\mathcal{A}$-superharmonic in the whole of B, and hence the functions u_i and u are (p, μ)-quasicontinuous in B by Theorem 10.9. Then fix $\varepsilon > 0$ and choose an open set $G \subset B$ such that

$$\mathrm{C}_{p,\mu}(G) < \varepsilon$$

and that the restrictions to $B \setminus G$ of functions u and u_i are continuous. Let

$$K = \frac{1}{2}\overline{B} \setminus G.$$

Then K is a compact subset of $B \setminus G$ and the continuous functions $u_i|_K$ decrease to the continuous function $u|_K$. Therefore the convergence is uniform on K. Since

$$u_i - u \ge \frac{1}{2}$$

on $D(2^{-i})$, it follows that for i large enough $D(2^{-i}) \cap K = \emptyset$, and so

$$D(2^{-i}) \subset G.$$

Now we have by Theorem 2.38 that

$$\mathrm{cap}_{p,\mu}(D(2^{-i}), B) \le c\,\mathrm{C}_{p,\mu}(G) < c\varepsilon,$$

where $c > 0$ is independent of i and ε. This completes the proof. $\square$

12.25. Lemma. *If the sets E_j, $j = 1, 2, \ldots$, are (p, μ)-thin at x_0, then there is a sequence of radii $r_j > 0$ such that*

$$\bigcup_{j=1}^{\infty} \left(E_j \cap B(x_0, r_j) \right)$$

is (p, μ)-thin at x_0.

PROOF: Fix j and choose i such that for $F_j = E_j \cap B_i$

$$W_{p,\mu}^{\Sigma}(F_j, x_0) < 2^{-j},$$

where $B_i = B(x_0, 2^{1-i})$; this is possible by Lemma 12.24. The Hölder inequality implies

$$\mathrm{cap}_{p,\mu} \left(\bigcup_j F_j \cap B_k, B_{k-1} \right)^{1/(p-1)} \leq \sum_{j=1}^{\infty} \mathrm{cap}_{p,\mu} \left(F_j \cap B_k, B_{k-1} \right)^{1/(p-1)}$$

if $p \geq 2$ and

$$\mathrm{cap}_{p,\mu} \left(\bigcup_j F_j \cap B_k, B_{k-1} \right)^{1/(p-1)}$$

$$\leq \left(\sum_{j=1}^{\infty} j^{-\frac{1}{2-p}} \right)^{\frac{2-p}{p-1}} \sum_{j=1}^{\infty} \left(j \, \mathrm{cap}_{p,\mu}(F_j \cap B_k, B_{k-1}) \right)^{1/(p-1)}$$

if $1 < p < 2$. Thus $r_j = 2^{1-i}$ gives the desired sequence and the lemma follows. $\square$

12.26. Theorem. *Suppose that a set E is not (p, μ)-thin at $x_0 \notin E$ and that $g \colon E \to [-\infty, \infty]$ is a function. Then*

$$\tau_{p,\mu}\text{-}\lim_{\substack{x \to x_0 \\ x \in E}} g(x) = \lambda$$

if and only if there is a (p, μ)-fine neighborhood V of x_0 such that

$$\lim_{\substack{x \to x_0 \\ x \in E \cap V}} g(x) = \lambda.$$

PROOF: For simplicity we assume that $\lambda \in \mathbf{R}$. If

$$\tau_{p,\mu}\text{-}\lim_{\substack{x \to x_0 \\ x \in E}} g(x) = \lambda,$$

then the set
$$E_j = \{x \in E \colon |g(x) - \lambda| \geq 1/j\}$$
is (p, μ)-thin at x_0 for each $j = 1, 2, \ldots.$ By Lemma 12.25 we find a
sequence of radii r_j such that

$$E_\infty = \bigcup_{j=1}^{\infty} (E_j \cap B(x_0, r_j))$$

is (p, μ)-thin at x_0. Then $V = \complement E_\infty$ is the desired (p, μ)-fine neighborhood.
To prove the converse, fix $\varepsilon > 0$. We can pick a $\delta > 0$ such that

$$|g(x) - \lambda| < \varepsilon$$

whenever $x \in E \cap V \cap B(x_0, \delta)$. The proof is complete as $V \cap B(x_0, \delta)$ is a
(p, μ)-fine neighborhood of x_0. $\square$

12.27. Corollary. *A function $g : \mathbf{R}^n \to [-\infty, \infty]$ is (p, μ)-finely continuous at x_0 if and only if there is a set E such that E is (p, μ)-thin at $x_0 \notin E$
and that $g|_{\complement E}$ is continuous at x_0.*

NOTES TO CHAPTER 12. The history of fine topology began in
a 1940 letter from Cartan to Brelot who had just initiated the study of
thinness of a set. For classical treatments we refer to Doob (1984), Helms
(1969), and Constantinescu and Cornea (1972). An excellent historical
survey can be found in Fuglede (1988). Density and fine topologies in
analysis are discussed in Lukeš *et al.* (1986) and Lukeš and Malý (1992).

The fine topology was first generalized to nonlinear theories by Meyers
(1975) and his definition was via Wiener type integrals; see also Adams
and Meyers (1972). Subsequent papers on fine topology and nonlinear potential theory include Adams and Hedberg (1984) and Hedberg and Wolff
(1983); see also an earlier work by Fuglede (1971a). Adams and Lewis
(1985) showed that in the unweighted case the fine topologies are arcwise
connected. Connectedness properties are discussed further in Heinonen *et
al.* (1990), where the quasi-Lindelöf property of (p, dx)-fine topologies is
proved; the latter follows also from Fuglede (1971a). The connection between the $\mathcal{A}$-fine and (p, dx)-fine topologies (Theorem 12.8) was established
in Heinonen *et al.* (1989a). The estimation techniques used in the proof
of Theorem 12.8 have their origin in Lindqvist and Martio (1985). The
weighted theory as presented here is all new.

In the linear case there exists a rich theory of harmonic functions on
finely open sets; the pioneering work is due to Fuglede (1971b, 1972, 1988).
Kilpeläinen and Malý (1992a) have initiated a study of supersolutions to
nonlinear equations in finely open sets.

One of the open problems in the weighted nonlinear potential theory is to find out whether an $\mathcal{A}$-finely isolated boundary point is irregular for the Dirichlet problem. This is known in the unweighted case (Heinonen *et al.* 1989a; Kilpeläinen and Malý 1992b); in particular, Theorem 12.21 holds in that case.

13

Harmonic morphisms

Many theorems about analytic functions or holomorphic mappings in several dimensions can be proved by only using the fact that these mappings preserve harmonic functions. This has led to an axiomatization of that property: a mapping $f : X \to Y$ between two harmonic spaces is called a harmonic morphism if $u \circ f$ is harmonic in X whenever u is harmonic in Y. In the present chapter we study a similar problem for $\mathcal{A}$-harmonic functions. We introduce $\mathfrak{A}_p$-harmonic morphisms in the general weighted theory and study potential theoretic properties of these mappings. In the next chapter we investigate a special class of morphisms, namely quasiregular mappings, and apply the results achieved in this chapter.

Let $\mathfrak{A}_p$ denote the class of all mappings $\mathcal{A} \colon \mathbf{R}^n \times \mathbf{R}^n \to \mathbf{R}^n$ that satisfy conditions (3.3)–(3.7) with common exponent $p \in (1, \infty)$. We understand throughout this chapter that $\mathcal{A}$ and $\mathcal{A}^*$ are mappings in $\mathfrak{A}_p$ and denote by w, w^* and μ, μ^* the corresponding weights and measures.

DEFINITION. A continuous mapping $f \colon \Omega \to \mathbf{R}^n$ is called an $(\mathcal{A}^*, \mathcal{A})$-*harmonic morphism* if $u \circ f$ is $\mathcal{A}^*$-harmonic in $f^{-1}(\Omega')$ whenever u is $\mathcal{A}$-harmonic in an open set Ω'. The mapping f is called an $\mathfrak{A}_p$-*harmonic morphism* if it is an $(\mathcal{A}^*, \mathcal{A})$-harmonic morphism for some $\mathcal{A}$ and $\mathcal{A}^*$ in $\mathfrak{A}_p$.

Recall that a mapping $f \colon \mathbf{R}^n \to \mathbf{R}^n$ is a *similarity* if there is a constant $\lambda \geq 0$ such that
$$|f(x) - f(y)| = \lambda |x - y|$$
for all $x, y \in \mathbf{R}^n$. A similarity f is always of the form $f(x) = \lambda \mathcal{O}(x) + b$, where $\mathcal{O} \colon \mathbf{R}^n \to \mathbf{R}^n$ is an orthogonal linear transformation and $b \in \mathbf{R}^n$. It is not difficult to see that every similarity is an $\mathfrak{A}_p$-harmonic morphism. Indeed, it is an $(\mathcal{A}, \mathcal{A})$-harmonic morphism for the (unweighted) p-Laplacian, i.e. $\mathcal{A}(x, \xi) = |\xi|^{p-2}\xi$. If $p = n$, then each Möbius transformation is an $(\mathcal{A}, \mathcal{A})$-harmonic morphism for the n-Laplacian. On the other hand, it is easy to construct examples of mappings $\mathcal{A}$ such that neither the dilation $f(x) = \lambda x$, $\lambda \in \mathbf{R} \setminus \{0\}$, nor the translation $f(x) = x + b$, $b \in \mathbf{R}^n \setminus \{0\}$, is an $(\mathcal{A}, \mathcal{A})$-harmonic morphism.

If we consider the Laplace equation, i.e. $\mathcal{A}(x, \xi) = \xi$, then $f \colon \Omega \to \mathbf{R}^n$ is an $(\mathcal{A}, \mathcal{A})$-harmonic morphism in a domain $\Omega \subset \mathbf{R}^n$ if and only if

(i) f is analytic or antianalytic in Ω for $n = 2$, and

(ii) f is the restriction to Ω of a similarity for $n > 2$

(Gehring and Haahti 1960; Fuglede 1978; Ishihara 1979). The difference between cases (i) and (ii) is preserved in the unweighted nonlinear theory: when $p = n$, the role of analytic functions is taken by quasiregular mappings, and when $p \neq n$, similarities are replaced by a subclass of quasiregular mappings, called mappings of bounded length distortion. This is the subject of Chapter 14.

13.1. Basic properties of $\mathfrak{A}_p$-harmonic morphisms

In this section we show that an $(\mathcal{A}^*, \mathcal{A})$-harmonic morphism $f \colon \Omega \to \mathbf{R}^n$ is, under some additional assumptions, either an open mapping or constant in each component of Ω. This leads to a maximum principle for $\mathfrak{A}_p$-harmonic morphisms. We show that morphisms preserve superharmonic functions as well. In the end we discuss Liouville type theorems for morphisms.

We begin with a result about homeomorphic $\mathfrak{A}_p$-harmonic morphisms.

13.2. Theorem. *Let f be a homeomorphism of Ω onto Ω'. If f is an $(\mathcal{A}^*, \mathcal{A})$-harmonic morphism, then f^{-1} is an $(\mathcal{A}, \mathcal{A}^*)$-harmonic morphism. In particular, f induces a one-to-one mapping $u \mapsto u \circ f$ of $\mathcal{H}_{\mathcal{A}}(\Omega')$ onto $\mathcal{H}_{\mathcal{A}^*}(\Omega)$.*

PROOF: Let $v \in \mathcal{H}_{\mathcal{A}^*}(\Omega_1)$, where Ω_1 is open in $\mathbf{R}^n$. To prove the first assertion, we must show that $u = v \circ f^{-1}$ is $\mathcal{A}$-harmonic in $\Omega'_1 = f(\Omega_1 \cap \Omega)$. (Note that Ω' and Ω'_1 are open because f is a homeomorphism.) To accomplish this, it suffices to verify that $u \in \mathcal{H}_{\mathcal{A}}(B')$ whenever B' is a ball whose closure is contained in Ω'_1. Since balls are regular and since u is continuous on $\partial B'$, we may choose an $\mathcal{A}$-harmonic function u_1 in B', continuous on $\overline{B'}$, with $u = u_1$ on $\partial B'$. Then $u_1 \circ f$ is $\mathcal{A}^*$-harmonic in $f^{-1}(B')$ and agrees with v on the boundary of $f^{-1}(B')$. It follows from the uniqueness that $u_1 \circ f = v$ in $f^{-1}(B')$ and, consequently, $u = v \circ f^{-1} = u_1$ is $\mathcal{A}$-harmonic in B' as desired. Thus f^{-1} is an $(\mathcal{A}, \mathcal{A}^*)$-harmonic morphism.

The second assertion is now immediate, and the theorem follows. □

Recall that a mapping $f \colon \Omega \to \mathbf{R}^n$ is *open* if $f(A)$ is open for every open $A \subset \Omega$. We also say that f is *open at a point* x if $f(U)$ is a neighborhood of $f(x)$ for every neighborhood U of x. Clearly f is open if and only if it is open at each point x in Ω.

DEFINITION. A mapping $f \colon \Omega \to \mathbf{R}^n$ has the *Radó property* if $f^{-1}(y)$ has no interior points whenever $y \in \mathbf{R}^n$.

Every open mapping has the Radó property but the converse is false; consider, for example, a projection onto a coordinate axis.

13.3. Theorem. *Suppose that $f \colon \Omega \to \mathbf{R}^n$ is an $(\mathcal{A}^*, \mathcal{A})$-harmonic morphism satisfying the Radó property. Let $x_0 \in \Omega$ and suppose that*

$f(x_0)$ *has a neighborhood U such that points in U are all $\mathcal{A}$-polar. Then f is open at x_0.*

PROOF: Assume, on the contrary, that f is not open at x_0 so that there are arbitrarily small balls B which are centered at x_0 and whose images $f(B)$ are not neighborhoods of $f(x_0)$. Choose a ball $B = B(x_0, r) \Subset \Omega$ such that $f(x_0) \in \partial f(B)$ and that

$$f(B) \Subset B' = B(f(x_0), r') \Subset U$$

for some $r' > 0$. Fix a point $z_0 \in B$ with $f(z_0) \neq f(x_0)$; note that by the Radó property f is not constant in B.

Let $y_i \in B' \setminus f(B)$, $i = 1, 2, \ldots$, be a sequence of points converging to $f(x_0)$ and let u_i be a singular $\mathcal{A}$-harmonic function in $B' \setminus \{y_i\}$. More precisely, let $u_i \in \mathcal{H}(B' \setminus \{y_i\})$ be nonnegative such that $\lim_{z \to y} u_i(z) = 0$ for each $y \in \partial B'$ and that

$$u_i(y_i) = \lim_{z \to y_i} u_i(z) = \infty.$$

For the construction of u_i, see Theorem 7.39. Since $f(z_0) \neq y_i$, we may assume that $u_i(f(z_0)) = 1$ for each i. It then follows from Harnack's inequality that for each compact set $K \subset B' \setminus \{f(x_0)\}$ there is an index i_0 such that the sequence u_i, $i \geq i_0$, is uniformly bounded, and hence equicontinuous on K (Corollary 6.12). By Ascoli's theorem we may select a subsequence of u_i which converges locally uniformly in $B' \setminus \{f(x_0)\}$ to an $\mathcal{A}$-harmonic function u_0 (Theorem 6.13). Because $u_i(f(z_0)) = 1$ for all i, we easily infer from the boundary estimate (6.37) that $u_0(z) \to 0$ as z tends to any boundary point of B'; therefore u_0 is not constant. Since u_0 is nonnegative, the limit

$$\lim_{z \to f(x_0)} u_0(z) = u_0(f(x_0))$$

exists (Lemma 6.15). If this limit were finite, u_0 would have an $\mathcal{A}$-harmonic extension to B' (Corollary 7.36) which is impossible because u_0 is nonconstant and tends to zero on the boundary. Thus $u_0(f(x_0)) = \infty$.

Now consider the functions $v_i = u_i \circ f$; they are $\mathcal{A}^*$-harmonic in B because $y_i \notin f(B)$. Since v_i is nonnegative and $v_i(z_0) = 1$, it again follows that a subsequence of v_i converges locally uniformly in B to an $\mathcal{A}^*$-harmonic function v_0. On the other hand, we have that $v_0 = u_0 \circ f$ in B and hence that $v_0(x_0) = u_0(f(x_0)) = \infty$, an obvious contradiction. Thus f is open at x_0 and the proof is complete. $\qquad\square$

To prove a version of Theorem 13.3 in the case when points are not necessarily $\mathcal{A}$-polar, we introduce a concept of strong nonpolarity.

DEFINITION. A point x_0 is said to be *strongly non-$\mathcal{A}$-polar* if

$$\text{cap}_{p,\mu}(\{x_j\}, \mathbf{R}^n \setminus \{x_0\}) \to \infty$$

as $j \to \infty$ whenever x_j is a sequence converging to x_0.

Of course, strong nonpolarity depends only on p and μ but we prefer the above terminology.

If x_0 is strongly non-$\mathcal{A}$-polar, then $\{x_0\}$ has positive (p, μ)-capacity and, moreover, x_0 cannot be a cluster point of $\mathcal{A}$-polar points. More generally, using Theorem 12.21 one can easily show that x_0 is in the $\mathcal{A}$-fine closure of any set E with $x_0 \in \overline{E}$, if x_0 is strongly non-$\mathcal{A}$-polar. If $d\mu = dx$, then points are strongly non-$\mathcal{A}$-polar if and only if $p > n$; this can be seen from the estimates in Chapter 2. It is possible, however, for a point x_0 to be nonpolar but not strongly nonpolar; for example, choose $1 < p \le n$, $w(x) = |x|^\gamma$ for $-n < \gamma < p - n$, and $x_0 = 0$ (see Example 2.22).

13.4. Theorem. *Suppose that $f: \Omega \to \mathbf{R}^n$ is an $(\mathcal{A}^*, \mathcal{A})$-harmonic morphism satisfying the Radó property. Let $x_0 \in \Omega$ and suppose that $f(x_0)$ is strongly non-$\mathcal{A}$-polar. Then f is open at x_0.*

PROOF: The proof is reminiscent of the proof of Theorem 13.3. Suppose that f is not open at x_0, and let $B = B(x_0, r) \Subset \Omega$, $B' = B(f(x_0), r')$, and $z_0 \in B$ be as in the proof of Theorem 13.3. Next choose a sequence $y_i \in B' \setminus f(B)$ converging to $f(x_0)$ and let u_i be the $\mathcal{A}$-potential of the singleton $\{y_i\}$ in $B' \setminus \{f(z_0)\}$. Note that $\{f(z_0)\}$ and $\{y_i\}$ are nonpolar if r' is sufficiently small. Hence for all $i = 1, 2, \ldots$

$$\lim_{z \to f(z_0)} u_i(z) = 0$$

and $u_i(y_i) = 1$.

We claim that u_i converges in B' to the $\mathcal{A}$-potential of the set $\{f(x_0)\}$ in $B' \setminus \{f(z_0)\}$. To see this, observe first that the quasiminimizing property of $\mathcal{A}$-harmonic functions together with the monotonicity of (p, μ)-capacities implies that

$$\int_{B'} |\nabla u_i|^p \, d\mu \le M < \infty,$$

where M is independent of i. Thus if $u_i(f(x_0)) < \lambda < 1$ for infinitely many i, the functions

$$s_i = \max \left(\frac{u_i - \lambda}{1 - \lambda}, 0 \right)$$

are admissible for the condenser $(\{y_i\}, B' \setminus \{f(x_0), f(z_0)\})$ and hence

$$M(1 - \lambda)^{-p} \geq \int_{B'} |\nabla s_i|^p \, d\mu \geq \mathrm{cap}_{p,\mu}(\{y_i\}, \mathbf{R}^n \setminus \{f(x_0)\}) \, .$$

Since the right hand side of this inequality tends to ∞ as i tends to infinity by the strong nonpolarity of $f(x_0)$, we have a contradiction. Therefore

$$\lim_{i \to \infty} u_i(f(x_0)) = 1 \, .$$

By using the equicontinuity and Ascoli's theorem we thus deduce that u_i converges to an $\mathcal{A}$-harmonic function u_0 uniformly in compact subsets of $B' \setminus \{f(x_0), f(z_0)\}$ and that

$$\lim_{y \to f(x_0)} u_0(y) = 1 \, .$$

We also infer from Theorem 6.33 and the boundary estimate (6.37) that $u_0(y)$ tends to 0 as y tends to $f(z_0)$ or to a point on the boundary of B'; so u_0 is the required $\mathcal{A}$-potential of $\{f(x_0)\}$ in $B' \setminus \{f(z_0)\}$.

Similarly, $u_i \circ f$ converges in $B \setminus f^{-1}(f(z_0))$ to an $\mathcal{A}^*$-harmonic function v_0. Since $v_0 \leq 1$ and $v_0(x_0) = 1$, the maximum principle implies $v_0 \equiv 1$ in the x_0-component of $B \setminus f^{-1}(f(z_0))$. This is, however, impossible because

$$\lim_{x \to y} v_0(x) = 0$$

for all $y \in f^{-1}(f(z_0))$. The theorem follows. $\square$

Now suppose that $f \colon \Omega \to \mathbf{R}^n$ is a nonconstant $(\mathcal{A}^*, \mathcal{A})$-harmonic morphism in a domain Ω and suppose that $x \in \Omega$ is such that either $f(x)$ is strongly non-$\mathcal{A}$-polar or for some neighborhood U of $f(x)$ points in U are all $\mathcal{A}$-polar. Then the proofs for Theorems 13.3 and 13.4 show that $f(x) \in \mathrm{int}\, f(A)$ for every neighborhood A of a point $y \in \partial f^{-1}(f(x)) \cap \Omega$. Note that the set $\partial f^{-1}(f(x)) \cap \Omega$ is nonempty because f is nonconstant. In other words, for each domain $A \subset \Omega$ either $f(A)$ is a point or an open set. This property of morphisms deserves to be formulated in the following theorem.

13.5. Theorem. *Suppose that f is a nonconstant $(\mathcal{A}^*, \mathcal{A})$-harmonic morphism in a domain Ω. If either all points are $\mathcal{A}$-polar or all points are strongly non-$\mathcal{A}$-polar, then $f(\Omega)$ is an open set.*

Theorem 13.5 does not assert that $f(U)$ is open whenever $U \subset \Omega$ is open; it says that $f(U)$ is either open or a point. In Theorem 13.10 below we show that if points are all $\mathcal{A}$-polar, then f is an open mapping. It is not known whether nonconstant $\mathfrak{A}_p$-harmonic morphisms are open mappings in general.

There exist nonconstant continuous mappings $f: \Omega \to \mathbf{R}^n$ such that for each open set $U \subset \Omega$, $f(U)$ is either open or a point and $f^{-1}(y)$ has nonempty interior for some $y \in \mathbf{R}^n$, i.e. f does not satisfy the Radó property. To display an example in $\mathbf{R}^2$, define first $g_1 : \mathbf{R} \to \mathbf{R}$ by

$$g_1(t) = \begin{cases} 1 + t & t \in (-\infty, -1) \\ 0 & t \in [-1, 0) \\ t & t \in [0, \infty). \end{cases}$$

Then define $g: \mathbf{R}^2 \to \mathbf{R}^2$ by $g(x) = \big(g_1(x_1), x_2\big)$ for $x = (x_1, x_2) \in \mathbf{R}^2$ and let $h(x) = (r, \varphi)$ be the mapping $r = |x_1|$, $\varphi = x_2/x_1$ in the polar coordinates (r, φ) of $\mathbf{R}^2$. Then $f = h \circ g: \mathbf{R}^2 \to \mathbf{R}^2$ has the required properties. Note that $h: \mathbf{R}^2 \to \mathbf{R}^2$ is a continuous, open mapping, which maps the x_2-axis to 0; thus f maps the strip $\{(x_1, x_2) \in \mathbf{R}^2 : -1 \leq x_1 \leq 0\}$ to 0.

Theorem 13.5 implies the following maximum principle for $\mathfrak{A}_p$-harmonic morphisms.

13.6. Theorem. *Suppose that* $f: \Omega \to \mathbf{R}^n$ *is an* $(\mathcal{A}^*, \mathcal{A})$-*harmonic morphism and that either all points are $\mathcal{A}$-polar or all points are strongly non-$\mathcal{A}$-polar. Then for each $x \in \Omega$ it holds that*

$$|f(x)| \leq \limsup_{y \to \partial\Omega} |f(y)| \,.$$

Moreover, if Ω is a domain and if f is nonconstant, then the above inequality is strict.

We next show that morphisms preserve superharmonic functions as well, excepting the trivial case when f is constant in a component. Namely, if $f \equiv b \in \mathbf{R}^n$ in Ω, an $\mathcal{A}$-superharmonic function u may satisfy $u(b) = \infty$, and then $u \circ f$ is identically ∞ in Ω.

13.7. Theorem. *Suppose that* $f: \Omega \to \mathbf{R}^n$ *is an* $(\mathcal{A}^*, \mathcal{A})$-*harmonic morphism and that u is $\mathcal{A}$-superharmonic in an open set Ω'. If either f is nonconstant in each component of Ω or u takes only finite values, then $u \circ f$ is $\mathcal{A}^*$-superharmonic in $f^{-1}(\Omega')$.*

PROOF: Let D be a domain, compactly contained in $f^{-1}(\Omega')$. It suffices to show that $u \circ f$ is $\mathcal{A}^*$-superharmonic in D. Choose an increasing sequence of continuous $\mathcal{A}$-superharmonic functions u_i converging to u in Ω' (Theorem

8.15). Let $s_i = u_i - 1/i$. Since s_i is continuous, there is $r_i > 0$ such that for each $x \in f(\overline{D})$ the ball $B(x, r_i)$ is compactly contained in Ω' and that

$$(13.8) \qquad \operatorname{osc}(s_i, B(x, r_i)) \leq \frac{1}{i(i+1)}.$$

Pick a finite cover $B_{i,j} = B(x_j, r_i)$, $j = 1, \ldots, k_i$, of $f(\overline{D})$ and let

$$s_{i,j} = P(s_i, B_{i,j})$$

be the Poisson modification of s_i with respect to $B_{i,j}$. Then

$$v_i = \min_{j=1,\ldots,k_i} s_{i,j}$$

is a continuous $\mathcal{A}$-superharmonic function in Ω'. We claim that v_i is an increasing sequence with limit u. To see this, note first that by the maximum principle and by (13.8) we have

$$0 \leq s_i - s_{i,j} \leq \operatorname{osc}(s_i, B_{i,j}) \leq \frac{1}{i(i+1)}.$$

Therefore, for each $j = 1, \ldots, k_i$ and $j' = 1, \ldots, k_{i+1}$ it holds that

$$\begin{aligned}
s_{i+1,j'} - s_{i,j} &\geq s_{i+1,j'} - s_i \\
&= s_{i+1,j'} - s_{i+1} + s_{i+1} - s_i \\
&\geq -\frac{1}{(i+1)(i+2)} + \frac{1}{i(i+1)} > 0.
\end{aligned}$$

Thus v_i is increasing, and since also $|v_i - s_i| \leq 1/i(i+1)$, we conclude that

$$\lim_{i \to \infty} v_i = u.$$

Fix i and $x \in D$. Let $B_1, \ldots, B_k$ be all the balls of the form $B_{i,j}$ that contain $f(x)$. By the continuity of f, the point x has a neighborhood $U \Subset D$ such that $f(U) \subset B_\ell$ for all $\ell = 1, \ldots, k$. Since $P(s_i, B_\ell) \leq s_i$, we obtain

$$v_i \circ f = \min_{\ell=1,\ldots,k} P(s_i, B_\ell) \circ f$$

in U. Since each $P(s_i, B_\ell)$ is $\mathcal{A}$-harmonic in B_ℓ and since f is an $(\mathcal{A}^*, \mathcal{A})$-harmonic morphism, $P(s_i, B_\ell) \circ f$ is $\mathcal{A}^*$-harmonic in U. As a minimum of a finite number of $\mathcal{A}^*$-harmonic functions, $v_i \circ f$ is $\mathcal{A}^*$-superharmonic in U. Thus $v_i \circ f$ is $\mathcal{A}^*$-superharmonic in D. Because $v_i \circ f$ increases to $u \circ f$ in

D, $u \circ f$ is $\mathcal{A}^*$-superharmonic in D provided it is not identically ∞ there (Lemma 7.3). Thus the proof is finished if u takes only finite values.

Now suppose that f is not constant in the D-component of Ω and that $u \circ f \equiv \infty$ in D. Then $f(D)$ is an $\mathcal{A}$-polar set. Because f assumes at least two distinct values in the D-component of Ω, we may assume, by choosing D large enough, that $f(D) \setminus \{y\}$ is nonempty for some $y \in f(D)$. By Theorem 10.2 there is an entire $\mathcal{A}$-superharmonic function s in $\mathbf{R}^n$ such that $s = \infty$ on $f(D) \setminus \{y\}$ but $s(y) < \infty$. Replacing u by s in the construction earlier in this proof, we find $\mathcal{A}^*$-superharmonic functions $s_i \circ f$ which increase to $s \circ f$ in D. Since $s(y) < \infty$ and $y \in f(D)$, $s \circ f$ is $\mathcal{A}^*$-superharmonic in D. On the other hand, we have that $s \circ f \equiv \infty$ in the nonempty open set $D \setminus f^{-1}(y)$, which is impossible because polar sets have no interior points. The theorem follows. $\qquad \square$

We have the following consequence of Theorem 13.7.

13.9. Theorem. *Suppose that $f \colon \Omega \to \mathbf{R}^n$ is an $(\mathcal{A}^*, \mathcal{A})$-harmonic morphism which is nonconstant in each component of Ω. If $E \subset \mathbf{R}^n$ is $\mathcal{A}$-polar, then $f^{-1}(E)$ is $\mathcal{A}^*$-polar. In particular, $f(U)$ is never $\mathcal{A}$-polar for open $U \subset \Omega$.*

PROOF: If E is $\mathcal{A}$-polar, there is an entire $\mathcal{A}$-superharmonic function u in $\mathbf{R}^n$ such that $u = \infty$ on E. Since $u \circ f = \infty$ on $f^{-1}(E)$ and since, by Theorem 13.7, $u \circ f$ is an $\mathcal{A}^*$-superharmonic function in Ω, the assertion follows. $\qquad \square$

Since an $\mathcal{A}$-polar set cannot contain any nonempty open set, Theorem 13.9 implies that a nonconstant $(\mathcal{A}^*, \mathcal{A})$-harmonic morphism has the Radó property whenever points are $\mathcal{A}$-polar. Thus we obtain from Theorem 13.3 the following result.

13.10. Theorem. *Suppose that Ω is a domain and that $f \colon \Omega \to \mathbf{R}^n$ is a nonconstant $(\mathcal{A}^*, \mathcal{A})$-harmonic morphism. If points are $\mathcal{A}$-polar, then f is an open mapping.*

In classical potential theory a function u in Ω is called *hyperharmonic* if in each component of Ω u is either superharmonic or identically infinite. Employing a similar definition for $\mathcal{A}$-superharmonic functions we may restate Theorem 13.7: *If $f \colon \Omega \to \mathbf{R}^n$ is an $(\mathcal{A}^*, \mathcal{A})$-harmonic morphism and if u is $\mathcal{A}$-hyperharmonic in Ω', then $u \circ f$ is $\mathcal{A}^*$-hyperharmonic in $f^{-1}(\Omega')$.* We do not employ the notion of hyperharmonicity in this book.

Theorem 13.9 states that the preimage of a polar set is polar under nonconstant morphisms. The following result demonstrates a similar phenomenon for nonpolar sets.

13.11. Theorem. *Suppose that* $f \colon \Omega \to \mathbf{R}^n$ *is an* $(\mathcal{A}^*, \mathcal{A})$-*harmonic morphism. If a Borel set* $E \subset \mathbf{R}^n$ *is not* $\mathcal{A}$-*polar, then* $f^{-1}(E)$ *is not* $\mathcal{A}^*$-*polar.*

PROOF: Since E is a Borel set of positive (p, μ)-capacity, it contains a compact set of positive (p, μ)-capacity, and we may assume that E itself is compact. Now by the Kellogg property (Theorem 9.13) $\Omega' = \mathbf{R}^n \setminus E$ has at least one regular boundary point x_0 with respect to $\mathcal{A}$. If $\{x_0\}$ is $\mathcal{A}$-polar, there must be another regular boundary point of Ω' and hence a positive $\mathcal{A}$-harmonic function u in Ω' such that $u(y)$ approaches 0 as y approaches x_0 inside Ω'. If $\{x_0\}$ is not $\mathcal{A}$-polar, the existence of a similar function u is guaranteed by Lemma 7.31.

Suppose that $f^{-1}(E)$ is $\mathcal{A}^*$-polar; then $f^{-1}(E)$ is of zero (p, μ^*)-capacity and hence removable for $\mathcal{A}^*$-superharmonic functions that are bounded below (Theorem 7.35). Because $v = u \circ f$ is a positive $\mathcal{A}^*$-harmonic function in the open set $\Omega \setminus f^{-1}(E)$, it can be extended to a nonnegative $\mathcal{A}^*$-superharmonic function v in Ω by setting

$$v(x) = \operatorname*{ess\,lim\,inf}_{y \to x} v(y)$$

for $x \in f^{-1}(E)$. It follows, however, that $v(x) = 0$ for $x \in f^{-1}(x_0)$, which violates the minimum principle. The theorem follows. $\square$

We close this section by establishing two nonexistence theorems for harmonic morphisms. The first is an immediate consequence of Theorem 13.9 which implies that images of nonpolar points cannot be polar. It is not known whether a nonconstant $(\mathcal{A}^*, \mathcal{A})$-harmonic morphism can map an $\mathcal{A}^*$-polar point to a point which is not $\mathcal{A}$-polar.

13.12. Theorem. *Suppose that points are* $\mathcal{A}$-*polar and that there is* $x \in \mathbf{R}^n$ *which is not* $\mathcal{A}^*$-*polar. If* Ω *is any connected neighborhood of* x, *then there are no nonconstant* $(\mathcal{A}^*, \mathcal{A})$-*harmonic morphisms* $f \colon \Omega \to \mathbf{R}^n$.

Theorem 13.12 is quite interesting in light of the following example. Let $\mathcal{A}$ be any mapping with constant weight $w = 1$, and let $\mathcal{A}^*$ be a mapping with weight $w^*(x) = |x|^\delta$, where $1 < p \le n$ and $-n < \delta < p - n$. Then *there are no nonconstant* $(\mathcal{A}^*, \mathcal{A})$-*harmonic morphisms* $f \colon \mathbf{R}^n \to \mathbf{R}^n$. Indeed, by Example 2.22, $\mathrm{cap}_{p, \mu^*}(\{0\}) > 0$, while every singleton is $\mathcal{A}$-polar. On the other hand, for each $1 < p < n$ and $\delta > p - n$ the quasiconformal mapping $f(x) = x|x|^{\delta/(n-p)}$ is an $(\mathcal{A}^*, \mathcal{A})$-harmonic morphism where $\mathcal{A}(y, \xi) = |\xi|^{p-2}\xi$ and $\mathcal{A}^*(x, \xi) = |x|^\delta |\xi|^{p-2}\xi$ (see Chapter 15).

Next we establish the following Picard type theorem for morphisms.

13.13. Theorem. *Suppose that* $f \colon \mathbf{R}^n \to \mathbf{R}^n$ *is an* $(\mathcal{A}^*, \mathcal{A})$-*harmonic morphism. If* f *omits a non-$\mathcal{A}$-polar set, then* f *is constant. In particular, if no point is $\mathcal{A}$-polar and* f *is nonconstant, then* f *is onto.*

PROOF: Suppose that f is nonconstant. We first claim that $E = f(\mathbf{R}^n)$ is dense in $\mathbf{R}^n$. If this is not the case, then elementary geometric considerations reveal that we can find two disjoint open balls B_1 and B_2 in the complement of E together with two distinct points x_1 and x_2 such that $x_i \in \partial E \cap \partial B_i$ for $i = 1, 2$. For $0 < t < 1$ let

$$\Omega_t = \mathbf{R}^n \setminus \{t\overline{B}_1 \cup t\overline{B}_2\}$$

and let, for $i = 1, 2$, x_i^t be the point on the boundary of tB_i closest to x_i. Then x_1^t and x_2^t both are regular boundary points of Ω_t (Theorem 9.10) and we can find $\mathcal{A}$-harmonic functions u_t in Ω_t such that $0 \le u_t < 1$ and that u_t has limit 0 at x_1^t and 1 at x_2^t. The uniform Wiener type boundary estimate of Theorem 6.18 guarantees that there is $t < 1$ such that $u_t(x_1) < 1/2$ and $u_t(x_2) > 1/2$. Consequently, with this choice of t, the function $v = u_t \circ f$ is nonconstant, bounded, and $\mathcal{A}^*$-harmonic in $\mathbf{R}^n$, contradicting Liouville's theorem. Thus E is dense in $\mathbf{R}^n$.

Next, since f is continuous, E as a countable union of compact sets is a Borel set, and the Choquet property of (p, μ)-capacities implies that there is a closed set $F \subset \mathbf{R}^n \setminus E$ such that F has positive (p, μ)-capacity. Thus by the Kellogg property (Theorem 9.13) $\Omega' = \mathbf{R}^n \setminus F$ has at least one regular boundary point x_0 with respect to $\mathcal{A}$. We have two possibilities: either $\{x_0\}$ is $\mathcal{A}$-polar or it is not. In the first case there must exist another regular boundary point y_0 of Ω', and hence also a bounded $\mathcal{A}$-harmonic function u in Ω' such that $u(y)$ approaches different limits when y approaches x_0 or y_0 inside Ω'. Because x_0 and y_0 lie on F, the composition $u \circ f$ is nonconstant, bounded, and $\mathcal{A}^*$-harmonic in $\mathbf{R}^n$, again a contradiction. Finally, if $\{x_0\}$ is not $\mathcal{A}$-polar, it follows from Lemma 7.31 and Theorem 6.33 that there is a positive $\mathcal{A}$-harmonic function u in the complement of x_0 with $\lim_{y \to x_0} u(y) = 0$. Consequently, $u \circ f$ is a nonconstant positive $\mathcal{A}^*$-harmonic function in $\mathbf{R}^n$, which is impossible by Harnack's inequality. The theorem follows.   □

As a corollary we obtain Liouville's theorem.

13.14. Liouville's theorem for harmonic morphisms. *If* $f \colon \mathbf{R}^n \to \mathbf{R}^n$ *is a bounded* $(\mathcal{A}^*, \mathcal{A})$-*harmonic morphism, then* f *is constant.*

13.15. Morphisms, Radó's theorem, and the $\mathcal{A}$-harmonic measure
A classical theorem of Radó says that if a continuous mapping $f \colon \Omega \to \mathbf{R}^2$ is analytic in $\Omega \setminus f^{-1}(x_0)$, then f is analytic in Ω. There is a plethora of

generalizations of this theorem in the literature; see Notes to this chapter. Before we present an extension of Radó's theorem we establish a removability theorem for $\mathfrak{A}_p$-harmonic morphisms.

13.16. Theorem. *Suppose that C is a relatively closed $\mathcal{A}^*$-polar set of Ω and that $f \colon \Omega \to \mathbf{R}^n$ is a continuous mapping which is an $(\mathcal{A}^*, \mathcal{A})$-harmonic morphism in $\Omega \setminus C$. Then f is an $(\mathcal{A}^*, \mathcal{A})$-harmonic morphism in Ω.*

PROOF: Let u be an $\mathcal{A}$-harmonic function in Ω' and consider its pull-back $v = u \circ f$. We know that v is $\mathcal{A}^*$-harmonic in $f^{-1}(\Omega') \setminus C$ and we want to show that v is $\mathcal{A}^*$-harmonic in $f^{-1}(\Omega')$. To do so, pick a ball $B \Subset f^{-1}(\Omega')$; it suffices to show that v is $\mathcal{A}^*$-harmonic in B. Since C is $\mathcal{A}^*$-polar, it is of zero (p, μ^*)-capacity, and hence removable for bounded $\mathcal{A}^*$-harmonic functions (Theorem 7.36). In particular, since v is locally bounded by continuity of f, v is $\mathcal{A}^*$-harmonic in B as desired. $\quad\square$

13.17. Theorem. *Suppose that $f \colon \Omega \to \mathbf{R}^n$ is continuous and that $C \subset \mathbf{R}^n$ is a closed $\mathcal{A}$-polar set. If f is an $(\mathcal{A}^*, \mathcal{A})$-harmonic morphism in $\Omega \setminus f^{-1}(C)$, then f is an $(\mathcal{A}^*, \mathcal{A})$-harmonic morphism in Ω.*

PROOF: We may assume that f is nonconstant and that Ω is connected. Since C is $\mathcal{A}$-polar, we may choose an $\mathcal{A}$-superharmonic function u in $\mathbf{R}^n$ such that $u = \infty$ on C. Because $v = u \circ f$ is $\mathcal{A}^*$-superharmonic in $\Omega \setminus C$ and f is continuous, we may extend v to be $\mathcal{A}^*$-superharmonic in Ω by setting $v = \infty$ on $f^{-1}(C)$. Hence $f^{-1}(C)$ is $\mathcal{A}^*$-polar and the assertion follows from the removability theorem above. $\quad\square$

In particular, we have:

13.18. Radó's theorem for harmonic morphisms. *Suppose that $f \colon \Omega \to \mathbf{R}^n$ is a continuous mapping which is an $(\mathcal{A}^*, \mathcal{A})$-harmonic morphism in $\Omega \setminus f^{-1}(x_0)$. If $\{x_0\}$ is $\mathcal{A}$-polar, then f is an $(\mathcal{A}^*, \mathcal{A})$-harmonic morphism in Ω.*

Next we illustrate the use of $\mathcal{A}$-harmonic measure in the theory morphisms. Recall that if Ω is an open set and $E \subset \partial\Omega$, then for $x \in \Omega$ the $\mathcal{A}$-harmonic measure $\omega = \omega(E, \Omega; \mathcal{A})$ is defined by

$$\omega(x) = \omega(E, \Omega; \mathcal{A})(x) = \inf u(x),$$

where the infimum is taken over all nonnegative $\mathcal{A}$-superharmonic functions u in Ω such that

$$\liminf_{x \to y} u(x) \geq 1$$

for all $y \in E$. See Chapter 11.

Before we present the principle of $\mathcal{A}$-harmonic measure for $\mathfrak{A}_p$-harmonic morphisms, we recall that if $f \colon \Omega \to \mathbf{R}^n$ is a mapping, the *cluster set* $\mathcal{C}(f, E)$ of f at $E \subset \partial\Omega$ is the set of all points $w \in \mathbf{R}^n \cup \{\infty\}$ for which there exists a sequence x_i in Ω converging to a point in E such that $f(x_i)$ converges to w.

Observe that $\partial\Omega$, and hence E, may contain the point at infinity. Also, $\mathcal{C}(f, E)$ may contain ∞ even if $\partial\Omega$ does not. If E is compact, the cluster set $\mathcal{C}(f, E)$ can alternatively be defined as the intersection of the sets $\overline{f(U \cap \Omega)}$, where U runs through all neighborhoods of E. If E is nonempty, the corresponding cluster set is also nonempty. Moreover, if E is compact, then $\mathcal{C}(f, E)$ is a compact set in $\mathbf{R}^n \cup \{\infty\}$.

13.19. Principle of $\mathcal{A}$-harmonic measure for $\mathfrak{A}_p$-harmonic morphisms. *Let Ω and Ω' be open sets in $\mathbf{R}^n$ and let $f \colon \Omega \to \Omega'$ be an $(\mathcal{A}^*, \mathcal{A})$-harmonic morphism. If $E \subset \partial\Omega$ is such that $\mathcal{C}(f, E) \cap \Omega'$ is a relatively closed subset of Ω', then*

$$\omega(E, \Omega; \mathcal{A}^*)(x) \le \omega(\mathcal{C}(f, E), \Omega'; \mathcal{A})\big(f(x)\big)$$

for all x in Ω.

Recall that

$$\omega(\mathcal{C}(f, E), \Omega'; \mathcal{A}) = \omega(\mathcal{C}(f, E) \cap \partial(\Omega' \setminus \mathcal{C}(f, E)), \Omega' \setminus \mathcal{C}(f, E); \mathcal{A}),$$

and the inequality in the theorem is understood to be the trivial inequality

$$\omega(E, \Omega; \mathcal{A}^*)(x) \le 1$$

if $f(x) \in \mathcal{C}(f, E) \cap \Omega'$.

PROOF OF 13.19: Suppose that u belongs to the upper class for the $\mathcal{A}$-harmonic measure $\omega(\mathcal{C}(f, E), \Omega'; \mathcal{A})$, i.e. u is nonnegative and $\mathcal{A}$-superharmonic in $\Omega' \setminus \mathcal{C}(f, E)$ with $\liminf_{y \to z} u(y) \ge 1$ for every $z \in \mathcal{C}(f, E) \cap \partial(\Omega' \setminus \mathcal{C}(f, E))$. We may assume that $u \le 1$, and by extending u as 1 to $\mathcal{C}(f, E) \cap \Omega'$, we may assume that u is $\mathcal{A}$-superharmonic in Ω'. Then $v = u \circ f$ is a nonnegative $\mathcal{A}^*$-superharmonic function in Ω by Theorem 13.7, and since $\liminf_{x \to z} v(x) = 1$ for every $z \in E$, v belongs to the upper class for $\omega = \omega(E, \Omega; \mathcal{A}^*)$. Thus

$$\omega(x) \le v(x) = u\big(f(x)\big),$$

and the result follows by taking the infimum over all such u. □

13.20. Principle of $\mathcal{A}$-harmonic measure for homeomorphic morphisms. *Suppose that f is a homeomorphic $(\mathcal{A}^*, \mathcal{A})$-harmonic morphism of Ω onto Ω'. If $E \subset \partial\Omega$, then*

$$(13.21) \qquad \omega(E, \Omega; \mathcal{A}^*)(x) \leq \omega\big(\mathcal{C}(f, E), \Omega'; \mathcal{A}\big)\big(f(x)\big)$$

for each $x \in \Omega$. If $\mathcal{C}\big(f^{-1}, \mathcal{C}(f, E)\big) = E$, then (13.21) holds as an equality. In particular, if f has a homeomorphic extension to $\overline{\Omega}$, then

$$(13.22) \qquad \omega(E, \Omega; \mathcal{A}^*)(x) = \omega\big(\mathcal{C}(f, E), \Omega'; \mathcal{A}\big)\big(f(x)\big) \, .$$

PROOF: Inequality (13.21) follows from 13.19 by observing that for homeomorphic mappings the cluster set $\mathcal{C}(f, E)$ is always contained in the boundary of Ω'.

Suppose next that $\mathcal{C}\big(f^{-1}, \mathcal{C}(f, E)\big) = E$. Let u be a function from the upper class for $\omega(E, \Omega; \mathcal{A}^*)$ and write $v = u \circ f^{-1}$. Now f^{-1} is an $(\mathcal{A}, \mathcal{A}^*)$-harmonic morphism from Ω' onto Ω (Theorem 13.2), whence v is $\mathcal{A}$-superharmonic and belongs to the upper class for $\omega\big(\mathcal{C}(f, E), \Omega'; \mathcal{A}\big)$. Thus we obtain the reverse inequality to (13.21) and the theorem follows. $\qquad\square$

The next theorem is a typical application of the principle of $\mathcal{A}$-harmonic measure.

13.23. Theorem. *Suppose that $f \colon \Omega \to \mathbf{R}^n$ is an $(\mathcal{A}^*, \mathcal{A})$-harmonic morphism in a domain Ω such that $\mathbf{R}^n \setminus f(\Omega)$ is not $\mathcal{A}$-polar. If $E \subset \partial\Omega$ has positive $\mathcal{A}^*$-harmonic measure in Ω and if $\mathcal{C}(f, E)$ is $\mathcal{A}$-polar, then f is constant.*

PROOF: Suppose that f is not constant. Because $f(\Omega)$ is a Borel set and because $\mathcal{C}(f, E)$ is contained in an $\mathcal{A}$-polar Borel set, there is a compact subset F of $\mathbf{R}^n \setminus \{f(\Omega) \cup \mathcal{C}(f, E)\}$ which has positive (p, μ)-capacity. By Theorem 10.12 we can find a positive $\mathcal{A}$-superharmonic function u in $\mathbf{R}^n \setminus F$ such that $u = \infty$ on $\mathcal{C}(f, E)$. Thus for all $\varepsilon > 0$ the function εv, where $v = u \circ f$, is a positive $\mathcal{A}^*$-superharmonic function in Ω with the property that

$$\liminf_{y \to x} \varepsilon\, v(y) = \infty$$

for all $x \in E$. Because v is finite at some point x_0, it follows that

$$\omega(x_0) = \omega(E, \Omega; \mathcal{A}^*)(x_0) \leq \varepsilon v(x_0) < \infty \, ,$$

and letting $\varepsilon \to 0$, we arrive at the contradiction $\omega = 0$. Thus f is constant and the theorem follows. $\qquad\square$

The following corollary illustrates the use of Theorem 13.23 in a special case.

13.24. Corollary. *Suppose that* $f\colon B \to \mathbf{R}^n$ *is a bounded* $(\mathcal{A}^*, \mathcal{A})$-*harmonic morphism in a ball* B. *If*

$$\lim_{x \to y} f(x) = b \in \mathbf{R}^n$$

for all points y *in a set* $E \subset \partial B$ *with nonempty interior on* ∂B *and if* $\{b\}$ *is* $\mathcal{A}$-*polar, then* f *is constant.*

PROOF: The assertion follows from Theorem 13.23 upon observing that E has positive $\mathcal{A}^*$-harmonic measure in B (Theorem 11.6).   □

NOTES TO CHAPTER 13. The study of harmonic morphisms for the Laplace equation can be traced back to a paper by Jacobi in 1848. Fuglede (1986) has a short historical account. The classification of morphisms between general Riemannian manifolds was completed by Fuglede (1978) and Ishihara (1979).

Radó's and Liouville's theorems have their origins in the corresponding results for analytic functions. Radó's theorem (Radó 1924) and its various extensions especially have attracted many authors; the papers by Král (1983) and Fuglede (1986) contain extensive bibliographies. The principle of $\mathcal{A}$-harmonic measure comes from the same source; see Nevanlinna (1953), Granlund et al. (1982). Fuglede (1979a) used singular harmonic functions to prove the openness of a harmonic morphism in the linear theory. Most of the results of this chapter were given by Heinonen et al. (1992) in the unweighted case. Theorem 13.7 is essentially due to Laine (1990); see also Constantinescu and Cornea (1965).

14

Quasiregular mappings

If u is a harmonic function in a planar open set Ω' and if $f\colon \Omega \to \Omega'$ is analytic, then the composed function $u \circ f$ is harmonic in Ω. This is an elementary fact in complex analysis. Conversely, one can show that if a mapping f pulls back harmonic functions in this fashion, then f is necessarily analytic or antianalytic. Quasiregular mappings constitute a natural generalization of analytic functions to higher dimensions and the important "pull-back principle" has its counterpart in the theory of these mappings. This is the subject of the present chapter. Throughout, our concern will be in the nonweighted case. We prove that quasiregular mappings $(p = n)$ and mappings of bounded length distortion (all $p > 1$) are $\mathfrak{A}_p$-harmonic morphisms. Several applications of this fact are given, including a proof of the Picard theorem for quasiregular mappings.

Many fundamental properties of quasiregular mappings lie deep and their thorough investigation would be unreasonable within this book. We present mostly without proof all the properties that are required for a geometric approach to the morphism property. Thus the exposition in the present chapter differs somewhat from the rest of the book which is more self-contained. For a more detailed study of quasiconformal and quasiregular mappings we refer to the monographs by Lehto and Virtanen (1973), Väisälä (1971), Reshetnyak (1989), Vuorinen (1988), and Rickman (In preparation).

In this chapter we assume that $w \equiv 1$ or $d\mu = dx$, and to emphasize this we let $W^{1,p}(\Omega)$, $W_0^{1,p}(\Omega)$, ... stand for the Sobolev spaces $H^{1,p}(\Omega; dx)$, $H_0^{1,p}(\Omega; dx)$, ...

14.1. Quasiregular mappings

DEFINITION. A continuous mapping $f\colon \Omega \to \mathbf{R}^n$ is said to be *quasiregular* if the coordinate functions of f belong to $W_{loc}^{1,n}(\Omega)$ and if there is $K \geq 1$ such that the inequality

$$(14.2) \qquad |f'(x)|^n \leq K\, J_f(x)$$

is satisfied for a.e. x in Ω.

Here $f'(x)$ denotes the formal derivative of f at x, i.e. the $n \times n$ matrix $\big(\partial_j f_i(x)\big)$ of the partial derivatives of the coordinate functions f_i of f. Further,

$$|f'(x)| = \max_{|h|=1} |f'(x)\, h|$$

250

and $J_f(x)$ is the Jacobian determinant of f at x, i.e. the determinant of $f'(x)$.

The smallest $K \geq 1$ for which inequality (14.2) is true is called the *outer dilatation* of f and denoted by $K_O(f)$. If f is quasiregular, then the smallest $K \geq 1$ for which the inequality

(14.3) $J_f(x) \leq K \, \ell(f'(x))$

holds for a.e. x in Ω is called the *inner dilatation* of f and denoted by $K_I(f)$. Here

$$\ell(f'(x)) = \min_{|h|=1} |f'(x)h|.$$

By elementary linear algebra, $K_I(f)$ is finite if f is quasiregular. The number

$$K(f) = \max\left(K_O(f), K_I(f)\right)$$

is the *maximal dilatation* of f and if $K(f) \leq K$, f is called K-*quasiregular*. By linear algebra it is not difficult to see that $K_O(f) \leq K_I(f)^{n-1}$ and $K_I(f) \leq K_O(f)^{n-1}$. Thus if $n = 2$, we have $K_I(f) = K_O(f) = K(f)$. One can also show that if $f_1 : \Omega \to \Omega'$ is K_1-quasiregular and if $f_2 : \Omega' \to \mathbf{R}^n$ is K_2-quasiregular, then $f_2 \circ f_1 : \Omega \to \mathbf{R}^n$ is $K_1 K_2$-quasiregular, but this is difficult to prove starting from the definition (Martio *et al.* 1969; Bojarski and Iwaniec 1983; Reshetnyak 1989; see also Theorem 14.28 below).

If $f \colon \Omega \to \mathbf{R}^n$ is a quasiregular homeomorphism onto $f(\Omega)$, then f is termed *quasiconformal*. In this case f^{-1} is a quasiconformal mapping in $f(\Omega)$ with $K_O(f^{-1}) = K_I(f)$, $K_I(f^{-1}) = K_O(f)$, and $K(f^{-1}) = K(f)$.

When $n = 2$ and $K = 1$ in (14.2) we recover analytic functions of one complex variable (Lehto and Virtanen 1973, Chapter VI). Especially, a 1-quasiconformal mapping is a conformal mapping. When $n \geq 3$, quasiregular mappings exhibit a remarkable rigidity property: a 1-quasiregular mapping $f \colon \Omega \to \mathbf{R}^n$ is either constant or the restriction to Ω of a Möbius transformation (Gehring 1962; Reshetnyak 1989; Bojarski and Iwaniec 1982). It is because of this rigidity that one has to allow a certain amount of distortion to obtain a nontrivial higher dimensional function theory.

Quasiregular mappings are usually constructed via explicit geometric procedures and existing examples show that quasiregularity allows exceedingly complicated behavior for the mapping (Rickman 1980a, 1985). As an elementary example we consider the *winding mapping* $f \colon \mathbf{R}^3 \to \mathbf{R}^3$ defined in cylindrical coordinates

$$(r, \varphi, x_3) \mapsto (r, k\varphi, x_3),$$

where $k \geq 2$ is an integer. This mapping is a k-times winding about the x_3-axis and can be viewed as a three-dimensional analogue of the analytic function $f(z) = z^k$. It is k^2-quasiregular and fails to be locally injective on the x_3-axis. Obviously, a similar construction can be done in every dimension $n \geq 2$. We use the winding map to produce a more interesting example. Fix $0 < \varepsilon < 1$ and consider a quasiconformal self-homeomorphism h of $\mathbf{R}^3$ which has the property that the preimage of the x_3-axis is a nonrectifiable curve with Hausdorff dimension $3 - \varepsilon$. For a construction of such a mapping h, see Gehring and Väisälä (1973). Now if f is a winding map, $h^{-1} \circ f \circ h$ is a quasiregular mapping which fails to be a local homeomorphism in a set of large Hausdorff dimension; it winds about an exotic curve in $\mathbf{R}^3$. Martio *et al.* (1971) and Martio and Srebro (1975a,b) have constructed further examples of quasiregular mappings.

We take for granted the basic topological and analytic properties of quasiregular mappings; these properties are well known for plane analytic functions. For future reference, we now present the essential facts without proof.

14.4. Discrete and open mappings

A mapping $f: \Omega \to \mathbf{R}^n$ is said to be *discrete* if $f^{-1}(y)$ consists of isolated points in Ω for every $y \in \mathbf{R}^n$. A nonconstant analytic function in a plane domain is discrete and open, and the same is true for general quasiregular mappings as we shall see in Theorem 14.14. Unfortunately, discrete, open maps are much more complicated in dimensions greater than two than what they are in the plane. The purpose of this section is to review the basic theory of these mappings. We first study various multiplicity functions associated with discrete and open mappings.

Let $f: \Omega \to \mathbf{R}^n$ be continuous. The *topological degree* $\mu(y, f, D)$ of f at y is defined whenever $D \Subset \Omega$ is a domain and $y \in \mathbf{R}^n \setminus f(\partial D)$. The degree $\mu(y, f, D)$ is integer valued and it has the following properties:

 (i) $y \mapsto \mu(y, f, D)$ is constant in each component of $\mathbf{R}^n \setminus f(\partial D)$.
 (ii) If $y \in f(D)$ and the restriction of f to $\overline{D}$ is one-to-one, then $|\mu(y, f, D)| = 1$.
 (iii) If $y \in D$ and id is the identity mapping, then $\mu(y, id, D) = 1$.
 (iv) If $\mu(y, f, D_i)$ is defined for $i = 1, \ldots, k$ and if $D_1, \ldots, D_k$ are mutually disjoint domains such that $f^{-1}(y) \cap D \subset \cup_{i=1}^k D_i \subset D$, then

$$\mu(y, f, D) = \sum_{i=1}^k \mu(y, f, D_i).$$

 (v) If f and g are connected with a homotopy h_t, $0 \leq t \leq 1$, such that $\mu(y, h_t, D)$ is defined for $0 \leq t \leq 1$, then $\mu(y, f, D) = \mu(y, g, D)$.

If Ω is a domain and if for all domains $D \Subset \Omega$ and $y \in f(D) \setminus f(\partial D)$ we have $\mu(y, f, D) > 0$, then f is called *sense-preserving*. If $\mu(y, f, D) < 0$ for all such y and D, then f is called *sense-reversing*.

The standard reference for the topological degree is the monograph by Radó and Reichelderfer (1955). Bojarski and Iwaniec (1983) and Reshetnyak (1989) present a different, more analytic approach to the topological degree. If $f\colon \Omega \to \mathbf{R}^n$ is differentiable at $x_0 \in \Omega$ and $J_f(x_0) \neq 0$, then there exists a neighborhood D of x_0 such that $\mu(y, f, D) = \operatorname{sign} J_f(x_0)$ for all $y \in f(D)$. Thus the above definition of a sense-preserving mapping is an extension of the more familiar case when f is differentiable.

For a continuous mapping f we let B_f denote the set of all points $x \in \Omega$, where f does not define a local homeomorphism; the set B_f is called the *branch set* of f. Clearly B_f is a relatively closed subset of Ω. If $n = 2$ and $f\colon \Omega \to \mathbf{R}^2$ is complex analytic, then

$$B_f = \{z \in \Omega \colon f'(z) = 0\}.$$

For discrete and open mappings $f\colon \Omega \to \mathbf{R}^n$ a theorem of Chernavskiĭ(1964, 1965) asserts that the topological dimension (Hurewicz and Wallman 1941) of both B_f and $f(B_f)$ is less than or equal to $n-2$. This assures that neither B_f nor $f(B_f)$ separates any domain, and using this the following theorem is easily proved.

14.5. Theorem. (Chernavskiĭ1964, 1965; Väisälä 1966) *Suppose that f is a discrete and open mapping of a domain Ω into $\mathbf{R}^n$. Then f is either sense-preserving or sense-reversing.*

Next we study the local behavior of discrete and open mappings $f\colon \Omega \to \mathbf{R}^n$. If $n = 2$ and f is analytic, then f behaves locally at $z_0 \in \Omega$ like the complex polynomial $z \mapsto (z - z_0)^k + f(z_0)$ for some positive integer k. Representation theorems like this are impossible in higher dimensional Euclidean spaces, but the topological degree enables us to obtain some useful results in this direction.

In the following discussion, we assume that $f\colon \Omega \to \mathbf{R}^n$ is discrete and open.

A domain $D \Subset \Omega$ is called a *normal domain* of f if $f(\partial D) = \partial f(D)$. Since f is open, we always have $\partial f(D) \subset f(\partial D)$. A normal domain U is a *normal neighborhood* of $x \in U$ if

$$\{x\} = U \cap f^{-1}\big(f(x)\big).$$

If U is a normal neighborhood of x, then $\mu\big(f(x), f, U\big)$ is defined and independent of U; this number is denoted by $i(x, f)$ and called the *local topological index* of f at x.

If we use the above local representation for a nonconstant plane analytic function f, we have $i(z_0, f) = k$.

If Ω is a domain, then $\Omega \setminus B_f$ is a domain and $i(x, f)$ has a constant value, either $+1$ or -1, in $\Omega \setminus B_f$. In the first case f is sense-preserving and in the second case sense-reversing (Theorem 14.5).

14.6. Lemma. (Martio *et al.* 1969, Lemma 2.5) *Let* $f: \Omega \to \mathbf{R}^n$ *be discrete and open. Suppose that V is a domain in $\mathbf{R}^n$ and that D is a component of $f^{-1}(V)$. If $D \Subset \Omega$, then D is a normal domain of f and $f(D) = V$.*

We let $U(x, f, r)$ denote the x-component of $f^{-1}\big(B(f(x), r)\big)$, $x \in \Omega$. If $U = U(x, f, r) \Subset \Omega$, then U is a normal domain of f (Lemma 14.6). The next lemma shows that normal neighborhoods always exist.

14.7. Lemma. (Martio *et al.* 1969, Lemma 2.9) *For every point $x \in \Omega$ there is $\sigma_x > 0$ such that for $0 < r \le \sigma_x$*

(a) $U(x, f, r)$ *is a normal neighborhood of x.*
(b) $U(x, f, r) = U(x, f, \sigma_x) \cap f^{-1}\big(B(f(x), r)\big)$.
(c) *If $0 < r < s \le \sigma_x$, then $\overline{U}(x, f, r) \subset U(x, f, s)$ and $A = U(x, f, s) \setminus \overline{U}(x, f, r)$ is a ring domain, i.e. its complement has exactly two components.*
(d) $\operatorname{diam} U(x, f, r) \to 0$ *as $r \to 0$.*

If $y \in \mathbf{R}^n$ and $A \subset \Omega$, then the (crude) *multiplicity* of f in A is

$$N(y, f, A) = \# \{ f^{-1}(y) \cap A \};$$

i.e. $N(y, f, A)$ is the number of points in the set $f^{-1}(y) \cap A$. The possibility that $N(y, f, A) = \infty$ is not excluded but if $A \Subset \Omega$, then $N(y, f, A) < \infty$ because f is discrete. We also define

$$N(f, A) = \sup_{y \in \mathbf{R}^n} N(y, f, A).$$

Relations between the crude multiplicity functions $N(y, f, A)$ and $N(f, A)$, and the topological degree and the local topological index, are described in the following lemma.

14.8. Lemma. (Martio *et al.* 1969, Lemma 2.12) *Suppose that $f: \Omega \to \mathbf{R}^n$ is discrete, open, and sense-preserving. Let D be a normal domain of f. Then*

(a) $N(f, D) = \mu(y, f, D)$ *for all $y \in f(D)$.*
(b) $N(y, f, D) = N(f, D)$ *for all $y \in f(D \setminus B_f)$.*
(c) $\mu(y, f, D) = \sum_{j=1}^{k} i(x_j, f)$ *for all $y \in f(D)$, where $k = N(y, f, D)$ and $\{x_1, \ldots, x_k\} = f^{-1}(y) \cap D$.*
(d) $i(x, f) = N(f, D)$ *for $x \in D$ if and only if D is a normal neighborhood of x.*

Since $i(x, f) = 1$ for $x \in \Omega \setminus B_f$, Lemma 14.8 implies that for discrete, open, and sense-preserving mappings $f \colon \Omega \to \mathbf{R}^n$ we have $i(x, f) \geq 2$ if and only if $x \in B_f$.

The monograph by Rickman (In preparation) contains a detailed account of the above facts.

14.9. Path lifting

Next we discuss the path lifting, which is an important tool in the theory of discrete and open mappings.

By a *path* we mean a continuous mapping $\beta \colon \Delta \to \mathbf{R}^n$, where Δ is an interval. We also denote by β the *locus* $\beta(\Delta)$ of β.

Suppose that $f \colon \Omega \to \mathbf{R}^n$ is discrete, open, and sense-preserving. Let $x \in \Omega$ and consider a path $\beta \colon [a, b] \to \mathbf{R}^n$ such that $\beta(a) = f(x)$. A path $\alpha \colon \Delta_c \to \Omega$, where $c \leq b$ and $\Delta_c = [a, c)$ or $\Delta_c = [a, b]$, is a *lift* of β (under f) starting at x if $\alpha(a) = x$ and $f \circ \alpha = \beta|_{[a,c)}$. We say that α is a *total lift* of β if $\Delta_c = [a, b]$ and α is a *maximal lift* of β if α is not a proper subpath of any lift of β starting at x.

There is always at least one maximal lift α of β starting at x (Martio *et al.* 1971, p. 12). Moreover, in normal domains it is possible to obtain total lifts with special properties.

14.10. Lemma. (Rickman 1973) *Suppose that D is a normal domain of f and that $y \in f(D)$. Let $f^{-1}(y) \cap D = \{x_1, \ldots, x_k\}$, where $k = N(f, D)$ and each point x in $f^{-1}(y) \cap D$ is counted according to its local topological index $i(x, f)$. If $\beta \colon [a, b] \to f(D)$ is a path with $\beta(a) = y$, then there are total lifts $\alpha_1, \ldots, \alpha_k$ of β such that α_j starts at x_j, $j = 1, 2, \ldots, k$. Moreover, whenever $t \in [a, b]$ and $1 \leq j \leq k$, we have*

$$(14.11) \qquad \#\{\ell \colon \alpha_\ell(t) = \alpha_j(t)\} = i\big(\alpha_j(t), f\big).$$

As an example, consider the winding mapping $f \colon \mathbf{R}^3 \to \mathbf{R}^3$ described in the beginning of this section. If $\beta \colon [0, 1] \to \mathbf{R}^3$ is the path $\beta(t) = t\, e_3$, then the only total lift of β starting at 0 is the path β itself. Note that β runs inside the branch set of f; a behavior like this is not possible when $n = 2$, for then the branch set consists of isolated points.

Condition (14.11) yields

$$(14.12) \qquad f^{-1}\big(\beta(t)\big) \cap D = \{\alpha_1(t), \ldots, \alpha_k(t)\}$$

for each $t \in [a, b]$. To prove this, fix $t \in [a, b]$ and note that

$$\{\alpha_1(t), \ldots, \alpha_k(t)\} \subset f^{-1}\big(\beta(t)\big) \cap D$$

by the definition of the lift. As for the reverse inclusion, write $\{y_1, \ldots, y_p\} = f^{-1}(\beta(t)) \cap D$ and let

$$S_j = \{i \colon \alpha_i(t) = y_j\}, \qquad j = 1, \ldots, p.$$

We want to show that $\# S_j \geq 1$ for all j. If $\# S_j \geq 1$, then (14.11) yields $\# S_j = i(y_j, f)$. On the other hand, there are k lifts α_j altogether and for each j, $\alpha_j(t) \in \{y_1, \ldots, y_p\}$, whence

$$k = N(f, D) = \sum_{j=1}^p i(y_j, f) \geq \sum_{j=1}^p \# S_j = k.$$

Hence $\# S_j = i(x_j, f) \geq 1$ for each $j = 1, \ldots, p$ and (14.12) follows.

14.13. Topological properties of quasiregular mappings

A nonconstant analytic function in a plane domain is sense-preserving, discrete, and open. By a theorem of Reshetnyak (1967a, 1968, 1989) the same is true for quasiregular mappings:

14.14. Theorem. *If f is a nonconstant quasiregular mapping of a domain Ω into $\mathbf{R}^n$, then f is discrete, open, and sense-preserving.*

We sketch a proof of this important result to the extent that it hinges on the nonlinear potential theory developed in the previous chapters. A full proof would require a thorough study of both analytic and topological properties of quasiregular mappings, and this would take us too far.

We begin with an elementary lemma on divergence free vector fields. Recall that the *adjoint matrix* $\operatorname{adj} A$ of an $n \times n$ matrix A satisfies the identity

$$A \operatorname{adj} A = (\det A)\, \mathbf{I},$$

where $\mathbf{I}$ is the identity matrix. Thus $\operatorname{adj} A = (\det A)A^{-1}$ if the matrix A is invertible. For a mapping $f \colon \Omega \to \mathbf{R}^n$, $f \in W^{1,1}_{loc}(\Omega)$, we write $\operatorname{adj} f$ for the a.e. defined adjoint matrix $\operatorname{adj} f'(x)$. Recall that $f'(x)$ is the Jacobian matrix of the partial derivatives of the coordinate functions of f.

14.15. Lemma. *Let $f \colon \Omega \to \mathbf{R}^n$ be a mapping in $W^{1,n-1}_{loc}(\Omega)$. Then each column vector v_i of $\operatorname{adj} f$, $i = 1, \ldots, n$, satisfies*

$$(14.16) \qquad \int_\Omega v_i(x) \cdot \nabla\varphi(x)\, dx = 0$$

for all $\varphi \in C^\infty_0(\Omega)$.

PROOF: When f is smooth, (14.16) is equivalent to

$$\operatorname{div} v_i = 0$$

which follows from a direct computation; for the details, see Bojarski and Iwaniec (1983). For the general case, fix $\varphi \in C_0^\infty(\Omega)$, choose open $U \Subset \Omega$ such that $\operatorname{spt} \varphi \subset U$, and let f_j be a sequence of C^∞ mappings converging to f in $W^{1,n-1}(U)$. Since

$$\int_\Omega v_i^j(x) \cdot \nabla \varphi(x)\, dx = 0$$

for each $j = 1, 2, \ldots$ and $i = 1, \ldots, n$, where v_i^j are the column vectors of $\operatorname{adj} f_j$, we only need to convince ourselves that $\operatorname{adj} f_j$ converges to $\operatorname{adj} f$ in $L^1(U)$. This follows easily from the fact that each entry of $\operatorname{adj} f_j$ is a homogeneous polynomial of degree $n - 1$ of entries of $f_j'(x)$; observe here that $f_j'(x)$ converges to $f'(x)$ in $L^{n-1}(U)$. $\square$

14.17. Corollary. *Let $f: \Omega \to \Omega'$ be a mapping in $W_{loc}^{1,n}(\Omega)$ and let $v: \Omega' \to \mathbf{R}^n$ be a C^∞-mapping in Ω' such that $\operatorname{div} v = 0$. Then*

$$\int_\Omega (\operatorname{adj} f)(v \circ f) \cdot \nabla \varphi\, dx = 0$$

for all $\varphi \in C_0^\infty(\Omega)$.

PROOF: By approximating we may assume $f \in C^\infty(\Omega)$, and then a computation gives

$$\operatorname{div}(\operatorname{adj} f)(x)(v \circ f)(x) = J_f(x) \operatorname{div} v(f(x)) = 0\,.$$

$\square$

Next, let $f: \Omega \to \mathbf{R}^n$ be a nonconstant K-quasiregular mapping and consider the function $u(x) = \log |f(x)|$. Because f is locally in $W^{1,n}(\Omega)$, it is easy to see that u is locally in $W^{1,n}(\Omega_f)$, where Ω_f is the open set $\Omega \setminus \{f^{-1}(0)\}$, and that

$$\nabla u(x) = \frac{f'(x)^* f(x)}{|f(x)|^2}\,.$$

Above, $*$ is a sign for transpose. If we then define

$$(14.18) \qquad \mathcal{A}(x, \xi) = J_f(x)|f'(x)^{-1*}\xi|^{n-2} f'(x)^{-1} f'(x)^{-1*}\xi$$

if $f'(x)$ exists and is invertible, and

$$\mathcal{A}(x, \xi) = |\xi|^{n-2}\xi$$

otherwise, we see that

$$A(x, \nabla u(x)) = \frac{\operatorname{adj} f(x) f(x)}{|f(x)|^n}$$

for a.e. $x \in \Omega_f$. Here we have used quasiregularity in an essential way: for a.e. $x \in \Omega$ such that $J_f(x) = 0$ we have $f'(x) = 0$, and hence $\nabla u(x) = 0$.

Because an easy computation shows that div $v(y) = 0$ for $v(y) = y/|y|^n$, $y \neq 0$, Corollary 14.17 implies that

$$\int_{\Omega_f} A(x, \nabla u) \cdot \nabla \varphi \, dx = 0$$

for all $\varphi \in C_0^\infty(\Omega)$. On the other hand, by using the K-quasiregularity of f, one can easily calculate that A satisfies structure conditions (3.3)–(3.7) with $p = n$, $\alpha = 1/K$, and $\beta = K$ (see Lemma 14.38). Since the same argument applies to the K-quasiregular mapping $f(x) - b$ for $b \in \mathbf{R}^n$, we have arrived at the following theorem.

14.19. Theorem. *Let $f: \Omega \to \mathbf{R}^n$ be a nonconstant K-quasiregular mapping and let $b \in \mathbf{R}^n$. Then the function $u(x) = \log|f(x) - b|$ is A-harmonic in the open set $\Omega \setminus f^{-1}(b)$, where A satisfies structure conditions (3.3)–(3.7) with $p = n$, $w = 1$, $\alpha = 1/K$, and $\beta = K$.*

We scrutinize quasiregular mappings and the pull-back equation (14.18) more closely in Section 14.35. At this point Theorem 14.19 is sufficient for our purposes.

The proof of the fact that a nonconstant quasiregular mapping is discrete and open rests on Theorem 14.19 and the potential theory of A-harmonic functions. Let f be a nonconstant K-quasiregular mapping in a domain Ω in $\mathbf{R}^n$ and let b be a point in $\mathbf{R}^n$. The function $u(x) = -\log|f(x) - b|$ can be extended continuously to the set $E(b) = \{x \in \Omega: f(x) = b\}$ by putting $u(x) = \infty$ there. This extension is A-superharmonic in Ω implying that $E(b)$ is A-polar, and hence of zero n-capacity by Theorem 10.1. We saw in Chapter 2 that sets of zero n-capacity have Hausdorff dimension zero, and hence each component of such a set is a point. In brief, we have shown that *the preimage of every point under a nonconstant quasiregular mapping is a totally disconnected set*. Mappings with that property are called *light*, and a theorem of Titus and Young (1962) asserts that sense-preserving and light mappings are discrete and open (see also Bojarski and Iwaniec 1983, p. 318).

Thus Theorem 14.14 follows if we could show that f is sense-preserving. This is usually done by first establishing the almost everywhere differentiability of f and then exploiting the fact that the topological degree is locally the sign of the Jacobian J_f if J_f is not zero. We refer to the discussion in Reshetnyak (1967a, 1968, 1989), Bojarski and Iwaniec (1983), or Rickman (In preparation), and conclude the proof of Theorem 14.14.

We end this section by indicating a fundamental difference between the plane and higher dimensional theory of quasiregular mappings. If $f: \Omega \to \mathbf{R}^2$ is a discrete and open mapping in a plane open set Ω, then the branch set B_f is either empty or consists of isolated points in Ω. Moreover, $f = g \circ h$ where h is a homeomorphism of Ω into $\mathbf{R}^2$ and $g: h(\Omega) \to \mathbf{R}^2$ is an analytic function. This result is known as Stoïlow's decomposition theorem (Lehto and Virtanen 1973). Thus from the topological point of view plane discrete and open mappings are not more general than analytic functions. In particular, we have:

14.20. Theorem. (Lehto and Virtanen 1973, Chapter VI) *Suppose that* $f: \Omega \to \mathbf{R}^2$ *is a plane K-quasiregular mapping. Then* $f = g \circ h$*, where h is a K-quasiconformal mapping and g is an analytic function in* $h(\Omega)$.

In dimension $n = 2$ Theorem 14.20 can be used on many occasions either to prove or disprove assertions which are not known in dimensions $n \geq 3$. The two-dimensional theory has another remarkable feature unknown in space: one can solve a Beltrami equation with measurable data to obtain quasiconformal mappings with prescribed dilatation. This powerful device is widely applied in complex analysis and elsewhere. We do not study this aspect here and refer to the monographs by Lehto and Virtanen (1973) and Lehto (1987) for further discussion.

14.21. Metric and analytic properties of quasiregular mappings
This brief section contains basic facts about the analytic properties of quasiregular mappings. Some of them are rather difficult to prove starting from the definition.

14.22. Lemma. *Suppose that* $f: \Omega \to \mathbf{R}^n$ *is a quasiregular mapping. Then*

(a) *f is differentiable a.e.*
(b) *f satisfies condition (N), i.e. if $A \subset \Omega$ and $|A| = 0$, then also $|f(A)| = 0$.*

In addition, if Ω is a domain and f nonconstant, then

(c) *$|B_f| = 0$;*
(d) *$|A| = 0$ if and only if $|f(A)| = 0$ whenever $A \subset \Omega$;*
(e) *$J_f(x) > 0$ a.e.*

The proofs for the assertions in Lemma 14.22 can be found in Reshetnyak (1967a, 1970, 1989), Martio *et al.* (1969), or Bojarski and Iwaniec (1983).

Conditions (b) and (c) imply that $|f(B_f)| = 0$. This fact seems to be more important than (c). The example in Section 14.1 shows that when $n \geq 3$, for a nonconstant quasiregular mapping f the Hausdorff dimension of B_f can be arbitrarily close to n.

Each function u in the Sobolev space $W_{loc}^{1,p}(\Omega)$ has an ACL-representative. This means that $u = v$ a.e. in Ω, where v is absolutely continuous on almost every line segment, parallel to the coordinate axes, in each closed n-interval

$$Q = \{x \in \mathbf{R}^n : a_i \leq x_i \leq b_i\}$$

in Ω, $i = 1, \ldots, n$. More precisely, if P_i is the orthogonal projection of $\mathbf{R}^n$ to the ith coordinate plane and if E_i is the set of all $x \in P_i(Q)$ such that the function $t \mapsto v(x + t\,e_i)$ is not absolutely continuous on $[a_i, b_i]$, then $m_{n-1}(E_i) = 0$. Conversely, if u is a measurable ACL-function in Ω such that the partial derivatives $\partial_i u$ of u are locally p-integrable in Ω, then u belongs to the Sobolev space $W_{loc}^{1,p}(\Omega)$. For this characterization, see Ziemer (1989, Theorem 2.1.4).

If $u \in C(\Omega) \cap W_{loc}^{1,p}(\Omega)$, then u itself is the ACL-representative v. In particular, since the coordinate functions f_i of a quasiregular mapping $f \colon \Omega \to \mathbf{R}^n$ belong to $C(\Omega) \cap W_{loc}^{1,n}(\Omega)$, each f_i, and hence also f, has the ACL-property. Absolute continuity of a vector-valued mapping $g \colon [a, b] \to \mathbf{R}^n$ is defined similarly to the real-valued case: g is absolutely continuous if and only if its coordinate functions have this property. However, in what follows we need an absolute continuity property of the set valued correspondence f^{-1}.

Suppose that $f \colon \Omega \to \mathbf{R}^n$ is quasiregular and that D is a normal domain of f. Let $Q = \{y \in \mathbf{R}^n : a_i \leq y_i \leq b_i\}$ be in $f(D)$. Fix $i = 1, \ldots, n$ and for $y \in P_i(Q)$ let

$$\beta_y \colon [a_i, b_i] \to Q$$

be the line segment $\beta_y(t) = y + t\,e_i$.

14.23. Lemma. (Martio *et al.* 1969, pp. 9–10; Martio 1970, pp. 9–10) *Let $\alpha_y \colon [a_i, b_i] \to D$ be a total lift of β_y starting at some point of $f^{-1}(y + a_i\,e_i)$. Then α_y is absolutely continuous for almost every $y \in P_i(Q)$.*

The proof for Lemma 14.23 uses the Lebesgue differentiation theorem for measures and the metric characterization of quasiregular mappings; it also shows that Lemma 14.23 equally holds if a radial projection to a sphere is considered.

14.24. Integral transformation formula

Next we consider integral transformation formulas for quasiregular mappings. Although an integral transformation formula

$$\int_A J_f(x)\,dx = \int_{\mathbf{R}^n} N(y, f, A)\,dy$$

exists for a quasiregular mapping $f\colon \Omega \to \mathbf{R}^n$ and a measurable $A \subset \Omega$ (Reshetnyak 1989, p. 99), we shall need it for quasiconformal mappings only. Then $N(y, f, A) = 1$ and the proof is more elementary.

14.25. Lemma. (Väisälä 1971, Section 33) *Suppose that* $f\colon \Omega \to \mathbf{R}^n$ *is a quasiconformal mapping. If u is a nonnegative measurable function in a measurable set $A \subset \Omega$, then*

$$(14.26) \qquad \int_A u\big(f(x)\big)\, J_f(x)\, dx = \int_{f(A)} u\, dx\,.$$

The transformation formula (14.26) will be used effectively for quasiregular mappings outside the branch set. This is illustrated in the next lemma.

14.27. Lemma. *Suppose that $f\colon \Omega \to \mathbf{R}^n$ is quasiregular and that D is a normal domain of f. Then*

$$\int_D u\big(f(x)\big)\, J_f(x)\, dx = N(f, D) \int_{f(D)} u\, dy$$

for each nonnegative measurable function u in $f(D)$.

PROOF: Since D is a normal domain of f, the set $f(B_f \cap D)$ is a relatively closed subset of $f(D)$, and the set $A = f(D \setminus B_f)$ is open, because f is open. By Lemma 14.8(b) each $y \in A$ has k distinct inverse images in D,

$$f^{-1}(y) \cap D = \{x_1, \ldots, x_k\}\,,$$

where $k = N(f, D)$. Now we can choose disjoint open balls $B_j = B(y_j, r_j) \subset A$, $j = 1, 2, \ldots$, such that

$$|A \setminus \cup B_j| = 0$$

and that f defines a quasiconformal homeomorphism $f_{j,i}$ from $U_{j,i} = U(x_{j,i}, f, r_j)$ onto B_j, where $i = 1, 2, \ldots, k$, $j = 1, 2, \ldots$, and

$$f^{-1}(y_j) \cap D = \{x_{j,1}, \ldots, x_{j,k}\}\,.$$

This construction is possible by Lemma 14.7 and by the Vitali covering theorem. Note that each x in $f^{-1}(y_j) \cap D$ is outside B_f, whence $i(x, f) = 1$, which means that f is injective in $U(x, f, r_j)$ whenever this is a normal neighborhood of x.

The sets $U_{j,i}$ are mutually disjoint and because $|B_f| = 0$ (Lemma 14.22), we obtain

$$\int_D u\big(f(x)\big)\, J_f(x)\, dx = \sum_{j=1}^{\infty} \sum_{i=1}^{k} \int_{U_{j,i}} u\big(f(x)\big)\, J_f(x)\, dx\,.$$

Now f is a quasiconformal mapping of $U_{j,i}$ onto B_j and the integral transformation formula (14.26) yields

$$\int_{U_{j,i}} u\big(f(x)\big)\, J_f(x)\, dx = \int_{B_j} u\, dx$$

for each i and j. Thus

$$\int_D u\big(f(x)\big)\, J_f(x)\, dx = k\sum_{j=1}^{\infty} \int_{B_j} u\, dx = k\int_{\cup B_j} u\, dx$$

$$= k\int_A u\, dx = k\int_{f(D)} u\, dx\,,$$

where the last equality follows from the fact that $|f(B_f)| = 0$. The lemma is proved. $\qquad\square$

We next show that quasiregular mappings preserve Sobolev functions. Observe that the proof of this result requires deep properties of quasiregular mappings.

14.28. Theorem. *Let f be a nonconstant quasiregular mapping in a domain Ω. If $u \in W^{1,n}_{loc}(\Omega')$, then $v = u \circ f$ belongs to $W^{1,n}_{loc}(A)$, where $A = f^{-1}(\Omega')$. Moreover,*

$$\nabla v(x) = f'(x)^* \,\nabla u\big(f(x)\big)$$

for a.e. x in A.

PROOF: Let $D \Subset A$ be a normal domain of f. It suffices to show that $v \in W^{1,n}_{loc}(D)$ (see Lemma 14.7). To this end, fix a ball $B \Subset D$ and let $\eta \in C^{\infty}_0(f(D))$ be a cut-off function such that $\eta = 1$ in $f(B)$. Then choose a sequence $\varphi_j \in C^{\infty}_0(f(D))$ converging to ηu in $W^{1,n}_0(f(D))$ and $\nabla\varphi_j \to \nabla(\eta\, u)$ a.e. in $f(D)$.

Set $v_j = \varphi_j \circ f$. Since f is a.e. differentiable, we have

$$\nabla v_j(x) = f'(x)^* \,\nabla\varphi_j\big(f(x)\big)$$

for a.e. x in D. Since D is a normal domain of f, Lemma 14.27 yields

$$\int_D |\nabla v_j|^n\, dx \le \int_D |f'(x)|^n\, |\nabla\varphi_j\big(f(x)\big)|^n\, dx$$

$$\le K \int_D J_f(x)\, |\nabla\varphi_j\big(f(x)\big)|^n\, dx$$

$$= N(f,D)\, K \int_{f(D)} |\nabla\varphi_j|^n\, dy\,.$$

The ACL-property of f implies that v_j has the ACL-property as well, and hence $v_j \in W^{1,n}(D)$ (see the discussion preceding Lemma 14.23). Because D is a normal domain of f, v_j has compact support in D, so that $v_j \in W_0^{1,n}(D)$. Now we obtain from the Poincaré inequality that the sequence v_j is bounded in $W_0^{1,n}(D)$ and hence bounded in $W^{1,n}(B)$. Passing to a subsequence we infer from Lemma 14.22(d) that $v_j \to v$ a.e. in B. Therefore the weak compactness of $W^{1,n}(B)$ yields $v \in W^{1,n}(B)$. The desired formula for ∇v holds because, similarly,

$$\nabla v_j(x) \to f'(x)^* \, \nabla u\big(f(x)\big)$$

for a.e. x in B. The theorem follows.   □

Next we show how functions ψ in Ω can be pushed forward to $f(\Omega)$. If f is a homeomorphism, this is easy: the function $\psi^* \colon f(\Omega) \to \mathbf{R}$ is given as $\psi^* = \psi \circ f^{-1}$. It is important to notice that if f is discrete and open, there is a possibility to push forward functions also when f is not injective.

Suppose that $f \colon \Omega \to \mathbf{R}^n$ is a discrete, open, and sense-preserving mapping in Ω. Let $\psi \in C_0(\Omega)$ and define $\psi^* \colon f(\Omega) \to \mathbf{R}$ by

$$(14.29) \qquad \psi^*(y) = \sum_{x \in f^{-1}(y)} i(x, f) \, \psi(x).$$

Here $i(x, f)$ is the local topological index of f at x as described in Section 14.4. The sum in (14.29) contains only a finite number of non-zero terms because f is discrete and because ψ is assumed to be compactly supported in Ω.

14.30. Lemma.   *The function ψ^* belongs to $C_0\big(f(\Omega)\big)$ and* $\operatorname{spt} \psi^* \subset f(\operatorname{spt} \psi)$.

PROOF: First we show that $\operatorname{spt} \psi^*$ is compact in $f(\Omega)$. Set $G = \{x \in \Omega \colon \psi(x) \neq 0\}$ and $G' = \{y \in f(\Omega) \colon \psi^*(y) \neq 0\}$. Then $G' \subset f(G)$ and because $\operatorname{spt} \psi$ is compact, the continuity of f implies $f(\operatorname{spt} \psi) = f(\overline{G}) \supset \overline{G'} = \operatorname{spt} \psi^*$. This implies that $\operatorname{spt} \psi^*$ is compact in $f(\Omega)$, which is open because f is open.

Next we show that ψ^* is continuous. Let $y \in f(\Omega)$ and $\varepsilon > 0$. We may assume that y belongs to $\operatorname{spt} \psi^*$. Choose a neighborhood U of $\operatorname{spt} \psi$ such that $U \Subset \Omega$ and $y \notin f(\partial U)$; this is possible because f is discrete. Let

$$\{x_1, \ldots, x_k\} = f^{-1}(y) \cap U.$$

By Lemma 14.7 and by the continuity of ψ, there exists a number r such that $0 < r < d\big(y, f(\partial U)\big)$ and such that the normal neighborhoods $U_i = U(x_i, f, r)$ of x_i, $i = 1, \ldots, k$, satisfy:

(i) $\overline{U}_i \subset U$ and
(ii) $|\psi(x) - \psi(x_i)| < \varepsilon$, $x \in U_i$.

Note that $f(U_i) = B(y, r)$ for each $i = 1, \ldots, k$. Let $z \in B(y, r)$. Using Lemma 14.6 one easily infers that

$$f^{-1}(z) \cap U \subset U_0 = \cup U_i \,.$$

Fix i, $1 \le i \le k$, and let

$$\{z_1^i, z_2^i, \ldots, z_{\ell(i)}^i\} = f^{-1}(z) \cap U_i \,.$$

We invoke Lemma 14.8 and infer that there exists $C_1 > 0$ such that

$$\sum_{x \in f^{-1}(w) \cap U_0} i(x, f) \le C_1$$

for all $w \in \mathbf{R}^n$; similarly, because U_i is a normal neighborhood of x_i, we find

$$i(x_i, f) = N(f, U_i) = \sum_{j=1}^{\ell(i)} i(z_j^i, f) \,.$$

This yields

$$\left| i(x_i, f)\, \psi(x_i) - \sum_{j=1}^{\ell(i)} i(z_j^i, f)\, \psi(z_j^i) \right|$$

$$= \left| \sum_{j=1}^{\ell(i)} i(z_j^i, f)\, \psi(x_i) - \sum_{j=1}^{\ell(i)} i(z_j^i, f)\, \psi(z_j^i) \right|$$

$$\le \sum_{j=1}^{\ell(i)} i(z_j^i, f)\, |\psi(x_i) - \psi(z_j^i)| < C_1\, \varepsilon \,,$$

and by summing over i we arrive at

$$|\psi^*(y) - \psi^*(z)| = \left| \sum_{i=1}^{k} \left(i(x_i, f)\, \psi(x_i) - \sum_{j=1}^{\ell(i)} i(z_j^i, f)\, \psi(z_j^i) \right) \right|$$

$$< k\, C_1\, \varepsilon \le C_1^2\, \varepsilon \,.$$

Since this holds for every $z \in B(y, r)$, we conclude that ψ^* is continuous at y, and the lemma follows. $\square$

If f is a nonconstant quasiregular mapping and if ψ is smooth, then also ψ^* is relatively smooth. This fact is more difficult to accomplish.

14.31. Lemma. *Suppose that $f\colon \Omega \to \mathbf{R}^n$ is a nonconstant quasiregular mapping in a domain Ω and that $\psi \in C_0^\infty(\Omega)$. Then $\psi^* \in W_0^{1,n}\big(f(\Omega)\big)$.*

PROOF: Because ψ^* belongs to $C_0\big(f(\Omega)\big)$ by Lemma 14.30, it remains to show that ψ^* is *ACL* and that

$$\int_{f(\Omega)} |\nabla \psi^*|^n \, dx < \infty .$$

It is clearly enough to prove that ψ^* is *ACL* in a neighborhood of each point of spt ψ^*. Fix $y_0 \in$ spt ψ^* and let

$$\{x_1, \dots, x_q\} = f^{-1}(y_0) \cap \operatorname{spt} \psi .$$

Choose $r_0 > 0$ such that the domains $U(x_j, f, r_0)$ are pairwise disjoint normal neighborhoods of x_j, $j = 1, \dots, q$, and choose a positive number $r' \le r_0$ such that

$$B(y_0, r') \cap f\big(\operatorname{spt} \psi \setminus \bigcup_{j=1}^{q} U(x_j, f, r_0)\big) = \emptyset .$$

The components of $f^{-1}(B(y_0, r'))$ which meet spt ψ are denoted by $U_j = U(x_j, f, r')$; also they are normal neighborhoods of x_j because $r' \le r_0$. Set $U = \cup U_j$. Then we have

$$\psi^*(y) = \sum_{x \in f^{-1}(y) \cap U} i(x, f)\, \psi(x)$$

for every $y \in B(y_0, r')$.

Next, fix a closed n-interval Q in $B(y_0, r')$. Write $Q = Q_0 \times J$, where Q_0 is an $(n-1)$-interval in $\mathbf{R}^{n-1}$ and $J = [a, b]$ is a closed segment on some x_i-axis, $1 \le i \le n$. For $z \in Q_0$ consider the line segment $\beta = \beta_z$,

$$\beta(t) = z + t\, e_i, \qquad t \in [a, b] ,$$

joining two faces of Q. We show that ψ^* is absolutely continuous on β for almost all z.

To this end, let $[y_1^1, y_1^2], \dots, [y_p^1, y_p^2]$ be a collection of disjoint closed intervals on β. Because U_j is a normal neighborhood of x_j, we have by Lemma 14.8 that

$$\sum_{x \in f^{-1}(y_r^\ell) \cap U_j} i(x, f) = N(f, U_j) = i(x_j, f) \equiv s(j) .$$

For each $j = 1, \ldots, q$ pick the total lifts $\alpha_{j,i}$, $i = 1, \ldots, s(j)$, of β in U_j given by Lemma 14.10. Then Lemma 14.23 implies that each $\alpha_{j,i}$ is absolutely continuous for almost every $z \in Q_0$. Fix such a point $z \in Q_0$. Then for every $\varepsilon > 0$ we can find $\delta > 0$ such that

$$(14.32) \qquad \sum_{j=1}^{q} \sum_{i=1}^{s(j)} \sum_{r=1}^{p} |\alpha_{j,i}(y_r^2) - \alpha_{j,i}(y_r^1)| < \varepsilon$$

whenever

$$\sum_{r=1}^{p} |y_r^2 - y_r^1| < \delta \,.$$

Now for fixed j, $1 \le j \le q$, and y_r^ℓ, $r = 1, 2, \ldots, p$ and $\ell = 1, 2$, we know the number of lifts going through each point in $f^{-1}(y_r^\ell) \cap U_j$; namely, it follows from (14.12) that

$$(14.33) \qquad \sum_{x \in f^{-1}(y_r^\ell) \cap U_j} i(x, f)\, \psi(x) = \sum_{i=1}^{s(j)} \psi\big(\alpha_{j,i}(y_r^\ell)\big) \,.$$

Finally, since $\psi \in C_0^\infty(\Omega)$, there is $C < \infty$ such that

$$|\psi(x) - \psi(y)| \le C\,|x - y|$$

for all $x, y \in \Omega$, and we obtain from (14.33) and (14.32) that

$$\sum_{r=1}^{p} |\psi^*(y_r^2) - \psi^*(y_r^1)| = \sum_{r=1}^{p} \left| \sum_{j=1}^{q} \sum_{i=1}^{s(j)} \big(\psi(\alpha_{j,i}(y_r^2)) - \psi(\alpha_{j,i}(y_r^1))\big) \right|$$

$$\le C \sum_{r=1}^{p} \sum_{j=1}^{q} \sum_{i=1}^{s(j)} |\alpha_{j,i}(y_r^2) - y_{j,i}(y_r^1)|$$

$$< C\varepsilon \,.$$

This shows that ψ^* is absolutely continuous on β.

It remains to show that the gradient of ψ^* is in $L^n(\Omega)$. Again fix $y_0 \in$ spt ψ^* and let $x_1, \ldots, x_q$, $U_1, \ldots, U_q$, $U = \cup U_j$, and $r' > 0$ be as in the first part of the proof. Suppose first that $y_0 \notin f(B_f \cap \mathrm{spt}\,\psi)$. Then $i(x_j, f) = 1$ for all $j = 1, \ldots, q$, and hence $f|_{U_j} = f_j$ is a quasiconformal mapping of U_j onto $B(y_0, r')$ (Lemma 14.8). Furthermore, for each $y \in B(y_0, r')$

$$\psi^*(y) = \sum_{x \in f^{-1}(y) \cap U} \psi(x) = \sum_{j=1}^{q} \psi\big(f_j^{-1}(y)\big) \,,$$

and invoking Theorem 14.28 we have

$$(14.34) \qquad \nabla \psi^*(y) = \sum_{j=1}^{q} (f_j^{-1})'(y)^* \, \nabla \psi \big(f_j^{-1}(y) \big)$$

for a.e. y in $B(y_0, r')$; here we also used the fact that the inverse of a quasiconformal mapping is quasiconformal. Thus

$$\int_{B(y_0, r')} |\nabla \psi^*|^n \, dy$$

$$\leq \sum_{j=1}^{q} q^{n-1} \int_{B(y_0, r')} |(f_j^{-1})'(y)|^n \, |\nabla \psi(f_j^{-1}(y))|^n \, dy$$

$$\leq q^{n-1} \sum_{j=1}^{q} K_O(f_j^{-1}) \int_{B(y_0, r')} J_{f_j^{-1}}(y) \, |\nabla \psi \big(f_j^{-1}(y) \big)|^n \, dy$$

$$\leq q^{n-1} \sum_{j=1}^{q} K_I(f) \int_{U_j} |\nabla \psi|^n \, dx = q^{n-1} \, K_I(f) \int_{U} |\nabla \psi|^n \, dx,$$

where we utilized the integral transformation formula (14.26) for homeomorphic mappings.

Finally, by the Vitali covering theorem, we have

$$\operatorname{spt} \psi^* \setminus f(B_f \cap \operatorname{spt} \psi) \subset \bigcup_{i=1}^{\infty} B(y_i, r_i) \cup A,$$

where $|A| = 0$ and $B(y_i, r_i)$, $i = 1, 2, \ldots$, are mutually disjoint with $y_i \notin f(B_f \cap \operatorname{spt} \psi)$ satisfying the conclusion obtained for $B(y_0, r')$. Since the multiplicity $N(f, \operatorname{spt} \psi)$ is finite and since $|f(B_f \cap \operatorname{spt} \psi)| = 0$ (Lemma 14.22), we have

$$\int_{\operatorname{spt} \psi^*} |\nabla \psi^*|^n \, dy \leq N(f, \operatorname{spt} \psi)^{n-1} K_I(f) \sum_{i=1}^{\infty} \int_{U^i} |\nabla \psi|^n \, dx$$

$$\leq N(f, \operatorname{spt} \psi)^{n-1} K_I(f) \int_{\Omega} |\nabla \psi|^n \, dx < \infty,$$

where U^i is the union of normal neighborhoods associated with y_i as before. This completes the proof. $\qquad \square$

Even if both f and ψ are smooth, ψ^* need not be smooth. This can be seen by considering a function $\psi \in C_0^\infty(\mathbf{R}^2)$ such that $\psi(z) = |z|^2$ in a neighborhood of 0 and the analytic function $f(z) = z^2$. If $z = r\,e^{i\theta}$ in the polar coordinates of $\mathbf{R}^2$, then $z_1 = \sqrt{r}\,e^{i\theta/2}$ and $z_2 = \sqrt{r}\,e^{i(2\pi+\theta)/2}$ are the points that are mapped to z under f. In a neighborhood of 0 we have

$$\psi^*(z) = \psi(z_1) + \psi(z_2) = 2\,r = 2\,|z|$$

which is not smooth at 0.

Note also that if $\psi \in C(\Omega)$ is not compactly supported in Ω, then ψ^* need not be continuous in Ω'.

14.35. Quasiregular mappings and $\mathcal{A}$-harmonic functions

As explained on several occasions, in nonlinear potential theory with $p = n$ quasiregular mappings often take the role of analytic functions in classical theory. We already cracked the surface when we demonstrated in Section 14.13 that if f is quasiregular, then there is a mapping $\mathcal{A}$ such that $u = \log|f|$ satisfies the equation

$$-\operatorname{div}A(x, \nabla u) = 0$$

outside the zero set of f. If $f \not\equiv 0$, we have the fundamental fact that $-\log|f|$ is $\mathcal{A}$-superharmonic. In this section we investigate relations between $\mathcal{A}$-harmonic functions and quasiregular mappings more generally.

Suppose that $\mathcal{A}$ satisfies assumptions (3.3)–(3.7) for $p = n$ in the unweighted case $w \equiv 1$. It will become obvious from the computations below that the "conformally invariant" case requires that the exponent p be equal to the dimension of the underlying space $\mathbf{R}^n$. Mappings associated with other values of p are studied in Section 14.78.

DEFINITION. Suppose that $f \colon \Omega \to \mathbf{R}^n$ is quasiregular. We define a new mapping $f^{\#}\mathcal{A} \colon \mathbf{R}^n \times \mathbf{R}^n \to \mathbf{R}^n$, the *pull-back of $\mathcal{A}$ under f*, as follows:

$$(14.36) \qquad f^{\#}\mathcal{A}(x, \xi) = J_f(x)\,f'(x)^{-1}\,\mathcal{A}(f(x), f'(x)^{-1^*}\xi)$$

whenever $x \in \Omega$ is such that $J_f(x) \neq 0$. If $x \notin \Omega$ or if $J_f(x)$ does not exist or if $J_f(x) = 0$, then we set

$$(14.37) \qquad f^{\#}\mathcal{A}(x, \xi) = \mathcal{A}(x, \xi)\,.$$

As usual, in (14.36) we have treated the (formal) derivative $f'(x)$ as a linear mapping $f'(x) \colon \mathbf{R}^n \to \mathbf{R}^n$. In particular, $f'(x)^{-1}$ is the inverse of $f'(x)$ which exists whenever $J_f(x) \neq 0$, and $f'(x)^*$ is the transpose of the linear mapping $f'(x)$. The definition in (14.37) does not play any role in our considerations, since the set of points $x \in \Omega$ to which it applies has measure zero (unless f is constant which is after all a rather uninteresting occurrence).

14.38. Lemma. *The pull-back $f^{\#}\mathcal{A}$ of $\mathcal{A}$ satisfies assumptions (3.3)–(3.7) with $p = n$, $w \equiv 1$, and structure constants*

$$\alpha' = \frac{\alpha}{K_O(f)}, \qquad \beta' = \beta\,K_I(f),$$

where α and β are the structure constants of $\mathcal{A}$.

PROOF: We may assume that Ω is a domain and that f is nonconstant. Since $|A| = 0$ if and only if $|f(A)| = 0$ (Lemma 14.22), it is not difficult to see that $f^{\#}\mathcal{A}$ satisfies (3.3). To prove (3.4), let $x \in \Omega$ be such that f is differentiable at x with $J_f(x) > 0$, that (14.2) is satisfied at x for $K = K_0(f)$, and that (3.4) holds for $\mathcal{A}$ at $f(x)$. Let $\xi \in \mathbf{R}^n$ and write $\xi^* = f'(x)^{-1^*}\xi$. Then

$$\begin{aligned}
f^{\#}\mathcal{A}(x,\xi)\cdot\xi &= J_f(x)\,\mathcal{A}\big(f(x),\xi^*\big)\cdot\xi^* \\
&\geq J_f(x)\,\alpha\,|\xi^*|^n \geq \alpha\,J_f(x)\,\ell\big(f'(x)^{-1}\big)^n\,|\xi|^n \\
&= \alpha\,J_f(x)\,|f'(x)|^{-n}\,|\xi|^n \geq \frac{\alpha}{K_O(f)}\,|\xi|^n,
\end{aligned}$$

and this holds for a.e. x in Ω. Hence (3.4) is true for $f^{\#}\mathcal{A}$ with constant $\alpha' = \alpha/K_O(f)$.

Similarly, we obtain for a.e. $x \in \Omega$

$$\begin{aligned}
|f^{\#}\mathcal{A}(x,\xi)| &\leq J_f(x)\,|f'(x)^{-1}|\,|\mathcal{A}\big(f(x),\xi^*\big)| \\
&\leq \beta\,J_f(x)\,|f'(x)^{-1}|\,|\xi^*|^{n-1} \leq \beta\,J_f(x)\,|f'(x)^{-1}|^n\,|\xi|^{n-1} \\
&= \beta\,J_f(x)\,\ell\big(f'(x)\big)^{-n}\,|\xi|^{n-1} \leq \beta\,K_I(f)\,|\xi|^{n-1}.
\end{aligned}$$

Hence (3.5) holds for $f^{\#}\mathcal{A}$ with constant $\beta' = \beta\,K_I(f)$.

Next observe that (3.6) for $f^{\#}\mathcal{A}$ follows from the corresponding property of $\mathcal{A}$ and from the fact that sets of zero measure are preserved under f. Finally, (3.7) is obvious, and the lemma follows. □

The following theorem is our main result in this section. It says that a quasiregular mapping f is an $(f^{\#}\mathcal{A}, \mathcal{A})$-harmonic morphism *for all $\mathcal{A}$* satisfying structure conditions (3.3)–(3.7) with $p = n$ and $w = 1$. If f has no branching, the theorem easily follows from the integral transformation formula (14.26); the general case is more involved.

14.39. Theorem. *Suppose that u is $\mathcal{A}$-harmonic in an open set $\Omega' \subset \mathbf{R}^n$ and that $f\colon \Omega \to \mathbf{R}^n$ is quasiregular. Then $u \circ f$ is $f^{\#}\mathcal{A}$-harmonic in $f^{-1}(\Omega')$.*

PROOF: There is no loss of generality in assuming that $f^{-1}(\Omega')$ is a domain and that f is nonconstant in $f^{-1}(\Omega')$. Because each $x_0 \in f^{-1}(\Omega')$ has arbitrary small normal neighborhoods $U = U(x_0, f, r) \Subset f^{-1}(\Omega')$ (Lemma 14.7) and because the problem is local, it suffices to verify that $v = u \circ f$ is $f^{\#}\mathcal{A}$-harmonic in such a neighborhood U. Moreover, since $v \in C(\overline{U}) \cap W^{1,n}(U)$ by Theorem 14.28, we only need to show that

$$\int_U f^{\#}\mathcal{A}(x, \nabla v) \cdot \nabla \psi \, dx = 0$$

whenever $\psi \in C_0^{\infty}(U)$.

Fix $\psi \in C_0^{\infty}(U)$. Write $B = B\big(f(x_0), r\big) = f(U)$ and for $y \in B$ define

$$\psi^*(y) = \sum_{x \in f^{-1}(y)} i(x, f) \, \psi(x)$$

as in (14.29). Let V be the open set $B \setminus f(U \cap B_f)$. If $B(y_0, r')$ is a ball in V, then $U \cap f^{-1}\big(B(y_0, r')\big)$ has exactly $q = N(f, U)$ components U_j, each of which is mapped homeomorphically onto $B(y_0, r')$ by the mapping $f_j = f|_{U_j}$, $j = 1, \ldots, q$. See Lemmas 14.7 and 14.8. Moreover, ψ^* is in $W_0^{1,n}(B)$ and its gradient $\nabla \psi^*$ has the expression

$$\nabla \psi^*(y) = \sum_{j=1}^q (f_j^{-1})'(y)^* \, \nabla \psi\big(f_j^{-1}(y)\big)$$

for a.e. $y \in B$; see Lemma 14.31 and (14.34).

We choose a sequence $B_1, B_2, \ldots$ of disjoint open balls from V which cover almost all of V, and hence almost all of B because $|f(B_f)| = 0$. Let $U_{i,1}, \ldots, U_{i,q}$ be the components of $U \cap f^{-1}(B_i)$ and let $f_{i,j} \colon U_{i,j} \to B_i$ be the quasiconformal restrictions of f to $U_{i,j}$. Put $g_{i,j} = f_{i,j}^{-1}$. Then the gradient of $\nabla \psi^*$ takes the form
(14.40)
$$\nabla \psi^*(y) = \sum_{j=1}^q g_{i,j}'(y)^* \, \nabla \psi\big(g_{i,j}(y)\big) = \sum_{j=1}^q f'\big(g_{i,j}(y)\big)^{-1^*} \nabla \psi\big(g_{i,j}(y)\big)$$

for a.e. $y \in B_i$, $i = 1, 2, \ldots$.

Now because $|B_f| = 0$ and because $\nabla v(x) = f'(x)^* \, \nabla u\big(f(x)\big)$ a.e. in U, the integral transformation formula (14.26) yields

$$\int_U f^{\#} \mathcal{A}(x, \nabla v) \cdot \nabla \psi \, dx$$

$$= \int_U J_f(x) \, f'(x)^{-1} \, \mathcal{A}\big(f(x), \nabla u(f(x))\big) \cdot \nabla \psi(x) \, dx$$

$$= \sum_{i=1}^{\infty} \sum_{j=1}^{q} \int_{U_{i,j}} J_f(x) \, \mathcal{A}\big(f(x), \nabla u(f(x))\big) \cdot f'(x)^{-1^*} \nabla \psi(x) \, dx$$

$$= \sum_{i=1}^{\infty} \sum_{j=1}^{q} \int_{B_i} \mathcal{A}\big(y, \nabla u(y)\big) \cdot f'\big(g_{i,j}(y)\big)^{-1^*} \nabla \psi\big(g_{i,j}(y)\big) \, dy \, ;$$

inserting (14.40) and using the fact that $|f(B_f)| = 0$, we obtain

$$\int_U f^{\#} \mathcal{A}(x, \nabla v) \cdot \nabla \psi \, dx = \int_B \mathcal{A}(y, \nabla u) \cdot \nabla \psi^* \, dy \, .$$

Because $u \in W^{1,n}(B)$ is $\mathcal{A}$-harmonic in B and $\psi^* \in W_0^{1,n}(B)$, the last integral is zero, as desired. This completes the proof. $\qquad\square$

14.41. Corollary. *Every quasiregular mapping* $f \colon \Omega \to \mathbf{R}^n$ *is an* $\mathfrak{A}_n$-*harmonic morphism.*

Combining Corollary 14.41 and Theorem 13.7 we obtain that quasiregular mappings "preserve" superharmonic functions as well. However, the following theorem could be proved directly since the proof of Theorem 14.39 applies almost verbatim when $\mathcal{A}$-harmonic functions are replaced by continuous supersolutions, and since an arbitrary $\mathcal{A}$-superharmonic function can be approximated by an increasing sequence of continuous supersolutions (Theorem 8.15). Note that the problem is trivial for quasiconformal mappings, for if f is a homeomorphism, the comparison definition for $\mathcal{A}$-superharmonic functions can be used.

14.42. Theorem. *Suppose that* $f \colon \Omega \to \mathbf{R}^n$ *is quasiregular and nonconstant in each component of* Ω. *Let* $\mathcal{A}$ *be as in Theorem 14.39. If* u *is* $\mathcal{A}$-*superharmonic in* $\Omega' \subset \mathbf{R}^n$, *then* $u \circ f$ *is* $f^{\#}\mathcal{A}$-*superharmonic in* $f^{-1}(\Omega')$.

The coordinate functions $u(x) = x_i$, $i = 1, \ldots, n$, trivially satisfy the n-Laplace equation

$$- \operatorname{div}(|\nabla u|^{n-2} \nabla u) = 0$$

in $\mathbf{R}^n$. Hence if $f \colon \Omega \to \mathbf{R}^n$ is a quasiregular mapping and $f^{\#}\mathcal{A}$ is the pullback of the n-Laplacian, the coordinate functions of f are $f^{\#}\mathcal{A}$-harmonic in Ω.

A computation shows that also the function $u(x) = \log|x|$ is n-harmonic in $\mathbf{R}^n \setminus \{0\}$ and it may seem as a consequence of Theorem 14.39 that the function

$$v(x) = \log|f(x) - b|, \qquad b \in \mathbf{R}^n,$$

is $f^\#\mathcal{A}$-harmonic in $\Omega \setminus f^{-1}(b)$. However, the fact that $\log|f(x) - b|$ is $f^\#\mathcal{A}$-harmonic for $\mathcal{A}(x,\xi) = |\xi|^{n-2}\xi$ was employed in the proof of Theorem 14.14, and the argument used there is essentially the only presently known route to the discreteness and openness of quasiregular mappings. The method in Theorem 14.39 seems necessary when dealing with more general mappings $\mathcal{A}$.

Since $t \mapsto e^{st}$, $s \geq 0$, is convex and increasing, it follows that $|f|^s$ is $f^\#\mathcal{A}$-subharmonic (Theorem 7.5).

If $f\colon \Omega \to \mathbf{R}^2$ is a plane analytic function, then the coordinate functions of f are harmonic. The above observation about the components of a quasiregular mapping is a counterpart of this result. However, the existence of a conjugate function $\tilde{u}$ such that $f = (u, \tilde{u})$ is analytic has no obvious extension in higher dimensions. In the plane the situation is different due to the existence theory of quasiconformal mappings with prescribed dilatation (Lehto and Virtanen 1973).

If $f\colon \Omega \to \mathbf{R}^n$ is quasiconformal, then $f^{-1}\colon f(\Omega) \to \Omega$ is also quasiconformal and it is easy to see by using Theorem 13.2 that

$$(f^{-1})^\# f^\# \mathcal{A} = \mathcal{A},$$
$$f^\# (f^{-1})^\# \mathcal{A} = \mathcal{A}.$$

Moreover, if $\mathcal{A}(x,\xi) = |\xi|^{n-2}\xi$ and if T is a sense-preserving Möbius transformation of $\mathbf{R}^n$, then the pull-back $T^\#\mathcal{A}$ is unchanged: $T^\#\mathcal{A}(x,\xi) = |\xi|^{n-2}\xi$. This follows from the fact that at each point x with $T(x) \neq \infty$ the derivative $T'(x)$ is of the form $J_T(x)^{1/n}\mathcal{O}_x$, where $\mathcal{O}_x$ is an orthogonal transformation of $\mathbf{R}^n$.

Heinonen et al. (1992) have proved that in the unweighted case Corollary 14.41 has a converse: every sense-preserving $\mathfrak{A}_n$-harmonic morphism is a quasiregular mapping. Manfredi (1991) has shown that every $(\mathcal{A},\mathcal{A})$-harmonic morphism with $\mathcal{A}(x,\xi) = |\xi|^{n-2}\xi$ is necessarily 1-quasiregular, and hence a Möbius transformation if $n \geq 3$.

14.43. Applications to quasiregular mappings
Potential theoretic methods are of great importance in the study of quasiregular mappings, even more so than in classical complex function theory where other means are often available. Because quasiregular mappings are $\mathfrak{A}_n$-harmonic morphisms, one can directly apply the results achieved in

the preceding chapter. In particular, we conclude that Liouville's theorem, Radó's theorem, the principle of $\mathcal{A}$-harmonic measure, etc. are true for quasiregular mappings. In this section we present these and some more advanced applications.

As explained in the previous section, each coordinate function of a quasiregular mapping $f \colon \Omega \to \mathbf{R}^n$ is $f^{\#}\mathcal{A}$-harmonic if $\mathcal{A}(x, \xi) = |\xi|^{n-2} \xi$. Thus we have by Theorem 6.6:

14.44. Theorem. *A quasiregular mapping $f \colon \Omega \to \mathbf{R}^n$ is Hölder continuous with exponent α on compact subsets of Ω, where $\alpha \in (0,1]$ depends only on n and $K(f)$.*

The use of $\mathcal{A}$-harmonic functions does not give the best possible Hölder exponent, which is known to be $K_I(f)^{1/(n-1)}$ (Martio et al. 1970; Vuorinen 1988).

Similarly, we apply Theorem 6.44 and record the following:

14.45. Theorem. *Suppose that f is a K-quasiregular mapping in a bounded open set Ω whose complement satisfies the capacity density condition (6.43) with $p = n$, $w = 1$, and constants c_0 and r_0. If f has a continuous extension to $\overline{\Omega}$ and if there are constants $M \geq 0$ and $0 < \delta \leq 1$ such that*

$$|f(x) - f(y)| \leq M |x - y|^\delta$$

for all $x, y \in \partial\Omega$, then

$$|f(x) - f(y)| \leq M_1 |x - y|^{\delta_1}$$

for all $x, y \in \overline{\Omega}$, where $\delta_1 > 0$ depends only on n, K, δ, and c_0, and

$$M_1 = \sqrt{n} 80 M r_0^{-2} \max\{1, (\operatorname{diam} \Omega)^2\}.$$

It was mentioned in Section 6.41 that a similar result for $\mathcal{A}$-harmonic functions, namely Theorem 6.44, is not necessarily true without the extra capacity density assumption on $\complement\Omega$. However, it was established by Gehring et al. (1982) that if an analytic function f is Hölder continuous with exponent α on the boundary of an *arbitrary bounded open set* Ω in the plane, then f is Hölder continuous with exponent α in the closure of Ω. Therefore, one might conjecture that if Ω is bounded and if $f \colon \overline{\Omega} \to \mathbf{R}^n$ is a continuous mapping which is K-quasiregular in Ω, then

$$|f(x) - f(y)| \leq M |x - y|^\delta$$

for all $x, y \in \partial\Omega$ implies

$$|f(x) - f(y)| \leq M_1 |x - y|^{\delta_1}$$

for all $x, y \in \overline{\Omega}$, where

$$\delta_1 = \min\{\delta, K_I(f)^{1/(1-n)}\}.$$

This is known to be true in the special case where f is quasiconformal (Martio and Näkki 1991), but seems to be an open question in general. Various forms of Theorem 14.45 had been known to several people but never published.

The removability result for morphisms (Theorem 13.16) can be sharpened in the quasiregular case. Recall that a subset E of $\mathbf{R}^n$ is of zero n-capacity if

$$\mathrm{cap}_n(E \cap \Omega', \Omega') = 0$$

for all open sets Ω'.

14.46. Theorem. *Suppose that C is a relatively closed subset of Ω and that C is of zero n-capacity. Then each bounded K-quasiregular mapping $f: \Omega \setminus C \to \mathbf{R}^n$ has a K-quasiregular extension $f^*: \Omega \to \mathbf{R}^n$.*

PROOF: Since the coordinate functions of f are $\mathcal{A}$-harmonic in $\Omega \setminus C$ for some $\mathcal{A}$ with $p = n$, where $w = 1$, and since sets of zero n-capacity are removable for bounded $\mathcal{A}$-harmonic functions by Theorem 7.36, f has a continuous extension f^* to Ω. Moreover, this extension belongs to $W_{loc}^{1,n}(\Omega)$. Because sets of zero capacity have Lebesgue measure zero, the required inequality $|f'(x)|^n \le K J_f(x)$ holds a.e. in Ω, and the theorem follows. $\quad\square$

The assumption that f is bounded in Theorem 14.46 is not necessary; it suffices that f omits a set of positive n-capacity. In this case f^* may take the value ∞ and the class of quasimeromorphic mappings must be considered. Quasimeromorphic mappings generalize quasiregular mappings the same way meromorphic functions generalize analytic functions. More precisely, if Ω is open in $\overline{\mathbf{R}}^n$, we say that a continuous mapping $f: \Omega \to \overline{\mathbf{R}}^n$ is K-*quasimeromorphic* if $f^{-1}(\infty)$ is a discrete set and if f is K-quasiregular in $\Omega \setminus \{f^{-1}(\infty), \infty\}$. The reader is invited to prove this more general removability theorem by using the fact that the complement of a nonpolar set carries nonconstant bounded $\mathcal{A}$-harmonic functions. Martio *et al.* (1970) have presented a different proof.

For a long time nothing better than Theorem 14.46 was known for quasiregular mappings in dimensions higher than two, but recently some new light has been shed on this quite difficult problem. In particular, Iwaniec (In press) has established that there is $\varepsilon = \varepsilon(n, K) > 0$ such that if f is a bounded K-quasiregular mapping in a neighborhood of a compact set E with Hausdorff dimension not exceeding ε, then f has a K-quasiregular extension to E; subsequently, Rickman (In press) constructed examples showing that this result is qualitatively the best possible. See also Koskela and Martio (1990), Iwaniec and Martin (In press), Järvi and Vuorinen (1992).

Radó's theorem for holomorphic functions has many applications, especially in the several variable theory, where it is often called the Radó–Cartan theorem. For a good account, see Narasimhan (1985, 11.8). We next restate Radó's theorem for quasiregular mappings and then discuss some applications.

14.47. Radó's theorem for quasiregular mappings. *Suppose that* $f: \Omega \to \mathbf{R}^n$ *is continuous and that* $C \subset \mathbf{R}^n$ *is a closed set of zero n-capacity. If f is K-quasiregular in* $\Omega \setminus f^{-1}(C)$, *then f is K-quasiregular in* Ω. *In particular, if* $f: \Omega \to \mathbf{R}^n$ *is continuous and K-quasiregular in* $\Omega \setminus f^{-1}(x_0)$, *then f is K-quasiregular in* Ω.

Theorem 14.47 was first proved by Kuusalo (1977). The removability theorem 14.46 allows a slightly more general formulation for Theorem 14.47 in the case when f is bounded; then it suffices to assume that f is quasiregular in $\Omega \setminus C$, where C is closed in Ω, and that the cluster set $\mathcal{C}(f, C)$ has zero n-capacity. Similar remarks hold for the following result. We leave such generalizations to the reader.

14.48. Theorem. *Let* $f: \Omega \to \mathbf{R}^n$ *be a quasiregular mapping in a domain* Ω *and let b be a nonisolated boundary point of* Ω. *If b has a neighborhood U such that* $\lim_{x \to y} f(x) = 0$ *for all* $y \in \partial\Omega \cap U$, *then* $f \equiv 0$ *in* Ω.

PROOF: Pick a ball B centered at b and contained in U. By setting $f = 0$ in $B \setminus \Omega$, we extend f continuously to $\Omega' = \Omega \cup B$. By Radó's theorem f is quasiregular in Ω'. Because Ω' is connected and because the zeros of f have a cluster point $b \in \Omega'$, the discreteness of nonconstant quasiregular mappings implies $f \equiv 0$, and the theorem follows. $\square$

A mapping $f: \Omega \to \Omega'$ is called *proper* if the preimage under f of every compact set in Ω' is compact. It is well known that $\mathbf{R}^n$ cannot be embedded quasiconformally onto its proper subset, and the following consequence of Radó's theorem can be viewed as a multivalued analogue of that result.

14.49. Theorem. *Suppose that* $f: \Omega \to \mathbf{R}^n$ *is a proper quasiregular mapping. Then* $\mathbf{R}^n \setminus \Omega$ *is a finite set and* $f(x) \to \infty$ *when* $x \to y$ *for all* $y \in \partial\Omega$. *In particular, f has a quasimeromorphic extension to* $\overline{\mathbf{R}}^n$ *with poles on* $\partial\Omega$.

PROOF: Since f is discrete and proper, the set $f^{-1}(0)$ is finite. Consider the mapping $g = T \circ f$, where T is a sense-preserving Möbius transformation which interchanges 0 and ∞. Then g is a quasiregular mapping in $\Omega \setminus f^{-1}(0)$. Let y be a boundary point of Ω and $x_j \in \Omega$ a sequence converging to y. Since f is proper, necessarily $f(x_j) \to \infty$. Consequently, $g(x) \to 0$ as $x \to y$ for all $y \in \partial\Omega$, and after setting $g(y) = 0$ for $y \in \complement\Omega \cap \mathbf{R}^n$, we deduce from Radó's theorem that g is quasiregular in $\mathbf{R}^n \setminus f^{-1}(0)$. In particular, it follows that $\mathbf{R}^n \setminus \Omega = \partial\Omega \cap \mathbf{R}^n$ is a discrete set, and putting $f(y) = \infty$ for $y \in \partial\Omega \cap \mathbf{R}^n$ defines a quasimeromorphic mapping $f \colon \mathbf{R}^n \to \overline{\mathbf{R}}^n$. Moreover, we have $\lim_{x \to \infty} f(x) = \infty$ so that f can be thought of as a continuous mapping of $\overline{\mathbf{R}}^n$, quasimeromorphic in $\mathbf{R}^n$. Using an appropriate inversion, it is clear that the removability theorem 14.46 can be used so as to conclude that f is quasimeromorphic in all of $\overline{\mathbf{R}}^n$. In particular, f is discrete which implies that $f^{-1}(0) = \partial\Omega$ is a finite set. The theorem is proved. □

The last topic in this section is the use of $\mathcal{A}$-harmonic measure in the study of quasiregular mappings. Let $\mathcal{A}$ satisfy the usual assumptions with $p = n$ and $w = 1$. The principle of $\mathcal{A}$-harmonic measure for quasiregular mappings reads as follows.

14.50. Principle of $\mathcal{A}$-harmonic measure for quasiregular mappings. *Let Ω and Ω' be open sets in $\mathbf{R}^n$ and let $f \colon \Omega \to \Omega'$ be a quasiregular mapping. If $E \subset \partial\Omega$ is such that $\mathcal{C}(f, E)$ is closed in Ω', then*

$$\omega(E, \Omega; f^{\#}\mathcal{A})(x) \le \omega(\mathcal{C}(f, E), \Omega'; \mathcal{A})(f(x))$$

for all $x \in \Omega$.

Recall the convention that $\omega(\mathcal{C}(f, E), \Omega'; \mathcal{A})(f(x)) = 1$ if $f(x) \in \mathcal{C}(f, E)$.

14.51. Principle of $\mathcal{A}$-harmonic measure for quasiconformal mappings. *Suppose that f is a quasiconformal mapping of Ω onto Ω'. If $E \subset \partial\Omega$, then*

$$(14.52) \qquad \omega(E, \Omega; f^{\#}\mathcal{A})(x) \le \omega(\mathcal{C}(f, E), \Omega'; \mathcal{A})(f(x))$$

for each $x \in \Omega$. If $\mathcal{C}(f^{-1}, \mathcal{C}(f, E)) = E$, then equality holds in (14.52).

Although an elementary inequality, the principle of $\mathcal{A}$-harmonic measure leads to many useful applications. Our first example extends the classical two-constant theorem for analytic functions.

14.53. Two-constant theorem for quasiregular mappings. *Suppose that $f \colon \Omega \to \mathbf{R}^n$ is quasiregular and that $E \subset \partial\Omega$. Let $\mathcal{A}(x, \xi) =$*

$|\xi|^{n-2}\xi$ and let $\omega = \omega(E, \Omega; f^{\#}\mathcal{A})$ denote the $f^{\#}\mathcal{A}$-harmonic measure of E in Ω. Suppose that there are $0 < m \leq M < \infty$ such that

$$(14.54) \qquad \limsup_{x \to y} |f(x)| \leq M$$

for all $y \in E$ and

$$(14.55) \qquad \limsup_{x \to y} |f(x)| \leq m$$

for all $y \in \partial\Omega \setminus E$. Then

$$|f(x)| \leq m \left(\frac{M}{m}\right)^{\omega(x)}$$

for all $x \in \Omega$.

PROOF: This follows by applying Theorem 11.9 to the $f^{\#}\mathcal{A}$-subharmonic function $\log|f|$.   □

As an example, let E be a porous Cantor set constructed on the boundary ∂B^n of the unit ball B^n in $\mathbf{R}^n$. Then E is of total harmonic measure zero by Theorem 11.27. If $f: B^n \to \mathbf{R}^n$ is a bounded quasiregular mapping, then Theorem 14.53 says that $|f|$ cannot "jump" on the set E. For analytic functions f in the unit disk Theorem 14.53 says that $|f|$ cannot jump on a set of linear measure zero in ∂B^2.

The two-constant theorem is often useful in a different formulation. Suppose that f is a quasiregular mapping in Ω such that $M = \sup|f| < \infty$. Suppose further that $E \subset \partial\Omega$ is such that

$$\limsup_{x \to y} f(x) \leq m < M$$

for all $y \in E$. Then

$$(14.56) \qquad |f(x)| \leq M \left(\frac{m}{M}\right)^{\omega(x)}$$

for all $x \in \Omega$, where ω is the $\mathcal{A}$-harmonic measure of E in Ω as in Theorem 14.53. The proof of this inequality is similar to the proof of Theorem 14.53, but it does not directly follow from that theorem, because it is not known whether $\omega(E, \Omega; \mathcal{A}) = 1 - \omega(\partial\Omega \setminus E, \Omega; \mathcal{A})$ in general.

Next we apply the Phragmén–Lindelöf principle 11.11 to obtain a growth estimate for quasiregular mappings in unbounded domains.

14.57. Theorem. Let $\mathcal{A}(x,\xi) = |\xi|^{n-2}\xi$. Suppose that $f \colon \Omega \to \mathbf{R}^n$ is a quasiregular mapping in an unbounded domain Ω containing 0. Denote by ω_R the $f^\# \mathcal{A}$-harmonic measure $\omega(E_R, \Omega_R; f^\# \mathcal{A})$, where $\Omega_R = \Omega \cap B(0, R)$ and $E_R = \partial B(0, R) \cap \partial \Omega_R$ for $R > 0$. If

$$\limsup_{x \to y} |f(x)| \leq 1$$

for all $y \in \partial\Omega$, then either $|f| \leq 1$ in Ω or

$$\liminf_{R \to \infty} \omega_R(x) \log M(R) > 0$$

for all $x \in \Omega$, where $M(R) = \max_{|x|=R} |f(x)|$.

PROOF: This follows from the Phragmén–Lindelöf principle 11.11 applied to the $f^\# \mathcal{A}$-subharmonic function $\log |f(x)|$. □

If the complement of Ω in Theorem 14.57 contains an unbounded continuum, it can be proved that asymptotically

$$\omega_R \lesssim R^{-\delta},$$

as $R \to \infty$, for some $\delta = \delta(n, K) > 0$. This follows from a more general Tsuji type estimate for ω_R given by Granlund et al. (1985). It follows, in particular, that if f is an entire nonconstant quasiregular mapping having at least one finite asymptotic value, then the lower order of f,

$$\liminf_{R \to \infty} \frac{\log \log M(R)}{\log R},$$

is positive. We recall that a mapping $f : \Omega \to \mathbf{R}^n$ is said to have an *asymptotic value* a at $y \in \partial\Omega$ if there is a path Γ in Ω ending at y such that $f(x) \to a$ as $x \to y$ along Γ. See also the papers by Rickman and Vuorinen (1982) and Heinonen et al. (1989b).

It is an open problem whether the Denjoy theorem holds for an entire quasiregular mapping in dimensions $n \geq 3$ even in the following weak form: if the lower order is finite, the number of distinct asymptotic values is finite.

The $\mathcal{A}$-harmonic measure can be used to prove a Lindelöf type theorem for quasiregular mappings. If f is bounded and analytic in the unit disk B and if f has an asymptotic value w_0 at $x_0 \in \partial B$, then f has the limit w_0 along each angle in B with vertex at x_0. In particular, f has the radial limit w_0 at x_0. This classical theorem of Lindelöf is best proved by using harmonic measure; see Ahlfors (1973). For $n \geq 3$ Lindelöf's theorem in the above formulation is false for quasiregular mappings (Rickman 1980a).

However, if B is the unit ball of $\mathbf{R}^n$ and if $f \colon B \setminus L \to \mathbf{R}^n$ is a bounded quasiregular mapping where L is a line, then the theorem holds at points $x_0 \in B \cap L$; this result (Granlund et al. 1985) is possible because L carries about the same amount of ($\mathcal{A}$-)harmonic measure as an asymptotic path. Similarly, Lindelöf's theorem in the ball is true if the asymptotic path is replaced by an asymptotic $(n-1)$-dimensional surface (Rickman 1980a). Lindelöf's theorem for quasiregular and quasimeromorphic mappings has also been studied in Vuorinen (1981) and Heinonen and Rossi (1990).

14.58. The Picard theorem for quasiregular mappings
Because quasiregular mappings are $\mathfrak{A}_n$-harmonic morphisms, we obtain from Theorem 13.13 that an entire quasiregular mapping $f : \mathbf{R}^n \to \mathbf{R}^n$ is either constant or assumes quasievery point in $\mathbf{R}^n$. It was an open problem for some time whether stronger Picard type theorems exist for quasiregular mappings. Zorich (1967) constructed a nonconstant quasiregular mapping of $\mathbf{R}^n$ that omits precisely one point, and he conjectured that omission of two points is not possible. Rickman (1980b) proved that a nonconstant K-quasiregular $f : \mathbf{R}^n \to \mathbf{R}^n$ assumes all but a finite number $q(n, K)$ of points, and later he showed via a deep construction that in 3-space any finite number of points can be omitted (Rickman 1985). Thus $q(n, K) \to \infty$ as $K \to \infty$ at least when $n = 3$, and probably the same is true for any $n > 3$ but there is no proof. Generally Rickman's methods are quite different from those presented in this book.

Our aim in this section is to present a potential theoretic proof of Rickman's version of the Picard theorem, recently given by Eremenko and Lewis (1991).

14.59. Picard's theorem for quasiregular mappings. *Let $f : \mathbf{R}^n \to \mathbf{R}^n$ be a K-quasiregular mapping. There is a number $q_0 = q_0(n, K) < \infty$ such that if f omits q_0 points, then f is constant.*

We prove a general statement about $\mathcal{A}$-harmonic functions from which Theorem 14.59 readily follows.

We suppose in this section that $\mathcal{A}$ is a mapping satisfying assumptions (3.3)–(3.7) with $p = n$ and $w = 1$, and, *in addition*, that there is a constant $\gamma > 0$ such that

$$(14.60) \qquad \gamma \left| \xi_1 - \xi_2 \right|^n \leq \big(\mathcal{A}(x, \xi_1) - \mathcal{A}(x, \xi_2) \big) \cdot \big(\xi_1 - \xi_2 \big)$$

holds for a.e. $x \in \mathbf{R}^n$ and all $\xi \in \mathbf{R}^n$. Then we prove:

14.61. Theorem. *Suppose that $\mathcal{A}$ satisfies assumptions (3.3)–(3.7) and (14.60) with $p = n$ and $w = 1$. Let $h_1, \ldots, h_q$ be nonconstant $\mathcal{A}$-harmonic functions in $\mathbf{R}^n$. Suppose that there is $\lambda > 0$ such that*

(i) $\{x : h_i(x) < -\lambda\} \cap \{x : h_j(x) < -\lambda\} = \emptyset$

(ii) $|h_i^+ - h_j^+| \leq \lambda$

(iii) $|h_i(0)| \leq \lambda$

whenever $1 \leq i, j \leq q$, $i \neq j$. Then q cannot exceed a number $q_0 < \infty$ which depends only on n, β/α, and γ.

Recall the notation $h^+ = \max(0, h)$ above.

Let us first show how Theorem 14.59 follows from Theorem 14.61. Suppose that f is a nonconstant K-quasiregular mapping in $\mathbf{R}^n$ omitting points $a_1, \ldots, a_q$. Then the functions

$$h_i(x) = \log |f(x) - a_i|$$

are $\mathcal{A}$-harmonic in $\mathbf{R}^n$ for $i = 1, \ldots, q$, where $\mathcal{A}$ is the pull-back of $|\xi|^{n-2}\xi$,

$$\mathcal{A}(x, \xi) = J_f(x)|f'(x)^{-1^*}\xi|^{n-2}f'(x)^{-1}f'(x)^{-1^*}\xi$$

whenever f' exists and is invertible, and $\mathcal{A}(x, \xi) = |\xi|^{n-2}\xi$ otherwise. By Lemma 14.38 the mapping $\mathcal{A}$ satisfies assumptions (3.3)–(3.7) with $p = n$, $w = 1$, $\alpha = 1/K_0(f)$, and $\beta = K_I(f)$. We compute that $\mathcal{A}$ satisfies assumption (14.60) with $\gamma = 1/2^{n-1}K_0$, where $K_0 = K_0(f)$. To do this, observe that $\mathcal{A}(x, \xi)$ is of the form

$$\mathcal{A}(x, \xi) = (\theta(x)\xi \cdot \xi)^{(n-2)/2}\theta(x)\xi,$$

where the matrix $\theta(x) = J_f(x)^{2/n}f'(x)^{-1}f'(x)^{-1^*}$ satisfies

(14.62) $$\frac{1}{K_0^{2/n}}|\xi|^2 \leq \theta(x)\xi \cdot \xi \leq K_I^{2/n}|\xi|^2$$

for a.e. x in $\mathbf{R}^n$. Fix x such that (14.62) is true and write for short $\theta(x) = T^*T$. Then

$$\left(\mathcal{A}(x, \xi_1) - \mathcal{A}(x, \xi_2)\right) \cdot (\xi_1 - \xi_2)$$
$$= \left(|T\xi_1|^{n-2}T\xi_1 - |T\xi_2|^{n-2}T\xi_2\right) \cdot (T\xi_1 - T\xi_2)$$
$$= \frac{1}{2}\left(|T\xi_1|^{n-2} + |T\xi_2|^{n-2}\right)\left(|T\xi_1 - T\xi_2|^2\right)$$
$$\quad + \frac{1}{2}\left(|T\xi_1|^2 - |T\xi_2|^2\right)\left(|T\xi_1|^{n-2} - |T\xi_2|^{n-2}\right)$$
$$\geq \frac{1}{2}\left(|T\xi_1|^{n-2} + |T\xi_2|^{n-2}\right)\left(|T\xi_1 - T\xi_2|^2\right)$$
$$= \frac{1}{2}\left((\theta(x)\xi_1 \cdot \xi_1)^{(n-2)/2} + (\theta(x)\xi_2 \cdot \xi_2)^{(n-2)/2}\right)\left(\theta(x)(\xi_1 - \xi_2) \cdot (\xi_1 - \xi_2)\right)$$
$$\geq \frac{1}{2K_0}\left(|\xi_1|^{n-2} + |\xi_2|^{n-2}\right)|\xi_1 - \xi_2|^2$$
$$\geq \frac{1}{2^{n-1}K_0}\left(|\xi_1| + |\xi_2|\right)^{n-2}|\xi_1 - \xi_2|^2 \geq \frac{1}{2^{n-1}K_0}|\xi_1 - \xi_2|^n,$$

as required.

Next, we assume without loss of generality that $f(0) = 0$ and that the points $a_1, \ldots, a_q$ all lie in the ball $B(0, 1/4)$. If we set

$$\delta = \frac{1}{2} \min_{i \neq j}\{|a_i|, \, |a_i - a_j|\}$$

and

$$\lambda = \log \frac{1}{\delta} \, ,$$

then it is an elementary task to check that (i), (ii), and (iii) are true for $h_i = \log |f - a_i|$ with this choice of λ. Consequently, Theorem 14.59 follows from Theorem 14.61.

We begin the proof of Theorem 14.61 with an auxiliary lemma which is interesting in its own right. The lemma as well as Theorem 14.61 could be presented in more generality (Eremenko and Lewis 1991) but we only give a version which is sufficient for the purposes of Theorem 14.59.

Suppose that u is $\mathcal{A}$-subharmonic in Ω and that $u \in W^{1,n}_{loc}(\Omega)$. Then $\operatorname{div} \mathcal{A}(x, \nabla u)$,

$$\operatorname{div} \mathcal{A}(x, \nabla u)(\varphi) = -\int_\Omega \mathcal{A}(x, \nabla u) \cdot \nabla \varphi \, dx \, ,$$

is a positive distribution, and the Riesz representation theorem guarantees that there is a unique positive measure ν on Ω such that

$$\int_\Omega \mathcal{A}(x, \nabla u) \cdot \nabla \varphi \, dx \ = \ -\int_\Omega \varphi \, d\nu$$

whenever $\varphi \in C_0^\infty(\Omega)$ (Schwartz 1966, p. 29). We call the measure ν the *Riesz measure associated with* u in Ω. The Riesz measure naturally exists for general $\mathcal{A}$-subharmonic functions which are locally in the Sobolev space $H^{1,p}(\Omega; \mu)$ although we only use it in the situation of Theorem 14.61. Moreover, we only consider continuous $\mathcal{A}$-subharmonic functions (see Corollary 7.20). The Riesz measure in connection with general $\mathcal{A}$-subharmonic functions has been studied by Kilpeläinen and Malý (1990, 1992b).

If u is continuous and $\mathcal{A}$-subharmonic in Ω with the associated Riesz measure ν and if $x \in \Omega$, $r > 0$, are such that $B(x, r) \Subset \Omega$, we write

$$\nu(x, r) = \nu\big(\overline{B}(x, r)\big)$$

and

$$M(x, r, u) \ = \ \max_{y \in \overline{B}(x, r)} u(y) \, .$$

Next we record an appropriate consequence of the local Hölder continuity estimate for $\mathcal{A}$-harmonic functions: let h be an $\mathcal{A}$-harmonic function in a ball $B(x_0, 2r)$ with $h(x_0) = 0$. Then for all $0 < t \leq 1$ we have that

$$(14.63) \qquad M(x_0, tr, h) \leq c\, t^\kappa M(x_0, 2r, h),$$

where $c > 0$ and $\kappa > 0$ depend only on n and β/α.

To achieve this, we first apply Harnack's inequality to the nonnegative $\mathcal{A}$-harmonic function $M(x_0, 2r, h) - h$ and obtain

$$(14.64) \qquad M(x_0, r, -h) \leq M(x_0, r, M(x_0, 2r, h) - h) \leq c\, M(x_0, 2r, h)$$

because $h(x_0) = 0$. Then the local Hölder continuity estimate (6.7) implies

$$M(x_0, tr, h) \leq \operatorname{osc}(h, B(x_0, tr)) \leq 2^\kappa t^\kappa \operatorname{osc}(h, B(x_0, r))$$
$$= 2^\kappa t^\kappa \big(M(x_0, r, h) + M(x_0, r, -h)\big) \leq c\, t^\kappa M(x_0, 2r, h),$$

as desired.

14.65. Lemma. *Suppose that $\mathcal{A}$ is as in Theorem 14.61 and that h is an $\mathcal{A}$-harmonic function in the ball $B(x_0, R)$ with $h(x_0) = 0$. Then*

$$\frac{1}{c}\nu(x_0, r/2) \leq M(x_0, r, h)^{n-1} \leq c\,\nu(x_0, 2r)$$

for all $0 < r < R/2$, where ν is the Riesz measure associated with the continuous $\mathcal{A}$-subharmonic function $u = \max(h, 0)$ and $c = c(n, \beta/\alpha, \gamma) > 0$.

PROOF: The first inequality is a consequence of the Caccioppoli estimate (3.28) and inequality (14.60) is not needed. Indeed, let $\eta \in C_0^\infty(B(x_0, 3r/4))$ be a nonnegative test function such that $\eta = 1$ on $B(x_0, r/2)$ and $|\nabla \eta| \leq 10/r$. Then

$$\nu(x_0, r/2) \leq \int_{B(x_0, 3r/4)} \eta \, d\nu$$

$$= -\int_{B(x_0, 3r/4)} \mathcal{A}(x, \nabla u) \cdot \nabla \eta \, dx \leq \beta \int_{B(x_0, 3r/4)} |\nabla u|^{n-1}|\nabla \eta| dx$$

$$\leq \beta \left(\int_{B(x_0, 3r/4)} |\nabla u|^n \, dx \right)^{(n-1)/n} \left(\int_{B(x_0, 3r/4)} |\nabla \eta|^n \, dx \right)^{1/n}$$

$$\leq c \left(\int_{B(x_0, 3r/4)} |\nabla u|^n \, dx \right)^{(n-1)/n}$$

On the other hand, choosing a slightly different test function, (3.28) implies

$$\int_{B(x_0,3r/4)} |\nabla u|^n \, dx \le c \, M(x_0, r, u)^n \,,$$

and we arrive at the first inequality.

In proving the second inequality, we invoke condition (14.60). Let h_u be the $\mathcal{A}$-harmonic function in $B(x_0, 2r)$ with boundary values u on $\partial B(x_0, 2r)$; then the comparison principle implies $0 \le u \le h_u$ in $B(x_0, 2r)$. By applying (14.63) and Harnack's inequality, we obtain for $0 < t \le 1/2$ that

$$\begin{aligned} M(x_0, tr, u) = M(x_0, tr, h) &\le c \, t^\kappa M(x_0, r, h) \\ = c \, t^\kappa M(x_0, r, u) &\le c \, t^\kappa M(x_0, r, h_u) \le c \, t^\kappa h_u(x_0) \,, \end{aligned}$$

where κ and c depend only on n and β/α. Using this and Harnack's inequality we may choose t and c', depending only on n and β/α, such that

$$c' h_u(x_0) \le h_u(x) - u(x)$$

for all $x \in B(x_0, tr)$. Now because the function

$$\varphi = \min(h_u - u, \, c' h_u(x_0))$$

is nonnegative and belongs to $W_0^{1,n}(B(x_0, 2r))$, we obtain by Poincaré's inequality 1.4 (with $d\mu = dx$), assumption (14.60), and the fact that h_u is $\mathcal{A}$-harmonic that

$$\begin{aligned} h_u(x_0)^n r^n &\le c \int_{B(x_0,2r)} \varphi^n \, dx \le c \, r^n \int_{B(x_0,2r)} |\nabla \varphi|^n \, dx \\ &= c \, r^n \int_{\{\nabla \varphi \ne 0\}} |\nabla h_u - \nabla u|^n \, dx \\ &\le c \, r^n \int_{\{\nabla \varphi \ne 0\}} \big(\mathcal{A}(x, \nabla h_u) - \mathcal{A}(x, \nabla u)\big) \cdot \big(\nabla h_u - \nabla u\big) \, dx \\ &= -c \, r^n \int_{B(x_0,2r)} \mathcal{A}(x, \nabla u) \cdot \nabla \varphi \, dx = c \, r^n \int_{B(x_0,2r)} \varphi \, d\nu \\ &\le c \, r^n h_u(x_0) \, \nu(x_0, 2r) \,, \end{aligned}$$

where c depends only on n, β/α, and γ. It follows that

$$M(x_0, r, h)^{n-1} \le M(x_0, r, h_u)^{n-1} \le c h_u(x_0)^{n-1} \le c\nu(x_0, 2r)$$

as desired. The lemma follows.   □

We require another lemma.

14.66. Lemma. *Let h be an $\mathcal{A}$-harmonic function in a ball $B = B(0, R)$ with $h(0) = 0$ and let ν be the Riesz measure associated with $u = \max(h, 0)$. Then there exists $x_1 \in \frac{1}{4}B$ and $0 < r < R/40$ such that $h(x_1) = 0$, that $B(x_1, 10r) \subset \frac{1}{4}B$, and that*

$$\max\{\nu(0, R/8),\ \nu(x_1, 10r)\} \le c\,\nu(x_1, r)$$

for some constant $c = c(n)$.

PROOF: Let $E = \{x : h(x) = 0\}$ and put

$$\tau = \sup\{\nu(x, \delta(x)) : x \in E \cap \frac{1}{4}B\},$$

where

$$\delta(x) = \frac{1}{20}\left(\frac{R}{4} - |x|\right).$$

Choose $x_1 \in E$ such that

$$\nu(x_1, \delta(x_1)) \ge \tau/2.$$

We claim that x_1 and $r = \delta(x_1)$ will do.

Clearly, $r = \delta(x_1) < R/40$ and $B(x_1, 10r) \subset \frac{1}{4}B$. Cover $E_1 = E \cap \overline{B}(x_1, 10r)$ by balls $B(y_j, \delta(y_j))$, $y_j \in E_1$, $j = 1, \ldots, \ell$, such that

$$B(y_j, \delta(y_j)/5) \cap B(y_k, \delta(y_k)/5) = \emptyset$$

for $j \ne k$ (see the covering theorem 2.28). Since

$$\frac{1}{2}\delta(x_1) \le \delta(y) \le \frac{3}{2}\delta(x_1)$$

for any $y \in E_1$, the number ℓ of the balls $B(y_j, \delta(y_j))$ does not exceed a constant $c(n)$. Thus

$$\nu(x_1, 10r) = \nu(E_1) \le \sum_{j=1}^{\ell} \nu(y_j, \delta(y_j)) \le \ell\tau \le 2c(n)\,\nu(x_1, r).$$

Similarly, $E_2 = E \cap \overline{B}(0, R/8)$ can be covered by $c(n)$ balls $B(y_k, \delta(y_k))$, $y_k \in E_2$, so that

$$\nu(0, R/8) = \nu(E_2) \le \sum_{k=1}^{\ell} \nu(y_k, \delta(y_k)) \le 2c(n)\,\nu(x_1, r).$$

The lemma follows. $\qquad\square$

PROOF OF THEOREM 14.61: We let c denote any positive constant that depends only on the parameters n, β/α, and γ. Without loss of generality we may assume that $h_1(0) = 0$, for otherwise h_1 can be replaced by $h_1(x) - h_1(0)$ and conditions (i)–(iii) are satisfied with 2λ in place of λ. Fix $R > 0$ and let $B = B(0, R)$. Then by Lemma 14.66 there is a point $x_1 \in \frac{1}{4}B$ and a radius $r < R/40$ such that $h_1(x_1) = 0$ and

$$\max\{\nu_1(0, R/8),\ \nu_1(x_1, 10r)\} \le c\,\nu_1(x_1, r),$$

where ν_1 is the Riesz measure associated with $u_1 = \max(h_1, 0)$. Consequently, Lemma 14.65 implies that

$$\max\{M(0, R/16, h_1),\ M(x_1, 5r, h_1)\} \le c\,M(x_1, 2r, h_1),$$

and because $|h_1^+ - h_j^+| \le \lambda$ for $j = 1, \ldots, q$, we find from this that all functions h_j satisfy

(14.67)
$$\begin{aligned}
&\max\{M(0, R/16, h_j),\ M(x_1, 5r, h_j)\} \\
&\le c\left(M(x_1, 2r, h_j) + \lambda\right) \\
&\le c\,M(x_1, 4r, -h_j) + c\,\lambda,
\end{aligned}$$

where the last inequality follows by applying (14.64) to the function $h_j(x_1) - h_j$ and noting that $h_j(x_1) \le \lambda$.

Fix j and choose $y_j \in \partial B(x_1, 4r)$ such that $-h_j(y_j) = M(x_1, 4r, -h_j)$. Consider the $\mathcal{A}$-harmonic function

$$g(x) = h_j(x) + M(x_1, 4r, -h_j).$$

The Hölder estimate (14.63) together with (14.67) implies for $0 < t \le 1/2$ that

$$\begin{aligned}
M(y_j, tr, g) &\le c\,t^\kappa M(y_j, r, g) \\
&\le c\,t^\kappa\left(M(x_1, 5r, h_j) + M(x_1, 4r, -h_j)\right) \\
&\le c\,t^\kappa\left(M(x_1, 4r, -h_j) + \lambda\right)
\end{aligned}$$

where $\kappa = \kappa(n, \beta/\alpha) > 0$. Therefore, choosing $t_0 = t_0(n, \beta/\alpha, \gamma)$ positive but small enough, we find that

$$M(y_j, t_0 r, g) \le \frac{1}{2}M(x_1, 4r, -h_j) + \lambda$$

which implies

(14.68)
$$h_j(x) \le -\frac{1}{2}M(x_1, 4r, -h_j) + \lambda$$

for all $x \in B(y_j, t_0 r)$. On the other hand, since $M(0, R/16, h_1) \to \infty$ as $R \to \infty$ and since $M(0, R/16, h_1) \leq M(0, R/16, h_j) + \lambda$, we deduce from (14.67) that R can be chosen so large that the right hand side in inequality (14.68) is less than $-\lambda$. In particular,

$$B(y_j, t_0 r) \subset \{x : h_j(x) < -\lambda\}.$$

Now the balls $B(y_j, t_0 r)$, $j = 1, \ldots, q$, are pairwise disjoint with centers y_j lying on $\partial B(x_1, 4r)$, and we conclude from this that q cannot exceed a finite number that depends only on n, β/α, and γ. The proof of Theorem 14.61 is complete.   □

14.69. Quasiregular mappings and variational integrals
Quasiregular mappings not only preserve $\mathcal{A}$-harmonic functions but they preserve variational integrals as well. The basic observation here is that the n-Dirichlet integral

$$\int |\nabla u|^n \, dx$$

is invariant under conformal mappings. In this section we briefly discuss variational integrals and quasiregular mappings. Granlund *et al.* (1983) and Reshetnyak (1989) have considered this topic more thoroughly.

Suppose that $F: \mathbf{R}^n \times \mathbf{R}^n \to \mathbf{R}^n$ is a variational kernel satisfying (5.2)–(5.5) with $p = n$ and $w = 1$, and let $f: \Omega \to \mathbf{R}^n$ be quasiregular. We define the *pull-back kernel* $f^\# F: \mathbf{R}^n \times \mathbf{R}^n \to \mathbf{R}$ of F as follows. If $x \in \Omega$ is such that $J_f(x) \neq 0$, then we set

$$f^\# F(x, \xi) = J_f(x) \, F\big(f(x), f'(x)^{-1^*} \xi\big);$$

otherwise

$$f^\# F(x, \xi) = F(x, \xi).$$

The proof for the following lemma is immediate (see Lemma 14.38).

14.70. Lemma.   The pull-back kernel $f^\# F$ satisfies the assumptions (5.2)–(5.5) for $p = n$ and $w = 1$. The structure constants of $f^\# F$ in (5.3) are

$$\delta' = \delta K_I(f), \qquad \gamma' = \gamma/K_O(f),$$

where δ and γ are the constants of F.

If u is in $W^{1,n}_{loc}(\Omega')$ and if $f: \Omega \to \mathbf{R}^n$ is quasiregular and nonconstant, then $v = u \circ f$ satisfies

$$\nabla v(x) = f'(x)^* \, \nabla u\big(f(x)\big)$$

a.e. in $f^{-1}(\Omega')$. Thus the integral transformation formulas (Section 14.24) imply:

14.71. Lemma. *Suppose that* $u \in W_{loc}^{1,n}(\Omega')$. *If* f *is a quasiconformal homeomorphism of* Ω *onto* Ω', *then*

$$(14.72) \qquad \int_A f^\# F\big(x, \nabla(u \circ f)\big) \, dx = \int_{f(A)} F(x, \nabla u) \, dx$$

whenever $A \subset \Omega$ *is measurable.*

If $f: \Omega \to \mathbf{R}^n$ *is quasiregular and if* D *is a normal domain of* f, *with* $f(D) \subset \Omega'$, *then*

$$(14.73) \qquad \int_D f^\# F\big(x, \nabla(u \circ f)\big) \, dx = N(f, D) \int_{f(D)} F(x, \nabla u) \, dx \,.$$

Since F-extremality and $\nabla_\xi F$-harmonicity are equivalent concepts (Theorem 5.18), Theorem 14.39 has an obvious reformulation for F-extremals.

Capacity inequalities are important tools in the theory of quasiconformal and quasiregular mappings. We present a short outline of the basic inequalities; they are not needed in the remainder of the book. First we recall the definition for the F-capacity (see Section 5.30). Suppose that the kernel F satisfies (5.2)–(5.4) with $p = n$ and $w = 1$. A pair $C = (E, \Omega)$ is called a condenser if E is a subset of an open set $\Omega \subset \mathbf{R}^n$. If E is compact, then the F-capacity of C is

$$\operatorname{cap}_F C = \inf_{u \in W(C)} \int_\Omega F(x, \nabla u) \, dx \,,$$

where

$$W(C) = \{u \in C_0^\infty(\Omega): u \geq 1 \text{ on } E\} \,.$$

For an arbitrary condenser $C = (E, \Omega)$ this is extended as follows:

$$\operatorname{cap}_F C = \inf_{\substack{E \subset G \subset \Omega \\ G \text{ open}}} \sup_{\substack{K \subset G \\ K \text{ compact}}} \operatorname{cap}_F(K, \Omega) \,.$$

If $f: \Omega \to \mathbf{R}^n$ is quasiconformal, then for all condensers $C = (E, D)$ with $D \subset \Omega$ we have

$$(14.74) \qquad \operatorname{cap}_F fC = \operatorname{cap}_{f\#F} C \,,$$

where fC is the condenser $\big(f(E), f(D)\big)$; this is readily inferred from the integral transformation formula (14.72).

If $f: \Omega \to \mathbf{R}^n$ is quasiregular and $C = (E, D)$ is a condenser with $D \subset \Omega$, then

$$(14.75) \qquad \operatorname{cap}_F fC \leq \operatorname{cap}_{f\#F} C$$

provided that f is nonconstant in every component of D; obviously,

$$fC = \big(f(E), f(D)\big)$$

is a condenser because f is open. The proof for (14.75) is more complicated than that of (14.74). If E is compact and $u \in W(C)$, then it can be shown that

$$v(y) = \sup_{x \in f^{-1}(y)} u(x)$$

is admissible for $W(fC)$. Of course, generally v is not in $C_0^\infty\big(f(D)\big)$, but it is in $C\big(f(D)\big) \cap W_0^{1,n}(D)$ with $v \geq 1$ on $f(E)$ (Martio $et\ al.$ 1969). Martio (1970) has presented a version of (14.75) which takes the multiplicity of f into account.

Finally note that capacity equations (14.74) and (14.75) can be written as capacity inequalities involving the usual n-capacity, i.e. $\mathrm{cap}_n C = \mathrm{cap}_F C$ with $F(x, \xi) = |\xi|^n$. Indeed, by Lemma 14.70 we have the double inequality

$$\frac{1}{K_O(f)}\, \mathrm{cap}_n C \leq \mathrm{cap}_{f\#F}\, C \leq K_I(f)\, \mathrm{cap}_n C$$

whenever $f\colon \Omega \to \mathbf{R}$ is quasiregular and $C = (E, D)$ is a condenser with $D \subset \Omega$. Thus from (14.74) and (14.75) we obtain the inequalities

$$(14.76) \qquad \frac{1}{K_O(f)}\, \mathrm{cap}_n C \leq \mathrm{cap}_n fC \leq K_I(f)\, \mathrm{cap}_n C$$

and

$$(14.77) \qquad \mathrm{cap}_n fC \leq K_I(f)\, \mathrm{cap}_n C$$

which hold under the same circumstances as (14.74) and (14.75), respectively.

14.78. Mappings of bounded length distortion

We have now studied at length quasiregular mappings, which are $\mathfrak{A}_n$-harmonic morphisms. In this section we investigate a class of mappings which are $\mathfrak{A}_p$-harmonic morphisms for all $p > 1$. Recall that for $n > 2$ all mappings that preserve the ordinary harmonic functions are similarities or constants. In the $\mathcal{A}$-harmonic setup bilipschitz mappings provide a natural extension of similarities. However, it is interesting that we need not confine ourselves to injective or even locally injective mappings.

DEFINITION. A continuous mapping $f \colon \Omega \to \mathbf{R}^n$ is of *L-bounded length distortion*, or *L-BLD*, if the coordinate functions of f are in $W^{1,1}_{loc}(\Omega)$, if $J_f(x) \geq 0$ a.e. in Ω, and if for some $L \geq 1$ the inequality

(14.79) $$|h|/L \leq |f'(x)\,h| \leq L\,|h|$$

holds for all $h \in \mathbf{R}^n$ and a.e. x in Ω.

We say that f is a *BLD* mapping if it is *L-BLD* for some L.

It is obvious that condition (14.79) together with $J_f(x) \geq 0$ is more restrictive than the condition for quasiregularity.

14.80. Lemma. *Suppose that* $f \colon \Omega \to \mathbf{R}^n$ *is L-BLD. Then*

(a) *f is locally L-Lipschitz; that is,*

$$|f(x) - f(y)| \leq L\,|x - y|$$

whenever the line segment $[x, y]$ lies in Ω.

(b) *f is K-quasiregular with $K = L^{2(n-1)}$.*

PROOF: As the coordinate functions of f belong to $W^{1,1}_{loc}(\Omega)$, they are *ACL*-functions (see Section 14.21). Suppose first that $\gamma = [x, y]$ is parallel to a coordinate axis, that the coordinate functions of f are absolutely continuous on γ, and that (14.79) holds a.e. with respect to the linear measure on γ. Then (a) follows by integration. Consequently, by continuity, (a) holds whenever γ is parallel to a coordinate axis, which implies that f is locally nL-Lipschitz. By the well-known theorem of Rademacher (Federer 1969, p. 216), Lipschitz functions are differentiable a.e. In particular, the coordinate functions of f are absolutely continuous on every line segment in Ω, and we can repeat the argument above to obtain (a) for all line segments in Ω.

To prove (b), observe that by (a) the partial derivatives of the coordinate functions of f are bounded a.e. In particular they belong to $W^{1,n}_{loc}(\Omega)$. Fix $x \in \Omega$ such that (14.79) holds and let $\lambda_1 \geq \lambda_2 \geq \cdots \geq \lambda_n \geq 0$ be the square roots of the eigenvalues of the symmetric $n \times n$ matrix $f'(x)^* f'(x)$. Then $\lambda_1 = |f'(x)|$, $\lambda_n = \ell(f'(x))$, and $J_f(x) = \prod_{i=1}^{n} \lambda_i$, and we see from (14.79) that

$$1/L \leq \lambda_n \leq \lambda_{n-1} \leq \cdots \leq \lambda_1 \leq L\,.$$

Thus

$$|f'(x)|^n = \lambda_1^n \leq \lambda_1\,L^{n-1} \leq \prod_{i=1}^{n} \lambda_i\,L^{2(n-1)} = L^{2(n-1)}\,J_f(x)\,,$$

and similarly
$$J_f(x) \leq L^{2(n-1)} \ell(f'(x)).$$

These hold for a.e. x in Ω and we conclude that f is $L^{2(n-1)}$-quasiregular.
□

14.81. Theorem. *A mapping $f \colon \Omega \to \mathbf{R}^n$ is L-BLD if and only if it is quasiregular and satisfies*

(14.82) $$|f'(x)| \leq L, \qquad \ell\big(f'(x)\big) \geq 1/L$$

a.e. in Ω.

PROOF: If f is L-BLD, then f is quasiregular by Lemma 14.80 while (14.82) follows from the BLD condition (14.79).

If f is quasiregular and (14.82) holds, then (14.79) is satisfied and the quasiregularity inequality $|f'(x)|^n \leq K J_f(x)$ guarantees that the Jacobian $J_f(x)$ is a.e. nonnegative. Hence f is L-BLD.
□

The winding map $f \colon \mathbf{R}^3 \to \mathbf{R}^3$, $(r, \varphi, x_3) \mapsto (r, 2\varphi, x_3)$ in cylindrical coordinates of $\mathbf{R}^3$ is 2-BLD and we see that a BLD mapping need not be locally injective. However, if f is a continuously differentiable BLD mapping, then f is locally injective because in this case $J_f(x) = 0$, and hence $\ell(f'(x)) = 0$, for all x in the branch set B_f. Let it be mentioned here that it is an open problem whether there can be branching for a C^1-smooth quasiregular mapping in dimensions $n \geq 3$.

Although a BLD mapping is quasiregular, the converse is obviously not true. A trivial counterexample is a constant mapping which cannot satisfy the left side of (14.79). A less trivial example is an analytic function with branching. On the other hand, $f(z) = e^z$ is e^c-BLD in each strip $\{z = (x, y) \in \mathbf{R}^2 \colon |x| < c\}$.

If $f \colon \Omega \to \mathbf{R}^n$ is L-bilipschitz, that is the double inequality

(14.83) $$\frac{1}{L} |x - y| \leq |f(x) - f(y)| \leq L |x - y|$$

is satisfied for all $x, y \in \Omega$, then f is L-BLD provided it is sense-preserving. Similarly, if f satisfies (14.83) in some neighborhood of each point in Ω, i.e. f is *locally L-bilipschitz*, then f is L-BLD if it is sense-preserving. Since the inverse of a quasiconformal mapping is quasiconformal, Theorem 14.81 implies that the inverse of a homeomorphic L-BLD mapping is L-BLD. This together with Lemma 14.80 and the above discussion means that f is a locally homeomorphic L-BLD mapping if and only if f is locally L-bilipschitz with $J_f(x) \geq 0$ a.e.

Finally, since a BLD mapping is nonconstant and quasiregular, we may record the following theorem (see Theorem 14.14 and Lemma 14.22).

14.84. Theorem. *Suppose that* $f \colon \Omega \to \mathbf{R}^n$ *is L-BLD. Then*

(a) f *is discrete, open, and sense-preserving;*
(b) *the restriction of* f *to* $\Omega \setminus B_f$ *is locally L-bilipschitz;*
(c) *if* f *is injective, then* $f^{-1} \colon f(\Omega) \to \Omega$ *is L-BLD;*
(d) $|B_f| = 0 = |f B_f|$.

Next we turn to the most useful characterization of *BLD* mappings. By the same token, the name "bounded length distortion" will be justified.

We recall that a path α is a continuous mapping $\alpha \colon [a, b] \to \mathbf{R}^n$ with length

$$\ell(\alpha) = \sup \sum_{i=1}^{k} |\alpha(t_i) - \alpha(t_{i-1})|,$$

where the supremum is taken over all subdivisions $a = t_0 \le t_1 \le \cdots \le t_k = b$ of $[a, b]$.

We first establish that *BLD* mappings possess an analogue of the bilipschitz property even at branch points. Recall that for discrete, open, and sense-preserving mappings there are arbitrary small normal neighborhoods $U = U(x, f, r)$ about each point $x \in \Omega$; U is the x-component of $f^{-1}\big(B(f(x), r)\big)$ (see Section 14.4).

14.85. Lemma. *If* $f \colon \Omega \to \mathbf{R}^n$ *is L-BLD and if* $U = U(x, f, r)$ *is a normal neighborhood of* $x \in \Omega$, *then*

$$\frac{1}{L} |y - x| \le |f(y) - f(x)| \le L |y - x|$$

whenever $|y - x| \le r/L$.

PROOF: Write

$$L^* = \sup\{|y - x| \colon y \in \partial U\}, \qquad \ell^* = \inf\{|y - x| \colon y \in \partial U\}.$$

We first show that

$$(14.86) \qquad\qquad L^* \le L r, \qquad \ell^* \ge r/L.$$

Choose $y \in \partial U$ with $|y - x| = \ell^*$. Since the line segment $[x, y]$ lies in U, we can use the Lipschitz character of f (Lemma 14.80) to conclude

$$r = |f(y) - f(x)| \le L \ell^*,$$

which is the second inequality in (14.86).

To prove the first inequality, we consider a point $z \in \partial B(f(x), r)$ and the radial path $\beta_z \colon [0, r) \to \mathbf{R}^n$ from $f(x)$ to z. Now $U \cap f^{-1}\beta_z$ can be covered by m maximal lifts of β_z starting at x, where $m = N(U, f) = i(x, f)$ (Lemma 14.10), and it follows from the remark after Lemma 14.23 that for almost every z on the boundary $\partial B(f(x), r)$, each lift α_z is absolutely continuous on every interval $[s, r]$, $s > 0$. Since $|f B_f| = 0$, almost every β_z meets $f B_f$ in a set of linear measure zero; moreover, because $f|_{G \setminus B_f}$ is locally L-bilipschitz, we then have that $|\alpha_z'(t)| \leq L$ for almost every $t \in (0, r)$. Thus

$$|\alpha_z(t) - x| = |\alpha_z(t) - \alpha_z(0)| \leq L r$$

for all $t \in (0, r)$ and for almost every $z \in B(f(x), r)$. Consequently, $|y - x| \leq L r$ for a dense set of points $y \in U$, and this implies $L^* \leq L r$.

The inequalities in (14.86) yield the double inequality of the lemma, for if $0 < |y - x| \leq r/L$, then $U(x, f, |f(y) - f(x)|)$ is a normal neighborhood of x in U (Lemma 14.6) and (14.86) can be used for this neighborhood in place of U. The lemma follows. $\square$

14.87. Theorem. *Let $f \colon \Omega \to \mathbf{R}^n$ be discrete, open, and sense-preserving. Then f is L-BLD if and only if*

$$(14.88) \qquad \frac{1}{L}\, \ell(\alpha) \leq \ell(f\alpha) \leq L\, \ell(\alpha)$$

for every path α in Ω.

PROOF: Assume first that f is L-BLD. Since f is locally L-Lipschitz, the second inequality in (14.88) is clear. The proof of the first inequality is less trivial. Write $\beta = f\alpha$. We may assume that $\ell(\beta) < \infty$ and that $\beta \colon [0, \ell(\beta)] \to \mathbf{R}^n$ is a parametrization by arc length. Let P be a partition of $[0, \ell(\beta)]$. It easily follows from Lemma 14.7 that there is a refinement $P' = \{t_0, \ldots, t_k\}$ of P and numbers $r_i > 0$ such that for each $i = 0, \ldots, k$, $U(\alpha(t_i), f, r_i)$ is a normal neighborhood of $\alpha(t_i)$ and $\beta[t_{i-1}, t_i] \subset B_{i-1} \cup B_i$, where $B_i = B(\beta(t_i), r_i)$. For each $i = 1, \ldots, k$ choose $s_i \in [t_{i-1}, t_i]$ with $\beta(s_i) \in B_{i-1} \cap B_i$. Then the preceding lemma implies

$$|\alpha(s_i) - \alpha(t_{i-1})| \leq L(s_i - t_{i-1}), \qquad |\alpha(t_i) - \alpha(s_i)| \leq L(t_i - s_i).$$

Thus

$$\sum_{i=1}^{k} |\alpha(t_i) - \alpha(t_{i-1})| \leq \sum_{i=1}^{k} \big(|\alpha(t_i) - \alpha(s_i)| + |\alpha(s_i) - \alpha(t_{i-1})| \big)$$

$$\leq L \sum_{i=1}^{k} (t_i - s_i) + (s_i - t_{i-1}) = L\, \ell(\beta).$$

This yields $\ell(\alpha) \leq L\,\ell(\beta)$ as required.

To prove the if part, observe first that (14.88) implies that f is locally L-Lipschitz, hence ACL. In consequence, f is differentiable and the derivative satisfies $|f'(x)| \leq L$ a.e. We invoke a result of Martio *et al.* (1969, Lemma 2.14) which is well known for homeomorphisms: for an almost everywhere differentiable sense-preserving discrete and open mapping f it holds that $J_f(x) \geq 0$ a.e.

It remains to verify the inequality

$$|f'(x)h| \geq |h|/L$$

whenever $h \in \mathbf{R}^n$ and $x \in \Omega$ is such that f is differentiable at x. Suppose that this is false for some $h = h_0$. Then

$$|f'(x)h_0|/|h_0| = \delta < 1/L.$$

To simplify notation, we assume that $x = 0 = f(x)$. Let $U = U(0, f, r)$ be a normal neighborhood of 0. Then f has the expansion

$$f(h) = f'(0)h + \varepsilon(h)|h|,$$

where $\varepsilon(h) \to 0$ as $h \to 0$. Pick $t > 0$ so small that $h = th_0 \in U$ and that

$$2|\varepsilon(h)| \leq 1/L - \delta.$$

Now the line segment γ from 0 to $f(h)$ lifts to a path γ^* from 0 to h, and assumption (14.88) implies

$$|f(h)| = \ell(\gamma) \geq \ell(\gamma^*)/L \geq |h|/L.$$

On the other hand, the previous inequalities and a computation yield

$$|f(h)| \leq |f'(0)h| + |\varepsilon(h)||h| \leq \delta|h| + (1/L - \delta)|h|/2$$
$$= (1/L + \delta)|h|/2 < |h|/L,$$

which is a contradiction. The theorem follows.   □

In (14.88) we used the length of a path instead of the length of its locus. Consider, for example, the 2-BLD mapping $f \colon \mathbf{R}^2 \to \mathbf{R}^2$, $(r, \varphi) \mapsto (r, 2\varphi)$ in polar coordinates of $\mathbf{R}^2$, and the path $\alpha(t) = (\cos t, \sin t)$, $t \in [0, 2\pi]$. Then $\beta = f\alpha$ is a path twice around $\partial B(0, 1)$ and $\ell(\beta) = 2\,\ell(\alpha)$, in consistency with (14.88).

Since a BLD mapping f is quasiregular, both the removability result (Theorem 14.46) and Radó's theorem 14.47 holds for BLD mappings. However, it is clear that BLD mappings should be much more rigid than quasiregular mappings. Here are some results in this direction (Martio and Väisälä 1988):

(i) If E is a relatively closed subset of Ω and if the $(n-1)$-dimensional Hausdorff measure of E is zero, then every L-BLD mapping $f: \Omega \setminus E \to \mathbf{R}^n$ has an L-BLD extension to Ω.

(ii) BLD mappings $f: \Omega \to \mathbf{R}^n$ preserve the Hausdorff dimension of any subset of Ω.

(iii) If $f: \Omega \to \mathbf{R}^n$ is L-BLD and $x \in \Omega$, then the local topological index of f at x does not exceed the number $L^{2(n-1)}$.

(iv) If $f: \mathbf{R}^n \to \mathbf{R}^n$ is L-BLD, then the maximal multiplicity $N(f, \mathbf{R}^n)$ does not exceed the number L^{2n}. Moreover, $f(x) \to \infty$ as $x \to \infty$.

None of assertions (i)–(iv) remain true for general quasiregular mappings.

14.89. BLD mappings are $\mathfrak{A}_p$-harmonic morphisms

Throughout this section we suppose that $\mathcal{A}: \mathbf{R}^n \to \mathbf{R}^n$ satisfies (3.3)–(3.7) with $1 < p < \infty$ and $w = 1$. If $f: \Omega \to \mathbf{R}^n$ is an L-BLD mapping, the pull-back $f^{\#}\mathcal{A}$ of $\mathcal{A}$ is

$$f^{\#}\mathcal{A}(x, \xi) = J_f(x) f'(x)^{-1} \mathcal{A}\big(f(x), f'(x)^{-1^*} \xi\big)$$

whenever this makes sense, and

$$f^{\#}\mathcal{A}(x, \xi) = \mathcal{A}(x, \xi)$$

otherwise. Note that the pull-back mapping $f^{\#}\mathcal{A}$ was defined in (14.36) for general quasiregular mappings by the same formula; here, however, f is BLD and we do not restrict the value of p.

The next lemma is similar to Lemma 14.38.

14.90. Lemma. *The mapping $f^{\#}\mathcal{A}$ satisfies assumptions (3.3)–(3.7) with $w = 1$ and with structure constants*

$$\alpha' = L^{2-n-p}\alpha, \qquad \beta' = L^{n+p-2}\beta,$$

where α, β are the structure constants of $\mathcal{A}$.

PROOF: The proof is almost identical to that of Lemma 14.38; the estimates

$$J_f(x)\,|f'(x)^{-1}\xi| \le L^{n-1}\,|\xi|$$

and

$$J_f(x)\,|f'(x)^{-1^*}\xi|^p \ge L^{2-n-p}\,|\xi|^p$$

are now employed to obtain the constants α' and β'. $\square$

14.91. Theorem. *Let $\mathcal{A}$ satisfy assumptions (3.3)–(3.7) with $p > 1$ and $w = 1$ and let $f \colon \Omega \to \mathbf{R}^n$ be a BLD mapping. If u is $\mathcal{A}$-harmonic in $\Omega' \subset \mathbf{R}^n$, then $u \circ f$ is $f^{\#}\mathcal{A}$-harmonic in $f^{-1}(\Omega')$.*

PROOF: The proof is similar to that of Theorem 14.39. Since f is L-BLD for some L, it is fairly easy to show that the function

$$\psi^*(y) = \sum_{x \in f^{-1}(y)} i(x, f)\, \psi(x)$$

is kLM-Lipschitz in a normal neighborhood U whenever $\psi \in C_0^\infty(U)$ is M-Lipschitz in U and $k = N(f, U)$; to establish this, Theorem 14.87 can be used. The integral transformation formulas involve only bilipschitz mappings in this case. The details are left to the reader, who may also consult Martio and Väisälä (1988, pp. 433–434). $\square$

14.92. Corollary. *A BLD mapping f is an $(f^{\#}\mathcal{A}, \mathcal{A})$-harmonic morphism. In particular, f is a $\mathfrak{A}_p$-harmonic morphism for each $p > 1$.*

Now Theorem 13.7 leads to the following result because a BLD mapping is never constant.

14.93. Theorem. *Suppose that $f \colon \Omega \to \mathbf{R}^n$ is BLD. If u is $\mathcal{A}$-superharmonic in an open set $\Omega' \subset \mathbf{R}^n$, then $u \circ f$ is $f^{\#}\mathcal{A}$-superharmonic in $f^{-1}(\Omega')$.*

As a final remark to this chapter we mention that a complete characterization of $\mathfrak{A}_p$-harmonic morphisms is not known. Heinonen *et al.* (1992) have obtained partial results in this direction; in particular they have proved that if $1 < p < n$, then a homeomorphic $\mathfrak{A}_p$-harmonic morphism is locally bilipschitz.

NOTES TO CHAPTER 14. Riemann founded his theory of analytic functions on potential theory, and now the connection between plane analytic and harmonic functions is explained in almost every textbook of complex variables. Tsuji (1959) is a classic reference to deeper applications of potential theory in complex analysis.

The analytic definition of plane quasiconformal mappings apparently emerged in a paper by Morrey (1938). He studied homeomorphic solutions to a Beltrami system. Morrey's idea was to reduce the study of the solutions v of a second order partial differential equation to the study of harmonic functions u via the representation $v = u \circ f$. This is exactly the morphism property of f. For further studies in this direction, see the monographs by Vekua (1962) and Bers and Schechter (1964).

Space quasiconformal mappings were first considered by Lavrentiev in 1938. Since 1959 these mappings have been studied extensively from the function theoretic point of view. Gehring, Väisälä, and the Russian school made groundbreaking contributions in this direction; see the bibliographies in the monographs by Väisälä (1971) and Vuorinen (1988). Systematic study of quasiregular mappings was initiated by Reshetnyak in the 1960s. He proved the $(f^\# \mathcal{A}, \mathcal{A})$-morphism property of a quasiregular mapping under slightly more restrictive assumptions on $\mathcal{A}$. The important result that a nonconstant quasiregular mapping is discrete and open is due to him (Reshetnyak 1967a, 1989). Martio et al. (1969, 1970, 1971) presented a more geometric approach to quasiregular mappings based on the method of extremal length. This approach has consequently resulted in a remarkable extension of classical Nevanlinna theory; see the monograph by Rickman (In preparation). The analytic–geometric approach followed in this book was initiated by Granlund et al. (1983). Bojarski and Iwaniec (1983) have presented still another approach.

There is a growing list of applications of nonlinear potential theory to quasiregular mappings (Martio 1992). Granlund et al. (1982, 1985), Heinonen and Martio (1987), Martio (1987), Heinonen et al. (1989b), and Heinonen and Rossi (1990, In press) have used the $\mathcal{A}$-harmonic measure in connection of quasiregular mappings. Heinonen et al. (1989a) have studied the fine limits of quasiregular and BLD mappings. Holopainen and Rickman (1992, In press) have extended the method of Eremenko and Lewis (1991) and proved Picard type theorems for quasiregular mappings in more general circumstances. Lewis (In press) has extended a conditional theorem of Littlewood for entire analytic functions to quasiregular mappings by using the $\mathcal{A}$-potential theory.

15

A_p-weights and
Jacobians of quasiconformal mappings

In this chapter we provide two important examples of p-admissible weights: A_p-weights and certain powers of the Jacobian of a quasiconformal mapping. To keep the presentation self-contained, we first establish some basic properties of A_p-weights, including Muckenhoupt's theorem. Then we prove that the two kinds of weights mentioned above are admissible. The present chapter is independent of the previous chapters.

15.1. A_p-weights

Let w be a locally integrable nonnegative function in $\mathbf{R}^n$ and assume that $0 < w < \infty$ almost everywhere. We say that w belongs to the *Muckenhoupt class A_p*, $1 < p < \infty$, or that w is an *A_p-weight*, if there is a constant $c_{p,w}$ such that

$$\fint_B w \, dx \leq c_{p,w} \Big(\fint_B w^{1/(1-p)} \, dx\Big)^{1-p}$$

for all balls B in $\mathbf{R}^n$; recall that for any measure ν,

$$\fint_E g \, d\nu = \frac{1}{\nu(E)} \int_E g \, d\nu$$

if $0 < \nu(E) < \infty$. We say that w belongs to A_1, or that w is an *A_1-weight*, if there is a constant $c_{1,w} \geq 1$ such that

$$\fint_B w \, dx \leq c_{1,w} \operatorname*{ess\,inf}_B w$$

for all balls B in $\mathbf{R}^n$. The union of all Muckenhoupt classes A_p is denoted by A_∞,

$$A_\infty = \bigcup_{p>1} A_p .$$

We also say that a weight w is an *A_∞-weight* if it belongs to some A_p. As customary, μ stands for the measure whose Radon–Nikodym derivative w is,

$$\mu(E) = \int_E w \, dx .$$

The following consequence of Hölder's inequality will be used repeatedly: for any measure ν and $-\infty < s \leq t < \infty$, with $s, t \neq 0$,

$$(15.2) \qquad \left(\fint_E g^s \, d\nu\right)^{1/s} \leq \left(\fint_E g^t \, d\nu\right)^{1/t}$$

whenever g is nonnegative and ν-measurable. In particular we have

$$\left(\fint_B w^{1/(1-p)} \, dx\right)^{1-p} \leq \fint_B w \, dx$$

so that $c_{p,w} \geq 1$ always. It also follows from (15.2) that

$$A_1 \subset A_q \subset A_p$$

for all $1 < q \leq p$. Moreover, if $w \in A_p$ and $0 < s \leq 1$, then

$$\left(\fint_B w^s \, dx\right) \leq \left(\fint_B w \, dx\right)^s \leq c_{p,w}^s \left(\fint_B w^{1/(1-p)} \, dx\right)^{s(1-p)}$$
$$\leq c_{p,w}^s \left(\fint_B w^{s/(1-p)} \, dx\right)^{1-p}$$

so that $w^s \in A_p$ with $c_{p,w^s} = c_{p,w}^s$.

As an example of A_p-weights we recall that the weight $w(x) = |x|^\delta$ is in A_p if and only if $-n < \delta < n(p-1)$. This is easily verified by a direct computation. Interesting examples are provided by ordinary superharmonic functions: if w is a positive superharmonic function in $\mathbf{R}^n$, then w is an A_1-weight. For this and more general examples, see the discussion after Theorem 3.59.

Our main goal in this section is to show that the definition for A_p-weights contains a remarkable self-improving property: if w is in A_p for some $p > 1$, then w is in $A_{p-\varepsilon}$ for some $\varepsilon > 0$. This *open-end property* is a consequence of the following *reverse Hölder inequality*.

15.3. Reverse Hölder inequality. *If $w \in A_p$, then there are numbers $r > 1$ and $c_r \geq 1$, depending only on n, p, and $c_{p,w}$, such that*

$$(15.4) \qquad \left(\fint_B w^r \, dx\right)^{1/r} \leq c_r \fint_B w \, dx$$

for all balls B.

To prove the reverse Hölder inequality (15.4), we require several preparatory results. The first asserts that the measure μ satisfies a *strong doubling property*.

15.5. Strong doubling of A_p-weights. *If $w \in A_p$, then*

$$(15.6) \qquad \qquad \left(\frac{|E|}{|B|}\right)^p \le c_{p,w} \frac{\mu(E)}{\mu(B)}$$

whenever B is a ball in $\mathbf{R}^n$ and E is a measurable subset of B.

PROOF: We have

$$
\begin{aligned}
|E| &= \int_E w^{1/p} w^{-1/p}\, dx \le \left(\int_E w\, dx\right)^{1/p} \left(\int_E w^{1/(1-p)}\, dx\right)^{(p-1)/p} \\
&\le \mu(E)^{1/p} \left(\fint_B w^{1/(1-p)}\, dx\right)^{(p-1)/p} |B|^{(p-1)/p} \\
&\le c_{p,w}^{1/p} \mu(E)^{1/p} \left(\fint_B w\, dx\right)^{-1/p} |B|^{(p-1)/p} \\
&= c_{p,w}^{1/p} \left(\frac{\mu(E)}{\mu(B)}\right)^{1/p} |B|,
\end{aligned}
$$

and the claim follows. $\qquad\qquad\qquad\qquad\qquad\qquad\qquad\qquad\qquad\qquad\square$

15.7. Corollary. *If $w \in A_p$, then μ is a doubling measure; that is,*

$$\mu(2B) \le c\, \mu(B)$$

for all balls B in $\mathbf{R}^n$, where $c = 2^{np} c_{p,w}$.

We next show that (15.6) leads to a similar reverse inequality.

15.8. Lemma. *If w is in A_p, then there are $0 < q \le 1$ and $c > 0$, depending only on n, p, and $c_{p,w}$, such that*

$$(15.9) \qquad \qquad \frac{\mu(E)}{\mu(B)} \le c \left(\frac{|E|}{|B|}\right)^q$$

whenever B is a ball in $\mathbf{R}^n$ and E is a measurable subset of B.

PROOF: It is more convenient to use cubes in place of balls. Indeed, since the doubling property implies $\mu(B) \approx \mu(Q)$ whenever Q is an open cube with $B \subset Q \subset \sqrt{n}B$, it suffices to verify (15.9) with B replaced by a cube Q. Similarly, it follows from the strong doubling property (15.6) that for $E \subset Q$ we have

$$\left(\frac{|E|}{|Q|}\right)^p \le c' \frac{\mu(E)}{\mu(Q)},$$

where $c' = c'(n, p, c_{p,w}) > 0$. Next observe that

(15.10) $\qquad |E| < \dfrac{1}{2}|Q| \qquad$ implies $\qquad \mu(E) < \delta\mu(Q)$

for all cubes Q and measurable $E \subset Q$, where $\delta = 1 - 1/(c'2^p)$; indeed, we have

$$\frac{1}{2^p} < \left(\frac{|Q \setminus E|}{|Q|}\right)^p \le c' \frac{\mu(Q \setminus E)}{\mu(Q)} = c'\left(1 - \frac{\mu(E)}{\mu(Q)}\right).$$

To prove (15.9), we may assume that

$$|E| < \frac{1}{2}|Q|.$$

Let $\mathcal{Q}$ be the collection of open dyadic subcubes of Q (see Appendix II). By the Lebesgue differentiation theorem (Stein 1970, p. 18), we have that

$$\lim_{\substack{|\tilde{Q}| \to 0 \\ x \in \tilde{Q} \in \mathcal{Q}}} \frac{|E \cap \tilde{Q}|}{|\tilde{Q}|} = 1$$

for a.e. $x \in E$. Thus for a.e. $x \in E$ there is a cube $\tilde{Q} \in \mathcal{Q}$ with $x \in \tilde{Q}$ and

(15.11) $\qquad\qquad |E \cap \tilde{Q}| \ge \dfrac{1}{2^{n+1}}|\tilde{Q}|;$

denote by Q_x the largest cube $\tilde{Q} \in \mathcal{Q}$ which contains x and satisfies (15.11). Then

(15.12) $\qquad\qquad \dfrac{1}{2^{n+1}}|Q_x| \le |E \cap Q_x| < \dfrac{1}{2}|Q_x|$

because $|E| < \frac{1}{2}|Q|$ and because $2^n|Q_x| = |Q'_x|$ for the smallest dyadic cube $Q'_x \neq Q_x$ containing Q_x. Since the cubes Q_x are dyadic subcubes of Q, we have that either $Q_x = Q_y$ or $Q_x \cap Q_y = \emptyset$. Let E_1 denote the union of all pairwise disjoint cubes of the form Q_x. Then E_1 contains almost every point of E. If

$$|E_1| \ge \frac{1}{2}|Q|,$$

set

$$E_2 = Q$$

and stop. Otherwise repeat the reasoning with E replaced by E_1. The process ends at some integer $k \ge 1$. More precisely, there are sets

$$E_0 \subset E_1 \subset \ldots \subset E_k = Q,$$

where $E_0 \subset E$ with $|E_0| = |E|$, and for each $1 \leq j \leq k$ the set E_j is a pairwise disjoint union of dyadic cubes $Q_{i,j}$ satisfying

$$\frac{1}{2^{n+1}}|Q_{i,j}| \leq |E_{j-1} \cap Q_{i,j}| < \frac{1}{2}|Q_{i,j}|$$

for $Q_{i,j} \subset E_j$. Then

$$\mu(E_{j-1}) = \sum_i \mu(E_{j-1} \cap Q_{i,j})$$
$$< \delta \sum_i \mu(Q_{i,j}) = \delta\mu(E_j)$$

and hence

$$\mu(E) = \mu(E_0) \leq \delta^k \mu(Q),$$

where δ is as in (15.10). On the other hand, for $j \geq 1$,

$$|E_j| = \sum_i |Q_{i,j}| \leq \sum_i 2^{n+1}|E_{j-1} \cap Q_{i,j}| = 2^{n+1}|E_{j-1}|$$

so that

$$|Q| \leq 2^{(n+1)k+1}|E_0| = 2^{(n+1)k+1}|E|.$$

Combining the above inequalities yields

$$\frac{\mu(E)}{\mu(Q)} \leq \left(\frac{|E|}{|Q|}\right)^q,$$

where $q = -\log\delta/\log 2^{n+2}$, and the lemma follows. $\qquad\square$

PROOF OF THE REVERSE HÖLDER INEQUALITY 15.3: For a ball B and $t > 0$ let $\lambda(t)$ be the distribution function of w; that is, $\lambda(t) = |E_t|$, where $E_t = \{x \in B : w(x) > t\}$. Then

$$t\lambda(t) \leq \int_{E_t} w(x)\,dx = \mu(E_t) \leq c\mu(B)\left(\frac{|E_t|}{|B|}\right)^q,$$

where c and q are as in (15.9). We may assume that $q < 1$ and so

$$\lambda(t) \leq \min\left\{|B|, \left(\frac{c}{t}\right)^{1/(1-q)}\left(\frac{\mu(B)}{|B|^q}\right)^{1/(1-q)}\right\}.$$

We obtain for $r > 1$

$$\int_B w(x)^r \, dx = r \int_0^\infty t^{r-1} \lambda(t) \, dt$$

$$\leq r \int_0^{t_0} t^{r-1} |B| \, dt + c^{1/(1-q)} r \int_{t_0}^\infty t^{r-1-1/(1-q)} \Big(\frac{\mu(B)}{|B|^q}\Big)^{1/(1-q)} dt \, ;$$

for the first equality see (3.31). We choose $t_0 = c \, \mu(B)/|B|$ and a computation then gives

$$\Big(\fint_B w(x)^r \, dx\Big)^{1/r} \leq c_1 \fint_B w(x) \, dx$$

for $r < 1/(1 - q)$ and $c_1 = c(1 - r(1 - q))^{-1/r}$, where $c = c(n, p, c_{p,w})$ is the constant in (15.9). This completes the proof of the reverse Hölder inequality (15.4). $\qquad\Box$

The arguments above show that for a nonnegative locally integrable function w the strong doubling property 15.5 implies (15.9), and that the reverse Hölder inequality follows from (15.9). It is well known that any of the inequalities (15.4), (15.6), and (15.9) is equivalent to w being an A_∞-weight. We shall not use this characterization of A_∞ here.

15.13. Open-end property. *If w is in A_p, $p > 1$, then w is in A_q for some $q < p$. Moreover, q and $c_{q,w}$ depend only on n, p, and $c_{p,w}$.*

PROOF: Since the A_p-condition equivalently reads

$$\fint_B u \, dx \leq c_{p,w}^{1/(p-1)} \Big(\fint_B u^{1/(1-p')} \, dx\Big)^{1-p'} \, ,$$

where $u = w^{1/(1-p)}$ and $p' = p/(p-1)$), we can apply the reverse Hölder inequality to u and obtain an $r > 1$ such that

$$\Big(\fint_B u^r \, dx\Big)^{1/r} \leq c \fint_B u \, dx$$

for all balls B. Here c and r depend only on n, p, and $c_{p,w}$. Since $r > 1$, we have $r/(p - 1) = 1/(q - 1)$ for some $1 < q < p$, and the claim follows from the computation

$$\Big(\fint_B w^{1/(1-q)} \, dx\Big)^{q-1} = \Big(\fint_B w^{r/(1-p)} \, dx\Big)^{(p-1)/r}$$

$$\leq c \Big(\fint_B w^{1/(1-p)} \, dx\Big)^{p-1} \leq c \Big(\fint_B w \, dx\Big)^{-1} \, .$$

$\qquad\Box$

15.14. Muckenhoupt's theorem

The celebrated theorem of Muckenhoupt asserts that the Hardy–Littlewood maximal function maps $L^p(B;\mu)$ continuously into itself if $w \in A_p$. This short section is devoted to a proof of Muckenhoupt's theorem.

For a measurable function f in $\mathbf{R}^n$ we define the Hardy–Littlewood maximal function

$$(15.15) \qquad Mf(x) = \sup \fint_B |f(y)|\, dy,$$

where the supremum is taken over all balls B that contain x. If f is defined only in a subset E of $\mathbf{R}^n$, we set $f(y) = 0$ for $y \in \mathbf{R}^n \setminus E$.

15.16. Theorem. *Suppose that $w \in A_p$ and let B_0 be a ball in $\mathbf{R}^n$. Then*

$$(15.17) \qquad \int_{B_0} |Mf|^p\, d\mu \le c \int_{B_0} |f|^p\, d\mu$$

for each $f \in L^p(B_0;\mu)$, where c depends only on n, p, and the A_p-constant $c_{p,w}$ of w.

PROOF: Set

$$(15.18) \qquad M_\mu f(x) = \sup \fint_B |f(y)|\, d\mu,$$

where again the supremum is taken over all balls B that contain x. The classical argument (Stein 1970, pp. 6–7) then shows that

$$(15.19) \qquad \int_{B_0} |M_\mu f|^s\, d\mu \le c \int_{B_0} |f|^s\, d\mu$$

for each $f \in L^s(B_0;\mu)$ and $1 < s < \infty$, where c depends only on n, s, and the A_p-constant $c_{p,w}$ of w. We demonstrate how the open-end property and (15.19) combined yield (15.17).

Let $x \in B_0$ and let B be a ball that contains x. Pick $q < p$ such that $w \in A_q$. Then Hölder's inequality gives

$$\fint_B |f|\, dy \le \left(\fint_B |f|^q w\, dy\right)^{1/q} \left(\fint_B w^{1/(1-q)}\, dy\right)^{(q-1)/q}$$

$$\le c_{q,w}^{1/q} \left(\frac{1}{|B|} \int_B |f|^q\, d\mu\right)^{1/q} \left(\frac{\mu(B)}{|B|}\right)^{-1/q}$$

$$= c_{q,w}^{1/q} \left(\fint_B |f|^q\, d\mu\right)^{1/q},$$

and hence
$$Mf(x) \leq c(M_\mu |f|^q(x))^{1/q}.$$

Therefore, using (15.19) with $s = p/q > 1$, we arrive at
$$\int_{B_0} |Mf|^p \, d\mu \leq c \int_{B_0} |M_\mu |f|^q|^{p/q} \, d\mu \leq c \int_{B_0} |f|^p \, d\mu,$$

as required.   □

It is not difficult to see that the A_p-condition is also necessary for (15.17) to hold. We do not need this fact here.

15.20. A_p-weights are p-admissible
In this section we establish that A_p-weights satisfy assumptions I–IV in Chapter 1, that is they are p-admissible.

15.21. Theorem. *A_p-weights are p-admissible. The constant c_μ depends only on n, p, and the A_p-constant $c_{p,w}$ of the weight w.*

Since $A_1 \subset A_p$, we obtain the following corollary.

15.22. Corollary. *A_1-weights are p-admissible for all $p > 1$.*

To prove Theorem 15.21, we observe first that trivially $0 < w < \infty$ a.e. in $\mathbf{R}^n$. Next, because requirement I (the doubling condition) is the content of Corollary 15.7 and because requirement II (the uniqueness of the gradient) was already established in Section 1.9, we only need to verify the weighted Sobolev and Poincaré inequalities.

For an A_∞-weight w define
$$p_0 = p_0(w) = \inf\{ p > 1 : w \in A_p \}.$$

15.23. Embedding theorem for A_p-weights. *Suppose that $w \in A_p$ and let $p_0 < q < p < nq$ and $\kappa = n/(n - p/q) > 1$. Then there exists a constant $c = c(n, p, c_{p,w}, q) > 0$ such that*

$$(15.24) \qquad \Big(\fint_B |\varphi|^{\kappa p} \, d\mu\Big)^{1/\kappa p} \leq c\, r \Big(\fint_B |\nabla\varphi|^p \, d\mu\Big)^{1/p}$$

whenever $B = B(x_0, r)$ is a ball and $\varphi \in C_0^\infty(B)$.

PROOF: Recall that the *Riesz potential* of a measurable function f in $\mathbf{R}^n$ is
$$If(x) = \int_{\mathbf{R}^n} \frac{|f(y)|}{|x - y|^{n-1}} \, dy.$$

Then we have the elementary pointwise inequality

$$|\varphi(x)| \le c(n) \int_{\mathbf{R}^n} \frac{|\nabla\varphi(y)|}{|x-y|^{n-1}}\, dy = c(n)\, I|\nabla\varphi|(x)$$

for $\varphi \in C_0^\infty(\mathbf{R}^n)$; see Gilbarg and Trudinger (1983, Lemma 7.14). In particular,

$$\Big(\fint_B |\varphi|^{\kappa p}\, d\mu\Big)^{1/\kappa p} \le c\,\Big(\fint_B |I|\nabla\varphi||^{\kappa p}\, d\mu\Big)^{1/\kappa p}$$

when φ is supported in a ball B. Therefore, to establish (15.24), we only need to verify the general inequality

$$(15.25) \qquad \Big(\fint_B |If|^{\kappa p}\, d\mu\Big)^{1/\kappa p} \le cr\,\Big(\fint_B |f|^p\, d\mu\Big)^{1/p}$$

whenever $f \in C(B) \cap L^p(B;\mu)$ and $B = B(x_0, r)$.

To this end, we set $f = 0$ outside B and put

$$I^\varepsilon f(x) = \int_{B(x,\varepsilon)} \frac{|f(y)|}{|x-y|^{n-1}}\, dy$$

for $\varepsilon > 0$. Since

$$Mf(x) \ge \fint_{B(x,\lambda)} |f(y)|\, dy \ge c\lambda^{-n} \int_{\{\lambda/2 \le |x-y| \le \lambda\}} |f(y)|\, dy$$

$$\ge c\lambda^{-1} \int_{\{\lambda/2 \le |x-y| \le \lambda\}} \frac{|f(y)|}{|x-y|^{n-1}}\, dy$$

for all $\lambda > 0$, we find that

$$I^\varepsilon f(x) = \sum_{j=1}^\infty \int_{\{\varepsilon/2^j \le |x-y| \le \varepsilon/2^{j-1}\}} \frac{|f(y)|}{|x-y|^{n-1}}\, dy$$

$$\le c\sum_{j=1}^\infty \varepsilon 2^{-j} Mf(x) = c(n)\varepsilon Mf(x)$$

for all $x \in \mathbf{R}^n$ and $\varepsilon > 0$.

Next, choose $p_0 < q < p$ such that $p < nq$ and use Hölder's inequality to estimate

$$\int_{\{|x-y|>\varepsilon\}} \frac{|f(y)|}{|x-y|^{n-1}}\, dy$$

$$\le \|f\|_{p,\mu} \Big(\int_{\{|x-y|>\varepsilon\}\cap B} |x-y|^{p(1-n)/(p-1)} w(y)^{1/(1-p)}\, dy\Big)^{(p-1)/p}$$

$$\le \|f\|_{p,\mu} \Big(\int_{\{|x-y|>\varepsilon\}} |x-y|^{p(1-n)/(p-q)}\, dy\Big)^{(p-q)/p}$$

$$\times \Big(\int_B w(y)^{1/(1-q)}\, dy\Big)^{(q-1)/p},$$

where

$$||f||_{p,\mu} = \big(\int_B |f|^p \, d\mu \big)^{1/p}.$$

Thus

$$If(x) \leq c_1 \varepsilon M f(x) + c_2 \varepsilon^{(p-nq)/p} ||f||_{p,\mu} \big(\int_B w^{1/(1-q)} dy \big)^{(q-1)/p},$$

and choosing

$$\varepsilon = \Big(\frac{||f||_{p,\mu}(\int_B w^{1/(1-q)} dy)^{(q-1)/p}}{M f(x)} \Big)^{p/nq}$$

gives

$$If(x) \leq c M f(x)^{(nq-p)/nq} ||f||_{p,\mu}^{p/nq} \big(\int_B w^{1/(1-q)} dy \big)^{(q-1)/nq}.$$

Therefore for $\kappa = n/(n - p/q)$ we have

$$\big(\int_B |If(x)|^{\kappa p} \, d\mu \big)^{1/\kappa p}$$

$$\leq c \big(\int_B |Mf(x)|^p \, d\mu \big)^{1/\kappa p} ||f||_{p,\mu}^{p/nq} \big(\int_B w^{1/(1-q)} dy \big)^{(q-1)/nq}$$

$$\leq c ||f||_{p,\mu} \big(\int_B w^{1/(1-q)} dy \big)^{(q-1)/nq},$$

where Muckenhoupt's theorem 15.16 was used in the last step. Since $w \in A_q$, we conclude

$$\big(\fint_B If(x)^{\kappa p} \, d\mu \big)^{1/\kappa p}$$

$$\leq c \mu(B)^{1/p - 1/\kappa p} \big(\fint_B |f|^p \, dy \big)^{1/p} \mu(B)^{-1/nq} |B|^{1/n}$$

$$\leq c r \big(\fint_B |f|^p \, dy \big)^{1/p}$$

as desired. The proof is complete.   □

In proving that A_p-weights are p-admissible we are left with the Poincaré inequality.

15.26. Poincaré inequality for A_p-weights. *Suppose that $w \in A_p$ and let $p_0 < q < p < nq$ and $\kappa = n/(n - p/q) > 1$. Then there exists a constant $c = c(n, p, c_{p,w}, q) > 0$ such that*

$$(15.27) \qquad \left(\fint_B |\varphi - \varphi_B|^{\kappa p} \, d\mu \right)^{1/\kappa p} \le c\, r \left(\fint_B |\nabla \varphi|^p \, d\mu \right)^{1/p}$$

whenever $B = B(x_0, r)$ is a ball and $\varphi \in C^\infty(B)$ is bounded. Here $\varphi_B = \fint_B \varphi \, d\mu$.

PROOF: Fix $B = B(x_0, r)$. We first prove that

$$(15.28) \qquad \fint_B |\varphi(x) - \varphi(y)| \, dy \le c(n) \int_B \frac{|\nabla \varphi(y)|}{|x - y|^{n-1}} \, dy$$

for all $x \in B$. Indeed, integration along the line segment from x to y yields

$$|\varphi(x) - \varphi(y)| \le \int_0^1 |\nabla \varphi(ty + (1 - t)x)| |x - y| \, dt$$

and hence

$$(15.29) \qquad \int_B |\varphi(x) - \varphi(y)| \, dy \le \int_B \int_0^1 |\nabla \varphi(ty + (1 - t)x)| \, |x - y| \, dt \, dy \, .$$

If we set $z = ty + (1 - t)x$ and change the variables, the integral on the right hand side becomes

$$\int_0^1 \int_{B(z_0, tr)} |\nabla \varphi(z)| \, |z - x| \, t^{-(n+1)} \, dz \, dt \, ,$$

where $z_0 = tx_0 + (1 - t)x$. Now $B(z_0, tr) \subset B(x, 2tr)$ since $x \in B(z_0, tr)$; hence (15.29) yields

$$\int_B |\varphi(x) - \varphi(y)| \, dy \le \int_0^1 \int_{B \cap B(x, 2tr)} |\nabla \varphi(z)| \, |z - x| \, t^{-(n+1)} \, dz \, dt$$

$$\le \int_B |\nabla \varphi(z)| \, |z - x| \int_{|z-x|/2r}^1 t^{-(n+1)} \, dt \, dz$$

$$= c \int_B |\nabla \varphi(z)| \, |z - x|^{1-n} r^n \, dz \, .$$

We arrive at (15.28) when we divide both sides by $|B|$.

To prove inequality (15.27), observe first that it is enough to verify it with φ_B replaced by

$$\gamma = \fint_B \varphi \, dy$$

because for any constant γ

$$\int_B |\varphi - \varphi_B|^{\kappa p} \, d\mu \le 2^{\kappa p} \int_B |\varphi - \gamma|^{\kappa p} \, d\mu.$$

Appealing to (15.28), we have

$$|\varphi(x) - \gamma| \le \fint_B |\varphi(x) - \varphi(y)| \, dy \le c(n) \int_B \frac{|\nabla\varphi(y)|}{|x - y|^{n-1}} \, dy,$$

whence by (15.25)

$$\left(\fint_B |\varphi - \gamma|^{\kappa p} \, d\mu\right)^{1/\kappa p} \le c\left(\fint_B |I|\nabla\varphi||^{\kappa p} \, d\mu\right)^{1/\kappa p}$$

$$\le c\,r\left(\fint_B |\nabla\varphi|^p \, d\mu\right)^{1/p}.$$

The proof of the Poincaré inequality 15.26 is complete. $\qquad\square$

We have now proved Theorem 15.21, the p-admissibility of A_p-weights. Take notice that if $1 < p \le p_0 n$ in Theorems 15.23 and 15.26, then any $1 < \kappa < n/(n - p/p_0)$ will do. In particular, we obtain the following interesting corollary.

15.30. Corollary. *Suppose that w is an A_1-weight and let $1 < p \le n$. Then for each $1 < \kappa < n/(n - p)$ there is a constant $c = c(n, p, c_{1,w}, \kappa) > 0$ such that*

$$\left(\fint_B |\varphi|^{\kappa p} \, d\mu\right)^{1/\kappa p} \le c\,r\left(\fint_B |\nabla\varphi|^p \, d\mu\right)^{1/p},$$

whenever $B = B(x_0, r)$ is a ball and $\varphi \in C_0^\infty(B)$, and that

$$\left(\fint_B |\varphi - \varphi_B|^{\kappa p} \, d\mu\right)^{1/\kappa p} \le c\,r\left(\fint_B |\nabla\varphi|^p \, d\mu\right)^{1/p}$$

whenever $B = B(x_0, r)$ is a ball and $\varphi \in C^\infty(B)$ is bounded; here $\varphi_B = \fint_B \varphi \, d\mu$.

15.31. Quasiconformal mappings and admissible weights

In Chapter 14 we studied in detail the role of quasiconformal mappings in nonlinear potential theory. However, one interesting aspect was left untouched: if f is a quasiconformal self-homeomorphism of $\mathbf{R}^n$, then certain powers of the Jacobian determinant of f are admissible weights. This is proved next.

Throughout this section *we assume that* $f \colon \mathbf{R}^n \to \mathbf{R}^n$ *is a K-quasiconformal mapping*. In particular, this means that f is a homeomorphism (necessarily onto), its coordinate functions belong to the Sobolev space $H^{1,p}_{loc}(\mathbf{R}^n; dx)$, and the inequality

$$|f'(x)|^n \leq K \, J_f(x)$$

is satisfied for a.e. x in $\mathbf{R}^n$.

The following result of Gehring is central to our investigation.

15.32. Theorem. *The Jacobian $J_f(x)$ of f is an A_∞-weight with constants depending only on n and K.*

We do not prove Theorem 15.32 here but refer to Gehring (1973).

The main result of this section is the following

15.33. Theorem. *The weight $w(x) = J_f(x)^{1-p/n}$ is p-admissible for all $1 < p < n$. The constant c_μ depends only on n, K, and p.*

Since p-admissible weights are q-admissible for $q \geq p$ (Theorem 1.8), we obtain the following corollary.

15.34. Corollary. *If $1 < p < n$, then $w(x) = J_f(x)^{1-p/n}$ is q-admissible for all $q \geq p$.*

As mentioned already in Section 1.6, Theorem 15.33 together with its corollary immediately provides interesting, concrete examples of admissible weights:

15.35. Corollary. *For each $\delta > -n$ and $p > 1$ the weight $w(x) = |x|^\delta$ is p-admissible with c_μ depending only on n, p, and δ.*

PROOF: The weight w is in A_p whenever $-n < \delta < n(p-1)$, so that the assertion holds whenever $-n < \delta \leq 0$. To treat the positive values of δ, we observe that the *radial stretching*,

$$g_\gamma(x) = x|x|^{\gamma-1}, \qquad \gamma > 0,$$

is K-quasiconformal in $\mathbf{R}^n$ with

$$K = \max(\gamma^{1/(n-1)}, \gamma^{1/(1-n)}).$$

See Väisälä (1971, p. 49). If $p < n$, choose $\gamma = \delta/(n-p) + 1 > 0$; then

$$J_{g_\gamma}(x)^{1-p/n} \approx |x|^\delta,$$

and thus the weight w is p-admissible. Finally, because p-admissible weights are also q-admissible for $q \geq p$, the theorem follows. □

We turn to the proof of Theorem 15.33. We first observe that $J_f(x)^{1-p/n}$ is an A_∞-weight, and hence doubling, so that the first requirement for admissible weights is satisfied.

Then we verify property **II**, the uniqueness of the gradient. We may assume that D is bounded. Let $\varphi_i \in C^\infty(D)$ be a sequence such that

$$\int_D |\varphi_i|^p J_f^{1-p/n}\, dx \to 0$$

and

$$\int_D |\partial_j \varphi_i - v_j|^p J_f^{1-p/n}\, dx \to 0$$

for all $j = 1, ..., n$ and for some vector function $v = (v_1, ..., v_n)$. Since $J_f > 0$ a.e. (Lemma 14.22), we may assume without loss of generality that $\varphi_i \to 0$ a.e. in D. Next consider the truncations η_i of φ_i,

$$\eta_i(x) = \begin{cases} \varphi_i(x) & \text{if } |\varphi_i(x)| \le 1 \\ \operatorname{sign} \varphi_i(x) & \text{otherwise}. \end{cases}$$

Then η_i is a sequence of uniformly bounded functions in D with $\eta_i \to 0$ a.e. in D. Since J_f and $J_f^{1-p/n}$ are locally integrable functions in $\mathbf{R}^n$ and since D is bounded, we have

$$\int_D |\eta_i|^p J_f\, dx \to 0$$

and

$$\int_D |\nabla \eta_i - v|^p J_f^{1-p/n}\, dx$$
$$\le \int_D |\nabla \varphi_i - v|^p J_f^{1-p/n}\, dx + \int_{\{\varphi_i > 1\}} |v|^p J_f^{1-p/n}\, dx \to 0.$$

Thus, by performing a change of variables we obtain

$$\int_{f(D)} |\psi_i(y)|^p\, dy \to 0$$

and

$$\int_{f(D)} |\nabla \psi_i(y) - u(y)|^p\, dy \to 0,$$

where $\psi_i(y) = \eta_i \circ f^{-1}(y)$ and $u = f^{-1'}(y)^* \circ v \circ f^{-1}(y)$; here we utilized the quasiconformality of f together with the well-known property of Jacobians:

$$J_{f^{-1}}(f(x)) = J_f(x)^{-1}$$

for a.e. x. It follows that $u = 0$. Since f^{-1} satisfies the condition (N) (see Lemma 14.22), we conclude that $v = 0$, as required.

To prove properties **III** and **IV**, we begin with:

15.36. Weighted embedding theorem for quasiconformal mappings. Let $1 < p < n$ and write $w(x) = J_f(x)^{1-p/n}$. There are constants $\kappa = \kappa(n, K) > 1$ and $c = c(n, K, p) > 0$ such that

$$(15.37) \qquad \Big(\fint_B |\varphi|^{\kappa p}\, d\mu \Big)^{1/\kappa p} \le c\, r \Big(\fint_B |\nabla\varphi|^p\, d\mu \Big)^{1/p}$$

whenever $B = B(x_0, r)$ is a ball and $\varphi \in C_0^\infty(B)$.

PROOF: Let $B = B(x_0, r)$ and $\varphi \in C_0^\infty(B)$. For $1 < \kappa < n/(n-p)$ we apply the Hölder inequality with the pair

$$q = \frac{n}{\kappa(n-p)}, \qquad q' = \frac{n}{\kappa p - (\kappa - 1)n}$$

to obtain

$$\int_B |\varphi|^{\kappa p} J_f^{1-p/n}\, dx = \int_B |\varphi|^{\kappa p} J_f^{1/q} J_f^{1-p/n-1/q}\, dx$$
$$\le \Big(\int_B |\varphi|^{np/(n-p)} J_f\, dx \Big)^{(n-p)\kappa/n} \Big(\int_B J_f^s\, dx \Big)^{(\kappa p - (\kappa-1)n)/n},$$

where

$$s = \frac{(1-\kappa)(n-p)}{\kappa p - (\kappa - 1)n} < 0.$$

By Gehring's theorem there is $q_0 > 1$ such that $J_f \in A_{q_0}$. In particular, we can find $\kappa = \kappa(n, K) > 1$ such that

$$\frac{1}{1 - q_0} \le s.$$

Hence

$$\Big(\fint_B J_f^{1-p/n}\, dx \Big)^{n/(n-p)} \le \fint_B J_f\, dx \le c \Big(\fint_B J_f^s\, dx \Big)^{1/s}.$$

Thus

$$\Big(\int_B J_f^s\, dx \Big)^{(\kappa p - (\kappa-1)n)/n} = |B|^{(\kappa p - (\kappa-1)n)/n} \Big(\fint_B J_f^s\, dx \Big)^{(\kappa p - (\kappa-1)n)/n}$$
$$\le c|B|^{(\kappa p - (\kappa-1)n)/n} \Big(\fint_B J_f^{1-p/n}\, dx \Big)^{ns(\kappa p - (\kappa-1)n)/n(n-p)}$$
$$= c|B|^{(\kappa p - (\kappa-1)n)/n} |B|^{\kappa-1} \Big(\int_B J_f^{1-p/n}\, dx \Big)^{1-\kappa} = c r^{\kappa p} \mu(B)^{1-\kappa}.$$

Combining the above inequalities we arrive at

$$\Big(\fint_B |\varphi|^{\kappa p}\, d\mu\Big)^{1/\kappa p} \le c\, r\, \mu(B)^{-1/p}\Big(\int_B |\varphi|^{np/(n-p)} J_f\, dx\Big)^{(n-p)/np}$$

and hence it suffices to show that

$$(15.38) \qquad \Big(\int_B |\varphi|^{np/(n-p)} J_f\, dx\Big)^{(n-p)/np} \le c\Big(\int_B |\nabla\varphi|^p J_f^{1-p/n}\, dx\Big)^{1/p}.$$

Towards this end, we use the notation $y = f(x)$ and $g = f^{-1}$ so that $g(y) = x$. Then also g is K-quasiconformal. Hence we have by the integral transformation formula (Lemma 14.25) that

$$\int_{f(B)} |\nabla(\varphi \circ g)(y)|^p\, dy \le \int_{f(B)} |\nabla\varphi(g(y))|^p |g'(y)|^p\, dy$$

$$\le K^{p/n} \int_{f(B)} |\nabla\varphi(g(y))|^p J_g(y)^{p/n}\, dy$$

$$= K^{p/n} \int_B |\nabla\varphi(x)|^p J_f(x)^{1-p/n}\, dx.$$

Similarly,

$$\int_B |\varphi(x)|^{np/(n-p)} J_f(x)\, dx = \int_{f(B)} |(\varphi \circ g)(y)|^{np/(n-p)}\, dy.$$

The above reasoning also shows that $\varphi \circ g$ belongs to $H_0^{1,p}(f(B); dx)$, and we can employ the usual Sobolev inequality (1.7) to conclude

$$\Big(\int_{f(B)} |(\varphi \circ g)(y)|^{np/(n-p)}\, dy\Big)^{(n-p)/np} \le c\Big(\int_{f(B)} |\nabla(\varphi \circ g)(y)|^p\, dy\Big)^{1/p}.$$

We have thereby accomplished (15.38) and the proof of the embedding theorem 15.36 is complete.   □

15.39.   Weighted Poincaré inequality for quasiconformal mappings.   Let $1 < p < n$ and write $w(x) = J_f(x)^{1-p/n}$. There are constants $\kappa = \kappa(n, K) > 1$ and $c = c(n, K, p) > 0$ such that

$$(15.40) \qquad \Big(\fint_B |\varphi - \varphi_B|^{\kappa p}\, d\mu\Big)^{1/\kappa p} \le c\, r\Big(\fint_B |\nabla\varphi|^p\, d\mu\Big)^{1/p}$$

whenever $B = B(x_0, r)$ is a ball and $\varphi \in C^\infty(B)$ is bounded.

PROOF: First we recall that it is enough to prove (15.40) with φ_B replaced by a constant γ (see the proof of Theorem 15.26). We reduce (15.40) to the usual Sobolev–Poincaré inequality much like in the proof of the weighted embedding theorem 15.36. Indeed, using Hölder's inequality and the A_∞-property of J_f exactly as in the proof of Theorem 15.36 we arrive at the inequality

$$\int_B |\varphi - \gamma|^{\kappa p} J_f^{1-p/n} \, dx \leq cr^{\kappa p} \mu(B)^{1-\kappa} \Big(\int_B |\varphi - \gamma|^{np/(n-p)} J_f \, dx \Big)^{(n-p)\kappa/n}$$

or

$$\Big(\fint_B |\varphi - \gamma|^{\kappa p} \, d\mu \Big)^{1/\kappa p} \leq cr \mu(B)^{-1/p} \Big(\int_B |\varphi - \gamma|^{np/(n-p)} J_f \, dx \Big)^{(n-p)/np},$$

where $B = B(x_0, r)$, $\varphi \in C^\infty(B)$ bounded, $\kappa = \kappa(n, K) > 1$, and γ is any constant. Moreover, performing the change of variables, we infer that only the inequality

$$
\begin{aligned}
(15.41) \quad & \Big(\int_{f(B)} |(\varphi \circ g)(y) - \gamma|^{np/(n-p)} \, dy \Big)^{(n-p)/np} \\
& \leq c \Big(\int_{f(B)} |\nabla(\varphi \circ g)(y)|^p \, dy \Big)^{1/p}
\end{aligned}
$$

needs to be proved. To accomplish this, we shall rely on the recent literature, for a complete treatment would take us too far from the subject.

Inequality (15.41) is the Sobolev–Poincaré inequality in the quasiball $f(B)$ (the image of an open ball under a quasiconformal self-homeomorphism of $\mathbf{R}^n$ is often termed a *quasiball*). This inequality is known to be valid, and we mention two different ways to obtain it. The first option is to adopt the proof in Martio (1988b), where a Poincaré inequality is established for the class of *John domains*. John domains are more general than quasiballs. The potential estimate (Martio 1988b, (3.10))

$$\Big(\int_\Omega \Big| \int_\Omega |x - y|^{1-n} u(x) \, dy \Big|^p \, dx \Big)^{1/p} \leq c |\Omega|^{1/n} \Big(\int_\Omega |u(x)|^p \, dx \Big)^{1/p},$$

can be replaced by

$$\Big(\int_\Omega \Big| \int_\Omega |x - y|^{1-n} u(x) \, dy \Big|^{np/(n-p)} \, dx \Big)^{(n-p)/np} \leq c |\Omega|^{1/n} \Big(\int_\Omega |u(x)|^p \, dx \Big)^{1/p},$$

see Ziemer (1989, p. 89), and then the rest of the proof in Martio (1988b) applies almost verbatim.

The second option is to employ an extension theorem of Jones (1981) which implies that there is a function ψ in $\mathbf{R}^n$ such that $\psi|_{f(B)} = \varphi \circ g$ and that

$$\int_{\mathbf{R}^n} |\nabla \psi|^p \, dy \leq c \int_{f(B)} |\nabla(\varphi \circ g)|^p \, dy \, ,$$

where the constant $c > 0$ depends only on n, p, and K. To be precise, Jones only established the theorem for $p = n$ whereas the extension for other values of p has been worked out by Herron and Koskela (1992, Section 3). Let B' be a ball that contains $f(B)$. We use the ordinary Sobolev–Poincaré inequality in B' and find a number γ, for example

$$\gamma = \fint_{B'} \psi \, dy \, ,$$

such that

$$\left(\int_{B'} |\psi(y) - \gamma|^{np/(n-p)} \, dy \right)^{(n-p)/np}$$
$$\leq c \left(\int_{B'} |\nabla \psi(y)|^p \, dy \right)^{1/p}$$
$$\leq c \left(\int_{f(B)} |\nabla(\varphi \circ g)(y)|^p \, dy \right)^{1/p} .$$

Hence (15.41) follows because

$$\int_{f(B)} |\varphi \circ g(y) - \gamma|^{np/(n-p)} \, dy \leq \int_{B'} |\psi(y) - \gamma|^{np/(n-p)} \, dy \, .$$

This concludes the proof of the weighted Poincaré inequality 15.40, and hence that of Theorem 15.33.   □

15.42. Quasiconformal mappings as $\mathfrak{A}_p$-harmonic morphisms

Suppose that $f \colon \mathbf{R}^n \to \mathbf{R}^n$ is a K-quasiconformal mapping and that $A \colon \mathbf{R}^n \times \mathbf{R}^n \to \mathbf{R}^n$ satisfies (3.3)–(3.7) with $w \equiv 1$ and $1 < p < n$. Let $f^\# A \colon \mathbf{R}^n \times \mathbf{R}^n \to \mathbf{R}^n$ be the pull-back of A under f as defined in Chapter 14. Then $f^\# A$ does not generally satisfy the assumptions (3.3)–(3.7) for $w \equiv 1$. However,

$$w(x) = J_f(x)^{1-p/n}$$

is a p-admissible weight and the proof of the following lemma is similar to Lemma 14.38.

15.43. Lemma. *The mapping* $f^\# \mathcal{A}$ *satisfies (3.3)–(3.7) with* $w(x) = J_f(x)^{1-p/n}$ *and*

$$\alpha' = \frac{\alpha}{K_O(f)^{p/n}} \qquad and \qquad \beta' = \beta\, K_I(f)^{p/n}.$$

15.44. Theorem. *If* u *is* A*-harmonic or* A*-superharmonic in* Ω*, then* $u \circ f$ *is* $f^\# \mathcal{A}$*-harmonic or* $f^\# \mathcal{A}$*-superharmonic in* $f^{-1}(\Omega)$*, respectively. In particular,* f *is an* $\mathfrak{A}_p$*-harmonic morphism.*

PROOF: If $\varphi \in C_0^\infty\big(f^{-1}(\Omega)\big)$, then by the integral transformation formula (Lemma 14.25) we have

$$\int_{f^{-1}(\Omega)} f^\# \mathcal{A}\big(x, \nabla(u \circ f)(x)\big) \cdot \nabla\varphi(x)\, dx$$

$$= \int_{f^{-1}(\Omega)} J_f(x)\, A\big(f(x), \nabla u(f(x))\big) \cdot \nabla(\varphi \circ f^{-1})\big(f(x)\big)\, dx$$

$$= \int_\Omega A(y, \nabla u(y)) \cdot \nabla(\varphi \circ f^{-1})(y)\, dy.$$

Since $f, f^{-1} \in H^{1,p}_{loc}(\mathbf{R}^n; dx)$, it is easily seen that $\varphi \circ f^{-1}$ belongs to $H^{1,p}(\Omega)$ and has compact support in Ω. The theorem follows since, by truncating, it suffices to show that $u \circ f$ is a supersolution if u is. $\qquad\square$

NOTES TO CHAPTER 15. A_p-weights were introduced by Muckenhoupt (1972) who characterized them as the weights w for which the Hardy–Littlewood maximal operator is bounded on $L^p(w(x)dx)$. Thereafter, weighted norm inequalities have been studied extensively. Several characterizations of A_p-weights were found by Coifman and Fefferman (1974); in particular they established the important open-end property from which Muckenhoupt's theorem easily follows. We refer to the monographs by García-Cuerva and Rubio De Francia (1985) and Torchinsky (1986) for a thorough discussion of the subject.

Weighted Sobolev inequalities in connection with A_p-weights were considered by Muckenhoupt and Wheeden (1974). The p-admissibility of A_p-weights was first proved by Fabes et al. (1982a). The short proof given here is due to Chiarenza and Frasca (1985).

That the Jacobian J_f of a quasiconformal mapping $f : \mathbf{R}^n \to \mathbf{R}^n$ is an A_∞-weight is the fundamental result of Gehring (1973). In a subsequent paper by Reimann (1974) it was established that $\log J_f$ is a BMO-function. These results opened up the investigation of sometimes deep connections between quasiconformal mappings and harmonic analysis. See the survey articles by Baernstein and Manfredi (1983), Iwaniec (1984), and the article by Iwaniec in Vuorinen (1992). For the connection between BMO, A_∞, and quasiconformal mappings, see also Reimann and Rychener (1975), Uchiyama (1975), and Astala (1983). More recent papers on A_∞ and quasiconformal mappings include David and Semmes (1990), Staples (1991, 1992), Heinonen and Koskela (In press), and Semmes (1992).

Fabes et al. (1982a) proved that $J_f^{1-2/n}$ is 2-admissible. Our more elementary proof is due to Pekka Koskela. Øksendal (1990) and Heinonen and Koskela (In preparation) have studied weighted Sobolev and Poincaré inequalities in connection with quasiregular mappings.

16
Axiomatic nonlinear potential theory

Classical potential theory is in a broad sense a study of the Laplace equation and its solutions, harmonic functions. In an axiomatic approach one starts with a few basic properties of harmonic functions and then builds a potential theory from these properties only. The linear axiomatic systems, pioneered by Tautz, Doob, Brelot, and Bauer between 1943 and 1963, have the following in common: on a locally compact space there is a linear sheaf of continuous real functions which solve the Dirichlet problem in a unique way in sufficiently many open sets and obey a minimum principle and the Harnack convergence principle. The monographs of Bauer (1966), Brelot (1960), and Constantinescu and Cornea (1972) are standard references to the linear axiomatic theory of harmonic functions.

In this chapter we introduce an axiomatic system which is applicable to nonlinear situations; this system is a slight modification of that introduced by Lehtola (1986) and is similar to the Brelot linear system. A somewhat different approach is presented by Laine (1985). Our principal model is the theory of the solutions to the equation

$$- \operatorname{div} \mathcal{A}(x, \nabla u) = 0$$

studied in the previous chapters. It turns out that in the nonlinear potential theory special attention must be paid to comparison principles rather than to various minimum principles.

After presenting the axioms for a quasilinear harmonic space we show that the Perron method is available. However, the nonlinearity prevents us from going too far; in the linear theories the harmonic measure provides a basic tool for the Dirichlet problem but that method cannot be used in general. Not much more than what we present here is known.

16.1. The axioms
Throughout this chapter we assume that X is a locally compact, locally connected, and connected Hausdorff space. To avoid certain unpleasant peculiarities we also require that X be noncompact.

Let $\mathfrak{H}$ be a sheaf of continuous real-valued functions, that is

(i) if $U \subset X$ is open, then each $u \in \mathfrak{H}(U)$ is a continuous real-valued function in U;

(ii) if $U' \subset U \subset X$ are open and $u \in \mathfrak{H}(U)$, then $u|_{U'} \in \mathfrak{H}(U')$;

(iii) if $U_j \subset X$ are open and $U = \cup_j U_j$, then $u|_{U_j} \in \mathfrak{H}(U_j)$ for all j implies $u \in \mathfrak{H}(U)$.

Given an open subset U of X, a function $u \in \mathfrak{H}(U)$ is called an $\mathfrak{H}$-*harmonic function*.

We say that a relatively compact open subset U of X is *regular* if

(a) (Dirichlet principle) for each $f \in C(\partial U)$ there is a unique $h \in \mathfrak{H}(U) \cap C(\overline{U})$ such that $h|_{\partial U} = f$ (we write $h = H_f = H_f^U$); and

(b) (comparison principle) for each pair $f, g \in C(\partial U)$ the condition $f \leq g$ implies $H_f \leq H_g$ in U.

The function H_f^U is called the *Dirichlet solution in U with boundary values f*.

After these preliminaries we are ready to list the axioms.

Axiom A. *For every open set $U' \subset X$ and for every compact set $C \subset U'$ there is a regular set U such that $C \subset U \Subset U'$.*

Axiom B. (Harnack's principle) *If h_j, $j = 1, 2, \ldots$, is an increasing sequence of functions in $\mathfrak{H}(U)$, where U is a domain in X, then the limit function $h = \lim h_j$ is in $\mathfrak{H}(U)$ if h is finite at some point of U.*

Axiom C. *If $h \in \mathfrak{H}(U)$ and $\lambda \in \mathbf{R}$, then both $h + \lambda$ and λh belong to $\mathfrak{H}(U)$.*

Given a space X as above and a sheaf $\mathfrak{H}$ satisfying Axioms A, B, and C, we call the pair $(X, \mathfrak{H})$ a *quasilinear harmonic space*. Of course, the sheaf may be linear but that is not required.

To compare our quasilinear harmonic space to the classical Brelot harmonic space, let X be as above and suppose that a linear sheaf $\mathcal{H}$ on X of continuous real-valued functions is given. In particular, for each open set $U \subset X$, $\mathcal{H}(U)$ is a vector subspace of $C(U)$. An open set U is called *Dirichlet regular* if U is relatively compact and if for every $f \in C(\partial U)$ there exists a unique $h \in C(\overline{U})$ such that $h|_U \in \mathcal{H}(U)$ and $h|_{\partial U} = f$. Moreover, if $f \geq 0$, then $h \geq 0$. Now the pair $(X, \mathcal{H})$ is called a *Brelot harmonic space*

(a) (Harnack's principle) if U is a domain, the limit function of any increasing sequence in $\mathcal{H}(U)$ belongs to $\mathcal{H}(U)$ if it is finite at some point of U; and

(b) if Dirichlet regular sets form a base of the topology of X.

The sheaf $\mathcal{H}$ is usually called a *harmonic sheaf* and its functions *harmonic functions*.

The basic difference between a Brelot harmonic space $(X, \mathcal{H})$ and a quasilinear harmonic space $(X, \mathfrak{H})$ is that in the latter case the sheaf $\mathfrak{H}$ is not assumed to be linear. There are other differences as well. In a Brelot harmonic space Axiom C need not hold since nonzero constants need not be harmonic. However, it should be remarked that an axiomatic theory can be developed without the requirement $h + \lambda \in \mathfrak{H}(U)$ whenever $h \in \mathfrak{H}(U)$ and $\lambda \in \mathbf{R}$ (Lehtola 1986). Similarly, Axiom A is not true in the general Brelot space (Hervé 1962, p. 440). Thus Brelot harmonic spaces are not included among the quasilinear harmonic spaces above; however, they are "locally quasilinear".

The most important tool in linear axiomatic potential theories is harmonic measure. If $(X, \mathcal{H})$ is a Brelot harmonic space and if U is a Dirichlet regular set, then for each $x \in U$ the mapping $f \mapsto h(x)$ is a positive linear functional on $C(\partial U)$, where h is the harmonic function in U with boundary values f. By the Riesz representation theorem there is a Radon measure ω_x on Borel subsets of ∂U such that

$$(16.2) \qquad h(x) = \int_{\partial U} f \, d\omega_x \,.$$

Here $\omega_x(A)$ is the *harmonic measure* of $A \subset \partial U$ at x. If $X = \mathbf{R}^n$ and $\mathcal{H}$ is the sheaf of ordinary harmonic functions, ω_x is just the classical harmonic measure. Because of the nonlinearity of the mapping $f \mapsto h(x)$ in the quasilinear case, no representation like (16.2) is possible. Hence there is no standard connection between the Dirichlet problem and harmonic measure although a counterpart for the latter exists in the quasilinear situation, as was seen in Chapter 11.

16.3. EXAMPLES. (a) A standard example of a quasilinear harmonic space is given by $(\mathbf{R}^n, \mathfrak{H})$, where $\mathfrak{H}$ is the sheaf of $\mathcal{A}$-harmonic functions, $\mathcal{A}$ satisfying the assumptions of Chapter 3. That $(\mathbf{R}^n, \mathfrak{H})$ satisfies Axioms A, B, and C has been established in the previous chapters.

(b) The next (somewhat pathological) example does not have its roots in the theory of differential equations.

Let

$$X = \{(x_1, x_2) \in \mathbf{R}^2 : x_1 x_2 = 0 \quad x_2 \geq 0\}$$

with the induced plane topology. The open intervals $I = (a, b) \times \{0\}$, $-\infty \leq a < b \leq \infty$ with $0 \notin I$, $I = \{0\} \times (a, b)$, $0 < a < b \leq \infty$, and the neighborhoods $T = (-t, t) \times \{0\} \cup \{0\} \times [0, t)$ of the origin $t > 0$ form a base for the topology of X. Now let $U \subset X$ be an open set and define $\mathfrak{H}(U)$ as follows:

(i) If U is of the form I above, then $\mathfrak{H}(U)$ consists of affine functions, i.e.

$$\mathfrak{H}(U) = \{h \colon h(x) = x_0 \cdot x + y \text{ for some } x_0 \in \mathbf{R}^2 \text{ and } y \in \mathbf{R}\} \,.$$

(ii) If U is of the form T above, then $\mathfrak{H}(U)$ consists of continuous functions h such that

$$(16.4) \qquad h(0) = \frac{1}{2}(\max_{\partial U} h + \min_{\partial U} h)$$

and h is affine on the intervals from the origin to the boundary points of U.

(iii) If $U = (a, b) \times [0, c) \cap X$ for $-\infty \leq a < 0 < b \leq \infty$ and $0 < c \leq \infty$, then each $h \in \mathfrak{H}(U)$ satisfies $h|_T \in \mathfrak{H}(T)$ for $T = (-t, t) \times \{0\} \cup \{0\} \times [0, t)$ with $t = \min(-a, b, c, 1)$, and h is affine on the intervals (respectively, half lines) from the origin to the boundary points (respectively, natural ideal boundary points) of U.

(iv) If U is an arbitrary open set in X, then each connected component U' of U is of the form described in (i) or in (iii) and we let $h \in \mathfrak{H}(U)$ if and only if $h \in \mathfrak{H}(U')$ for each such component U'.

It is clear that $\mathfrak{H}$ is a sheaf of continuous functions on X. Note that each relatively compact open set $U \subset X$ is regular and it is easy to check that Axioms A, B, and C are satisfied; here the only point worth observing is that if $\lambda \in \mathbf{R}$ is negative, then in case (ii), $\max_{\partial U} \lambda h = \lambda \min_{\partial U} h$ and $\min_{\partial U} \lambda h = \lambda \max_{\partial U} h$. Thus the family $\mathfrak{H}(U)$ is stable under multiplication by scalars. The sheaf $\mathfrak{H}$ is not linear because of the definition of $h(0)$ in case (ii). Indeed, let U be an open set of the form T with $\partial U = \{a, b, c\}$. If $f, g \in C(\overline{U}) \cap \mathfrak{H}(U)$ are such that

$$f(a) = 0 = g(b)$$
$$f(b) = f(c) = 1 = g(a) = g(c),$$

then

$$f(0) = \frac{1}{2} = g(0)$$

by (16.4). Then $f + g$ is not an $\mathfrak{H}$-harmonic function because by (16.4) we should have

$$(f + g)(0) = \frac{3}{2}$$

and that is not the case. Hence $\mathfrak{H}$ is not linear.

If, instead of (16.4) in case (ii), h is defined via the formula

$$h(0) = \frac{1}{3}(h(x_{-t}) + h(y_t) + h(x_t)),$$

where $x_{-t} = (-t, 0)$, $y_t = (0, t)$, $x_t = (t, 0)$, and $t > 0$, then a linear Brelot space is obtained. In fact, this space is an extension of the space of harmonic functions on a line to the space formed by three rays starting from the origin. Recall that ordinary harmonic functions on a line are just affine functions.

The quasilinear space $(X, \mathfrak{H})$ in Example (b) does not satisfy the strict comparison principle: "If U is a regular domain and $f, g \in C(\partial U)$ are such that $f \leq g$ and $f \neq g$, then $H_f < H_g$ in U." Moreover, two $\mathfrak{H}$-harmonic functions f and g in a domain U may coincide in an open nonempty subset of U without being identical. For example, let f be the function as in the example above, and let $g \in C(\overline{U}) \cap \mathfrak{H}(U)$ be any function such that

$$g(a) = f(a), \quad g(b) = f(b), \quad \text{and} \quad 0 < g(c) < 1.$$

Thus that harmonic space does not possess the unique continuation property; a quasilinear harmonic space $(X, \mathfrak{H})$ is said to have the *unique continuation property* if for each domain U of X and for each $h, g \in \mathfrak{H}(U)$ the interior of the set

$$\{x \in U : h(x) = g(x)\}$$

is either empty or all of U. Note that the classical harmonic functions in $\mathbf{R}^n$ have the unique continuation property because they are real analytic. It is not known which operators div $\mathcal{A}$ produce quasilinear harmonic spaces $(\mathbf{R}^n, \mathcal{H})$ with the unique continuation property. See Notes to Chapter 6.

16.5. Properties of $\mathfrak{H}$-harmonic functions
For the rest of this chapter we let $(X, \mathfrak{H})$ be a quasilinear harmonic space and U an open subset of X.

The next simple lemma is often employed without reference.

16.6. Lemma. *If U is a regular open set, then each component of U is also regular.*

PROOF: Let U' be a component of U. Since X is locally connected, U' is open. If $f \in C(\partial U')$, then by the Tietze extension theorem there is $f^* \in C(\partial U)$ such that $f^* = f$ on $\partial U'$. Let $u = H_{f^*}^U$. Now the restriction $h = u|_{U'}$ is an $\mathfrak{H}$-harmonic function in U' with boundary values f. Moreover, the sheaf property of $\mathfrak{H}$ and the uniqueness of the Dirichlet solution in U imply that h is unique. Since the comparison principle is clearly satisfied, the lemma follows.   □

It follows from Lemma 16.6 and Axiom A that X has a base consisting of regular domains.

The proof for the following lemma is immediate.

16.7. Lemma. *Let U be regular. If $f \in C(\partial U)$ and $\lambda \in \mathbf{R}$, then $H_{f+\lambda} = H_f + \lambda$ and $H_{\lambda f} = \lambda H_f$.*

If the sheaf $\mathfrak{H}$ is linear, the following properties of $\mathfrak{H}$-harmonic functions are well known.

16.8. Minimum zero principle. *Let U be a domain in X. If $h \in \mathfrak{H}(U)$ is nonnegative, then either $h \equiv 0$ or $h > 0$ in U.*

PROOF: If $h(x_0) > 0$ for some $x_0 \in U$, then the limit function of the increasing sequence $h_j = jh$, $j = 1, 2, \ldots$, of $\mathfrak{H}$-harmonic functions is identically ∞ by Harnack's principle, Axiom B. Hence $h(x) > 0$ for each $x \in U$. $\quad\square$

16.9. Strong maximum principle. *A nonconstant $\mathfrak{H}$-harmonic function h in a domain U cannot attain its supremum or infimum.*

PROOF: Axiom C allows us to apply Theorem 16.8 to the $\mathfrak{H}$-harmonic functions $\sup_U h - h$ and $h - \inf_U h$. $\quad\square$

The sheaf of $\mathfrak{H}$-harmonic functions is closed under locally uniform convergence.

16.10. Theorem. *Suppose that $h_i \in \mathfrak{H}(U)$ is a sequence which converges to h uniformly on compact subsets of U. Then $h \in \mathfrak{H}(U)$.*

PROOF: Since X is locally compact and since $\mathfrak{H}$ is a sheaf, we may assume that $h_i \to h$ uniformly in U. Fix then an integer $j \geq 3$ and choose h_{i_j} such that $\sup_U |h_{i_j} - h| < j^{-3}$. Write

$$u_j = h_{i_j} - 1/j .$$

Thereby we obtain an increasing sequence $u_j \in \mathfrak{H}(U)$, $j = 3, 4, \ldots$, with the finite limit function h. Hence $h \in \mathfrak{H}(U)$ by Harnack's principle. $\quad\square$

Recall that a family $\mathcal{F}$ of functions $f \colon A \to [-\infty, \infty]$ is upper directed if for each pair $f, g \in \mathcal{F}$ there is $u \in \mathcal{F}$ such that $u \geq \max(f, g)$.

16.11. Theorem. *Suppose that U is a domain in X and that $\mathcal{F}$ is an upper directed family of $\mathfrak{H}$-harmonic functions in U. If $h = \sup \mathcal{F}$, then either h is an $\mathfrak{H}$-harmonic function or $h \equiv \infty$ in U.*

For the proof we invoke two topological results. Because of the first lemma there is no need to assume that X has a countable base.

16.12. Lemma. *(Cornea 1968, Proposition 1.1) If $\mathcal{F}$ is an upper directed family of continuous functions on a locally compact space Y such that the limit of any increasing sequence in $\mathcal{F}$ is continuous, then $\sup \mathcal{F}$ is continuous.*

The next result is the well-known Dini convergence theorem.

16.13. Lemma. *(Bourbaki 1961, Ch. 4, no 1, Théorème 1) Let $\mathcal{F}$ be an upper directed family of continuous functions on a compact space Y. If $f = \sup \mathcal{F}$ is continuous, then f can be approximated uniformly by functions in $\mathcal{F}$.*

PROOF FOR THEOREM 16.11: If $h(x_0) < \infty$ for some $x_0 \in U$, then Harnack's principle and Lemma 16.12 ensure that h is continuous. Fix a domain $V \Subset U$. The Dini convergence theorem 16.13 and the upper directedness of $\mathcal{F}$ give a sequence of functions in $\mathcal{F}$ converging to h uniformly on $\overline{V}$. Thus h belongs to $\mathfrak{H}(V)$ by Theorem 16.10. Consequently, $h \in \mathfrak{H}(U)$ because X is locally compact and because $\mathfrak{H}$ is a sheaf.   □

We recall that a function $u: Y \to \mathbf{R} \cup \{\infty\}$ in a topological space Y is *lower semicontinuous* if the sets $\{y \in Y : u(y) > \lambda\}$ are open in Y for all $\lambda \in \mathbf{R}$; a function $v: Y \to \mathbf{R} \cup \{-\infty\}$ is *upper semicontinuous* if $-v$ is lower semicontinuous. If Y is compact, then for every lower semicontinuous function u in Y there is an increasing sequence of continuous functions f_i in Y such that $\lim f_i = u$.

For the next result let U be regular and $u: \partial U \to \mathbf{R} \cup \{\infty\}$ be lower semicontinuous. Define

$$\underline{h}_u = \underline{h}_u^U = \sup\{H_f : f \in C(\partial U) \text{ and } f \leq u\}.$$

16.14. Lemma. *In each component of U, $\underline{h}_u$ is either an $\mathfrak{H}$-harmonic function or identically ∞. Moreover, if $h \in \mathfrak{H}(U) \cap C(\overline{U})$ and $h \leq u$ on ∂U, then $h \leq \underline{h}_u$.*

PROOF: The first claim follows from Theorem 16.11 since, by the comparison principle, the set

$$\{H_f : f \in C(\partial U) \text{ and } f \leq u\}$$

is upper directed. The second assertion follows since $H_h = h$ by the uniqueness of the Dirichlet solution in regular open sets.   □

If $\underline{h}_u \in \mathfrak{H}(U)$, then it is called the *best $\mathfrak{H}$-minorant* of u in U.

16.15. Lemma. *If U is regular and $u: \partial U \to \mathbf{R} \cup \{\infty\}$ is lower semicontinuous, then*

$$\liminf_{x \to y} \underline{h}_u(x) \geq u(y)$$

for all $y \in \partial U$.

PROOF: Fix $y \in \partial U$. Then for each $f \in C(\partial U)$ with $f \leq u$ we have

$$(16.16) \qquad \liminf_{x \to y} \underline{h}_u(x) \geq \lim_{x \to y} H_f(x) = f(y).$$

Since u is lower semicontinuous, there exists an increasing sequence of functions $f_j \in C(\partial U)$ such that $f_j \to u$ on ∂U. Thus the lemma follows from (16.16).   □

16.17. Lemma. *If $u \in C(\partial U)$, then $\underline{h}_u = H_u$.*

PROOF: By the comparison principle $\underline{h}_u \leq H_u$ and the reverse inequality is trivial. □

Let U be a regular open set. If we endow the spaces $C(\overline{U})$ and $C(\partial U)$ with sup-norms, then the mapping $C(\partial U) \to C(\overline{U})$,

$$(16.18) \qquad\qquad f \mapsto H_f^U,$$

is continuous. Indeed, if $\sup_{\partial U} |f - g| \leq \varepsilon$, then $f - \varepsilon \leq g \leq f + \varepsilon$, and hence Lemma 16.7 and the comparison principle imply

$$H_f - \varepsilon = H_{f-\varepsilon} \leq H_g \leq H_{f+\varepsilon} = H_f + \varepsilon$$

or

$$\sup |H_f - H_g| \leq \varepsilon.$$

Note that in general the mapping in (16.18) is not linear.

16.19. Superharmonic functions

In the quasilinear harmonic space $(X, \mathfrak{H})$ we define superharmonic functions via the comparison principle in a similar fashion as $\mathcal{A}$-superharmonic functions are defined in Chapter 7.

If U is an open set in X and if $u: U \to \mathbf{R} \cup \{\infty\}$ is lower semicontinuous, not identically ∞ in any component of U, then u is an $\mathfrak{H}$-*superharmonic function*, or a *superharmonic function* for short, if it satisfies the following comparison principle: if $V \Subset U$ is open and $h \in C(\overline{V}) \cap \mathfrak{H}(V)$ is such that $h \leq u$ on ∂V, then $h \leq u$ in V.

We let $\mathfrak{S}(U)$ denote the family of all superharmonic functions in U. Functions in $-\mathfrak{S}(U)$ are called $(\mathfrak{H}\text{-})$*subharmonic functions*; these functions have a natural interpretation as upper semicontinuous functions which obey the reverse comparison principle.

The next lemma follows from the comparison principle and the uniqueness of the Dirichlet solution.

16.20. Lemma. *If $U \subset X$ is open, then $\mathfrak{H}(U) = \mathfrak{S}(U) \cap -\mathfrak{S}(U)$.*

In axiomatic theories it is customary to employ hyperfunctions; these are defined as superharmonic functions except that they are allowed to be ∞ in a component of U. We do not use the class of hyperfunctions here.

The following theorem is immediate in light of Axiom C.

16.21. Theorem. *Suppose that $u, v \in \mathfrak{S}(U)$, $\sigma \in \mathbf{R}$, and $\lambda \geq 0$. Then $\min(u,v)$, λu, and $u + \sigma$ all belong to $\mathfrak{S}(U)$.*

Moreover, if $\mathcal{F}$ is an upper directed family of superharmonic functions in U, then $\sup \mathcal{F}$ is a superharmonic function if it is not identically ∞ in any component of U.

The lattice property of superharmonic functions is used in Perron's method which we develop in the next section.

If $\mathcal{F} \subset \mathfrak{S}(U)$, then $\inf \mathcal{F}$ need not be lower semicontinuous. However, if $\mathcal{F} \subset \mathfrak{S}(U)$ is locally uniformly bounded from below, then it is easy to see that the lower semicontinuous regularization of $\inf \mathcal{F}$ belongs to $\mathfrak{S}(U)$ (see Lemma 7.4).

Let $\dot{X} = X \cup \{\infty\}$ denote the one-point compactification of the space X. From now on the topological notions are taken in $\dot{X}$. For example, if U is an open set in X which is not relatively compact, we consider ∞ a boundary point of U.

16.22. Generalized comparison principle. *Let $u \in \mathfrak{S}(U)$ and $v \in -\mathfrak{S}(U)$ be such that*

$$(16.23) \qquad \liminf_{x \to y} u(x) \geq \limsup_{x \to y} v(x)$$

for every $y \in \partial U$. If both sides of (16.23) are not simultaneously $-\infty$ or ∞, then $u \geq v$ in U.

PROOF: Let $\varepsilon > 0$. Then (16.23) takes the form

$$\liminf_{x \to y}(u(x) + \varepsilon) > \limsup_{x \to y} v(x)$$

for all $y \in \partial U$ and the set

$$C = \{x \in U : u(x) + \varepsilon \leq v(x)\}$$

is a compact subset of U. By Axiom A we may choose a regular open set V with $C \subset V \Subset U$. Pick an increasing sequence of continuous functions f_j on ∂V such that $f_j \to u + \varepsilon$ on ∂V. Since v is upper semicontinuous and ∂V is compact, there is an index j such that $f_j \geq v$ on ∂V. Because $u + \varepsilon \in \mathfrak{S}(U)$, we have

$$u + \varepsilon \geq H^V_{f_j} \geq v$$

in V. Hence $u + \varepsilon \geq v$ in C and, consequently, $u + \varepsilon \geq v$ in U. By letting $\varepsilon \to 0$ we obtain the desired inequality $u \geq v$ in U. $\qquad \square$

The next theorem shows that in the definition of superharmonic functions it suffices to test the comparison principle in regular open sets only.

16.24. Theorem. *Suppose that $u: U \to \mathbf{R} \cup \{\infty\}$ is a lower semicontinuous function which is not identically ∞ in any component of U. Then $u \in \mathfrak{S}(U)$ if and only if for each regular set $V \Subset U$ and each $f \in C(\partial V)$ the condition $u \geq f$ on ∂V implies $u \geq H_f$ in V.*

PROOF: The necessity part of the theorem is trivial. For the converse it suffices to establish the comparison principle in an arbitrary open set $W \Subset U$. Let $h \in \mathfrak{H}(W) \cap C(\overline{W})$ be such that $u \geq h$ on ∂W and fix $\varepsilon > 0$. Since the set

$$C = \{x \in W : u(x) \leq h(x) - \varepsilon\}$$

is compact, we may pick a regular open set U such that $C \subset U \subset W$ (Axiom A). Now the assumption implies that

$$u \geq H_{h-\varepsilon} = H_h - \varepsilon = h - \varepsilon$$

in U (Lemma 16.7). Thus $u \geq h - \varepsilon$ in W and, by letting $\varepsilon \to 0$, we obtain $u \geq h$ in W as desired.   □

In the definition of a superharmonic function we required that it not be identically infinite in any component. It follows, in fact, that a superharmonic function cannot be infinite in an open set.

16.25. Theorem. *Each superharmonic function u in U is finite in a dense subset of U.*

PROOF: We may assume that u is nonnegative in U. Suppose that the set

$$\check{V} = \text{int}\{x \in U : u(x) = \infty\}$$

is nonempty. Let V be a component of $\check{V}$ and $y \in \partial V \cap U$. An easy topological argument shows that we may pick a connected regular open neigborhood $W \Subset U$ of y such that $\partial W \cap V \neq \emptyset$. Fix then $z \in \partial W \cap V$ and choose a nonnegative function $f \in C(\partial W)$ such that $f(z) = 1$ and $f(x) = 0$ for $x \in \partial W \setminus V$; note that such a function exists because X, and hence ∂W, being a locally compact Hausdorff space, is regular.

Next write $h_j = jH_f^W$, $j = 1, 2, \ldots$. Then $h_j \leq u$ in W as the same inequality holds on ∂W. Then, since W is a neighborhood of a boundary point of V, W contains a point x where u is finite. Consequently,

$$h(x) = \lim_{j \to \infty} h_j(x) \leq u(x) < \infty.$$

On the other hand, the minimum zero principle 16.8 implies that $H_f^W > 0$ in W and, therefore, $h = \lim_{j \to \infty} jH_f^W \equiv \infty$ in W. This contradiction shows that the assumption $\check{V} \neq \emptyset$ is absurd and the lemma is proved.   □

16.26. Corollary. *If $u \in \mathfrak{S}(U)$ and $V \subset U$ is open, then $u \in \mathfrak{S}(V)$.*

16.27. Perron's method

We next discuss the generalized Dirichlet problem with arbitrary boundary functions in general open sets. We start with the Poisson modification of a superharmonic function.

Let $V \Subset U$ be regular. If $u \in \mathfrak{S}(U)$, then the function

$$u_V = \begin{cases} \underline{h}_u^V & \text{in } V \\ u & \text{in } U \setminus V \end{cases}$$

is called the *Poisson modification of u in V.* We recall that $\underline{h}_u^V$ is the best $\mathfrak{H}$-minorant of u in V defined by

$$\underline{h}_u^V = \sup\{H_f^V : f \in C(\partial V) \text{ and } f \leq u \text{ in } V\}.$$

16.28. Lemma. *If $u \in \mathfrak{S}(U)$ and $V \Subset U$ is regular, then the Poisson modification u_V is superharmonic in U and an $\mathfrak{H}$-harmonic in V. Moreover, $u_V \leq u$ in U.*

PROOF: Clearly $u_V \leq u$ in U and hence Lemma 16.14 yields that $u_V \in \mathfrak{H}(V)$. We next show that u_V is lower semicontinuous in U. Since this is clearly the case in $U \setminus \overline{V}$ and in V, it suffices to verify the lower semicontinuity on ∂V. Fix $y \in \partial V$. Lemma 16.15 implies that

$$\liminf_{\substack{x \to y \\ x \in V}} u_V(x) \geq u(y)$$

and since the same inequality is trivially true when x approaches y along $U \setminus V$, u_V is lower semicontinuous in U.

The following pasting lemma completes the proof for Lemma 16.28.

16.29. Lemma. *Suppose that $V \subset U$ are open sets in X. If $u \in \mathfrak{S}(U)$ and $v \in \mathfrak{S}(V)$, then the function*

$$s = \begin{cases} \min(u,v) & \text{in } V \\ u & \text{in } U \setminus V \end{cases}$$

belongs to $\mathfrak{S}(U)$ if it is lower semicontinuous.

PROOF: Let $W \Subset U$ be an open set and $h \in C(\overline{W}) \cap \mathfrak{H}(W)$ be such that $h \leq s$ in ∂W. Then $h \leq u$ in $\overline{W}$. In particular, for all $x \in \partial V \cap W$

$$\lim_{\substack{y \to x \\ y \in V \cap W}} h(y) \leq u(x) = s(x) \leq \liminf_{\substack{y \to x \\ y \in V \cap W}} v(y) .$$

since s is lower semicontinuous. Thus

$$\lim_{\substack{y \to x \\ y \in V \cap W}} h(y) \le s(x) \le \liminf_{\substack{y \to x \\ y \in V \cap W}} s(y)$$

whenever $x \in \partial(V \cap W)$, and since $s = \min(u, v)$ is in $\mathfrak{S}(V \cap W)$, the generalized comparison principle 16.22 implies $h \le s$ in $V \cap W$. Therefore $h \le s$ in W and the lemma is proved. $\qquad\square$

We recall that a family $\mathcal{P}$ is lower directed if for each pair $u, v \in \mathcal{P}$ there is a $w \in \mathcal{P}$ such that $w \le \min(u, v)$. A lower directed family $\mathcal{P} \subset \mathfrak{S}(U)$ is called a *Perron family* if the Poisson modification $u_V \in \mathcal{P}$ whenever $u \in \mathcal{P}$ and $V \Subset U$ is regular.

The next theorem is the main step in Perron's process; we use the convention $\inf \emptyset = \infty$.

16.30. Theorem. *Let $\mathcal{P}$ be a Perron family in a domain U. Then $\inf \mathcal{P}$ is either $\mathfrak{H}$-harmonic or identically ∞ or $-\infty$ in U.*

PROOF: If $\mathcal{P} = \emptyset$, there is nothing to prove. Thus suppose that $\mathcal{P} \ne \emptyset$ and let $v = \inf \mathcal{P}$. If $V \Subset U$ is a regular domain, then $u_V \le u$ for each $u \in \mathcal{P}$, and therefore

$$v = \inf\{u_V : u \in \mathcal{P}\}.$$

Since the family

$$\mathcal{P}_V = \{u_V|_V : u \in \mathcal{P}\}$$

is a lower directed family of $\mathfrak{H}$-harmonic functions in V (Lemma 16.28), we deduce from the lower directed version of Theorem 16.11 that $v|_V = \inf \mathcal{P}_V$ is either $\mathfrak{H}$-harmonic or identically $-\infty$ in V. Because regular domains cover U and U is connected, we conclude that v is either $\mathfrak{H}$-harmonic or identically $-\infty$ in U. The theorem follows. $\qquad\square$

After this preparation we are ready to solve the generalized Dirichlet problem by using Perron's method. Let U be an open set in X and let $f \colon \partial U \to \overline{\mathbf{R}}$ be any function. The *upper class* $\mathcal{U}_f = \mathcal{U}_f^U$ of f in U consists of all functions $u \in \mathfrak{S}(U)$ such that u is bounded below and

$$\liminf_{x \to y} u(x) \ge f(y)$$

for each $y \in \partial U$. We emphasize that if U is not relatively compact, then we consider ∞ as a boundary point of U; therefore ∂U is always nonempty and compact.

The function $\overline{H}_f = \overline{H}_f^U = \inf \mathcal{U}_f$ is called the *upper Perron solution in U with boundary values f*. Of course, it may happen that $\mathcal{U}_f = \emptyset$, for example if $f \equiv \infty$. Then $\overline{H}_f = \inf \emptyset = \infty$.

The *lower class* $\mathcal{L}_f = \mathcal{L}_f^U$ of f in U consists of all functions $v \in -\mathfrak{S}(U)$ such that v is bounded above and

$$\limsup_{x \to y} v(x) \leq f(y)$$

for each $y \in \partial U$ and the function $\underline{H}_f = \underline{H}_f^U = \sup \mathcal{L}_f$ is called the *lower Perron solution in U with boundary values f*. Then $\mathcal{L}_f = -\mathcal{U}_{-f}$ and $\underline{H}_f = -\overline{H}_{-f}$.

Since the functions in $\mathcal{U}_f$ are bounded below and the functions in $\mathcal{L}_f$ bounded above, the generalized comparison principle 16.22 implies:

16.31. Theorem. $\underline{H}_f \leq \overline{H}_f$.

The next result culminates our development of axiomatic quasilinear potential theory. It demonstrates that not much structure is needed to solve the generalized Dirichlet problem.

16.32. Theorem. *Let U be an open set in X and $f : \partial U \to \overline{\mathbf{R}}$. If V is a component of U, then $\overline{H}_f$ is either $\mathfrak{H}$-harmonic or identically ∞ or $-\infty$ in V. The similar assertion is true if $\overline{H}_f$ is replaced by $\underline{H}_f$.*

PROOF: The proof follows from Theorem 16.30 because $\{u|_V : u \in \mathcal{U}_f\}$ is a Perron family in V. $\qquad\square$

It is desirable that the Perron solution coincides with the Dirichlet solution H_f when U is regular and $f \in C(\partial U)$. This is established in the next theorem.

16.33. Theorem. *If U is a regular open set and $f \in C(\partial U)$, then $H_f = \underline{H}_f = \overline{H}_f$.*

PROOF: Clearly, $H_f \in \mathcal{U}_f$ and $H_f \in \mathcal{L}_f$. Thus by Theorem 16.31 we have

$$H_f \leq \underline{H}_f \leq \overline{H}_f \leq H_f,$$

as desired. $\qquad\square$

If $\underline{H}_f = \overline{H}_f$ and if both are $\mathfrak{H}$-harmonic in U, then the function $h = \overline{H}_f = \underline{H}_f$ is called the *Perron solution with boundary values f* and the boundary function f is called *resolutive*. As a result of Theorem 16.33 we can write $H_f = H_f^U = \overline{H}_f = \underline{H}_f$ whenever f is resolutive.

It is trivial that constant functions are resolutive. More generally, if there is a bounded $\mathfrak{H}$-harmonic function h in U such that

$$f(y) = \lim_{x \to y} h(x)$$

for all $y \in \partial U$, then it is quickly seen that f is resolutive and $H_f = h$. Theorem 16.36 below provides a slight extension of this result but a characterization of resolutive functions remains open.

In the remaining theorems in this chapter we state some simple properties of Perron solutions. Suppose that $U \subset X$ is open. The first theorem follows at once.

16.34. Theorem. *Let $f, g \colon \partial U \to \overline{\mathbf{R}}$ be functions with $f \le g$ and $\lambda \in \mathbf{R}$. Then*

(i) $\overline{H}_{\lambda f} = \lambda \overline{H}_f$ *if $\lambda \ge 0$; here $\lambda f = 0$ if $\lambda = 0$,*

(ii) $\overline{H}_{-f} = -\underline{H}_f$,

(iii) $\overline{H}_{f+\lambda} = \overline{H}_f + \lambda$, *and*

(iv) $\overline{H}_f \le \overline{H}_g$.

The same is true for lower solutions and, if f and g are resolutive, for Perron solutions.

16.35. Theorem. *Let $f_j \colon \partial U \to \mathbf{R}$, $j = 1, 2, \ldots$, be a sequence of resolutive functions. If $f_j \to f$ uniformly on ∂U, then f is resolutive.*

PROOF: Fix $\varepsilon > 0$ and pick j such that $|f - f_j| < \varepsilon$ on ∂U. Then

$$\overline{H}_f - \varepsilon \le \overline{H}_{f_j} = \underline{H}_{f_j} \le \underline{H}_f + \varepsilon \le \overline{H}_f + \varepsilon,$$

and letting $\varepsilon \to 0$ we obtain $\overline{H}_f = \underline{H}_f$; moreover, $\overline{H}_f$ is $\mathfrak{H}$-harmonic because $\overline{H}_{f_j} \to \overline{H}_f$ uniformly in U (Theorem 16.34). Hence f is resolutive. □

16.36. Theorem. *Suppose that $u \in \mathfrak{S}(U)$ is such that*

$$f(y) = \lim_{x \to y} u(x)$$

for $y \in \partial U$ defines a bounded function on ∂U. Then f is resolutive.

PROOF: Since f is bounded, $\overline{H}_f$ is a bounded $\mathfrak{H}$-harmonic function and it remains to show that $\overline{H}_f \le \underline{H}_f$. First note that $u \ge \inf f > -\infty$ and hence $u \in \mathcal{U}_f$. Thus $u \ge \overline{H}_f$ and hence

$$\limsup_{x \to y} \overline{H}_f(x) \le \lim_{x \to y} u(x) = f(y)$$

for each $y \in \partial U$. Therefore $\overline{H}_f \in \mathcal{L}_f$ and so $\overline{H}_f \le \underline{H}_f$ as required. □

If $u\colon U \to \mathbf{R}$ is any function and

$$\lim_{x \to y} u(x) = f(y) \in \mathbf{R}$$

for all $y \in \partial U$, then it is easy to see that f is continuous. Thus the function in Theorem 16.36 is necessarily continuous on ∂U; the main difference to Theorem 16.33 is that we do not require that U be regular in Theorem 16.36.

NOTES TO CHAPTER 16. The axiomatic theory presented here was basically developed by Lehtola (1986); he has a slightly more general setup where he does not assume that constants belong to the harmonic sheaf. Lehtola (1986) also develops a theory of regular boundary points for Perron solutions through the barrier approach (see Chapter 9).

The obstacle problem finds its natural interpretation in quasilinear potential theory by using balayage; cf. Lemma 9.26 and Theorem 3.71.

In modern linear axiomatic potential theories axioms are sometimes stated for superharmonic functions instead of for harmonic functions as in the Brelot axiomatic theory. Such an axiomatic quasilinear theory has been considered by Laine (1985, 1990).

17

Appendix I: The existence of solutions

Let w be a p-admissible weight and let μ be the corresponding measure,

$$d\mu(x) = w(x)\,dx\,.$$

Suppose that $\mathcal{A}: \mathbf{R}^n \times \mathbf{R}^n \to \mathbf{R}^n$ satisfies the assumptions (3.3)–(3.6). Let Ω be a bounded open set in $\mathbf{R}^n$, $\vartheta \in H^{1,p}(\Omega; \mu)$, and let $\psi : \Omega \to [-\infty, \infty]$ be any function. Suppose that the set

$$\mathcal{K}_{\psi,\vartheta} = \{v \in H^{1,p}(\Omega; \mu) : v \geq \psi \text{ a.e. in } \Omega \text{ and } v - \vartheta \in H_0^{1,p}(\Omega; \mu)\}$$

is not empty. We prove:

17.1. Theorem. *There exists a solution u to the obstacle problem in $\mathcal{K}_{\psi,\vartheta}$. That is, there is a function u in $\mathcal{K}_{\psi,\vartheta}$ such that*

$$\int_\Omega \mathcal{A}(x, \nabla u) \cdot \nabla(v - u)\,dx \geq 0$$

whenever $v \in \mathcal{K}_{\psi,\vartheta}$.

We deduce Theorem 17.1 from a general result in the theory of monotone operators. Let X be a reflexive Banach space with dual X' and let $\langle \cdot, \cdot \rangle$ denote a pairing between X' and X. If $\mathbf{K} \subset X$ is a closed convex set, then a mapping $\mathfrak{A} : \mathbf{K} \to X'$ is called *monotone* if

$$\langle \mathfrak{A}u - \mathfrak{A}v, u - v \rangle \geq 0$$

for all u, v in $\mathbf{K}$. Further, $\mathfrak{A}$ is called *coercive on* $\mathbf{K}$ if there exists $\varphi \in \mathbf{K}$ such that

$$\frac{\langle \mathfrak{A}u_j - \mathfrak{A}\varphi, u_j - \varphi \rangle}{\|u_j - \varphi\|} \to \infty$$

whenever u_j is a sequence in $\mathbf{K}$ with $\|u_j\| \to \infty$.

For the following proposition see Kinderlehrer and Stampacchia (1980, Corollary III.1.8, p. 87).

17.2. Proposition. *Let* **K** *be a nonempty closed convex subset of* X *and let* $\mathfrak{A} : \mathbf{K} \to X'$ *be monotone, coercive, and weakly continuous on* **K**. *Then there exists an element* u *in* **K** *such that*

$$\langle \mathfrak{A}u, v - u \rangle \geq 0$$

whenever $v \in \mathbf{K}$.

Now let $X = L^p(\Omega; \mu; \mathbf{R}^n)$ and let $\langle \cdot, \cdot \rangle$ be the usual pairing between X and X',

$$\langle f, g \rangle = \int_\Omega f \cdot g \, d\mu,$$

where g is in X and f in $X' = L^{p/(p-1)}(\Omega; \mu; \mathbf{R}^n)$. Write

$$\mathbf{K} = \{ \nabla v : v \in \mathcal{K}_{\psi,\vartheta} \}.$$

Then $\mathbf{K} \subset X$ is convex. To show that $\mathbf{K}$ is closed in X, let $\nabla v_i \in \mathbf{K}$ be a sequence converging to $\tilde{v}$ in X. The Poincaré inequality implies

$$\int_\Omega |v_i - \vartheta|^p \, d\mu \leq c \, (\mathrm{diam}\,\Omega)^p \int_\Omega |\nabla v_i - \nabla \vartheta|^p \, d\mu \leq M < \infty$$

so that v_i is a bounded sequence in $H^{1,p}(\Omega; \mu)$. Since $\mathcal{K}_{\psi,\vartheta}$ is a convex and closed subset of $H^{1,p}(\Omega; \mu)$, there is a function $v \in \mathcal{K}_{\psi,\vartheta}$ such that $\nabla v = \tilde{v}$ (Theorems 1.30 and 1.31). Thus $\tilde{v} \in \mathbf{K}$ and hence $\mathbf{K}$ is closed.

Next we define a mapping $\mathfrak{A} : \mathbf{K} \to X'$ by

$$\langle \mathfrak{A}v, u \rangle = \int_\Omega \mathcal{A}\big(x, v(x)\big) \cdot u(x) \, dx$$

for $u \in X$. Because

$$\Big| \int_\Omega \mathcal{A}\big(x, v(x)\big) \cdot u(x) \, dx \Big| \leq \beta \Big(\int_\Omega |v|^p \, d\mu \Big)^{(p-1)/p} \Big(\int_\Omega |u|^p \, d\mu \Big)^{1/p}$$

by (3.5), we have that $\mathfrak{A}v \in X'$ whenever $v \in \mathbf{K}$. Moreover, it follows from (3.6) that $\mathfrak{A}$ is monotone.

To show that $\mathfrak{A}$ is coercive on $\mathbf{K}$, fix $\varphi \in \mathbf{K}$. Then

$$\langle \mathfrak{A}u - \mathfrak{A}\varphi, u - \varphi \rangle = \int_\Omega \big(\mathcal{A}(x, u) - \mathcal{A}(x, \varphi) \big) \cdot \big(u - \varphi \big) \, dx$$

$$\geq \alpha(\|u\|^p + \|\varphi\|^p) - \beta(\|u\|^{p-1}\|\varphi\| + \|u\|\|\varphi\|^{p-1})$$

$$\geq \|u - \varphi\| \alpha \, 2^{-p} \|u - \varphi\|^{p-1}$$

$$- \beta 2^{p-1} \|\varphi\| (\|\varphi\| + \|u - \varphi\|^{p-1}) - \beta \|\varphi\|^{p-1} (\|\varphi\| + \|u - \varphi\|),$$

where $\| \cdot \|$ is the usual $L^p(\Omega; \mu)$-norm

$$\|f\| = \left(\int_\Omega |f|^p \, d\mu \right)^{1/p}.$$

It follows that $\mathfrak{A}$ is coercive on $\mathbf{K}$.

Finally, we show that $\mathfrak{A}$ is weakly continuous on $\mathbf{K}$. Let $u_i \in \mathbf{K}$ be a sequence that converges to an element $u \in \mathbf{K}$ in $L^p(\Omega; \mu)$. Pick a subsequence u_{i_j} such that $u_{i_j} \to u$ a.e. in Ω. Since the mapping $\xi \mapsto \mathcal{A}(x, \xi)$ is continuous for a.e. x, we have that

$$\mathcal{A}\big(x, u_{i_j}(x)\big) \, w^{-1/p}(x) \to \mathcal{A}\big(x, u(x)\big) \, w^{-1/p}(x)$$

a.e. in Ω. Because the $L^{p/(p-1)}(\Omega; dx)$-norms of $\mathcal{A}(x, u_{i_j}) \, w^{-1/p}$ are uniformly bounded, we have that

$$\mathcal{A}(x, u_{i_j}) \, w^{-1/p} \to \mathcal{A}(x, u) \, w^{-1/p}$$

weakly in $L^{p/(p-1)}(\Omega; dx)$. Since the weak limit is independent of the choice of the subsequence, it follows that

$$\mathcal{A}(x, u_i) \, w^{-1/p} \to \mathcal{A}(x, u) \, w^{-1/p}$$

weakly in $L^{p/(p-1)}(\Omega; dx)$. Consequently we have for all $v \in L^p(\Omega; \mu)$ that

$$\langle \mathfrak{A}u_i, v \rangle = \int_\Omega \mathcal{A}(x, u_i) \cdot v \, dx = \int_\Omega \mathcal{A}(x, u_i) \, w^{-1/p} \cdot v \, w^{1/p} \, dx$$
$$\to \int_\Omega \mathcal{A}(x, u) \, w^{-1/p} \cdot v \, w^{1/p} \, dx = \langle \mathfrak{A}u, v \rangle.$$

Hence $\mathfrak{A}$ is weakly continuous on $\mathbf{K}$.

Therefore we can apply Proposition 17.2 and find an element $\tilde{u}$ in $\mathbf{K}$ such that

$$\langle \mathfrak{A}\tilde{u}, \tilde{v} - \tilde{u} \rangle \geq 0$$

for all $\tilde{v} \in \mathbf{K}$. But this means that there is a function $u \in \mathcal{K}_{\psi, \vartheta}$ such that $\nabla u = \tilde{u}$ and

$$\int_\Omega \mathcal{A}(x, \nabla u) \cdot (\nabla v - \nabla u) \, dx = \langle \mathfrak{A} \nabla u, \nabla v - \nabla u \rangle \geq 0$$

whenever $v \in \mathcal{K}_{\psi, \vartheta}$. Theorem 17.1 is thereby proved.

17.3. Corollary. *Let Ω be bounded and $\vartheta \in H^{1,p}(\Omega; \mu)$. There is a function $u \in H^{1,p}(\Omega; \mu)$ with $u - \vartheta \in H_0^{1,p}(\Omega; \mu)$ such that*

$$-\operatorname{div} \mathcal{A}(x, \nabla u) = 0$$

weakly in Ω, that is

$$\int_\Omega \mathcal{A}(x, \nabla u) \cdot \nabla \varphi \, dx = 0$$

whenever $\varphi \in H_0^{1,p}(\Omega; \mu)$.

PROOF: Choose $\psi \equiv -\infty$ and let u be the solution to the obstacle problem in $\mathcal{K}_{\psi,\vartheta}$. Let $\varphi \in H_0^{1,p}(\Omega; \mu)$. Since the functions $u + \varphi$ and $u - \varphi$ both belong to $\mathcal{K}_{\psi,\vartheta}$ we have

$$-\int_\Omega \mathcal{A}(x, \nabla u) \cdot \nabla \varphi \, dx \geq 0$$

and

$$\int_\Omega \mathcal{A}(x, \nabla u) \cdot \nabla \varphi \, dx \geq 0 \,.$$

Thus

$$\int_\Omega \mathcal{A}(x, \nabla u) \cdot \nabla \varphi \, dx = 0 \,,$$

as desired.   □

NOTES TO APPENDIX I. The standard reference here is Kinderlehrer and Stampacchia (1980). An important early work is Leray and Lions (1965). For existence results for refined obstacle problems, see Michael and Ziemer (1991).

18

Appendix II : The John–Nirenberg lemma

In this section we establish a weighted version of the John–Nirenberg lemma. We assume here that μ is a doubling measure in $\mathbf{R}^n$ with locally integrable density w; then $d\mu(x) = w(x)\,dx$. We also assume that $0 < w < \infty$ a.e. in $\mathbf{R}^n$. The doubling constant of μ is denoted by c_μ. For all functions $f \in L^1_{loc}(\mathbf{R}^n; \mu)$ it holds that

$$(18.1) \qquad \lim_{\substack{\mu(B) \to 0 \\ x \in B}} \fint_B f(y)\,d\mu(y) = f(x)$$

for a.e. $x \in \mathbf{R}^n$, where B designates a ball (Stein 1970, pp. 4–9; Ziemer 1989, p. 14). We use the familiar notation

$$f_E = \fint_E f\,d\mu = \fint_E f(y)\,d\mu(y) = \frac{1}{\mu(E)} \int_E f(y)\,d\mu(y)$$

for any set E with positive and finite measure.

A function $f \in L^1_{loc}(\Omega; \mu)$ is said to be in $BMO(\Omega; \mu)$ if there is a constant C such that

$$(18.2) \qquad \fint_B |f - f_B|\,d\mu \le C$$

for all balls $B \Subset \Omega$. The least C for which (18.2) holds is denoted by $\|f\|_*$ and called the *BMO-norm* of f. The letters BMO stand for *bounded mean oscillation*.

One of the most useful theorems in analysis is the John–Nirenberg lemma which asserts that (18.2) is in fact equivalent to a seemingly much stronger statement.

18.3. John–Nirenberg lemma for doubling weights. *A function f is in $BMO(\Omega; \mu)$ if and only if for each ball $B \Subset \Omega$ and $t > 0$ it holds that*

$$\mu(\{x \in B : |f(x) - f_B| > t\}) \le c_1 e^{-c_2 t}\mu(B).$$

The positive constants c_1, c_2 and the BMO norm $\|f\|_*$ depend only on each other, the dimension n, and c_μ.

The proof of Theorem 18.3 is usually based on the *Calderón–Zygmund decomposition* performed for dyadic subcubes of $\mathbf{R}^n$. Throughout this book, however, we have used balls instead of cubes in our estimates. This deviation from the standard proof causes only minor technical complications.

18.4. Calderón–Zygmund decomposition for doubling weights
Let B_0 be a fixed ball in $\mathbf{R}^n$ and let f be a nonnegative function in $L^1(B_0; \mu)$. Assume that

$$\fint_{B_0} f \, d\mu \leq t.$$

Then there are balls B_1, B_2, $\ldots$, contained in B_0, such that

$$(18.5) \qquad \sum_{i=1}^{\infty} \chi_{B_i}(x) \leq N < \infty$$

for all $x \in B_0$, that

$$(18.6) \qquad t < \fint_{B_i} f \, d\mu \leq c_0 t$$

for all $i = 1, 2, \ldots$, and that

$$(18.7) \qquad f(x) \leq t$$

for a.e. x in $B_0 \setminus \cup_i B_i$. The constants $N \geq 4$ and c_0 depend only on the doubling constant c_μ and the dimension n.

To prove (18.5)–(18.7), we need to find a sequence of balls in B_0 which substitute for the dyadic subdivision.

In the *dyadic subdivision* of an open cube Q we first divide Q into 2^n disjoint open subcubes $\{Q_i\}$ such that their sides are parallel to the sides of Q, their union covers Q up to a set of measure zero, and $|Q| = 2^n|Q_i|$ for all i. Then we repeat the construction in each cube Q_i, and continue in this manner. Cubes that are obtained in this process are called the *dyadic subcubes* of Q. Then for a.e. x in Q there is a unique descending chain of dyadic subcubes $Q \supset Q_{1,x} \supset Q_{2,x} \supset \cdots \ni x$ such that $|Q_{i,x}| = 2^n|Q_{i+1,x}|$.

Let Q_0 be the smallest cube with sides parallel to the coordinate axes that contains B_0 and has the same center as B_0. Let $\mathcal{Q} = \{Q\}$ be the collection of the dyadic subcubes of Q_0 such that $Q \in \mathcal{Q}$ if and only if B_Q, the smallest ball containing Q and centered at the center of Q, is contained in B_0. Then for a.e. $x \in B_0$ there is a descending chain of balls $B_0 = B_{0,x} \supset B_{1,x} \supset B_{2,x} \supset \cdots \ni x$ such that $|B_{i,x}| \leq c(n)|B_{i+1,x}|$; namely, put $B_{i,x} = B_{Q_{i,x}}$. Denote the set of all such x by B_0'. Then $\mu(B_0') = \mu(B_0)$, and because μ is doubling, we have

$$\mu(B_{i,x}) \leq c_0 \mu(B_{i+1,x}),$$

where c_0 depends only on n and c_μ. Next, let

$$A = \{x \in B_0': f(x) > t \text{ and } (18.1) \text{ holds}\};$$

then $\mu(A) = \mu(\{x \in B_0 : f(x) > t\})$. For each $x \in A$ let $k = k_x$ be the unique positive integer such that

$$t < \fint_{B_{k,x}} f \, d\mu$$

but

$$\fint_{B_{j,x}} f \, d\mu \leq t \text{ for all } j \leq k-1.$$

Hence the balls $\{B_{k_x,x}\}$ cover A, and by the Besicovitch covering theorem (Besicovitch 1945; Ziemer 1989, p. 9; de Guzmán 1975) we can select a countable subfamily $\{B_1, B_2, \ldots\}$ that covers A and satisfies

$$\sum_{i=1}^{\infty} \chi_{B_i}(x) \leq N < \infty$$

for all $x \in B_0$. Moreover, for any ball $B_i = B_{k,x}$ we have

$$t < \fint_{B_i} f \, d\mu \leq \frac{c_0}{\mu(B_{k-1,x})} \int_{B_{k-1,x}} f \, d\mu \leq c_0 \fint_{B_{k-1,x}} f \, d\mu \leq c_0 \, t.$$

PROOF OF THEOREM 18.3: The if part follows immediately from the formula

$$\int_B \varphi \circ f(x) \, d\mu(x) = \int_0^{\infty} \varphi'(t) \mu(\{x \in B : |f(x)| > t\}) \, dt$$

for any nonnegative and continuously differentiable φ with $\varphi(0) = 0$, which is a simple consequence of Fubini's theorem. Indeed, choose $\varphi(t) = e^{c_2 t/2} - 1$ and compute

$$\int_B e^{\frac{c_2}{2}|f-f_B|} \, d\mu - \mu(B)$$

$$= \frac{c_2}{2} \int_0^{\infty} e^{c_2 t/2} \mu(\{x \in B : |f(x) - f_B| > t\}) \, dt$$

$$\leq \frac{c_1 c_2}{2} \mu(B) \int_0^{\infty} e^{-c_2 t/2} \, dt = c_1 \mu(B).$$

This implies

$$\fint_B |f - f_B| \, d\mu \leq \frac{2}{c_2} \fint_B e^{\frac{c_2}{2}|f-f_B|} \, d\mu \leq \frac{2(c_1 + 1)}{c_2} < \infty$$

as desired.

The only if part requires a repeated application of the Calderón–Zygmund decomposition. Suppose that $f \in BMO(\Omega; \mu)$ and fix $B \Subset \Omega$. Replacing f by $(f - f_B)/N\|f\|_*$ we may assume that $f_B = 0$ and $\|f\|_* = 1/N$. Then, since

$$\fint_B |f| \, d\mu \le \|f\|_* = 1/N < 2 \,,$$

there are balls B_j^1, $j = 1, 2, \ldots$, in B such that

$$\sum_j \chi_{B_j^1} \le N$$

and

$$2 < \fint_{B_j^1} |f| \, d\mu \le 2c_0 \,,$$

$$|f| \le 2 \qquad \text{a.e. in } B \setminus \cup_j B_j^1 \,.$$

Hence

$$\sum_j \mu(B_j^1) < \frac{1}{2} \sum_j \int_{B_j^1} |f| \, d\mu$$

$$\le \frac{N}{2} \int_B |f| \, d\mu \le \frac{1}{2}\mu(B) \,.$$

Next fix one of the balls $B_j^1 = B_{j_1}$ and apply the Calderón–Zygmund decomposition with function $f - f_{B_{j_1}}$. Since

$$\fint_{B_{j_1}} |f - f_{B_{j_1}}| \, d\mu \le \|f\|_* = 1/N < 2 \,,$$

we again find balls $B_{j_1,j}^2$, $j = 1, 2, \ldots$, in B_{j_1} such that

$$\sum_j \chi_{B_{j_1,j}^2} \le N$$

and

$$2 < \fint_{B_{j_1,j}^2} |f - f_{B_{j_1}}| \, d\mu \le 2c_0 \,,$$

$$|f - f_{B_{j_1}}| \le 2 \qquad \text{a.e. in } B_{j_1} \setminus \cup_j B_{j_1,j}^2 \,.$$

Hence

$$\sum_j \mu(B_{j_1,j}^2) < \frac{1}{2} \sum_j \int_{B_{j_1,j}^2} |f - f_{B_{j_1}}| \, d\mu$$

$$\le \frac{N}{2} \int_{B_{j_1}} |f - f_{B_{j_1}}| \, d\mu$$

$$\le \frac{1}{2} \mu(B_{j_1}) \, .$$

Thus if we denote the balls $B_{j_1,j}^2$ by B_{j_1,j_2}, we see that for all balls B_{j_1} it holds that

$$|f(x)| \le |f(x) - f_{B_{j_1}}| + |f_{B_{j_1}}| \le 2 + 2c_0 \le 2 \cdot 2c_0$$

for a.e. $x \in B_{j_1} \setminus \cup_{j_2} B_{j_1,j_2}$ and

$$\sum_{j_1,j_2} \mu(B_{j_1,j_2}) \le (\frac{1}{2})^2 \mu(B) \, .$$

Continuing in this way, we apply at the $(k+1)$st step the Calderón–Zygmund decomposition to the ball $B_{j_1,\ldots,j_k}$ and to the function $f - f_{B_{j_1,\ldots,j_k}}$ and obtain a collection $\{B_{j_1,\ldots,j_k,j_{k+1}}\}$ of balls in $B_{j_1,\ldots,j_k}$ such that

$$\sum_{j_{k+1}} \chi_{B_{j_1,\ldots,j_k,j_{k+1}}} \le N \, ,$$

that

$$|f(x)| \le 2(k+1)c_0$$

for a.e. $x \in B_{j_1,\ldots,j_k} \setminus \cup_j B_{j_1,\ldots,j_k,j_{k+1}}$, and that

$$\sum_{j_1,\ldots,j_{k+1}} \mu(B_{j_1,\ldots,j_{k+1}}) \le (\frac{1}{2})^{k+1} \mu(B).$$

Now if $t > 4c_0$, let k be the first integer for which

$$2kc_0 < t \le 2(k+1)c_0 \, .$$

Then almost every point of $\{x \in B: |f(x)| > t\}$ belongs to some of the balls $B_{j_1,\ldots,j_k}$ so that

$$\mu(\{x \in B: |f(x)| > t\}) \le (\frac{1}{2})^k \mu(B) \le e^{-c_2 t} \mu(B) \, ,$$

where $c_2 = (\log 2)/4c_0$. If $0 < t \le 4c_0$, then

$$\mu(\{x \in B: |f(x)| > t\}) \le \mu(B) \le c_1 e^{-c_2 t} \mu(B)$$

where $c_1 = e^{4c_0 c_2} = 2$. This proves the theorem.   □

The proof of the lemma of John and Nirenberg (Theorem 18.3) implies:

18.8. Corollary. *A function f is in $BMO(\Omega; \mu)$ if and only if there are positive constants c and C such that*

$$(18.9) \qquad \int_B e^{c|f - f_B|} \, d\mu \leq C$$

for all balls $B \in \Omega$. If (18.9) holds, then $\|f\|_* \leq C/c$. Conversely, if f is in $BMO(\Omega; \mu)$, then (18.9) holds with $C = 3$ and $c = (\log 2)/(8 c_0 N \|f\|_*)$, where N and c_0 are as in (18.5) and (18.6).

NOTES TO APPENDIX II. The BMO space was introduced by John and Nirenberg (1961). "Bounded mean oscillation" soon became one of the central concepts in harmonic analysis, complex analysis, and partial differential equations. A weighted John–Nirenberg lemma was proved in Muckenhoupt and Wheeden (1976). Today more abstract versions of it are available; see e.g. Strömberg and Torchinsky (1989). For the connection between BMO and quasiconformal mappings, see Reimann (1974) and Reimann and Rychener (1975).

Bibliography

Adams, D. R. (1981). *Lectures on L^p-potential theory*. Department of Mathematics, University of Umeå.

Adams, D. R. (1986). Weighted nonlinear potential theory. *Transactions of the American Mathematical Society* **297**, 73–94.

Adams, D. R. (1992). L^p potential theory techniques and nonlinear PDE. In *Potential theory*. (ed. M. Kishi). pp. 1–15. Walter de Gruyter & Co, Berlin.

Adams, D. R. and Hedberg, L. I. (1984). Inclusion relations among fine topologies in non-linear potential theory. *Indiana University Mathematics Journal* **33**, 117–126.

Adams, D. R and Hedberg, L. I. *Function spaces and potential theory*. (In preparation.).

Adams, D. R. and Lewis, J. L. (1985). Fine and quasi connectedness in nonlinear potential theory. *Annales de l'Institut Fourier* **35**.1, 53–73.

Adams, D. R. and Meyers, N. G. (1972). Thinness and Wiener criteria for non-linear potentials. *Indiana University Mathematics Journal* **22**, 169–197.

Adams, D. R. and Meyers, N. G. (1973). Bessel potentials. Inclusion relations among classes of exceptional sets. *Indiana University Mathematics Journal* **22**, 873–905.

Adams, R. A. (1975). *Sobolev spaces*. Academic Press, London.

Ahlfors, L. V. (1973). *Conformal invariants*. McGraw-Hill, New York.

Ahlfors, L. V. and Sario, L. (1960). *Riemann surfaces*. Princeton University Press.

Aronsson, G. and Lindqvist, P. (1988). On p-harmonic functions in the plane and their stream functions. *Journal of Differential Equations* **74**, 157–178.

Aronszajn, N. and Smith, K. T. (1956). Functional spaces and functional completion. *Annales de l'Institut Fourier* **6**, 125–185.

Aronszajn, N. and Smith, K. T. (1961). Theory of Bessel potentials, I. *Annales de l'Institut Fourier* **11**, 385–475.

Aronszajn, N., Krzywicki, A., and Szarski, J. (1962). A unique continuation theorem for the exterior differential forms on Riemannian manifolds. *Arkiv för Matematik* **4**, 417–453.

Aronszajn, N., Mulla, F., and Szeptycky, P. (1963). On spaces of potentials connected with L^p-classes. *Annales de l'Institut Fourier* **13**.2, 211–306.

Astala, K. (1983). A remark on quasi-conformal mappings and BMO functions. *The Michigan Mathematical Journal* **30**, 209–212.

Avilés, P. and Manfredi, J. J. On null sets of p-harmonic measures. In *Partial differential equations with minimal smoothness and applications* (ed. B. Dahlberg, E. Fabes, R. Fefferman, D. Jerison, C. Kenig, J. Pipher), pp. 33–36. Springer-Verlag, Berlin.

Baernstein, A. and Manfredi, J. J. (1983). Topics in quasiconformal mappings. In *Topics in harmonic analysis*. Proceedings of seminar held in Torino and Milano vol. II, pp. 849–862. Istituto Nazionale di Alta Matematica Francesco Severi, Roma.

Bagby, T. (1972). Quasi topologies and rational approximation. *Journal of Functional Analysis* **10**, 259–268.

342

Bauer, H. (1966). *Harmonische Räume und ihre Potentialtheorie.* Lecture Notes in Mathematics 22. Springer-Verlag, Berlin.

Bauer, H. (1990). *Mass- und Integrationstheorie.* Walter de Gruyter & Co, Berlin.

Beckenbach, E. F. and Jackson, L. K. (1953). Subfunctions of several variables. *Pacific Journal of Mathematics* **3**, 291–313.

Bers, L. and Schechter, M. (1964). Elliptic equations. In *Partial differential equations.* Lectures in Applied Mathematics 3, pp. 131–300. Interscience Publishers, New York.

Besicovitch, A. (1945). A general form of the covering principle and relative differentiation of additive functions. *Proceedings of the Cambridge Philosophical Society* **41**, 103–110.

Beurling, A. and Ahlfors, L. V. (1956). The boundary correspondence under quasiconformal mappings. *Acta Mathematica* **96**, 125–142.

Biroli, M. and Marchi, S. (1986). Wiener estimates at boundary points for degenerate elliptic equations. *Bollettino Unione Matematica Italiana* (6) **5–B**, 689–706.

Bojarski, B. and Iwaniec, T. (1982). Another approach to Liouville theorem. *Mathematische Nachrichten* **107**, 253–262.

Bojarski, B. and Iwaniec, T. (1983). Analytical foundations of the theory of quasiconformal mappings in $\mathbb{R}^n$. *Annales Academiae Scientiarum Fennicae. Series A I. Mathematica* **8**, 257–324.

Bojarski, B. and Iwaniec, T. (1987). p-harmonic equation and quasiregular mappings. In *Partial differential equations.* Banach Center Publications, vol. 19, pp. 25–38. PWN-Polish Scientific Publishers, Warsaw.

Bourbaki, N. (1961). *Topologie générale* (2nd edn). Hermann, Paris.

Bourgain, J. (1987). On the Hausdorff dimension of harmonic measure in higher dimension. *Inventiones Mathematicae* **87**, 477–483.

Brelot, M. (1938). Sur le potentiel et les suites de fonctions sous-harmoniques. *Comptes Rendus de l'Académie des Sciences. Série I. Mathématique* **207**, 836–839.

Brelot, M. (1939). Familles de Perron et problème de Dirichlet. *Acta Universitatis Szegediensis. Acta Scientiarum Mathematicarum* **9**, 133–153.

Brelot, M. (1941). Sur la théorie autonome des fonctions sous-harmoniques. *Bulletin des Sciences Mathématiques* **65**, 72–98.

Brelot, M. (1945). Minorantes sousharmoniques, extrémales et capacilés. *Journal de Mathématiques Pures et Appliquées. Neuviéme Série* **24**, 1–32.

Brelot, M. (1960). *Lectures on potential theory..* Part IV (rev. edn 1967). Tata Institute of Fundamental Research, Bombay.

Brelot, M. (1971). *On topologies and boundaries in potential theory.* Lecture Notes in Mathematics 175. Springer-Verlag, Berlin.

Caffarelli, L., Fabes, E. B., and Kenig, C. (1981). Completely singular elliptic harmonic measures. *Indiana University Mathematics Journal* **30**, 917–924.

Calderón, C. P. (1961). Lebesgue spaces of differentiable functions and distributions. *Proceedings of Symposia in Pure Mathematics* **4**, 33–49.

Carleson, L. (1967). *Selected problems on exceptional sets.* Van Nostrand, Princeton, NJ.

Carleson, L. (1985). On the support of harmonic measure for sets of Cantor type. *Annales Academiae Scientiarum Fennicae. Series A I. Mathematica* **10**, 113–123.

Cartan, H. (1945). Théorie du potentiel newtonian: énergie, capacité, suites de potentiels. *Bulletin de la Société Mathématique de France* **73**, 74–106.

Chabrowski, J. (1991). *The Dirichlet problem with L^2-boundary data for elliptic linear equations.* Lecture Notes in Mathematics 1482. Springer-Verlag, Berlin.

Chanillo, S. and Wheeden, R. L. (1985). Weighted Poincaré and Sobolev inequalities and estimates for weighted Peano maximal functions. *American Journal of Mathematics* **107**, 1191–1226.

Chanillo, S. and Wheeden, R. L. (1986). Harnack's inequality and mean-value inequalities for solutions of degenerate elliptic equations. *Communications in Partial Differential Equations* **11**, 1111–1134.

Chanillo, S. and Wheeden, R. L. Poincaré inequalities for a class of non-A_p weights. *Indiana University Mathematics Journal*. (In press.).

Chernavskiĭ, A. V. (1964). Discrete and open mappings on manifolds. *Matematicheskiĭ Sbornik* **65**, 357–369.

Chernavskiĭ, A. V. (1965). Continuation to "Discrete and open mappings on manifolds". *Matematicheskiĭ Sbornik* **66**, 471–472.

Chiarenza, F. and Frasca, M. (1985). A note on a weighted Sobolev inequality. *Proceedings of the American Mathematical Society* **93**, 703–704.

Chiarenza, F., Rustichini, A., and Serapioni, R. (1989). De Giorgi–Moser theorem for a class of degenerate non-uniformly elliptic equations. *Communications in Partial Differential Equations* **14**, 635–662.

Choquet, G. (1953–54). Theory of capacities. *Annales de l'Institut Fourier* **5**, 131–295.

Coifman, R. R. and Fefferman, C. (1974). Weighted norm inequalities for maximal functions and singular integrals. *Studia Mathematica* **51**, 241–250.

Constantinescu, C. and Cornea, A. (1965). Compactifications of harmonic spaces. *Nagoya Mathematical Journal* **25**, 1–57.

Constantinescu, C. and Cornea, A. (1972). *Potential theory on harmonic spaces.* Springer-Verlag, Berlin.

Cornea, A. (1968). *Weakly compact sets in vector lattices and convergence theorems in harmonic spaces.* In Seminar über Potentialtheorie (ed. H. Bauer), pp. 173–180. Lecture Notes in Mathematics 69. Springer-Verlag, Berlin.

Dacorogna, B. (1982). *Weak continuity and weak lower semicontinuity of nonlinear functionals.* Lecture Notes in Mathematics 922. Springer-Verlag, Berlin.

David, G. and Semmes, S. (1990). Strong A_∞-weights, Sobolev inequalities, and quasiconformal mappings. In *Analysis and partial differential equations.* (ed. C. Sadosky). Lecture Notes in Pure and Applied Mathematics 122. Marcel Dekker.

De Giorgi, E. (1957). Sulla differenziabilità e l'analiticità delle estremali degli integrali multipli regolari. *Memorie della Reale Accademia delle Science di Torino. Classe di Science Fisiche, Matematiche e Naturali* **3**, 25–43.

Deny, J. (1950). Les potentiels d'énergie finie. *Acta Mathematica* **82**, 107–183.

Deny, J. (1954). Sur la convergence de certaines integrales de la théorie du potentiel. *Archiv der Mathematik* **5**, 367–370.

Deny, J. and Lions, J. L. (1953–54). Les espaces du type de Beppo Levi. *Annales de l'Institut Fourier* **5**, 305–370.

Di Benedetto, E. and Trudinger, N. S. (1984). Harnack inequalities for quasi-minima of variational integrals. *Annales de l'Institut Henri Poincaré, Analyse non linéaire* **1**, 295–308.

Donaldson, S. K. and Sullivan, D. P. (1989). Quasiconformal 4-manifolds. *Acta Mathematica* **163**, 181–252.

Doob, J. L. (1984). *Classical potential theory and its probabilistic counterpart.* Springer-Verlag, New York.

Dunford, N. and Schwartz, J. T. (1958). *Linear operators.* Part I. Interscience, New York.

Eremenko, A. and Lewis, J. L. (1991). Uniform limits of certain A-harmonic functions with applications to quasiregular mappings. *Annales Academiae Scientiarum Fennicae. Series A I. Mathematica* **16**, 361–375.

Evans, L. C. (1990). Weak convergence methods for nonlinear partial differential equations. *Regional Conference Series in Mathematics* **74**, 1–80.

Evans, L. C. and Gariepy, R. F. (1992). *Measure theory and fine properties of functions.* Studies in Advanced Mathematics. CRC Press, Boca Raton, FL.

Fabes, E. B., Kenig, C. E., and Serapioni, R. P. (1982a). The local regularity of solutions of degenerate elliptic equations. *Communications in Partial Differential Equations* **7**, 77–116.

Fabes, E. B., Jerison, D., and Kenig, C. E. (1982b). The Wiener test for degenerate elliptic equations. *Annales de l'Institut Fourier* **21.1**, 123–169.

Fabes, E. B, Jerison, D., and Kenig, C. E. (1983). *Boundary behavior of solutions to degenerate elliptic equations.* In Conference on harmonic analysis in honor of Antoni Zygmund. Vol. 2. Wadsworth, Belmont, CA.

Fabes, E., Garofalo, N, Marín-Malave, S., and Salsa, S. (1988). Fatou theorems for some nonlinear elliptic equations. *Revista Matemática Iberoamericana* **4**, 227–251.

Falconer, K. J. (1985). *The geometry of fractal sets.* Cambridge University Press, Cambridge.

Federer, H. (1969). *Geometric measure theory.* Springer-Verlag, Berlin.

Federer, H. and Ziemer, W. P. (1972). The Lebesgue set of a function whose distribution derivatives are p-th power summable. *Indiana University Mathematics Journal* **22**, 139–158.

Fefferman, R. A., Kenig, C. E., and Pipher, J. (1991). The theory of weights and the Dirichlet problem for elliptic equations. *Annals of Mathematics* **134**, 65–124.

Franchi, B. (1991). Weighted Sobolev–Poincaré inequalities and pointwise estimates for a class of degenerate elliptic equations. *Transactions of the American Mathematical Society* **327**, 125–158.

Franchi, B. and Serapioni, R. (1987). Pointwise estimates for a class of strongly degenerate elliptic operators: a geometrical approach. *Annali della Scuola Normale Superiore di Pisa.* Serie IV. **14**, 527–568.

Frehse, J. (1982). Capacity methods in the theory of partial differential equations. *Jahresbericht der Deutschen Mathematiker-Vereinigung* **84**, 1–44.

Fuglede, B. (1957). Extremal length and functional completion. *Acta Mathematica* **98**, 171–219.

Fuglede, B. (1960). On the theory of potentials in locally compact spaces. *Acta Mathematica* **103**, 139–215.

Fuglede, B. (1971a). The quasi topology associated with a countable subadditive set function. *Annales de l'Institut Fourier* **21.1**, 123–169.

Fuglede, B. (1971b). Connexion en topologie fine et balayage des mesures. *Annales de l'Institut Fourier* **21.3**, 227–244.

Fuglede, B. (1972). *Finely harmonic functions.* Lecture Notes in Mathematics 289. Springer-Verlag, Berlin.

Fuglede, B. (1978). Harmonic morphisms between Riemannian manifolds. *Annales de l'Institut Fourier* **28.2**, 107–144.

Fuglede, B. (1979a). *Harmonic morphisms.* In Proceedings of the colloquium on complex analysis, Joensuu 1978, pp. 123–131. Lecture Notes in Mathematics 747. Springer-Verlag, Berlin.

Fuglede, B. (1979b). Harnack sets and openness of harmonic morphisms. *Mathematische Annalen* **241**, 181–186.

Fuglede, B. (1986). Value distribution of harmonic morphisms and applications in complex analysis. *Annales Academiae Scientiarum Fennicae. Series A I. Mathematica* **11**, 111–135.

Fuglede, B. (1988). Fine potential theory. In *Potential theory, surveys and problems,* pp. 82–97. Lecture Notes in Mathematics 1344. Springer-Verlag, Berlin.

García-Cuerva, J. and Rubio de Francia, J. L. (1985). *Weighted norm inequalities and related topics.* North-Holland, Amsterdam.

Gariepy, R. and Ziemer, W. P. (1977). A regularity condition at the boundary for solutions of quasilinear elliptic equations. *Archive for Rational Mechanics and Analysis* **67**, 25–39.

Gardiner, S. J. and Klimek, M. (1986). Convexity and subsolutions of partial differential equations. *Bulletin of the London Mathematical Society* **18**, 41–43.

Garofalo, N. and Lin Fang-Hua (1986). Monotonicity properties of variational integrals, A_p weights and unique continuation. *Indiana University Mathematics Journal* **35**, 245–268.

Gatto, A. E. and Wheeden, R. L. (1989). Sobolev inequalities for products of powers. *Transactions of the American Mathematical Society* **314**, 727–743.

Gehring, F. W. (1962). Rings and quasiconformal mappings in space. *Transactions of the American Mathematical Society* **103**, 353–393.

Gehring, F. W. (1973). The L^p-integrability of the partial derivatives of quasiconformal mappings. *Acta Mathematica* **130**, 265–277.

Gehring, F. W. (1987). Topics in quasiconformal mappings. In Proceedings of the international congress of mathematicians, Berkeley, CA 1986. American Mathematical Society, Providence, RI.

Gehring, F. W. and Haahti, H. (1960). The transformations which preserve the harmonic functions. *Annales Academiae Scientiarum Fennicae. Series A I. Mathematica* **293**, 1–12.

Gehring, F. W. and Palka, B. P. (1976). Quasiconformally homogeneous domains. *Journal d'Analyse Mathématique* **30**, 172–199.

Gehring, F. W. and Väisälä, J. (1973). Hausdorff dimension and quasiconformal mappings. *The Journal of the London Mathematical Society* (2) **6**, 504–512.

Gehring, F. W., Hayman, W. K., and Hinkkanen, A. (1982). Analytic functions satisfying Hölder conditions on the boundary. *Journal of Approximation Theory* **35**, 243–249.

Giaquinta, M. (1983). *Multiple integrals in the calculus of variations and nonlinear elliptic systems.* Annals of Mathematics Studies 105. Princeton University Press.

Giaquinta, M. and Giusti, E. (1982). On the regularity of the minima of variational integrals. *Acta Mathematica* **148**, 31–46.

Giaquinta, M. and Giusti, E. (1984). Quasi-minima. *Annales de l'Institut Henri Poincaré, Analyse non linéaire* **1**, 79–107.

Gilbarg, D. and Trudinger, N. S. (1983). *Elliptic partial differential equations of second order* (2nd edn). Die Grundlehren der mathematischen Wissenschaften 224. Springer-Verlag, Berlin.

Granlund, S. (1982). An L^p-estimate for the gradient of extremals. *Mathematica Scandinavica* **50**, 66–72.

Granlund, S., Lindqvist, P., and Martio, O. (1982). F-harmonic measure in space. *Annales Academiae Scientiarum Fennicae. Series A I. Mathematica* **7**, 233–247.

Granlund, S., Lindqvist, P., and Martio, O. (1983). Conformally invariant variational integrals. *Transactions of the American Mathematical Society* **277**, 43–73.

Granlund, S., Lindqvist, P., and Martio, O. (1985). Phragmén-Lindelöf's and Lindelöf's theorems. *Arkiv för Matematik* **23**, 103–128.

Granlund, S., Lindqvist, P., and Martio, O. (1986). Note on the PWB-method in the non-linear case. *Pacific Journal of Mathematics* **125**, 381–395.

de Guzmán, M. (1975). *Differentiation of integrals in* $\mathbb{R}^n$. Lecture Notes in Mathematics 481. Springer-Verlag, Berlin.

Hayman, W. K. (1990). *Subharmonic functions.* Vol. 2. London Mathematical Society Monographs 20. Academic Press, London.

Hayman, W. K. and Kennedy, P. B. (1976). *Subharmonic functions.* Vol. 1. London Mathematical Society Monographs 9. Academic Press, London.

Hedberg, L. I. (1972). Non-linear potentials and approximation in the mean by analytic functions. *Mathematische Zeitschrift* **129**, 299–319.

Hedberg, L. I. (1987). Nonlinear potential theory and Sobolev spaces. In *Nonlinear analysis, function spaces and applications*, vol. 3. Proceedings of the Spring School, Litomyšl 1986. Teubner, Leipzig.

Hedberg, L. I. and Wolff, Th. H. (1983). Thin sets in nonlinear potential theory. *Annales de l'Institut Fourier* **33.4**, 161–187.

Heinonen, J. (1988a). Asymptotic paths for subsolutions of quasilinear elliptic equations. *Manuscripta mathematica* **62**, 449–465.

Heinonen, J. (1988b). Boundary accessibility and elliptic harmonic measures. *Complex Variables* **10**, 273–282.

Heinonen, J. and Kilpeläinen, T. (1988a). $\mathcal{A}$-superharmonic functions and supersolutions of degenerate elliptic equations. *Arkiv för Matematik* **26**, 87–105.

Heinonen, J. and Kilpeläinen, T. (1988b). Polar sets for supersolutions of degenerate elliptic equations. *Mathematica Scandinavica* **63**, 136–150.

Heinonen, J. and Kilpeläinen, T. (1988c). On the Wiener criterion and quasilinear obstacle problems. *Transactions of the American Mathematical Society* **310**, 239–255.

Heinonen, J. and Koskela, P. A_∞-condition for the Jacobian of a quasiconformal mapping. *Proceedings of the American Mathematical Society.* (In press.).

Heinonen, J. and Koskela, P. Weighted Sobolev inequalities and quasiregular mappings. (In preparation.).

Heinonen, J. and Martio, O. (1987). Estimates for F-harmonic measures and Øksendal's theorem for quasiconformal mappings. *Indiana University Mathematics Journal* **36**, 659–683.

Heinonen, J. and Rossi, J. (1990). Lindelöf's theorem for normal quasimeromorphic mappings. *The Michigan Mathematical Journal* **37**, 219–226.

Heinonen, J. and Rossi, J. Remarks on the value distribution of quasimeromorphic mappings. *Complex Variables.* (In press.).

Heinonen, J., Kilpeläinen, T., and Martio, O. (1989a). Fine topology and quasilinear elliptic equations. *Annales de l'Institut Fourier* **39.2**, 293–318.

Heinonen, J., Kilpeläinen, T., and Rossi, J. (1989b). The growth of $\mathcal{A}$-subharmonic functions and quasiregular mappings along asymptotic paths. *Indiana University Mathematics Journal* **38**, 581–601.

Heinonen, J., Kilpeläinen, T., and Malý, J. (1990). Connectedness in fine topologies. *Annales Academiae Scientiarum Fennicae. Series A I. Mathematica* **15**, 107–123.

Heinonen, J., Kilpeläinen, T., and Martio, O. (1992). Harmonic morphisms in nonlinear potential theory. *Nagoya Mathematical Journal* **125**, 115–140.

Helms, L. L. (1969). *Introduction to potential theory.* Wiley-Interscience, New York.

Herron, D. and Koskela, P. (1990). Quasiextremal distance domains and extendability of quasiconformal mappings. *Complex Variables* **15**, 167–179.

Herron, D. and Koskela, P. (1992). Uniform, Sobolev extension and quasiconformal circle domains. *Journal d'Analyse Mathématique*. (In press.).

Hervé, R.-M. (1962). Recherches axiomatiques sur la théorie des fonctions surharmoniques et du potentiel. *Annales de l'Institut Fourier* **12**, 415–571.

Hewitt, E. and Stromberg, K. (1965). *Real and abstract analysis*. Springer-Verlag, Berlin.

Hinkkanen, A. (1988). Modulus of continuity of harmonic functions. *Journal d'Analyse Mathématique* **51**, 1–29.

Holopainen, I. (1990). Nonlinear potential theory and quasiregular mappings on Riemannian manifolds. *Annales Academiae Scientiarum Fennicae. Series A I. Mathematica Dissertationes* **74**, 1–45.

Holopainen, I. (1992). Positive solutions of quasilinear elliptic equations on Riemannian manifolds. *Proceedings of the London Mathematical Society*. (In press.).

Holopainen, I. and Rickman, S. (1991). Classification of Riemannian manifolds in nonlinear potential theory. Preprint.

Holopainen, I. and Rickman, S. (1992). A Picard type theorem for quasiregular mappings of $\mathbf{R}^n$ into n-manifolds with many ends. *Revista Matemática Iberoamericana* (In press.).

Holopainen, I. and Rickman, S. Quasiregular mappings of the Heisenberg group. *Mathematische Annalen* (In press.).

Hurewicz, W. and Wallman, H. (1941). *Dimension theory*. Princeton University Press.

Hurri, R. (1988). Poincaré domains in $\mathbf{R}^n$. *Annales Academiae Scientiarum Fennicae. Series A I. Mathematica Dissertationes* **71**, 1–42.

Ishihara, T. (1979). A mapping of Riemannian manifolds which preserves harmonic functions. *Journal of Mathematics of Kyoto University* **19**, 215–229.

Iwaniec, T. (1984). Some aspects of partial differential equations and quasiregular mappings. In *Proceedings of the international congress of mathematicians, Warszawa 1983*, vol. 2, pp. 1193–1208. Polish Scientific Publishers, Warszawa.

Iwaniec, T. p-harmonic tensors and quasiregular mappings. *Annals of Mathematics*. (In press.).

Iwaniec, T. and Manfredi, J. J. (1989). Regularity of p-harmonic functions on the plane. *Revista Matemática Iberoamericana* **5**, 1–19.

Iwaniec, T. and Martin, G. Quasiregular mappings in even dimensions. *Acta Mathematica*. (In press.).

Iwaniec, T. and Nolder, C. (1985). Hardy–Littlewood inequality for quasiregular mappings in certain domains in $\mathbf{R}^n$. *Annales Academiae Scientiarum Fennicae. Series A I. Mathematica* **10**, 267–282.

Jacobi, C. G. J. (1848). Über eine particuläre Lösung der partiellen Differentialgleichung $\partial^2 V/\partial x^2 + \partial^2 V/\partial y^2 + \partial^2 V/\partial z^2 = 0$. *Journal für die reine und angewandte Mathematik* **36**, 113–134.

Jackson, L. K. (1955). On generalized subharmonic functions. *Pacific Journal of Mathematics* **5**, 215–228.

Järvi, P. and Vuorinen, M. (1992). Self-similar Cantor sets and quasiregular mappings. *Journal für die reine und angewandte Mathematik* **424**, 31–45.

John, F. and Nirenberg, L. (1961). On functions of bounded mean oscillation. *Communications on Pure and Applied Mathematics* **14**, 415–426.

Jones, P. W. (1981). Quasiconformal mappings and extendability of functions in Sobolev spaces. *Acta Mathematica* **147**, 71–88.

Jones, P. W. and Wolff, T. H. (1988). Hausdorff dimension of harmonic measures in the plane. *Acta Mathematica* **161**, 131–144.

Kellogg, O. D. (1929). *Foundations of potential theory.* Springer-Verlag, Reprinted (1967), Berlin.

Kilpeläinen, T. (1985). Homogeneous and conformally invariant variational integrals. *Annales Academiae Scientiarum Fennicae. Series A I. Mathematica Dissertationes* **57**, 1–37.

Kilpeläinen, T. (1987). Convex increasing functions preserve the sub-F-extremality. *Annales Academiae Scientiarum Fennicae. Series A I. Mathematica* **12**, 55–60.

Kilpeläinen, T. (1989). Potential theory for supersolutions of degenerate elliptic equations. *Indiana University Mathematics Journal* **38**, 253–275.

Kilpeläinen, T. (1992a). Nonlinear potential theory and PDEs. *Potential Analysis.* (In press.).

Kilpeläinen, T. (1992b). Weighted Sobolev spaces and capacity. (In press.).

Kilpeläinen, T. and Malý, J. (1989). Generalized Dirichlet problem in nonlinear potential theory. *Manuscripta Mathematica* **66**, 25-44.

Kilpeläinen, T. and Malý, J. (1990). Degenerate elliptic equations with measure data and nonlinear potentials. Preprint, University of Jyväskylä.

Kilpeläinen, T. and Malý, J. (1992a). Supersolutions to degenerate elliptic equations on quasi open sets. *Communications in Partial Differential Equations* **17**, 371–405.

Kilpeläinen, T. and Malý, J. (1992b). The Wiener test and potential estimates for quasilinear elliptic equations. Preprint, University of Jyväskylä.

Kilpeläinen, T. and Ziemer, W. P. (1991). Pointwise regularity of solutions to nonlinear double obstacle problems. *Arkiv för Matematik* **29**, 83–106.

Kinderlehrer, D. and Stampacchia, G. (1980). *An introduction to variational inequalities and their applications.* Academic Press, New York.

Koskela, P. and Martio, O. (1990). Removability theorems for quasiregular mappings. *Annales Academiae Scientiarum Fennicae. Series A I. Mathematica* **15**, 381–399.

Koskela, P. and Martio, O. Removability theorems for solutions of degenerate elliptic partial differential equations. *Arkiv för Matematik.* (In press.).

Král, J. (1983). Some extension results concerning harmonic functions. *Journal of the London Mathematical Society* **28**, 62–70.

Krol', I. N. and Maz'ya, V. G. (1972). On the absence of continuity and Hölder continuity of solutions of quasilinear elliptic equations near a nonregular boundary. *Transactions of the Moscow Mathematical Society* **26**, 73–93.

Kufner, A. (1985). *Weighted Sobolev spaces.* Wiley, New York.

Kufner, A., John, O., and Fučík, S. (1977). *Function spaces.* Noordhoff International Publishing, Leyden.

Kuusalo, T. (1977). Radó's theorem for quasiregular mappings. *Reports of the Department of Mathematics, University of Helsinki, Series A* **12**, 1–4.

Ladyzhenskaya, O. A. and Ural'tseva, N. N. (1968). *Linear and quasilinear elliptic equations.* Mathematics in Science and Engineering 46. Academic Press, New York.

Laine, I. (1985). Introduction to a quasi-linear potential theory. *Annales Academiae Scientiarum Fennicae. Series A I. Mathematica* **10**, 339–348.

Laine, I. (1990). Harmonic morphisms and non-linear potential theory. Preprint.

Landkof, N. S. (1972). *Foundations of modern potential theory.* Springer-Verlag, Berlin.

Lehto, O. (1987). *Univalent functions and Teichmüller spaces.* Graduate Texts in Mathematics 109. Springer-Verlag, New York.

Lehto, O. and Virtanen, K. I. (1973). *Quasiconformal mappings in the plane* (2nd edn). Die Grundlehren der mathematischen Wissenschaften 126.. Springer-Verlag, Berlin.

Lehtola, P. (1986). An axiomatic approach to non-linear potential theory. *Annales Academiae Scientiarum Fennicae. Series A I. Mathematica Dissertationes* **62**, 1–40.

Leray, J. and Lions, J.-L. (1965). Quelques résultats de Višik sur les problèmes elliptiques non linéaires par les méthodes de Minty–Browder. *Bulletin de la Societé Mathématique de France* **93**, 97–107.

Lewis, J. L. (1980). Smoothness of certain degenerate elliptic equations. *Proceedings of the American Mathematical Society* **80**, 259–265.

Lewis, J. L. (1983). Regularity of the derivatives of solutions to certain degenerate elliptic equations. *Indiana University Mathematics Journal* **32**, 849–858.

Lewis, J. L. (1988). Uniformly fat sets. *Transactions of the American Mathematical Society* **308**, 177–196.

Lewis, J. L. On a conditional theorem of Littlewood for quasiregular entire functions. *Journal d'Analyse Mathématique.* (In press.).

Lindqvist, P. (1985). On the growth of the solutions of the differential equation $\mathrm{div}(|\nabla u|^{p-2}\nabla u) = 0$ in n-dimensional space. *Journal of Differential Equations* **58**, 307–317.

Lindqvist, P. (1986). On the definition and properties of p-superharmonic functions. *Journal für die reine und angewandte Mathematik* **365**, 67–79.

Lindqvist, P. Global integrability and degenerate quasilinear elliptic equations. *Journal d'Analyse Mathématique.* (In press.).

Lindqvist, P. and Martio, O. (1985). Two theorems of N. Wiener for solutions of quasilinear elliptic equations. *Acta Mathematica* **155**, 153–171.

Lindqvist, P. and Martio, O. (1988). Regularity and polar sets of supersolutions of certain degenerate elliptic equations. *Journal d'Analyse Mathématique* **50**, 1–17.

Littman, W., Stampacchia, G., and Weinberger, H. F. (1963). Regular points for elliptic equations with discontinuous coefficients. *Annali della Scuola Normale Superiore di Pisa. Serie III.* **17**, 43–77.

Lukeš, J. and Malý, J. (1992). Thinness, Lebesgue density and fine topology (an interplay between real analysis and potential theory). In *Summer School in Potential Theory, Joensuu 1990.* (Ed. I. Laine), pp. 35–70. University of Joensuu, Publications in Sciences.

Lukeš, J., Malý, J., and Zajíček, L. (1986). *Fine topology methods in real analysis and potential theory.* Lecture Notes in Mathematics 1189. Springer-Verlag, Berlin.

Maeda, F.-Y. (1980). *Dirichlet integrals on harmonic spaces.* Lecture Notes in Mathematics 803. Springer-Verlag, Berlin.

Makarov, N. G. (1985). Distortion of boundary sets under conformal mappings. *Proceedings of the London Mathematical Society* **51**, 369–384.

Manfredi, J. J. (1988). p-harmonic functions in the plane. *Proceedings of the American Mathematical Society* **103**, 473–479.

Manfredi, J. J. (1991). N-harmonic morphisms are Möbius transformations. *Preprint.*

Manfredi, J. J. and Weitsman, A. (1988). On the Fatou theorem for p-harmonic functions. *Communications in Partial Differential Equations* **13**, 651–668.

Martio, O. (1970). A capacity inequality for quasiregular mappings. *Annales Academiae Scientiarum Fennicae. Series A I. Mathematica* **474**, 1–18.

Martio, O. (1975). Equicontinuity theorem with an application to variational integrals. *Duke Mathematical Journal* **42**, 569–581.

Martio, O. (1978/79). Capacity and measure densities. *Annales Academiae Scientiarum Fennicae. Series A I. Mathematica* **4**, 109–118.

Martio, O. (1983). *Non-linear potential theory*, In *Summer school in potential theory*. (ed. I. Laine and O. Martio), pp. 65–104. University of Joensuu, Joensuu.

Martio, O. (1987). *F*-harmonic measures, quasihyperbolic distance and Milloux's problem. *Annales Academiae Scientiarum Fennicae. Series A I. Mathematica* **12**, 151–162.

Martio, O. (1988a). Counterexamples for unique continuation. *Manuscripta Mathematica* **60**, 21–47.

Martio, O. (1988b). John domains, bilipschitz balls and Poincaré inequality. *Revue Roumaine de Mathématiques Pures et Appliqueés* **33**, 107–112.

Martio, O. (1989a). Sets of zero elliptic harmonic measures. *Annales Academiae Scientiarum Fennicae. Series A I. Mathematica* **14**, 47–55.

Martio, O. (1989b). Harmonic measures for second order non-linear partial differential equations. In *Function spaces, differential operators and nonlinear analysis* (ed. L. Päivärinta), pp. 271–279.. Pitman Research Notes in Mathematics Series 211, Longman Scientific & Technical, Harlow.

Martio, O. (1992). Potential theory and quasiconformal mappings. In *Potential theory* (ed. M. Kishi) , pp. 55–64. Walter de Gruyter & Co, Berlin.

Martio, O. and Näkki, R. (1991). Boundary Hölder continuity and quasiconformal mappings. *The Journal of the London Mathematical Society* (2) **44**, 339–350.

Martio, O. and Srebro, U. (1975a). Periodic quasimeromorphic mappings. *Journal d'Analyse Mathématique* **28**, 20–40.

Martio, O. and Srebro, U. (1975b). Automorphic quasimeromorphic mappings in $\mathbb{R}^n$. *Acta Mathematica* **135**, 221–247.

Martio, O. and Väisälä, J. (1988). Elliptic equations and maps of bounded length distortion. *Mathematische Annalen* **282**, 423–443.

Martio, O., Rickman, S., and Väisälä, J. (1969). Definitions for quasiregular mappings. *Annales Academiae Scientiarum Fennicae. Series A I. Mathematica* **448**, 1–40.

Martio, O., Rickman, S., and Väisälä, J. (1970). Distortion and singularities of quasiregular mappings. *Annales Academiae Scientiarum Fennicae. Series A I. Mathematica* **465**, 1–13.

Martio, O., Rickman, S., and Väisälä, J. (1971). Topological and metric properties of quasiregular mappings. *Annales Academiae Scientiarum Fennicae. Series A I. Mathematica* **488**, 1–31.

Mattila, P. (1986). *Lecture notes on geometric measure theory*. Departamento de Matemáticas, Universidad de Extremadura, Salamanca.

Maz'ya, V. G. (1976). On the continuity at a boundary point of solutions of quasi-linear elliptic equations (English translation). *Vestnik Leningrad University. Mathematics* **3**, 225–242. Original in *Vestnik Leningradskogo Universiteta* **25** (1970), 42–55 (in Russian).

Maz'ya, V. G. (1985). *Sobolev spaces*. Springer-Verlag, Berlin.

Maz'ya, V. G. and Khavin, V. P. (1972). Non-linear potential theory. *Russian Mathematical Surveys* **27**, 71–148.

Meyers, N. G. (1970). A theory of capacities for potentials of functions in Lebesgue classes. *Mathematica Scandinavica* **26**, 255–292.

Meyers, N. G. (1975). Continuity properties of potentials. *Duke Mathematical Journal* **42**, 157–166.

Michael, J. H. and Ziemer, W. P. (1986). Interior regularity for solutions to obstacle problems. *Nonlinear Analysis. Theory, Methods and Applications* **10**, 1427–1448.

Michael, J. H. and Ziemer, W. P. (1991). Existence of solutions to obstacle problems. *Nonlinear Analysis. Theory, Methods and Applications* **17**, 45–71.

Miklyukov, V. M. (1980). Asymptotic properties of subsolutions of quasilinear equations of elliptic type and mappings with bounded distortion. *Matematicheskiĭ Sbornik* **111**, 42–66.

Miller, K. (1974). Nonunique continuation for uniformly parabolic and elliptic equations in self-adjoint divergence forms with Hölder continuous coefficients. *Archive for Rational Mechanics and Analysis* **54**, 105–117.

Miller, N. (1982). Weighted Sobolev spaces and pseudodifferential operators with smooth symbols. *Transactions of the American Mathematical Society* **269**, 91–109.

Monna, A. F. (1975). *Dirichlet's principle*. Oosthock, Scheltema and Holkema, Utrecht.

Morrey, C. B. (1938). On the solutions of quasilinear elliptic partial differential equations. *Transactions of the American Mathematical Society* **43**, 120–166.

Morrey, C. B. (1966). *Multiple integrals in the calculus of variations*. Die Grundlehren der mathematischen Wissenschaften 130. Springer-Verlag, Berlin.

Mosco, U. (1987). Wiener criterion and potential estimates for the obstacle problem. *Indiana University Mathematics Journal* **36**, 455–494.

Moser, J. (1961). On Harnack's theorem for elliptic differential equations. *Communications on Pure and Applied Mathematics* **14**, 577–591.

Muckenhoupt, B. (1972). Weighted norm inequalities for the Hardy maximal function. *Transactions of the American Mathematical Society* **165**, 207–226.

Muckenhoupt, B. and Wheeden, R. L. (1974). Weighted norm inequalities for fractional integrals. *Transactions of the American Mathematical Society* **192**, 261–274.

Muckenhoupt, B. and Wheeden, R. L. (1976). Weighted bounded mean oscillation and the Hilbert transform. *Studia Mathematica* **54**, 221–237.

Narasimhan, R. (1985). *Complex analysis in one variable*. Birkhäuser, Boston.

Nash, J. (1958). Continuity of solutions of parabolic and elliptic equations. *American Journal of Mathematics* **80**, 931–954.

Nevanlinna, R. (1953). *Eindeutige analytische Funktionen* (2nd edn). Die Grundlehren der mathematischen Wissenschaften 46.. Springer-Verlag, Berlin.

Nieminen, E. (1991). Hausdorff measures, capacities, and Sobolev spaces with weights. *Annales Academiae Scientiarum Fennicae. Series A I. Mathematica Dissertationes* **81**, 1–39.

Nolder, C. (1991). Hardy–Littlewood theorems for solutions of elliptic equations in divergence form. *Indiana University Mathematics Journal* **40**, 149–160.

Ohtsuka, M. (1970). *Dirichlet problem, extremal length, and prime ends*. Van Nostrand Reinhold Company, New York.

Øksendal, B. (1988). Dirichlet forms, quasiregular functions and Brownian motion. *Inventiones Mathematicae* **91**, 273–297.

Øksendal, B. (1990). Weighted Sobolev inequalities and harmonic measure associated with quasiregular functions. *Communications in Partial Differential Equations* **15**, 1447–1459.

Pfluger, A. (1955). Extremallängen und Kapazität. *Commentarii Mathematici Helvetici* **29**, 120–131.

DuPlessis, N. (1970). *An introduction to potential theory*. Oliver and Boyd, Edinburgh.

Pliš, A. (1963). On non-uniqueness in Cauchy problem for an elliptic second order differential equation. *Bulletin of the Polish Academy of Sciences* **11**, 95–100.

Pommerenke, Ch. (1984). On uniformly perfect sets and Fuchsian groups. *Analysis* **4**, 299–321.

Radó, T. (1924). Über eine nicht fortsetzbare Riemannsche Mannigfaltigkeit. *Mathematische Zeitschrift* 20, 1–6.

Radó, T. (1949). *Subharmonic functions*. Chelsea Publishing Company, New York.

Radó, T. and Reichelderfer, P. V. (1955). *Continuous transformations in analysis*. Die Grundlehren der mathematischen Wissenschaften 75. Springer-Verlag, Berlin.

Reimann, H. M. (1974). Functions of bounded mean oscillation and quasiconformal mappings. *Commentarii Mathematici Helvetici* 49, 260–276.

Reimann, H. M. and Rychener, T. (1975). *Funktionen beschränkter mittlerer Oszillation*. Lecture Notes in Mathematics 487. Springer-Verlag, Berlin.

Reshetnyak, Yu. G. (1966). Estimates of the modulus of continuity for certain mappings. *Sibirskiĭ Matematicheskiĭ Zhurnal* 7, 1106–1114.

Reshetnyak, Yu. G. (1967a). Space mappings with bounded distortion. *Sibirskiĭ Matematicheskiĭ Zhurnal* 8, 629–659.

Reshetnyak, Yu. G. (1967b). The Liouville theorem with minimal regularity conditions. *Sibirskiĭ Matematicheskiĭ Zhurnal* 8, 835–840.

Reshetnyak, Yu. G. (1967c). General theorems on semicontinuity and on convergence with a functional. *Sibirskiĭ Matematicheskiĭ Zhurnal* 8, 1051–1069.

Reshetnyak, Yu. G. (1968). On the condition of the boundedness of index for mappings with bounded distortion. *Sibirskiĭ Matematicheskiĭ Zhurnal* 9, 368–374.

Reshetnyak, Yu. G. (1969). The concept of capacity in the theory of functions with generalized derivatives. *Sibirskiĭ Matematicheskiĭ Zhurnal* 10, 1109–1138.

Reshetnyak, Yu. G. (1970). On the branch set of mappings with bounded distortion. *Sibirskiĭ Matematicheskiĭ Zhurnal* 11, 1333–1339.

Reshetnyak, Yu. G. (1989). *Space mappings with bounded distortion*. Translations of Mathematical Monographs 73. American Mathematical Society, Providence, RI.

Rickman, S. (1973). Path lifting for discrete open mappings. *Duke Mathematical Journal* 40, 187–191.

Rickman, S. (1980a). Asymptotic values and angular limits of quasiregular mappings of a ball. *Annales Academiae Scientiarum Fennicae. Series A I. Mathematica* 5, 185–196.

Rickman, S. (1980b). On the number of omitted values of entire quasiregular mappings. *Journal d'Analyse Mathématique* 37, 100–117.

Rickman, S. (1985). The analogue of Picard's theorem for quasiregular mappings in dimension three. *Acta Mathematica* 154, 195–242.

Rickman, S. Nonremovable Cantor sets for bounded quasiregular mappings. *Annales Academiae Scientiarum Fennicae. Series A I. Mathematica*. (In press.).

Rickman, S. *Quasiregular mappings.*. (In preparation.).

Rickman, S. and Vuorinen, M. (1982). On the order of quasiregular mappings. *Annales Academiae Scientiarum Fennicae. Series A I. Mathematica* 7, 221–231.

Riesz, F. (1926). Sur les fonctions subharmoniques et leur rapport à la théorie du potentiel I. *Acta Mathematica* 48, 329–343.

Riesz, F. (1930). Sur les fonctions subharmoniques et leur rapport à la théorie du potentiel II. *Acta Mathematica* 54, 321–360.

Roberts, A. W. and Varberg, D. E. (1973). *Convex functions*. Academic Press, London.

Rogers, C. A. (1970). *Hausdorff measures*. Cambridge University Press, London.

Sawyer, E. and Wheeden, R. L. Weighted inequalities for fractional integrals on euclidean and homogeneous spaces. *American Journal of Mathematics*. (In press.).

Schwartz, L. (1966). *Théorie des distributions*. Hermann, Paris.

Semmes, S. (1992). Bilipschitz mappings and strong A_∞-weights. Preprint.

Serrin, J. (1963). A Harnack inequality for nonlinear equations. *Bulletin of American Mathematical Society* **69**, 481–486.

Serrin, J. (1964). Local behavior of solutions of quasi-linear equations. *Acta Mathematica* **111**, 247–302.

Serrin, J. (1965). Isolated singularities of solutions of quasilinear equations. *Acta Mathematica* **113**, 219–240.

Staples, S. (1991). Maximal functions, A_∞-measures, and quasiconformal maps. *Proceedings of the American Mathematical Society* **113**, 689–700.

Staples, S. (1992). Doubling measures and quasiconformal maps. *Commentarii Mathematici Helvetici* **67**, 119–128.

Stein, E. M. (1970). *Singular integrals and differentiability properties of functions.* Princeton University Press.

Stredulinsky, E. W. (1984). *Weighted inequalities and degenerate elliptic partial differential equations.* Lecture Notes in Mathematics 1074. Springer-Verlag, Berlin.

Strömberg, J.-D. and Torchinsky, A. (1989). *Weighted Hardy spaces.* Lecture Notes in Mathematics 1381. Springer-Verlag, Berlin.

Titus, C. J. and Young, G. S. (1962). The extension of interiority with some applications. *Transactions of the American Mathematical Society* **103**, 329–340.

Tolksdorf, P. (1984). Regularity for a more general class of quasi-linear elliptic equations. *Journal of Differential Equations* **51**, 126–150.

Torchinsky, A. (1986). *Real-variable methods in harmonic analysis.* Academic Press, Orlando, FL.

Trudinger, N. S. (1967). On Harnack type inequalities and their application to quasilinear elliptic equations. *Communications on Pure and Applied Mathematics* **20**, 721–747.

Tsuji, M. (1959). *Potential theory in modern function theory.* Maruzen, Tokyo.

Tukia, P. (1989). Hausdorff dimension and quasisymmetric mappings. *Mathematica Scandinavica* **65**, 152–160.

Uchiyama, A. (1975). Weight functions of the class (A_∞) and quasiconformal mappings. *Japan Academy. Proceedings. Series A. Mathematical Sciences* **50**, 811–814.

Väisälä, J. (1966). Discrete open mappings on manifolds. *Annales Academiae Scientiarum Fennicae. Series A I. Mathematica* **392**, 1–10.

Väisälä, J. (1971). *Lectures on n-dimensional quasiconformal mappings.* Lecture Notes in Mathematics 229. Springer-Verlag, Berlin.

Väisälä, J. (1975). Capacity and measure. *The Michigan Mathematical Journal* **22**, 1–3.

Väisälä, J. (1980). *A survey of quasiregular mappings in* $\mathbf{R}^n$, In *Proceedings of the international congress of mathematicians, Helsinki 1978.* Academia Scientiarum Fennica, Helsinki.

Vekua, I. N. (1962). *Generalized analytic functions.* Pure and Applied Mathematics 25. Pergamon Press, Oxford.

Vodop'yanov, S. K. (1989). Potential theory on homogeneous groups. *Matematicheskiĭ Sbornik* **180**, 57–77.

Vodop'yanov, S. K. (1990). L_p-theory of potentials for general kernels and applications. *Mathematics Institute, Siberian Branch of the Academy of Sciences of USSR, Preprint* **6**, 1–47.

Vuorinen, M. (1981). Capacity densities and angular limits of quasiregular mappings. *Transactions of the American Mathematical Society* **263**, 343–354.

Vuorinen, M. (1988). *Conformal geometry and quasiregular mappings.* Lecture Notes in Mathematics 1319. Springer-Verlag, Berlin.

Vuorinen, M. (ed.) (1992). *Quasiconformal space mappings, A collection of surveys 1960–90.* Lecture Notes in Mathematics 1508. Springer-Verlag, Berlin.

Wallin, H. (1963). Continuous functions and potential theory. *Arkiv för Matematik* **5**, 55–84.

Wermer, J. (1974). *Potential theory.* Lecture Notes in Mathematics 408. Springer-Verlag, Berlin.

Wiener, N. (1924a). Certain notions in potential theory. *Journal of Mathematics and Physics* **3**, 24–51. Reprinted in *Norbert Wiener: Collected works.* Vol. 1 (1976), pp. 364–391, MIT Press.

Wiener, N. (1924b). The Dirichlet problem. *Journal of Mathematics and Physics* **3**, 127–147. Reprinted in *Norbert Wiener: Collected works.* Vol. 1 (1976), pp. 394–413, MIT Press.

Wu, Jang-Mei (1992). Null sets for doubling and dyadic doubling measures. *Annales Academiae Scientiarum Fennicae. Series A I. Mathematica.* (In press.).

Yosida, K. (1980). *Functional analysis* (6th edn). Springer-Verlag, Berlin.

Ziemer, W. P. (1983). Mean values of subsolutions of elliptic and parabolic equations. *Transactions of the American Mathematical Society* **279**, 555–568.

Ziemer, W. P. (1989). *Weakly differentiable functions: Sobolev spaces and functions of bounded variation.* Graduate Texts in Mathematics 120. Springer-Verlag, New York.

Zorich, V. A. (1967). The theorem of M. A. Lavrent'ev on quasiconformal mappings in space. *Matematicheskiĭ Sbornik* **74**, 417–433.

List of symbols

We list the symbols defined in the text in the order of their appearance, see also pp. 5–6.

Index

19. The John–Nirenberg lemma

Here we present an argument for a slightly modified version of the John–Nirenberg lemma 18.3 that serves the theory as presented in this book. More specifically, we require that the $BMO(\Omega; \mu)$ condition (19.4) and the exponential measure decay estimate (19.6) hold only for balls B such that $2B \subset \Omega$. The John–Nirenberg lemma is used in this book only in the proof of Theorem 3.51 on p. 71, where the hypotheses allow for this change in the formulation. The required modification to the hypotheses of the John–Nirenberg lemma 3.46, p. 69, were indicated in the preface to the Dover edition, p. II.

Our discussion of the John–Nirenberg lemma here is self-contained. In particular, no reference to Chapter 18 is necessary; when convenient, we repeat the arguments given there. The proof is standard except for some technical details.

We note that various abstract versions of the John–Nirenberg lemma have appeared recently in [18], [64], [69].

We assume that μ is a *doubling measure* on $\mathbf{R}^n$; that is, μ is a nontrivial Radon measure on $\mathbf{R}^n$ such that

$$(19.2) \qquad \mu(2B) \le \gamma_\mu \, \mu(B)$$

for all balls $B \subset \mathbf{R}^n$ and for some constant $\gamma_\mu > 1$ independent of the ball. In particular, we need not assume that $d\mu = w \, dx$ for some locally integrable nonnegative function w. In this chapter, the phrase *almost every*, or *a.e.*, refers to the measure μ. By a *ball* we mean an open ball. The Lebesgue differentiation theorem for doubling measures gives that

$$(19.3) \qquad \lim_{r \to 0} \fint_{B(x,r)} f(y) \, d\mu(y) = f(x)$$

for a.e. $x \in \mathbf{R}^n$, whenever $f \in L^1_{loc}(\mathbf{R}^n; \mu)$. (See, for example, [81, p. 13], [40, p. 4].) We use the familiar notation

$$f_E = \fint_E f \, d\mu = \fint_E f(y) \, d\mu(y) = \frac{1}{\mu(E)} \int\limits_E f(y) \, d\mu(y)$$

for any measurable set E with positive and finite measure.

A function $f \in L^1_{loc}(\Omega; \mu)$ is said to be in $BMO(\Omega; \mu)$ if there is a constant $C > 0$ such that

$$(19.4) \qquad \fint_B |f - f_B| \, d\mu \leq C$$

for every ball B with $2B \subset \Omega$. The least C for which (19.4) holds is denoted by $\|f\|_*$ and called the BMO-norm of f. The letters BMO stand for *bounded mean oscillation*.

19.5.   John–Nirenberg lemma for doubling measures. *A function f is in $BMO(\Omega; \mu)$ if and only if for every ball B such that $2B \subset \Omega$ and for every $t > 0$ it holds that*

$$(19.6) \qquad \mu(\{x \in B : |f(x) - f_B| > t\}) \leq c_1 e^{-c_2 t} \mu(B) \, .$$

The positive constants c_1, c_2 and the BMO norm $\|f\|_*$ depend only on each other, the dimension n, and γ_μ.

The proof will show the following: if f is in $BMO(\Omega; \mu)$, then (19.6) holds for $c_1 = \sqrt{2}$ and $c_2 = \log 2/(8\gamma_\mu^{11}\|f\|_*)$, while if (19.6) holds, then $\|f\|_* \leq 2(c_1 + 1)/c_2$.

19.7.   Calderón–Zygmund type decomposition for doubling measures. *Let B be a ball in $\mathbf{R}^n$ and let f be a nonnegative function in $L^1(B; \mu)$. Assume that*

$$\fint_B f \, d\mu \leq s \, .$$

Then there is a countable collection of disjoint balls $\{B_i\}$ with centers in B such that $10B_i \subset 2B$, that

$$(19.8) \qquad s < \fint_{B_i \cap B} f \, d\mu \leq \gamma_\mu^6 s \, ,$$

and that

$$(19.9) \qquad f(x) \leq s$$

for a.e. x in $B \setminus \bigcup_i 5B_i$.

PROOF: Let

$$A = \{x \in B : s < \lim_{r \to 0} \fint_{B(x,r)} f(y)\, d\mu(y) = f(x)\}.$$

Let $x \in A$ and consider the balls $B(x, 2^{-j}R)$ for $j = 4, 5, \ldots$, where R is the radius of B. Observe that $B(x, 10 \cdot 2^{-j}R) \subset 2B$. Let $k_x \geq 4$ be the smallest positive integer such that

$$s < \fint_{B(x, 2^{-k_x}R) \cap B} f\, d\mu.$$

From the doubling property of μ, and from elementary geometry, we obtain that

$$\mu(B(x, 2^{-j+1}R) \cap B) \leq \gamma_\mu^3\, \mu(B(x, 2^{-j}R) \cap B)$$

for every $j \geq 1$. Thus

$$s < \fint_{B(x, 2^{-k_x}R) \cap B} f\, d\mu \leq \gamma_\mu^3 \fint_{B(x, 2^{-k_x+1}R) \cap B} f\, d\mu \leq \gamma_\mu^3 s$$

if $k_x \geq 5$. If $k_x = 4$, we similarly have that

$$s < \fint_{B(x, 2^{-k_x}R) \cap B} f\, d\mu = \fint_{B(x, R/16) \cap B} f\, d\mu \leq \gamma_\mu^6 \fint_B f\, d\mu \leq \gamma_\mu^6 s.$$

By the covering theorem 2.28, we can pick from the collection $\{B(x, 2^{-k_x}R)\}$ a countable pairwise disjoint subcollection $\{B_i\}$ such that

$$A \subset \bigcup_i 5B_i.$$

These balls B_i have the required properties. □

PROOF OF THEOREM 19.5: The if part follows immediately from the formula

$$\int_B \varphi \circ g(x)\, d\mu(x) = \int_0^\infty \varphi'(t)\mu(\{x \in B : g(x) > t\})\, dt,$$

which is valid for every nonnegative and continuously differentiable φ with $\varphi(0) = 0$ and for every measurable nonnegative g. Indeed, we choose $\varphi(t) = e^{c_2 t/2} - 1$ and find that

$$\int_B e^{\frac{c_2}{2}|f-f_B|} d\mu - \mu(B)$$

$$= \frac{c_2}{2} \int_0^\infty e^{c_2 t/2} \mu(\{x \in B : |f(x) - f_B| > t\}) dt$$

$$\leq \frac{c_1 c_2}{2} \mu(B) \int_0^\infty e^{-c_2 t/2} dt = c_1 \mu(B),$$

which implies

$$\fint_B |f - f_B| d\mu \leq \frac{2}{c_2} \fint_B e^{\frac{c_2}{2}|f-f_B|} d\mu \leq \frac{2(c_1 + 1)}{c_2} < \infty.$$

The only if part requires a repeated application of the Calderón–Zygmund type decomposition as in 19.7. Suppose that $f \in BMO(\Omega; \mu)$ and fix a ball B with $2B \subset \Omega$. Replacing f by $(f - f_B)/\|f\|_*$ we may assume that $f_B = 0$ and $\|f\|_* = 1$.

We claim that for each $k = 1, 2, \ldots$ there is a countable collection of balls $\{B_{j_1 \ldots j_k}\}$ satisfying the following properties (with the convention that $B_{j_1 \ldots j_0} = 5B_{j_1 \ldots j_0} = B$):

(i) each $B_{j_1 \ldots j_{k-1} j_k}$ is centered at a point in $5B_{j_1 \ldots j_{k-1}}$;

(ii) $10B_{j_1 \ldots j_k} \subset 2B$;

(iii) $\sum_{j_1 \ldots j_k} \mu(5B_{j_1 \ldots j_k}) \leq (\frac{1}{2})^k \mu(B)$;

(iv) $|f(x)| \leq k\lambda$ for a.e. $x \in 5B_{j_1 \ldots j_{k-1}} \setminus \bigcup_{j_k} 5B_{j_1 \ldots j_{k-1} j_k}$, where $\lambda = 4\gamma_\mu^{11}$.

In the proof of this claim, we will repeatedly use the following fact: if B' is a ball that is centered in a ball B'' such that $\operatorname{diam} B' \leq \operatorname{diam} B''$, then

$$\mu(B') \leq \gamma_\mu^2 \mu(B' \cap B'').$$

To prove the case $k = 1$, we first observe that

$$\fint_B |f| d\mu \leq \|f\|_* = 1 < 2\gamma_\mu^5 = s.$$

The Calderón–Zygmund type decomposition in 19.7 implies that there is a collection of disjoint balls $\{B_{j_1}\}$ with centers in B such that $10B_{j_1} \subset 2B$, that

$$s < \fint_{B_{j_1} \cap B} |f| d\mu \leq \gamma_\mu^6 s,$$

and that

$$|f| \leq s \qquad \text{a.e. in } B \setminus \bigcup_{j_1} 5B_{j_1}.$$

In particular, (i), (ii), and (iv) hold for $k = 1$. We also have that

$$\sum_{j_1} \mu(5B_{j_1}) \leq \gamma_\mu^3 \sum_{j_1} \mu(B_{j_1}) \leq \gamma_\mu^5 \sum_{j_1} \mu(B_{j_1} \cap B)$$

$$< \frac{\gamma_\mu^5}{s} \sum_{j_1} \int_{B_{j_1} \cap B} |f| \, d\mu \leq \frac{1}{2} \int_B |f| \, d\mu \leq \frac{1}{2} \mu(B),$$

so that (iii) holds as well for $k = 1$.

Suppose now that we have proved the claim for some $k \geq 1$. Fix a ball $B'_{j_1 \ldots j_k} = 5B_{j_1 \ldots j_k}$. Because $2B'_{j_1 \ldots j_k} \subset 2B \subset \Omega$, we have that

$$\fint_{B'_{j_1 \ldots j_k}} |f - f_{B'_{j_1 \ldots j_k}}| \, d\mu \leq \|f\|_* = 1 < 2\gamma_\mu^5 = s,$$

and again the Calderón–Zygmund type decomposition in 19.7 provides us with a disjoint collection of balls $\{B_{j_1 \ldots j_k j_{k+1}}\}$ with centers in $B'_{j_1 \ldots j_k}$ such that $10B_{j_1 \ldots j_k j_{k+1}} \subset 2B'_{j_1 \ldots j_k}$, that

$$s < \fint_{B_{j_1 \ldots j_k j_{k+1}} \cap B'_{j_1 \ldots j_k}} |f - f_{B'_{j_1 \ldots j_k}}| \, d\mu \leq \gamma_\mu^6 s,$$

and that

$$|f - f_{B'_{j_1 \ldots j_k}}| \leq s \qquad \text{a.e. in } B'_{j_1 \ldots j_k} \setminus \bigcup_{j_{k+1}} 5B_{j_1 \ldots j_k j_{k+1}}.$$

Hence

$$\sum_{j_{k+1}} \mu(5B_{j_1 \ldots j_k j_{k+1}}) \leq \gamma_\mu^3 \sum_{j_{k+1}} \mu(B_{j_1 \ldots j_k j_{k+1}})$$

$$\leq \gamma_\mu^5 \sum_{j_{k+1}} \mu(B_{j_1 \ldots j_k j_{k+1}} \cap B'_{j_1 \ldots j_k})$$

$$< \frac{\gamma_\mu^5}{s} \sum_{j_{k+1}} \int_{B_{j_1 \ldots j_k j_{k+1}} \cap B'_{j_1 \ldots j_k}} |f - f_{B'_{j_1 \ldots j_k}}| \, d\mu$$

$$\leq \frac{1}{2} \int_{B'_{j_1 \ldots j_k}} |f - f_{B'_{j_1 \ldots j_k}}| \, d\mu \leq \frac{1}{2} \mu(B'_{j_1 \ldots j_k}).$$

Therefore (i), (ii), and (iii) hold for $k + 1$, for the collection $\{B_{j_1 \ldots j_{k+1}}\}$. It remains to show that (iv) also holds.

To this end, write $B'_{j_1 \ldots j_m} = 5 B_{j_1 \ldots j_m}$, and let $x \in B'_{j_1 \ldots j_{k-1} j_k}$. Then

$$
\begin{aligned}
|f(x)| \leq\ & |f(x) - f_{B'_{j_1 \ldots j_{k-1} j_k}}| + |f_{B'_{j_1 \ldots j_{k-1} j_k}} - f_{B_{j_1 \ldots j_{k-1} j_k}}| \\
& + |f_{B_{j_1 \ldots j_{k-1} j_k}} - f_{B_{j_1 \ldots j_{k-1} j_k} \cap B'_{j_1 \ldots j_{k-1}}}| + |f_{B_{j_1 \ldots j_{k-1} j_k} \cap B'_{j_1 \ldots j_{k-1}}} - f_{B'_{j_1 \ldots j_{k-1}}}| \\
& + |f_{B'_{j_1 \ldots j_{k-1}}} - f_{B_{j_1 \ldots j_{k-1}}}| \\
& \cdots \\
& + |f_{B_{j_1 \ldots j_{l-1} j_l}} - f_{B_{j_1 \ldots j_{l-1} j_l} \cap B'_{j_1 \ldots j_{l-1}}}| + |f_{B_{j_1 \ldots j_{l-1} j_l} \cap B'_{j_1 \ldots j_{l-1}}} - f_{B'_{j_1 \ldots j_{l-1}}}| \\
& + |f_{B'_{j_1 \ldots j_{l-1}}} - f_{B_{j_1 \ldots j_{l-1}}}| \\
& \cdots \\
& + |f_{B_{j_1}} - f_{B_{j_1} \cap B}| + |f_{B_{j_1} \cap B}| .
\end{aligned}
$$

We have bounds

$$
\begin{aligned}
|f_{B_{j_1 \ldots j_{l-1} j_l}} - f_{B_{j_1 \ldots j_{l-1} j_l} \cap B'_{j_1 \ldots j_{l-1}}}| & \leq \fint_{B_{j_1 \ldots j_{l-1} j_l} \cap B'_{j_1 \ldots j_{l-1}}} |f - f_{B_{j_1 \ldots j_{l-1} j_l}}| \, d\mu \\
& \leq \gamma_\mu^2 \fint_{B_{j_1 \ldots j_{l-1} j_l}} |f - f_{B_{j_1 \ldots j_{l-1} j_l}}| \, d\mu \leq \gamma_\mu^2 ,
\end{aligned}
$$

$$
|f_{B_{j_1 \ldots j_{l-1} j_l} \cap B'_{j_1 \ldots j_{l-1}}} - f_{B'_{j_1 \ldots j_{l-1}}}| \leq \fint_{B_{j_1 \ldots j_{l-1} j_l} \cap B'_{j_1 \ldots j_{l-1}}} |f - f_{B'_{j_1 \ldots j_{l-1}}}| \, d\mu \leq \gamma_\mu^6 s ,
$$

and

$$
|f_{B'_{j_1 \ldots j_l}} - f_{B_{j_1 \ldots j_l}}| \leq \gamma_\mu^3 \fint_{B'_{j_1 \ldots j_l}} |f - f_{B'_{j_1 \ldots j_l}}| \, d\mu \leq \gamma_\mu^3
$$

for every $l = 1, \ldots, k$. It follows that

$$
|f(x)| \leq s + k\gamma_\mu^2 + k\gamma_\mu^6 s + k\gamma_\mu^3 \leq (k+1)\lambda
$$

for a.e.

$$
x \in B'_{j_1 \ldots j_k} \setminus \bigcup_{j_{k+1}} 5 B_{j_1 \ldots j_k j_{k+1}} .
$$

This proves the induction step, and the claim follows.

Now if $t > \lambda$, let $k \geq 1$ be the integer for which

$$
k\lambda < t \leq (k+1)\lambda .
$$

Then almost every point of $\{x \in B : |f(x)| > t\}$ belongs to the union of the balls $5B_{j_1\ldots j_k}$, so that

$$\mu(\{x \in B : |f(x)| > t\}) \leq (\frac{1}{2})^k \mu(B) \leq e^{-c_2 t} \mu(B),$$

where $c_2 = \log 2/2\lambda$. If $0 < t \leq \lambda$, then

$$\mu(\{x \in B : |f(x)| > t\}) \leq \mu(B) \leq c_1 e^{-c_2 t} \mu(B)$$

where $c_1 = e^{c_2 \lambda} = \sqrt{2}$. This proves the theorem. □

The proof of the John–Nirenberg lemma also gives the following (see the comments after the statement of 19.5):

19.10. Corollary. *A function f is in $BMO(\Omega; \mu)$ if and only if there are positive constants c and C such that*

(19.11) $$\fint_B e^{c|f - f_B|} \, d\mu \leq C$$

for all balls B with $2B \subset \Omega$. If (19.11) holds, then $\|f\|_* \leq C/c$. Conversely, if f is in $BMO(\Omega; \mu)$, then (19.11) holds with $C = \sqrt{2} + 1$ and $c = c_2/2 = \log 2/(16\gamma_\mu^{11} \|f\|_*)$.

20. Admissible weights

In this chapter we discuss some recent developments related to p-admissible weights. First, the four conditions **I** – **IV** on pp. 7–8 describing the admissibility have been reduced to two; in fact, conditions **I** and **IV** suffice. Second, many new examples of admissible weights have been found.

Throughout this chapter, we let w be a weight in $\mathbf{R}^n$, $n \geq 2$, and let $d\mu = w\,dx$ be the associated Radon measure as in (1.2). We also assume that $0 < w < \infty$ almost everywhere. Thus the term "almost everywhere" can unambiguously be used; it may refer either to μ or to Lebesgue measure.

All Lipschitz functions are assumed to be real-valued, and all balls are assumed to be open. A function $u : E \to \mathbf{R}$, $E \subset \mathbf{R}^n$, is L-Lipschitz if $|u(x) - u(y)| \leq L|x - y|$ for every $x, y \in E$. A function $u : E \to \mathbf{R}$ is *locally Lipschitz* in E if for each compact subset C of E there is L such that the restriction of u to C is L-Lipschitz.

20.1. Doubling weights and Poincaré inequality

We recall that μ is doubling if there is a constant $\gamma_\mu > 0$ such that

$$(20.2) \qquad \mu(2B) \leq \gamma_\mu\, \mu(B)$$

whenever B is a ball in $\mathbf{R}^n$.

Next, we say that μ admits a *p-Poincaré inequality*, $1 \leq p < \infty$, if there are constants $C_p > 0$ and $\lambda \geq 1$ such that

$$(20.3) \qquad \fint_B |\varphi - \varphi_B|\, d\mu \leq C_p\, r\, \Big(\fint_{\lambda B} |\nabla\varphi|^p\, d\mu\Big)^{1/p}$$

whenever $B = B(x_0, r)$ is a ball in $\mathbf{R}^n$ and $\varphi : \lambda B \to \mathbf{R}$ is a bounded locally Lipschitz function. Here φ_B denotes the integral average of φ on B and $\nabla\varphi$ denotes the almost everywhere defined gradient of φ (see pp. 8 and 15). In recent literature, condition (20.3) is sometimes called a *weak $(1, p)$-Poincaré inequality*.

In particular, for every weight w that satisfies conditions **I** and **IV**, pp. 7–8, the associated measure $d\mu = w\,dx$ is both doubling and admits a p-Poincaré inequality. To see this, we only need to observe that the requirement that $\varphi \in C^\infty(B)$ in **IV** can be changed to the requirement that φ is a bounded locally Lipschitz function (Lemma 1.11, p. 15, and Lemma 1.15, p. 17).

Note that we allow the value $p = 1$ in (20.3).

If μ satisfies (20.2) and (20.3), we refer to the constants γ_μ, C_p, and λ, along with n and p, as the *data (associated with μ)*.

We now describe the main results in this section.

20.4. Theorem. *Assume that μ is doubling and admits a p-Poincaré inequality for some $1 \le p < \infty$. If D is an open set in $\mathbf{R}^n$ and φ_i a sequence of locally Lipschitz functions in D such that $\int_D |\varphi_i|^p \, d\mu \to 0$ and $\int_D |\nabla\varphi_i - v|^p \, d\mu \to 0$ as $i \to \infty$, where v is a vector-valued measurable function in $L^p(D; \mu; \mathbf{R}^n)$, then $v = 0$.*

20.5. Theorem. *Assume that μ is doubling and admits a p-Poincaré inequality for some $1 \le p < \infty$. There are constants $\varkappa > 1$ and $C > 0$, only depending on the data associated with μ, such that*

$$(20.6) \qquad \left(\fint_B |\varphi|^{\varkappa p} d\mu \right)^{1/\varkappa p} \le C \, r \left(\fint_B |\nabla\varphi|^p \, d\mu \right)^{1/p}$$

whenever $B = B(x_0, r)$ is a ball in $\mathbf{R}^n$ and $\varphi : B \to \mathbf{R}$ is a Lipschitz function with compact support in B.

20.7. Theorem. *Assume that μ is doubling and admits a p-Poincaré inequality for some $1 \le p < \infty$. There are constants $\varkappa > 1$ and $C > 0$, only depending on the data associated with μ, such that*

$$(20.8) \qquad \left(\fint_B |\varphi - \varphi_B|^{\varkappa p} d\mu \right)^{1/\varkappa p} \le C \, r \left(\fint_B |\nabla\varphi|^p \, d\mu \right)^{1/p}$$

whenever $B = B(x_0, r)$ is a ball in $\mathbf{R}^n$ and $\varphi : B \to \mathbf{R}$ is a bounded locally Lipschitz function.

20.9. Corollary. *A weight w in $\mathbf{R}^n$ is p-admissible if and only if the associated measure is doubling and admits a p-Poincaré inequality. In particular, a weight w in $\mathbf{R}^n$ is p-admissible if and only if conditions **I** and **IV** hold. The statement is quantitative in the sense that the constants in the conclusion only depend on the data in the hypotheses.*

Theorem 20.4 is due to Semmes; it first appeared in [41]. We will follow the streamlined later proof of [29]. Theorem 20.5 is credited to Saloff-Coste [72] and Grigor'yan [33] in [37, p. 79], where also further references to numerous related works can be found. Theorem 20.7 was first proved by Jerison [45] in a slightly weaker formulation. Here we follow Hajłasz and Koskela [36] with arguments that can be applied in great generality; see [37].

In preparation for the proofs, we require a lemma on maximal functions of gradients of Lipschitz functions. Recall ((15.18) p. 303) that

$$(20.10) \qquad M_\mu f(x) = \sup \fint_B |f| \, d\mu$$

for $f \in L^1_{loc}(\mathbf{R}^n; \mu)$, where the supremum is taken over all balls B that contain x. If f is a priori defined only on some measurable subset $A \subset \mathbf{R}^n$, and is locally integrable there, the definition in (20.10) applies with the understanding that f is defined to be zero outside A.

20.11. Lemma. *Assume that μ is doubling and admits a p-Poincaré inequality for some $1 \le p < \infty$. There is a constant $C > 0$ depending only on the data such that*

$$(20.12) \qquad |\varphi(x) - \varphi(y)| \le C \, |x - y| \left((M_\mu |\nabla \varphi|^p(x))^{1/p} + (M_\mu |\nabla \varphi|^p(y))^{1/p} \right)$$

whenever $\varphi : B \to \mathbf{R}$ is a locally Lipschitz function in a ball B and $x, y \in (2\lambda)^{-1}B$, where $\lambda \ge 1$ is as in (20.3).

Proof. Let $\varphi : B \to \mathbf{R}$ be a locally Lipschitz function in a ball $B = B(x_0, r)$, and let $x, y \in (2\lambda)^{-1}B$, $x \ne y$. Put $z_0 = (x + y)/2$ and on the line segment $[z_0, x]$ choose consecutive points $z_0, z_1, \ldots$ converging to x such that $|z_k - z_{k+1}| = 2^{-k-1}d$, where $d = |z_0 - x|$. Write $B_k = B(z_k, 2^{-k-1}d)$ for $k \ge 0$. Because $\varphi_{B_k} \to \varphi(x)$ as $k \to \infty$, and because $B_{k+1} \subset 2B_k \subset \lambda^{-1}B$ for every $k \ge 0$, we conclude from the doubling property (20.2) and from the Poincaré inequality (20.3) that

$$|\varphi(x) - \varphi_{B_0}| \le \sum_{k=0}^{\infty} |\varphi_{B_k} - \varphi_{B_{k+1}}| \le \sum_{k=0}^{\infty} |\varphi_{B_k} - \varphi_{2B_k}| + |\varphi_{2B_k} - \varphi_{B_{k+1}}|$$

$$\le c \sum_{k=0}^{\infty} \fint_{2B_k} |\varphi - \varphi_{2B_k}| \, d\mu \le c \sum_{k=0}^{\infty} 2^{-k} |x - y| \left(\fint_{2\lambda B_k} |\nabla \varphi|^p \, d\mu \right)^{1/p}.$$

Noting that $2\lambda B_k \subset B$ and that $x \in 3\lambda B_k$ for every $k \ge 0$, we obtain from the preceding that

$$|\varphi(x) - \varphi_{B_0}| \le c \, |x - y| \, (M_\mu |\nabla \varphi|^p(x))^{1/p}.$$

By symmetry, we also have

$$|\varphi(y) - \varphi_{B_0}| \le c \, |x - y| \, (M_\mu |\nabla \varphi|^p(y))^{1/p}.$$

Inequality (20.12) follows from the preceding two estimates and the triangle inequality, and the lemma is thereby proved. $\qquad \square$

Proof of Theorem 20.4. Let φ_i and v be functions as in the theorem. Fix a ball B such that $2\lambda B \subset D$. It suffices to show that $v = 0$ in B. By passing to a subsequence, we may assume that $\varphi_i \to 0$ and $\nabla\varphi_i \to v$ pointwise almost everywhere in B, and that

$$(20.13) \qquad \int_B |\nabla\varphi_{i+1} - \nabla\varphi_i|^p \, d\mu \leq 4^{-ip}.$$

Write $u_i = \varphi_{i+1} - \varphi_i$. Then Lemma 20.11 gives that

$$|u_i(x) - u_i(y)| \leq c\,|x - y| \left((M_\mu|\nabla u_i|^p(x))^{1/p} + (M_\mu|\nabla u_i|^p(y))^{1/p} \right)$$

for every $x, y \in B$. Hence we obtain that

$$(20.14) \qquad |\varphi_i(x) - \varphi_i(y)| \leq c\,|x - y|\,(v_k(x) + v_k(y))$$

for every $x, y \in B$ and $i \geq k$, where

$$v_k(z) = \sum_{j=k}^\infty (M_\mu|\nabla u_j|^p(z))^{1/p}.$$

Then for $0 < t = \sum_{j=k}^\infty t2^{k-j-1}$ we have

$$\{z \in B : v_k(z) > t\} \subset \bigcup_{j=k}^\infty \{z \in B : M_\mu|\nabla u_j|^p(z) > (t2^{k-j-1})^p\},$$

whence

$$\mu(\{z \in B : v_k(z) > t\}) \leq \sum_{j=k}^\infty \mu(\{z \in B : M_\mu|\nabla u_j|^p(z) > (t2^{k-j-1})^p\})$$
$$\leq c \sum_{j=k}^\infty (t2^{k-j-1})^{-p} \int_B |\nabla u_j|^p \, d\mu$$
$$\leq c\, t^{-p} 4^{-kp},$$

where we used the weak estimate [81, Theorem 1 (b), p. 13] and (20.13). Write $A_t^k = B \setminus \{z \in B : v_k(z) > t\}$. Then the preceding inequality gives that

$$(20.15) \qquad \mu(B \setminus A_t^k) \to 0$$

as $k \to \infty$, while we obtain from (20.14) that the restriction of φ_i to A_t^k is ct-Lipschitz whenever $i \geq k$. By the standard Lipschitz extension lemma [27, 2.10.44], we may extend φ_i to a ct-Lipschitz function $\Phi_i : B \to \mathbf{R}$. In particular, we have that $|\nabla\Phi_i| \leq ct$ almost everywhere in B, and hence that $|\nabla\varphi_i| \leq ct$ almost everywhere in A_t^k whenever $i \geq k$. This implies that $|v| \leq ct$ almost everywhere in A_t^k, and hence by (20.15) that $|v| \leq ct$ almost everywhere in B. Finally, by letting $t \to 0$, we have that $v = 0$ almost everywhere in B. The theorem follows. $\qquad \square$

For the proof of Theorem 20.5, we require the following two lemmas.

20.16.   Lemma. *Assume that μ satisfies (20.2). Let $B(x_0, r) \subset \mathbf{R}^n$ be a ball, $x \in B(x_0, r)$, and $0 < s < 2r$. Then*

$$(20.17) \qquad \left(\frac{s}{r}\right)^\alpha \leq 4^\alpha \frac{\mu(B(x,s))}{\mu(B(x_0,r))},$$

where $\alpha = \log_2 \gamma_\mu > 0$.

Proof. Let $k \geq 1$ be the integer such that $2^{k-1}s < 2r \leq 2^k s$. Then

$$\mu(B(x_0, r)) \leq \mu(B(x, 2^k s)) \leq \gamma_\mu^k \mu(B(x, s)),$$

and (20.17) follows from the choice of k.   □

20.18.   Lemma. *Let (X, ν) be a finite measure space and let $u : X \to \mathbf{R}$ be a measurable function. Assume that there are constants $\gamma > 1$ and $C_0 > 0$ such that*

$$\nu(\{x \in X : |u(x)| > t\}) \leq C_0 t^{-\gamma}$$

for every $t > 0$. Then for each $0 < \beta < \gamma$ we have

$$(20.19) \qquad \int_X |u|^\beta \, d\nu \leq \left(\frac{\gamma}{\gamma - \beta}\right) C_0^{\beta/\gamma} \nu(X)^{(\gamma-\beta)/\gamma}.$$

Proof. We have

$$\int_X |u|^\beta \, d\nu = \beta \int_0^\infty \nu(\{x \in X : |u(x)| > t\}) t^{\beta-1} \, dt$$

$$\leq \beta \int_0^A \nu(X) t^{\beta-1} \, dt + C_0 \beta \int_A^\infty t^{-\gamma+\beta-1} \, dt$$

for $A > 0$. By setting $A = (C_0/\nu(X))^{1/\gamma}$ and computing the integrals, we arrive at (20.19).   □

Proof of Theorem 20.5. Let φ be a Lipschitz function with compact support in a ball $B = B(x_0, r)$. We treat φ as a globally defined Lipschitz function that is zero outside B. Let $\alpha > 0$ be as in (20.17). Fix $0 < \varepsilon < 1$, depending only on p and α, such that $q = p(1 - \varepsilon)/\alpha < 1$. We claim that

$$(20.20) \qquad \mu(\{x \in B : |\varphi(x)| > t\}) \leq c \left(r^p \mu(B)\right)^{-q} \int_B |\nabla\varphi|^p \, d\mu)^{1/(1-q)} t^{-p/(1-q)}$$

for every $t > 0$, where $c > 0$ depends only on the data. Indeed, (20.6) follows from (20.20) and Lemma 20.18 by a direct computation; we can choose any $\varkappa < 1/(1-q)$. It therefore suffices to prove the claim (20.20).

To this end, fix $t > 0$ and fix $x \in B$ such that $|\varphi(x)| > t$. We slightly modify the argument in the proof of Lemma 20.11. Set $y_0 = x_0 - 3r(x - x_0)/|x - x_0|$, and on the line segment $[y_0, x]$ choose consecutive points $y_0, y_1, \ldots$ converging to x such that $|y_k - y_{k+1}| = 2^{-k-1}d$, where $d = |y_0 - x|$. Then put $B_k = B(y_k, 2^{-k-1}d)$ for $k \geq 0$. Note that $\varphi = 0$ in B_0. As in the proof of Lemma 20.11, we obtain that

$$(20.21) \qquad t < |\varphi(x)| = |\varphi(x) - \varphi_{B_0}| \leq c \sum_{k=0}^{\infty} d\, 2^{-k} \left(\fint_{2\lambda B_k} |\nabla\varphi|^p \, d\mu \right)^{1/p}.$$

Because

$$t = c(\varepsilon) \sum_{k=0}^{\infty} t\, 2^{-k\varepsilon},$$

we find from (20.21) that there must be a ball $B_{k_x} = B(y_{k_x}, 2^{-k_x-1}d)$ such that

$$t\, 2^{-k_x\varepsilon} \leq c\, d\, 2^{-k_x} \left(\fint_{2\lambda B_{k_x}} |\nabla\varphi|^p \, d\mu \right)^{1/p}.$$

The doubling condition (20.2), inequality (20.17), and the choice for q then give

$$t \leq c\, d^\varepsilon (d\, 2^{-k_x})^{1-\varepsilon} \mu(B_{k_x})^{-1/p} \left(\int_{2\lambda B_{k_x}} |\nabla\varphi|^p \, d\mu \right)^{1/p}$$

$$\leq c\, d^\varepsilon r^{\alpha q/p} \mu(B_{k_x})^{(q-1)/p} \mu(B)^{-q/p} \left(\int_{2\lambda B_{k_x}} |\nabla\varphi|^p \, d\mu \right)^{1/p},$$

whence

$$(20.22) \qquad \mu(B_{k_x}) \leq c \left(r^p \mu(B)\right)^{-q} \int_{2\lambda B_{k_x}} |\nabla\varphi|^p \, d\mu \right)^{1/(1-q)} t^{-p/(1-q)}.$$

Note that $x \in 3\lambda B_{k_x}$. It follows from the covering theorem 2.28, p. 43, that there exists a countable pairwise disjoint collection of balls $B_i = 2\lambda B_{k_{x_i}}$ of the type in (20.22) such that

$$\{x \in B : |\varphi(x)| > t\} \subset \bigcup_i 10B_i \,.$$

In particular,

$$
\begin{aligned}
\mu(\{x \in B : |\varphi(x)| > t\}) &\leq \sum_i \mu(10B_i) \leq c \sum_i \mu(B_{k_{x_i}}) \\
&\leq c \sum_i (r^p \mu(B)^{-q} \int_{B_i} |\nabla\varphi|^p \, d\mu)^{1/(1-q)} t^{-p/(1-q)} \\
&\leq c \, (r^p \mu(B)^{-q} \int_B |\nabla\varphi|^p \, d\mu)^{1/(1-q)} t^{-p/(1-q)} \,,
\end{aligned}
$$

where we also used the fact $0 < 1 - q < 1$. Thus (20.20) follows, and by what was said earlier, the proof is complete.   □

Before the proof of Theorem 20.7, we require the following lemma, whose proof is left to the reader.

20.23.   Lemma. *Let $B = B(x_0, r)$ be a ball in $\mathbf{R}^n$, $\lambda \geq 1$, and $x \in B$. Then there exists a sequence of balls $B_0 = (2\lambda)^{-1}B, B_1, B_2, \ldots$ with following properties for every $k \geq 0$: (i) $B_k \cap B_{k+1} \neq \emptyset$, (ii) $B_{k+1} \subset 2B_k$, (iii) $2\lambda B_k \subset B$, (iv) $x \in 4\lambda B_k$, (v) $b^{-1}2^{-k}r \leq \operatorname{diam} B_k \leq b \, 2^{-k}r$ for some constant $b > 1$ depending only on λ, and (vi) $\operatorname{dist}(B_k, x) \to 0$ as $k \to \infty$.*

Proof of Theorem 20.7. The proof here is similar to the proof of Theorem 20.5. Indeed, fix a bounded locally Lipschitz function $\varphi : B \to \mathbf{R}$, where $B = B(x_0, r)$. Let $B_0 = (2\lambda)^{-1}B$. Because

$$\left(\fint_B |\varphi - \varphi_B|^{\varkappa p} \, d\mu\right)^{1/\varkappa p} \leq c \left(\fint_B |\varphi - \varphi_{B_0}|^{\varkappa p} \, d\mu\right)^{1/\varkappa p},$$

it suffices to show that

$$(20.24) \qquad \left(\fint_B |\varphi|^{\varkappa p} d\mu\right)^{1/\varkappa p} \leq C r \left(\fint_B |\nabla\varphi|^p \, d\mu\right)^{1/p}$$

under the assumption that $\varphi_{B_0} = 0$.

The preceding understood, fix $t > 0$. Then fix $x \in B$ such that $|\varphi(x)| > t$. Let $B_0 = (2\lambda)^{-1}B, B_1, B_2, \ldots$ be a sequence of balls associated with x as in Lemma 20.23. As in the proof of Theorem 20.5, we obtain

$$t < |\varphi(x)| = |\varphi(x) - \varphi_{B_0}| \leq \sum_{k=0}^{\infty} |\varphi_{B_k} - \varphi_{2B_k}| + |\varphi_{2B_k} - \varphi_{B_{k+1}}|$$

$$\leq c \sum_{k=0}^{\infty} r \, 2^{-k} \Big(\fint_{2\lambda B_k} |\nabla\varphi|^p \, d\mu \Big)^{1/p}.$$

Using the this inequality, a covering argument, and Lemma 20.18, exactly as in the proof of Theorem 20.5, we arrive at (20.24). Note again that we can pick any $\varkappa < 1/(1 - q)$, where $q = p(1 - \varepsilon)/\alpha$ as in the proof of Theorem 20.5. The theorem follows.   □

20.2. Strong A_∞-weights as admissible weights

The main new source of examples of p-admissible weights is the class of strong A_∞-weights. We next describe this class. (See Chapter 15 for the basic theory of A_∞-weights.)

Let w be a weight in $\mathbf{R}^n$, $n \geq 2$, and assume that the measure $d\mu = w \, dx$ is doubling. Associated with w is a quasi-distance $d_w : \mathbf{R}^n \times \mathbf{R}^n \to [0, \infty)$ defined by

$$(20.25) \qquad d_w(x, y) = \mu\big(B_{x,y}\big)^{1/n},$$

where $B_{x,y} = B(\frac{x+y}{2}, \frac{|x-y|}{2})$ if $x \neq y$ and $B_{x,y} = \emptyset$ if $x = y$. Thus, $d_w(x, y) = d_w(y, x)$, $d_w(x, x) = 0$, and

$$d_w(x, y) \leq c \big(d_w(x, z) + d_w(z, y) \big)$$

for all $x, y, z \in \mathbf{R}^n$ and for some $c \geq 1$ depending only on the doubling constant of μ and on n.

We call w a *strong A_∞-weight* if the associated quasi-distance d_w is equivalent to a distance, i.e., if there is a metric ρ in $\mathbf{R}^n$ such that

$$(20.26) \qquad C^{-1}\rho(x, y) \leq d_w(x, y) \leq C \, \rho(x, y)$$

for every $x, y \in \mathbf{R}^n$, where $C \geq 1$ is a constant that is independent of x and y. Alternatively, a weight w is a strong A_∞-weight if there exists a constant $C \geq 1$ such that

$$(20.27) \qquad d_w(x, y) \leq C \inf \sum_{i=1}^{k} d_w(x_i, x_{i+1})$$

for every $x, y \in \mathbf{R}^n$, where the infimum is taken over all finite sequences of points such that $x_1 = x$ and $x_{k+1} = y$.

Every A_1-weight is a strong A_∞-weight. On the other hand, the weight $w(x_1, x_2) = |x_1|$ in $\mathbf{R}^2$ is an A_p-weight for every $p > 1$, but not a strong A_∞-weight.

Strong A_∞-weights were introduced by David and Semmes [21], [74]. The initial motivation was to define a class of weights that would potentially characterize Jacobians of quasiconformal mappings (up to comparability).

If $f : \mathbf{R}^n \to \mathbf{R}^n$ is a quasiconformal mapping, then the Jacobian determinant J_f (see Section 15.31) is a strong A_∞-weight. In fact, one can set $\rho(x, y) = |f(x) - f(y)|$ in (20.26). This follows from basic distortion and change of variables properties of quasiconformal mappings. However, there are strong A_∞ weights w in $\mathbf{R}^n$ for each $n \geq 2$ that are not comparable to quasiconformal Jacobians, i.e., we have that

$$(20.28) \qquad \max \left\{ \operatorname{ess\,sup} \frac{J_f(x)}{w(x)}, \ \operatorname{ess\,sup} \frac{w(x)}{J_f(x)} \right\} = \infty$$

for every quasiconformal mapping $f : \mathbf{R}^n \to \mathbf{R}^n$. All hitherto known examples of strong A_∞-weights w with property (20.28) are somewhat nontrivial [76], [62], [4].

The following theorem generalizes Corollary 15.33.

20.29. Theorem. *Let w be a strong A_∞-weight in $\mathbf{R}^n$. Then the measure $d\mu = w^{1-p/n}dx$ is doubling and admits a p-Poincaré inequality for every $1 \leq p < n$.*

20.30. Corollary. *Let w be a strong A_∞-weight in $\mathbf{R}^n$. Then $w^{1-p/n}$ is p-admissible whenever $1 < p < n$. Moreover, $w^{1-1/n}$ is p-admissible for every $p > 1$.*

Theorem 20.29 was first proved in [41] for $p > 1$, by using results from [21], [28]. A different proof covering also the case $p = 1$ was given by Björn [10]. The article [10] contains further examples of p-admissible weights, obtained by using weighted A_p-conditions (against a strong A_∞-weight).

Using Corollary 20.30 we obtain many nontrivial and concrete examples of p-admissible weights. Geometric examples can be found in [74], [60], [16]. (See also [46], where strong A_∞-weights are not used.) For analytic examples, several subspaces of $BMO(\mathbf{R}^n)$ are known to have the property that exponentials of functions with small enough BMO-norm are strong A_∞-weights. See [15], [14], [20].

21. The Riesz measure of an $\mathcal{A}$-superharmonic function

In this chapter we discuss the Riesz measure associated with an $\mathcal{A}$-superharmonic function. First we study existence questions, and then discuss the relationship between the Riesz measure and capacity. The main result is Theorem 21.21, where estimate (21.22) resembles the classical Riesz decomposition theorem (cf. [43, Theorem 6.18, p. 116] or [25, 1.I.8]). The theory is applied to prove the necessity of the Wiener test for the Dirichlet problem. The results here also provide converse statements to those in Lemma 9.10, Theorem 9.18, and Theorem 12.5. All in all these results give a complete characterization of the boundary regularity for the Dirichlet problem associated with $\mathcal{A}$-harmonic functions.

We follow the discussion in [53], [54]. In Notes to Chapter 9, p. 192, it is stated that the methods in Kilpeläinen and Malý (1992b) do not extend to the weighted situation. This statement was formally correct at the time, for an early version of the paper [54] did not allow for such extension. However, the argument in the final published version of [54] applies so as to cover the weighted equations considered in this book. See [70] for a complete account.

We use the notation and terminology of the main text. In particular, $1 < p < \infty$, and $\mathcal{A}$ is a mapping satisfying (3.3)–(3.7) for some constants $0 < \alpha \leq \beta < \infty$ and for some p-admissible weight w in $\mathbf{R}^n$, $n \geq 2$. Also recall that the notation c_μ stands for constants appearing in connection with conditions **I**, **III**, and **IV** (see p. 9) and that Ω is an open set in $\mathbf{R}^n$.

Here a *Radon measure* in Ω means a locally finite nonnegative Borel regular measure defined on all subsets of Ω [27, 2.2.5].

21.1. Riesz measures

The *weak gradient* Du of an $\mathcal{A}$-superharmonic function u in Ω was defined on p. 150 by

$$Du = \lim_{k \to \infty} \nabla \min(u, k).$$

By Theorem 7.46, we have that $Du \in L^{p-1}_{loc}(\Omega; \mu)$, and hence

$$\mathcal{A}(x, Du) \in L^1_{loc}(\Omega; dx; \mathbf{R}^n)$$

determines a vector-valued distribution. As mentioned on p. 151, if $u \in H^{1,p}_{loc}(\Omega; \mu)$, then $Du = \nabla u$. Conversely, it is easy to see (by using the Poincaré inequality **IV** and Theorem 1.32) that $Du \in L^p_{loc}(\Omega; \mu)$ implies $u \in H^{1,p}_{loc}(\Omega; \mu)$ and $Du = \nabla u$.

21.2. Theorem. *Let u be an $\mathcal{A}$-superharmonic function in Ω. Then there is a unique Radon measure ν in Ω such that*

$$(21.3) \qquad\qquad -\operatorname{div}\mathcal{A}(x, Du) = \nu$$

in the sense of distributions; that is,

$$\int_\Omega \mathcal{A}(x, Du) \cdot \nabla\varphi\, dx = \int_\Omega \varphi\, d\nu$$

for every $\varphi \in C_0^\infty(\Omega)$.

The measure ν as in Theorem 21.2 is called the *Riesz measure*, or the *Riesz mass*, associated with u. Note that Riesz measures were introduced and used earlier in the book in an unweighted context on p. 281.

PROOF: Let $\varphi \in C_0^\infty(\Omega)$ be nonnegative and choose an open set $\Omega' \Subset \Omega$ so that $\operatorname{spt}\varphi \subset \Omega'$. Then for each $k = 1, 2, \dots$ the truncation $\min(u, k)$ is a supersolution of (3.8) in Ω (Corollary 7.20). Moreover, by Theorem 7.46 there is $q > 1$ such that $|Du|^{p-1} \in L^q(\Omega'; \mu)$. Hence the functions $\mathcal{A}(x, \nabla\min(u, k))$ form a bounded sequence in $L^q(\Omega'; \mu; \mathbf{R}^n)$ converging weakly to $\mathcal{A}(x, Du)$ in $L^q(\Omega'; dx; \mathbf{R}^n)$ (cf. the discussion on p. 25). We collect these observations to deduce

$$\int_\Omega \mathcal{A}(x, Du) \cdot \nabla\varphi\, dx = \lim_{k\to\infty} \int_{\Omega'} \mathcal{A}(x, \nabla\min(u, k)) \cdot \nabla\varphi\, dx \geq 0\,.$$

It follows that the assignment

$$\varphi \mapsto \int_\Omega \mathcal{A}(x, Du) \cdot \nabla\varphi\, dx$$

is a nonnegative distribution and hence it can be represented by a measure ν as claimed in the theorem. (See [73, p. 29].) The proof is complete.   □

We have the following converse to Theorem 21.2.

21.4. Theorem. *Let ν be a finite Radon measure in $\mathbf{R}^n$. Then there exists an $\mathcal{A}$-superharmonic function u in $\mathbf{R}^n$ such that $-\operatorname{div}\mathcal{A}(x, Du) = \nu$.*

If, moreover, ν is a finite Radon measure in a bounded open set Ω, then there exists an $\mathcal{A}$-superharmonic function u in Ω such that $-\operatorname{div}\mathcal{A}(x, Du) = \nu$ and that $\min(u, k) \in H_0^{1,p}(\Omega; \mu)$ for every $k \geq 1$.

For a proof of Theorem 21.4, see [53, Theorem 2.4], [49, Theorem 2.10]. For related existence results, see also [3], [13], [23], [24]. Theorem 21.4 fails to hold in general for Radon measures with infinite total mass. The uniqueness of the $\mathcal{A}$-superharmonic solution is also not known in general. We will not discuss these issues further here, but refer to [50].

Radon measures that give rise to $\mathcal{A}$-superharmonic functions in the Sobolev space $H_0^{1,p}(\Omega; \mu)$, with Ω bounded, can be characterized as measures in the dual of the Dirichlet space $L_0^{1,p}(\Omega; \mu)$, defined on p. 13. To say that a Radon measure ν in Ω belongs to the dual of $L_0^{1,p}(\Omega; \mu)$ means, by definition, that the action of ν on the dense subspace $C_0^\infty(\Omega)$ of $L_0^{1,p}(\Omega; \mu)$ is given by

$$(21.5) \qquad \langle \nu, \varphi \rangle = \int_\Omega \varphi \, d\nu \, .$$

In other words, a Radon measure ν in Ω belongs to the dual of $L_0^{1,p}(\Omega; \mu)$ if and only if there exists a constant $C > 0$ such that

$$\left| \int_\Omega \varphi \, d\nu \right| \leq C \left(\int_\Omega |\nabla \varphi|^p \, d\mu \right)^{1/p}$$

whenever $\varphi \in C_0^\infty(\Omega)$. Later we will see that the integral action (21.5) extends as such to all (quasicontinuous) functions in $L_0^{1,p}(\Omega; \mu)$.

We have a continuous embedding $H_0^{1,p}(\Omega; \mu) \to L_0^{1,p}(\Omega; \mu)$, so that every element in the dual of $L_0^{1,p}(\Omega; \mu)$ naturally defines an element in the dual of $H_0^{1,p}(\Omega; \mu)$. If Ω is bounded, then the Sobolev space $H_0^{1,p}(\Omega; \mu)$ is isomorphic to the space $L_0^{1,p}(\Omega; \mu)$ modulo functions that are constant in a component of Ω by the Sobolev inequality **III**. Indeed,

$$\|\nabla \varphi\|_p \leq \|\varphi\|_{1,p} \leq (C_{\mathbf{III}} \operatorname{diam} \Omega + 1) \|\nabla \varphi\|_p$$

for $\varphi \in C_0^\infty(\Omega)$. In particular, if Ω is bounded, then a measure ν belongs to the dual of $L_0^{1,p}(\Omega; \mu)$ as defined above if and only if it belongs to the dual of $H_0^{1,p}(\Omega; \mu)$ by using the action (21.5) in the dense subspace.

21.6. Theorem. *Let u be an $\mathcal{A}$-superharmonic function in Ω such that $Du \in L^p(\Omega; \mu)$. Then the Riesz mass ν of u belongs to the dual of $L_0^{1,p}(\Omega; \mu)$.*

Conversely, if Ω is bounded and ν is a Radon measure in the dual of $L_0^{1,p}(\Omega; \mu)$, then there is a unique $\mathcal{A}$-superharmonic function $u \in H_0^{1,p}(\Omega; \mu)$ with Riesz mass ν.

Note that there is no finiteness assumption for the measure in Theorem 21.6 (cf. Theorem 21.4).

PROOF: Assume u is $\mathcal{A}$-superharmonic with ν its Riesz mass, and assume that $Du \in L^p(\Omega; \mu)$. Then $\nabla u = Du$ by the remarks made before Theorem 21.2, so that

$$(21.7) \quad \int_\Omega \varphi \, d\nu = \int_\Omega \mathcal{A}(x, \nabla u) \cdot \nabla \varphi \, dx \le \beta \left(\int_\Omega |\nabla u|^p \, d\mu \right)^{(p-1)/p} \left(\int_\Omega |\nabla \varphi|^p \, d\mu \right)^{1/p}$$

whenever $\varphi \in C_0^\infty(\Omega)$. Hence ν belongs to the dual of $L_0^{1,p}(\Omega; \mu)$.

Conversely, let Ω be bounded and let ν be a measure in the dual of $L_0^{1,p}(\Omega; \mu)$. The existence of an $\mathcal{A}$-superharmonic function u with Riesz mass ν follows by applying Proposition 17.2, Appendix I, p. 333. Indeed, choose $\mathbf{K} = X = H_0^{1,p}(\Omega; \mu)$ and let $\mathfrak{A}_\nu : X \to X'$ be defined by

$$\langle \mathfrak{A}_\nu v, \varphi \rangle = \int_\Omega \mathcal{A}(x, \nabla v) \cdot \nabla \varphi \, dx - \langle \nu, \varphi \rangle.$$

Moreover, since $\mathfrak{A}_\nu$ is of the form $\mathfrak{A}_\nu = \mathfrak{A} - \nu$, where $\mathfrak{A}$ is defined on p. 333, we have that the hypotheses of Proposition 17.2 are all satisfied. It follows that there is an element $u \in H_0^{1,p}(\Omega; \mu)$ such that

$$\int_\Omega \mathcal{A}(x, \nabla u) \cdot \nabla \varphi \, dx = \int_\Omega \varphi \, d\nu$$

for every $\varphi \in C_0^\infty(\Omega)$. In particular, u is an $\mathcal{A}$-supersolution and we can take its $\mathcal{A}$-superharmonic representative. The uniqueness follows as in Lemma 3.22.   □

We will not address the existence of an $\mathcal{A}$-superharmonic solution in $L_0^{1,p}(\Omega; \mu)$ if Ω is not bounded. The issue is more complicated than in Theorem 21.6. For example, in many cases there are no nontrivial $\mathcal{A}$-superharmonic functions in $L_0^{1,p}(\mathbf{R}^n; \mu)$ (cf. Section 9.21).

We say that a Radon measure ν in Ω is *absolutely continuous with respect to the* (p, μ)-*capacity* if $\mathrm{cap}_{p,\mu} E = 0$ implies $\nu(E) = 0$ for every $E \subset \Omega$.

21.8.   Proposition. *The Riesz measure ν of an $\mathcal{A}$-superharmonic function $u \in H_{loc}^{1,p}(\Omega; \mu)$ is absolutely continuous with respect to the (p, μ)-capacity.*

In particular, every Radon measure in the dual of $L_0^{1,p}(\Omega; \mu)$ is absolutely continuous with respect to the (p, μ)-capacity.

PROOF: Let $E \subset \Omega$ be of (p, μ)-capacity zero. By Section 2.7, we may assume that E is a Borel set. Therefore, because ν is a Radon measure, it suffices to show that $\nu(K) = 0$ for every compact set $K \subset E$. Fix a compact set $K \subset E$ and fix an open set $\Omega' \Subset \Omega$ containing K. For every function $\varphi \in C_0^\infty(\Omega')$ satisfying $\varphi \geq 0$ and $\varphi = 1$ on K, we have that

$$
\nu(K) \leq \int_{\Omega'} \varphi \, d\nu = \int_{\Omega'} \mathcal{A}(x, \nabla u) \cdot \nabla \varphi \, dx
$$
$$
\leq \beta \Big(\int_{\Omega'} |\nabla u|^p \, d\mu \Big)^{(p-1)/p} \Big(\int_{\Omega'} |\nabla \varphi|^p \, d\mu \Big)^{1/p}.
$$

Since $\mathrm{cap}_{p,\mu}(K, \Omega') = 0$, taking the infimum over all functions φ as above yields $\nu(K) = 0$.

To prove the second assertion, we simply observe that every member in the dual of $L_0^{1,p}(\Omega; \mu)$ lies naturally in the dual of $L_0^{1,p}(\Omega'; \mu)$ for every $\Omega' \Subset \Omega$. The claim then follows from the first part of the theorem, from the subadditivity of capacity, and from Theorem 21.6.   □

Next, recall the definition for quasicontinuous functions from Section 4.1. The uniqueness, up to a set of (p, μ)-capacity zero, of the (p, μ)-quasicontinuous representative of a Sobolev function was established in Theorem 4.14. There is an obvious local version of this result, which is not explicitly recorded in the main text. We next give a simple proof of the uniqueness found in [48]. We consider only a special case of the general statement in [48] that is relevant in the present context.

21.9.   Theorem. *Let u and v be two (p, μ)-quasicontinuous functions in Ω such that $u = v$ almost everywhere. Then $u = v$ (p, μ)-quasieverywhere.*

PROOF: We have to show that $C_{p,\mu}(N) = 0$, where $N = \{x \in \Omega : u(x) \neq v(x)\}$. Fix $\varepsilon > 0$. Let $G \subset \Omega$ be an open set such that $C_{p,\mu}(G) < \varepsilon$ and such that the restrictions of both u and v to $\Omega \setminus G$ are continuous. It follows that we can find an open set $U \subset \Omega$ such that

$$
(21.10) \qquad\qquad (\Omega \setminus G) \cap U = (\Omega \setminus G) \cap N \subset N.
$$

Since $(G \cup U) \setminus (U \setminus G) = G$, and since $G \cup U$ is open and $U \setminus G$ has measure zero by (21.10), the definition for the Sobolev capacity gives that

$$
C_{p,\mu}(G \cup U) = C_{p,\mu}(G) < \varepsilon.
$$

Finally, since $N \subset G \cup U$, we have that $C_{p,\mu}(N) < \varepsilon$. The theorem follows.   □

21.11. Remark. It follows from the definitions that every (p, μ)-quasicontinuous function has a pointwise defined (p, μ)-quasicontinuous representative that is a Borel function. Moreover, every two such representatives differ only in a set of (p, μ)-capacity zero by Theorem 21.9. In particular, if ν is a Radon measure in Ω that is in the dual of $L_0^{1,p}(\Omega; \mu)$, then it follows from Proposition 21.8 that the expression

$$(21.12) \qquad \int_\Omega v \, d\nu$$

makes sense for every function v that is (p, μ)-quasicontinuous; we can unambiguously use a Borel representative of v.

In the next theorem we show that the dual action of a finite measure ν in $L_0^{1,p}(\Omega; \mu)$ is exactly given by formula (21.12) if Ω is bounded and $v \in H_0^{1,p}(\Omega; \mu)$ (cf. the discussion before Theorem 21.6).

21.13. Theorem. *Let Ω be bounded and let ν be a finite Radon measure in the dual of $L_0^{1,p}(\Omega; \mu)$. Then*

$$\langle \nu, v \rangle = \int_\Omega v \, d\nu$$

for every (p, μ)-quasicontinuous function $v \in H_0^{1,p}(\Omega; \mu)$.

We require the following lemma.

21.14. Lemma. *Let $u \in H_0^{1,p}(\Omega; \mu)$ be such that $|u| \leq M$ almost everywhere. Then there exists a sequence $\varphi_j \in C_0^\infty(\Omega)$ such that $|\varphi_j| \leq M$ for every j and that $\varphi_j \to u$ in $H_0^{1,p}(\Omega; \mu)$.*

Proof. Let $\varphi_j \in C_0^\infty(\Omega)$ be a sequence such that $\varphi_j \to u$ in $H_0^{1,p}(\Omega; \mu)$. Then the sequence $\psi_j = \min(\max(\varphi_j, -M), M)$ consists of compactly supported Lipschitz functions in Ω such that $\psi_j \to u$ in $H_0^{1,p}(\Omega; \mu)$ (Lemma 1.22). It therefore suffices to show that every compactly supported Lipschitz function ψ in Ω can be approximated in $H_0^{1,p}(\Omega; \mu)$ by functions $\varphi \in C_0^\infty(\Omega)$ such that $||\varphi||_\infty \leq ||\psi||_\infty$. This is achieved by considering convolutions ψ_ε of ψ, for they satisfy

$$||\psi_\varepsilon||_\infty \leq ||\psi||_\infty, \qquad ||\nabla \psi_\varepsilon||_\infty \leq ||\nabla \psi||_\infty, \qquad \varepsilon > 0.$$

(See also Lemma 1.11.) Indeed, since we also have that $\psi_\varepsilon \to \psi$ and $\nabla \psi_\varepsilon \to \nabla \psi$ almost everywhere, as $\varepsilon \to 0$, the claim follows from the dominated convergence theorem.   □

PROOF OF THEOREM 21.13: Let $v \in H_0^{1,p}(\Omega; \mu)$ be (p, μ)-quasicontinuous and Borel. By considering the positive and the negative part of v separately, if necessary, we may assume that $v \geq 0$. Moreover, since the truncations $v_k = \min(v, k)$ converge to v in $H_0^{1,p}(\Omega; \mu)$, as $k \to \infty$, we may assume by the monotone convergence theorem that v is bounded. Let $\varphi_j \in C_0^\infty(\Omega)$ be a sequence of uniformly bounded functions that converges to v both in $H_0^{1,p}(\Omega; \mu)$ and (p, μ)-quasieverywhere (Lemma 21.14, Theorems 4.3 and 4.5). Since $\nu(\Omega) < \infty$, we have that

$$\langle \nu, v \rangle = \lim_{j \to \infty} \langle \nu, \varphi_j \rangle = \lim_{j \to \infty} \int_\Omega \varphi_j \, d\nu = \int_\Omega v \, d\nu$$

by the dominated convergence theorem. The theorem follows.   □

21.15.   Theorem. *Let Ω be bounded, let ν be a finite Radon measure in the dual of $L_0^{1,p}(\Omega; \mu)$, and let $u \in H_0^{1,p}(\Omega; \mu)$ be the unique $\mathcal{A}$-superharmonic function with Riesz mass ν guaranteed by Theorem 21.6. Then*

$$(21.16) \qquad \int_\Omega \mathcal{A}(x, \nabla u) \cdot \nabla u \, dx = \int_\Omega u \, d\nu$$

and the dual norm $\|\nu\|$ of ν satisfies

$$(21.17) \qquad \alpha \|\nabla u\|_p^{p-1} \leq \|\nu\| \leq \beta \|\nabla u\|_p^{p-1}.$$

Proof. The inequality on the right in (21.17) follows from (21.7). To prove the inequality on the left in (21.17), as well as (21.16), we recall that u is (p, μ)-quasicontinuous and Borel (Theorem 10.9), and that by the definition for the mapping $\mathfrak{A}_\nu$ in the proof of Theorem 21.6 we have

$$0 = \langle \mathfrak{A}_\nu u, u \rangle = \int_\Omega \mathcal{A}(x, \nabla u) \cdot \nabla u \, dx - \langle \nu, u \rangle.$$

Theorem 21.13 then gives that (21.16) holds. Moreover, we have that

$$\alpha \int_\Omega |\nabla u|^p \, d\mu \leq \int_\Omega \mathcal{A}(x, \nabla u) \cdot \nabla u \, dx = \langle \nu, u \rangle \leq \|\nu\| \, \|\nabla u\|_p.$$

The proof of the theorem is complete.   □

The following corollary is noteworthy.

21.18. Corollary. *Let Ω be bounded, let ν be a finite Radon measure in the dual of $L_0^{1,p}(\Omega; \mu)$, and let $u \in H_0^{1,p}(\Omega; \mu)$ be the unique $\mathcal{A}$-superharmonic function with $\mathcal{A}(x, \xi) = w(x)|\xi|^{p-2}\xi$ and with Riesz mass ν guaranteed by Theorem 21.6. Then the dual norm $\|\nu\|$ of ν satisfies*

$$\|\nu\| = \|\nabla u\|_p^{p-1}.$$

21.19. Remark. A Radon measure ν with compact support in a bounded open set Ω is in the dual of $H_0^{1,p}(\Omega; \mu)$ if and only if

$$(21.20) \qquad \int\limits_{\Omega} \int\limits_0^r \left(\frac{t^p \, \nu(B(x,t))}{\mu(B(x,t))} \right)^{1/(p-1)} \frac{dt}{t} \, d\nu(x) < \infty$$

for some $r > 0$. This result in the unweighted case is due to Hedberg and Wolff [39]. A different proof can be found in [1], [84, Theorem 4.7.5]. For the weighted case, see [70, Theorem 6.1]. Indeed, the result follows fairly directly from (21.22).

By using condition (21.20), one can exhibit many concrete and nontrivial examples of measures in the dual of $L_0^{1,p}(\Omega; \mu)$.

21.2. The Wolff potential

The main new concept in the proof of the necessity of the Wiener test for the regularity of a boundary point is the Wolff potential of a measure. Let ν be a Radon measure in Ω. *The Wolff potential of ν* is the expression

$$\mathbf{W}_{1,p}^\nu(x, r) = \int\limits_0^r \left(\frac{t^p \, \nu(B(x,t))}{\mu(B(x,t))} \right)^{1/(p-1)} \frac{dt}{t}$$

defined for every $x \in \Omega$ and $0 < r < \text{dist}(x, \partial\Omega)$. Note that $\mathbf{W}_{1,p}^\nu(x, r)$ is comparable to the integral

$$\int\limits_0^r \left(\frac{\nu(B(x,t))}{\text{cap}_{p,\mu}(B(x,t), B(x,2t))} \right)^{1/(p-1)} \frac{dt}{t}$$

with constants of comparability only depending on n, p, and c_μ (Lemma 2.14). Also note that the Wolff potential already appeared in condition (21.20).

The main theorem of this chapter is the following.

21.21.   Theorem. *Let u be a nonnegative $\mathcal{A}$-superharmonic function in $B(x_0, 3r)$ with ν its Riesz mass. Then*

$$(21.22) \qquad c_1 \mathbf{W}^{\nu}_{1,p}(x_0, r) \le u(x_0) \le c_2 \inf_{B(x_0,r)} u + c_3 \mathbf{W}^{\nu}_{1,p}(x_0, 2r) \,,$$

where c_1, c_2, and c_3 are positive constants, depending only on n, p, c_μ, and the structural constants α and β.

In particular, $u(x_0) < \infty$ if and only if $\mathbf{W}^{\nu}_{1,p}(x_0, r) < \infty$.

Theorem 21.21 was first established by Kilpeläinen and Malý in [53], [54]; its weighted version is included in Mikkonen's thesis [70]. Malý [65], [66] extended the estimate for equations depending also on u; for a complete record of this see the book [67] by Malý and Ziemer. Later Trudinger and Wang [82] gave a different proof for Theorem 21.21 that works for a wider class of subelliptic operators. Labutin [63] extended the result for Hessian equations. For further information about the Wolff potential and its extensions, see [2].

We will not prove Theorem 21.21 here. Instead, we employ the Wolff potential estimate (21.22) to establish the necessity of the Wiener test for the regularity of a boundary point. First we present estimates that connect capacities and Riesz masses of capacitary potentials. Recall the definitions for the balayage $\hat{R}^u_E = \hat{R}^u_E(\Omega; \mathcal{A})$ of u relative to $E \subset \Omega$, and for the $\mathcal{A}$-potential $\hat{R}^1_E = \hat{R}^1_E(\Omega; \mathcal{A})$ of a set $E \subset \Omega$ (Chapter 8, pp. 157–8).

The next lemma is not explicitly stated in the main text, although it can be derived easily from the results therein. For simplicity, we outline an argument.

21.23.   Lemma. *Suppose that Ω is bounded, that $E \Subset \Omega$, and that u is a bounded nonnegative $\mathcal{A}$-superharmonic function in Ω. Then $\hat{R}^u_E(\Omega; \mathcal{A}) \in H^{1,p}_0(\Omega; \mu)$.*

Proof. Let $v = \hat{R}^u_E(\Omega; \mathcal{A})$ and choose an open set $D \Subset \mathbf{R}^n \setminus \overline{E}$ that contains $\partial\Omega$. Let $\theta \in H^{1,p}(D \cap \Omega; \mu) \cap C(\overline{D \cap \Omega})$ be such that $\theta = v$ in $\partial D \cap \Omega$ and that $\theta = 0$ in $\partial\Omega$. We note that v is $\mathcal{A}$-harmonic in $\Omega \setminus \overline{E}$ by Lemma 8.4. If h is the $\mathcal{A}$-harmonic function in $D \cap \Omega$ with $h - \theta \in H^{1,p}_0(D \cap \Omega; \mu)$, we infer from the comparison with exceptional set (Lemma 7.37, Theorem 9.11, and Theorem 9.34) that $h = v$ in $D\cap\Omega$. The lemma follows.   □

21.24.   Lemma. *Suppose that Ω is bounded and $E \Subset \Omega$. Let $u = \hat{R}^1_E(\Omega; \mathcal{A})$ be the $\mathcal{A}$-potential of E in Ω with Riesz mass ν. Then*

$$\nu(U) \le c \operatorname{cap}_{p,\mu}(E \cap U, \Omega)$$

whenever $U \subset \Omega$ is open, where $c = c(p, \alpha, \beta) > 0$.

PROOF: Let $U \subset \Omega$ be open, and let $G \subset \Omega$ be an open set containing $E \cap U$. Choose an increasing sequence of compact sets K_j such that $G = \cup_j K_j$. Fix j and let u_j be the $\mathcal{A}$-potential of $(E \cap K_j) \cup (E \setminus U)$ in Ω with ν_j the associated Riesz measure. Then let $\sigma_j = \nu_j \lfloor U$ be the restriction of ν_j to U. The support of σ_j is contained in K_j, and hence in G. It follows from Lemma 21.23 and from Theorem 21.6 that σ_j is in the dual of $L_0^{1,p}(\Omega; \mu)$. Let $v_j \in H_0^{1,p}(\Omega; \mu)$ be the unique $\mathcal{A}$-superharmonic function with Riesz mass σ_j guaranteed by Theorem 21.6. We see similarly as in Lemma 3.22 that

$$0 \le v_j \le u_j \le 1 .$$

Next, let $\varphi \in C_0^\infty(\Omega)$, $0 \le \varphi \le 1$, be such that $\varphi = 1$ in K_j. We obtain from the Caccioppoli inequality (3.29) (applied to the function $v_j - 1$) that

$$
\begin{aligned}
\nu_j(U) = \sigma_j(K_j) &= \int_\Omega \mathcal{A}(x, \nabla v_j) \cdot \nabla \varphi^p \, dx \\
&\le c \Big(\int_\Omega |\nabla v_j|^p \varphi^p \, d\mu \Big)^{(p-1)/p} \Big(\int_\Omega |\nabla \varphi|^p \, d\mu \Big)^{1/p} \\
&\le c \int_\Omega |\nabla \varphi|^p \, d\mu .
\end{aligned}
$$

By taking the infimum over all φ in the preceding inequality, we find that

$$\nu_j(U) \le c \operatorname{cap}_{p,\mu}(G, \Omega) ,$$

and hence that

$$\nu_j(U) \le c \operatorname{cap}_{p,\mu}(E \cap U, \Omega) .$$

Next, by following the proof of Theorem 3.75, we infer that, for each $\psi \in C_0^\infty(\Omega)$,

$$\int_\Omega \psi \, d\nu_j = \int_\Omega \mathcal{A}(x, \nabla u_j) \cdot \nabla \psi \, dx \to \int_\Omega \mathcal{A}(x, \nabla u) \cdot \nabla \psi \, dx = \int_\Omega \psi \, d\nu$$

as $j \to \infty$. Hence in particular

$$\nu(U) \le \liminf_{j \to \infty} \nu_j(U) \le c \operatorname{cap}_{p,\mu}(E \cap U, \Omega) ,$$

as required. The lemma follows. □

21.25. Lemma. *Suppose that Ω is bounded, that ν is a Radon measure in Ω, and that $u \in H_0^{1,p}(\Omega; \mu)$ is $\mathcal{A}$-superharmonic with the Riesz measure ν. Then*

$$\lambda^{p-1} \operatorname{cap}_{p,\mu}(\{x \in \Omega : u(x) > \lambda\}, \Omega) \leq \frac{\nu(\Omega)}{\alpha}$$

for every $\lambda > 0$.

PROOF: We may assume that $\nu(\Omega) < \infty$. Then we deduce from Theorems 21.6 and 21.13 that

$$\alpha \int_\Omega |\nabla \min(u,\lambda)|^p \, d\mu \leq \int_\Omega \mathcal{A}(x, \nabla u) \cdot \nabla \min(u,\lambda) \, dx$$
$$= \int_\Omega \min(u,\lambda) \, d\nu \leq \lambda \nu(\Omega) \, .$$

The lemma follows from this, for $\min(u,\lambda)/\lambda$ is admissible to test the capacity (see the proofs of Theorems 4.3 and Lemma 4.7). $\square$

We now provide a converse to Theorem 12.12. Recall the definition for (p,μ)-thinness from 12.7.

21.26. Theorem. *Suppose that E is (p,μ)-thin at $x_0 \notin E$. Then E is $\mathcal{A}$-thin at x_0, i.e., there is an $\mathcal{A}$-superharmonic function u in a neighborhood of x_0 such that*

$$\liminf_{\substack{x \to x_0 \\ x \in E}} u(x) > u(x_0) \, .$$

PROOF: We may assume that E is open (Lemma 12.11). For $j = 0, 1, 2, \ldots$, write $B_j = B(x_0, 2^{-j})$ and $E_j = E \cap B_j$. Fix an integer $k \geq 2$, and let $u = \hat{R}_{E_k}^1(B_{k-2}; \mathcal{A})$ be the $\mathcal{A}$-potential of E_k in B_{k-2} with ν its Riesz mass. Then $u = 1$ on E_k (Lemma 8.4). We will show that when k is chosen large enough, we also have that $u(x_0) < 1$.

To do so, write $\lambda = \inf_{B_k} u$. By the minimum principle $u > \lambda$ on B_k, and Lemmas 2.14, 21.23, and 21.25 give that

$$\lambda^{p-1} \mu(B_k) 2^{pk} \leq c \, \lambda^{p-1} \operatorname{cap}_{p,\mu}(\{u > \lambda\}, B_{k-2}) \leq c\nu(B_{k-2}) = c\nu(B_{k-1}) \, .$$

From this we obtain that

$$(21.27) \qquad \inf_{B_k} u \leq c \left(2^{-(k-1)p} \frac{\nu(B_{k-1})}{\mu(B_{k-1})} \right)^{1/(p-1)} \, .$$

Moreover, it follows from Lemma 21.24 that for $j \geq k - 1$ we have

$$(21.28) \qquad \nu(B_j) \leq c \, \mathrm{cap}_{p,\mu}(E_j, B_{k-2}) \leq c \, \mathrm{cap}_{p,\mu}(E_j, B_{j-1}) \,.$$

Hence, using (21.27), Lemma 2.14, and (21.28), we obtain from Theorem 21.21 that

$$u(x_0) \leq c \inf_{B_k} u + c \, \mathbf{W}_{1,p}^{\nu}(x_0, 2^{-(k-1)})$$

$$\leq c \sum_{j=k-1}^{\infty} \left(\frac{\mathrm{cap}_{p,\mu}(E_j, B_{j-1})}{\mathrm{cap}_{p,\mu}(B_j, B_{j-1})} \right)^{1/(p-1)} \,,$$

where $c = c(n, p, \alpha, \beta, c_\mu) > 0$. Since E is (p, μ)-thin at x_0, it follows (see Lemma 12.10) that the sum on the right in the preceding inequality is at most $1/2$, provided we choose k large enough. This completes the proof. $\qquad \square$

Theorem 21.26 implies the following general form of Theorem 12.21.

21.29. Theorem. *Suppose that $E \subset \mathbf{R}^n$ is a set with $x_0 \notin E$. Then the following are equivalent:*

(i) *E is $\mathcal{A}$-thin at x_0.*

(ii) *$\complement E$ is an $\mathcal{A}$-fine neighborhood of x_0.*

(iii) *E is (p, μ)-thin at x_0.*

It follows from Theorems 9.18, 6.33, and 21.26 that $x_0 \in \partial\Omega \setminus \{\infty\}$ is irregular if $\complement\Omega$ is (p, μ)-thin at x_0. Hence the list of different characterizations for regularity in Theorem 9.17 can be continued by the Wiener test (cf. Section 12.7):

21.30. Theorem. *Suppose that $x_0 \in \partial\Omega \setminus \{\infty\}$ is such that $\mathrm{cap}_{p,\mu}\{x_0\} = 0$. Then the following are equivalent:*

(i) *x_0 is regular.*

(ii) *There is a barrier at x_0 relative to Ω.*

(iii) *If $U \Subset V$ are bounded open sets with $x_0 \in U$, then*

$$\hat{R}_{U \setminus \Omega}^u (V)(x_0) = u(x_0)$$

whenever $u \in \mathcal{S}(V)$ is nonnegative.

(iv) *For all balls B containing x_0 it holds that*

$$\hat{R}^1_{B\setminus\Omega}(2B)(x_0) = 1.$$

(v) *The complement $\complement\Omega$ is (p,μ)-thick at x_0.*

21.31. Remark. (a) Using the Wolff potential estimate (21.22), Kilpeläinen [49] constructed for any given G_δ set $E \subset \mathbf{R}^n$ of p-capacity zero an $\mathcal{A}$-superharmonic function u in $\mathbf{R}^n$ such that

$$E = \{x : u(x) = \infty\}.$$

This was raised as an open question in Notes to Chapter 10, p. 200. (Note that there is an error in this question: one should obviously require that E is of p-capacity zero.) The proof in [49] can be extended to a weighted case as well.

(b) Using the results in [54], one can argue as in [52] to show that Theorem 9.34 has a converse, too. Namely, if $u \in \mathcal{S}(\Omega)$ is positive, $E \Subset \Omega$ with $\mathrm{cap}_{p,\mu} E > 0$, and

$$\lim_{x \to x_0} \hat{R}^u_E(x) = 0$$

for a point $x_0 \in \partial\Omega$, then x_0 is a regular boundary point of Ω.

22. Generalizations

In this final chapter, we discuss recent generalizations of the nonlinear potential theory as developed in this book.

22.1. Potential theory in metric measure spaces

Axiomatic linear or nonlinear potential theories specify the properties of harmonic functions without any reference to differential equations or to variational integrals, and these theories can be applied in a wide range of topological spaces. The Dirichlet integral and the Dirichlet principle have been fundamental in the applications of potential theory. Since many spaces carry natural measures, it is desirable to have a theory where Dirichlet type integrals are taken with respect to a given measure. In this book such a theory is developed in the framework of weighted Sobolev spaces in $\mathbf{R}^n$ and, at least from a local point of view, an extension of the theory to Riemannian manifolds is straightforward.

During the past ten years several variants for the first order classical Sobolev space $H^{1,p}$ have appeared in the general context of metric measure spaces. Here by a *metric measure space* we mean a metric space X equipped with a locally finite Borel regular measure μ. In 1996 Hajlasz [34], and in 1999 Cheeger [19] and Shanmugalingam [78], [79], introduced their counterparts for $H^{1,p}$. In many situations these new function spaces coincide [79], [37], [35], [47].

For Cheeger and Shanmugalingam the starting point was the concept of an upper gradient introduced by Heinonen and Koskela in [42]. Namely, consider the standard inequality

$$|u(x) - u(y)| \leq \int_\gamma |\nabla u|\, ds$$

satisfied by every C^1-function u and every rectifiable path γ joining x to y. Accordingly, a nonnegative Borel measurable function g on a metric measure space X is said to be an *upper gradient* of a function $u : X \to \mathbf{R}$ if for every pair of points $x, y \in X$ and for every rectifiable path γ joining x and y in X the inequality

$$|u(x) - u(y)| \leq \int_\gamma g\, ds$$

holds. The function g is called a *p-weak upper gradient* of u if the above inequality holds for p-almost all paths; that is, if the family of paths for which the property

394

fails to hold has zero p-modulus [30], [40]. The space $\widetilde{N}^{1,p}(X)$ is now defined to be the collection of all p-integrable functions u on X that have a p-integrable p-weak upper gradient g on X. This space is equipped with the seminorm

$$\|u\|_{\widetilde{N}^{1,p}(X)} = \|u\|_{L^p(X)} + \inf \|g\|_{L^p(X)},$$

where the infimum is taken over all p-weak upper gradients g of u. Finally, the *Newtonian-Sobolev space* $N^{1,p}(X)$ is obtained after an identification of functions u and v with $\|u - v\|_{\widetilde{N}^{1,p}(X)} = 0$. One can show that $N^{1,p}(X)$ is a Banach space. In an open set $\Omega \subset \mathbf{R}^n$, equipped with the Lebesgue measure, the space $N^{1,p}(\Omega)$ coincides with the classical Sobolev space $H^{1,p}(\Omega)$. More generally, if w is a p-admissible weight in $\mathbf{R}^n$ and μ the associated measure, then $N^{1,p}(\Omega)$ agrees with $H^{1,p}(\Omega; \mu)$, where in the definition of the former space we consider Ω equipped with Euclidean distance and measure μ [19], [71]. A function $u \in N^{1,p}(\Omega)$ is automatically a quasicontinuous representative of a function in $H^{1,p}(\Omega)$ as in Chapter 4 in this book; see [6]. In general, a function in $N^{1,p}(X)$ is absolutely continuous on p-almost all paths in X. One can also define the local spaces $N^{1,p}_{loc}(\Omega)$, and the spaces $N^{1,p}_0(\Omega)$ of functions with zero boundary values (in the Sobolev sense) whenever $\Omega \subset X$ is open [51].

If u belongs to $N^{1,p}(X)$, or to the local space $N^{1,p}_{loc}(X)$, with $1 \leq p < \infty$, then u has a unique minimal p-weak upper gradient g_u. That is, g_u is a p-weak upper gradient of u and if g is any other p-weak upper gradient of u, then $g_u \leq g$ almost everywhere. For $1 < p < \infty$, the concept and existence of a minimal weak upper gradient is due to Cheeger [19] (who used a different but equivalent definition both for the Sobolev space $N^{1,p}$ and for g_u). For the case $p = 1$, see [35]. The fact that such a minimal upper gradient exists, makes it possible to employ not only the unique solutions to boundary value problems (the Dirichlet principle), but also solutions to various obstacle problems involving the p–Dirichlet integral of a p-weak upper gradient [55].

The definition for a Newtonian-Sobolev function u provides control for u on curves in X, but it does not impose any control on the L^p-norm of u in terms of its upper gradient. Extensive research has shown that a right condition for this purpose is that the space X supports a Poincaré inequality [42], [19], [79], [37], [75], [77], [47]. Note that such an inequality is used in this book in the context of weighted Sobolev spaces in $\mathbf{R}^n$. Nowadays many nontrivial examples of spaces supporting a Poincaré inequality are known. Among them are all compact Riemannian manifolds, complete Riemannian manifolds of nonnegative Ricci curvature and their limit spaces, Carnot–Carathéodory spaces, some classes of metric topological manifolds, as well as certain exotic fractals. See [42], [37], [75], [61], [17], [38], [19], [46], [59], and the references therein.

For most potential theoretical applications the following (weak) q-Poincaré inequality suffices, although there are many refinements (cf. Section 20.1). A metric measure space X is said to support a *q-Poincaré inequality*, $1 \leq q < \infty$, if the measure of every ball is positive and finite and if there are constants $C > 0$ and $\lambda \geq 1$ such that

(22.2) $$\fint_{B(z,r)} |u - u_{B(z,r)}| \, d\mu \leq C\,r \left(\fint_{B(z,\lambda r)} g^q \, d\mu \right)^{1/q}$$

for all balls $B(z,r)$ in X, for all integrable functions u in $B(z, \lambda r)$, and for all q-weak upper gradients g of u. Inequality (22.2) corresponds to inequality (20.3) in the (weighted) Euclidean case.

For a p-potential theory, $1 < p < \infty$, in a metric measure space X, the standard assumptions are that X is complete as a metric space, that the underlying measure μ is a doubling measure, and that the space supports a q-Poincaré inequality for some q with $1 < q < p$. In fact, a recent remarkable theorem of Keith and Zhong [47] implies that one only needs to assume that X admits a p-Poincaré inequality; then, under the assumptions of completeness and doubling measure, X also admits a q-Poincaré inequality for some $q < p$.

In this general context, a natural replacement of a p-harmonic function (i.e., a solution to the p-harmonic equation in an open subset of $\mathbf{R}^n$) is a function $u \in N_{loc}^{1,p}(\Omega)$ which minimizes the p-Dirichlet integral

$$\int_{\Omega'} g_u^p \, d\mu$$

with its own boundary values in each open subset Ω' that is compactly contained in Ω, where $\Omega \subset X$ is open. Such a function is called a *p-minimizer* in Ω.

The De Giorgi method can be used to show that a p-minimizer is locally Hölder continuous in Ω [55], [80]; see [9] for the use of the Moser method in this context. Moreover, nonnegative p-minimizers satisfy the Harnack inequality [57], and the important Harnack principle is then an easy consequence.

A function $u \in N_{loc}^{1,p}(\Omega)$ is called a *p-superminimizer* if for all open $\Omega' \Subset \Omega$ and for all functions $v \in N_{loc}^{1,p}(\Omega)$ such that $v - u \in N_0^{1,p}(\Omega')$ and that $v \geq u$ μ-almost everywhere we have

$$\int_{\Omega'} g_u^p \, d\mu \leq \int_{\Omega'} g_v^p \, d\mu \, .$$

Like classical superharmonic functions p-superminimizers can be defined pointwise and they are lower semicontinuous [55]. Finally a *p-superharmonic function* can be defined as a limit of an increasing sequence of p-superminimizers. See [5] for definitions based on comparison conditions. Boundary regularity, removable sets, the Perron method, polar sets, and p-harmonic measures in this setup have been studied in [8], [7], [12], [11], [55], [58].

Potential theory on metric measure spaces provides a continuous approach to the discrete p-Laplace equation and to the treatment of these equations on graphs [44]. Harmonic maps between general spaces [26] also have connections to metric measure spaces and their potential theory.

22.2. Quasiminimizers

Quasiminimizers can be viewed as perturbations of the minimizers of variational integrals. More precisely, let $\Omega \subset \mathbf{R}^n$ be an open set, $K \geq 1$ and $1 \leq p < \infty$. In the case of the p-Dirichlet integral, a function u belonging to the local (unweighted) Sobolev space $H^{1,p}_{loc}(\Omega)$ is said to be a (p, K)-*quasiminimizer* or simply a K-*quasiminimizer*, if

$$\int_{\Omega'} |\nabla u|^p \, dx \leq K \int_{\Omega'} |\nabla v|^p \, dx$$

for all functions $v \in H^{1,p}(\Omega')$ with $v - u \in H^{1,p}_0(\Omega')$ and for all open sets Ω' with compact closure in Ω.

Quasiminimizers have been used as a tool to study regularity of minimizers of variational integrals [31], [32]. The advantage of this approach is that it covers a wide range of applications and that it is based only on the minimization of the variational integrals instead of the corresponding Euler equation. Regularity properties (Hölder continuity, L^p-estimates) as well as the Harnack inequality are consequences of the quasiminimizing property [22]. Quasiminimizers form a wide and flexible class of functions in the calculus of variations under very general circumstances. A natural environment for them is a metric measure space [57], [56].

From the potential theoretic point of view K-quasiminimizers have several drawbacks; they do not provide unique solutions of the Dirichlet problem, they do not obey the comparison principle, they do not form a sheaf, and they do not have a linear structure even when the corresponding Euler equation is linear. In spite of this, quasiminimizers offer an interesting potential theoretic model. In metric measure spaces, quasisuperminimizers, quasisuperharmonic functions, Poisson modifications, and polar sets were studied in [56], and boundary regularity properties in [83], [11].

One-dimensional quasiminimizers are characterized by a reverse type inequality [32]. Quasisuperminimizers, integrability conditions for the derivatives, and connections to A_p-weights were considered in [68] in this case.

New Bibliography

[1] ADAMS, D. R. Weighted nonlinear potential theory. *Trans. Amer. Math. Soc.* *297*, 1 (1986), 73–94.

[2] ADAMS, D. R., AND HEDBERG, L. I. *Function spaces and potential theory*, vol. 314 of *Grundlehren der Mathematischen Wissenschaften*. Springer-Verlag, Berlin, 1996.

[3] BÉNILAN, P., BOCCARDO, L., GALLOUËT, T., GARIEPY, R., PIERRE, M., AND VÁZQUEZ, J. L. An L^1-theory of existence and uniqueness of solutions of nonlinear elliptic equations. *Ann. Scuola Norm. Sup. Pisa Cl. Sci. (4) 22*, 2 (1995), 241–273.

[4] BISHOP, C. J. An A_1-weight not comparable to any quasiconformal Jacobian. To appear in *Contemporary Mathematics*. Proceedings of the Ahlfors-Bers Colloquium, Ann Arbor, 2005.

[5] BJÖRN, A. Characterizations of p-superharmonic functions on metric spaces. *Studia Math. 169*, 1 (2005), 45–62.

[6] BJÖRN, A., BJÖRN, J., AND SHANMUGALINGAM, N. Quasicontinuity of Newton–Sobolev functions and density of Lipschitz functions on metric spaces. Preprint, 2006.

[7] BJÖRN, A., BJÖRN, J., AND SHANMUGALINGAM, N. The Dirichlet problem for p-harmonic functions on metric spaces. *J. Reine Angew. Math. 556* (2003), 173–203.

[8] BJÖRN, A., BJÖRN, J., AND SHANMUGALINGAM, N. The Perron method for p-harmonic functions in metric spaces. *J. Differential Equations 195*, 2 (2003), 398–429.

[9] BJÖRN, A., AND MAROLA, N. Moser iteration for (quasi)minimizers on metric spaces. Preprint, 2005.

[10] BJÖRN, J. Poincaré inequalities for powers and products of admissible weights. *Ann. Acad. Sci. Fenn. Math. 26*, 1 (2001), 175–188.

[11] BJÖRN, J. Boundary continuity for quasiminimizers on metric spaces. *Illinois J. Math. 46*, 2 (2002), 383–403.

[12] BJÖRN, J., MACMANUS, P., AND SHANMUGALINGAM, N. Fat sets and pointwise boundary estimates for p-harmonic functions in metric spaces. *J. Anal.
Math. 85* (2001), 339–369.

[13] BOCCARDO, L., AND GALLOUËT, T. Nonlinear elliptic equations with right-
hand side measures. *Comm. Partial Differential Equations 17*, 3-4 (1992), 641–
655.

[14] BONK, M., HEINONEN, J., AND SAKSMAN, E. Logarithmic potentials,
quasiconformal flows, and the Q-curvature. Preprint, 2006.

[15] BONK, M., HEINONEN, J., AND SAKSMAN, E. The quasiconformal Jacobian
problem. In *In the tradition of Ahlfors and Bers, III*, vol. 355 of *Contemp.
Math.* Amer. Math. Soc., Providence, RI, 2004, pp. 77–96.

[16] BONK, M., AND LANG, U. Bi-Lipschitz parameterization of surfaces. *Math.
Ann. 327*, 1 (2003), 135–169.

[17] BOURDON, M., AND PAJOT, H. Poincaré inequalities and quasiconformal
structure on the boundaries of some hyperbolic buildings. *Proc. Amer. Math.
Soc. 127*, 8 (1999), 2315–2324.

[18] BUCKLEY, S. M. Inequalities of John-Nirenberg type in doubling spaces. *J.
Anal. Math. 79* (1999), 215–240.

[19] CHEEGER, J. Differentiability of Lipschitz functions on metric measure spaces.
Geom. Funct. Anal. 9 (1999), 428–517.

[20] COSTEA, S. *Strong A_∞-weights and scaling invariant Besov and Sobolev-
Lorentz capacities.* PhD thesis, University of Michigan, 2006.

[21] DAVID, G., AND SEMMES, S. Strong A_∞ weights, Sobolev inequalities and
quasiconformal mappings. In *Analysis and partial differential equations*, vol. 122
of *Lecture Notes in Pure and Appl. Math.* Marcel Dekker, 1990, pp. 101–111.

[22] DIBENEDETTO, E., AND TRUDINGER, N. S. Harnack inequalities for
quasiminima of variational integrals. *Ann. Inst. H. Poincaré Anal. Non Linéaire
1*, 4 (1984), 295–308.

[23] DOLZMANN, G., HUNGERBÜHLER, N., AND MÜLLER, S. Non-linear elliptic
systems with measure-valued right hand side. *Math. Z. 226*, 4 (1997), 545–574.

[24] DOLZMANN, G., HUNGERBÜHLER, N., AND MÜLLER, S. Uniqueness and
maximal regularity for nonlinear elliptic systems of n-Laplace type with measure
valued right hand side. *J. Reine Angew. Math. 520* (2000), 1–35.

[25] DOOB, J. L. *Classical potential theory and its probabilistic counterpart*, vol. 262 of *Grundlehren der Mathematischen Wissenschaften*. Springer-Verlag, New York, 1984.

[26] EELLS, J., AND FUGLEDE, B. *Harmonic maps between Riemannian polyhedra*, vol. 142 of *Cambridge Tracts in Mathematics*. Cambridge University Press, Cambridge, 2001. With a preface by M. Gromov.

[27] FEDERER, H. *Geometric Measure Theory*, vol. 153 of *Die Grundlehren der mathematischen Wissenschaften*. Springer-Verlag, New York, 1969.

[28] FRANCHI, B., GUTIÉRREZ, C. E., AND WHEEDEN, R. L. Weighted Sobolev-Poincaré inequalities for Grushin type operators. *Comm. Partial Differential Equations 19*, 3-4 (1994), 523–604.

[29] FRANCHI, B., HAJŁASZ, P., AND KOSKELA, P. Definitions of Sobolev classes on metric spaces. *Ann. Inst. Fourier (Grenoble) 49*, 6 (1999), 1903–1924.

[30] FUGLEDE, B. Extremal length and functional completion. *Acta Math. 98* (1957), 171–219.

[31] GIAQUINTA, M., AND GIUSTI, E. On the regularity of the minima of variational integrals. *Acta Math. 148* (1982), 31–46.

[32] GIAQUINTA, M., AND GIUSTI, E. Quasiminima. *Ann. Inst. H. Poincaré Anal. Non Linéaire 1*, 2 (1984), 79–107.

[33] GRIGOR'YAN, A. A. The heat equation on noncompact Riemannian manifolds. *Mat. Sb. 182*, 1 (1991), 55–87.

[34] HAJŁASZ, P. Sobolev spaces on an arbitrary metric space. *Potential Anal. 5* (1996), 403–415.

[35] HAJŁASZ, P. Sobolev spaces on metric-measure spaces. In *Heat kernels and analysis on manifolds, graphs, and metric spaces (Paris, 2002)*, vol. 338 of *Contemp. Math.* Amer. Math. Soc., Providence, RI, 2003, pp. 173–218.

[36] HAJŁASZ, P., AND KOSKELA, P. Sobolev meets Poincaré. *C. R. Acad. Sci. Paris Sér. I Math. 320* (1995), 1211–1215.

[37] HAJŁASZ, P., AND KOSKELA, P. Sobolev met Poincaré. *Memoirs Amer. Math. Soc. 145*, 688 (2000).

[38] HANSON, B., AND HEINONEN, J. An *n*-dimensional space that admits a Poincaré inequality but has no manifold points. *Proc. Amer. Math. Soc. 128*, 11 (2000), 3379–3390.

[39] HEDBERG, L. I., AND WOLFF, T. H. Thin sets in nonlinear potential theory. *Ann. Inst. Fourier (Grenoble) 33*, 4 (1983), 161–187.

[40] HEINONEN, J. *Lectures on analysis on metric spaces.* Springer-Verlag, New York, 2001.

[41] HEINONEN, J., AND KOSKELA, P. Weighted Sobolev and Poincaré inequalities and quasiregular mappings of polynomial type. *Math. Scand. 77*, 2 (1995), 251–271.

[42] HEINONEN, J., AND KOSKELA, P. Quasiconformal maps in metric spaces with controlled geometry. *Acta Math. 181* (1998), 1–61.

[43] HELMS, L. L. *Introduction to potential theory.* Pure and Applied Mathematics, Vol. XXII. Wiley-Interscience A Division of John Wiley & Sons, New York-London-Sydney, 1969.

[44] HOLOPAINEN, I., AND SOARDI, P. M. *p*-harmonic functions on graphs and manifolds. *Manuscripta Math. 94*, 1 (1997), 95–110.

[45] JERISON, D. The Poincaré inequality for vector fields satisfying Hörmander's condition. *Duke Math. J. 53* (1986), 503–523.

[46] KEITH, S. Modulus and the Poincaré inequality on metric measure spaces. *Math. Z. 245*, 2 (2003), 255–292.

[47] KEITH, S., AND ZHONG, X. The Poincaré inequality is an open ended condition. Preprint, 2003.

[48] KILPELÄINEN, T. A remark on the uniqueness of quasi continuous functions. *Ann. Acad. Sci. Fenn. Math. 23*, 1 (1998), 261–262.

[49] KILPELÄINEN, T. Singular solutions to *p*-Laplacian type equations. *Ark. Mat. 37*, 2 (1999), 275–289.

[50] KILPELÄINEN, T. *p*-Laplacian type equations involving measures. In *Proceedings of the International Congress of Mathematicians, Vol. III* (Beijing, 2002), Higher Ed. Press, pp. 167–176.

[51] KILPELÄINEN, T., KINNUNEN, J., AND MARTIO, O. Sobolev spaces with zero boundary values on metric spaces. *Potential Anal. 12*, 3 (2000), 233–247.

[52] KILPELÄINEN, T., AND LINDQVIST, P. Nonlinear ground states in irregular domains. *Indiana Univ. Math. J. 49*, 1 (2000), 325–331.

[53] KILPELÄINEN, T., AND MALÝ, J. Degenerate elliptic equations with measure data and nonlinear potentials. *Ann. Scuola Norm. Sup. Pisa Cl. Sci. (4) 19*, 4 (1992), 591–613.

[54] KILPELÄINEN, T., AND MALÝ, J. The Wiener test and potential estimates for quasilinear elliptic equations. *Acta Math. 172*, 1 (1994), 137–161.

[55] KINNUNEN, J., AND MARTIO, O. Nonlinear potential theory on metric spaces. *Illinois J. Math. 46*, 3 (2002), 857–883.

[56] KINNUNEN, J., AND MARTIO, O. Potential theory of quasiminimizers. *Ann. Acad. Sci. Fenn. Math. 28*, 2 (2003), 459–490.

[57] KINNUNEN, J., AND SHANMUGALINGAM, N. Regularity of quasi-minimizers on metric spaces. *Manuscripta Math. 105*, 3 (2001), 401–423.

[58] KINNUNEN, J., AND SHANMUGALINGAM, N. Polar sets on metric spaces. *Trans. Amer. Math. Soc. 358*, 1 (2006), 11–37.

[59] KOSKELA, P. Upper gradients and Poincaré inequalities. In *Lecture notes on analysis in metric spaces (Trento, 1999)*. Appunti Corsi Tenuti Docenti Sc. Scuola Norm. Sup., Pisa, 2000, pp. 55–69.

[60] KOVALEV, L. V., AND MALDONADO, D. Mappings with convex potentials and the quasiconformal Jacobian problem. *Ilinois J. Math. 49*, 4 (2005), 1039–1060.

[61] LAAKSO, T. J. Ahlfors Q-regular spaces with arbitrary $Q > 1$ admitting weak Poincaré inequality. *Geom. Funct. Anal. 10*, 1 (2000), 111–123.

[62] LAAKSO, T. J. Plane with A_∞-weighted metric not bi-Lipschitz embeddable to $\mathbb{R}^N$. *Bull. London Math. Soc. 34*, 6 (2002), 667–676.

[63] LABUTIN, D. A. Potential estimates for a class of fully nonlinear elliptic equations. *Duke Math. J. 111*, 1 (2002), 1–49.

[64] MACMANUS, P., AND PÉREZ, C. Trudinger inequalities without derivatives. *Trans. Amer. Math. Soc. 354*, 5 (2002), 1997–2012.

[65] MALÝ, J. Nonlinear potentials and quasilinear PDEs. In *Potential theory—ICPT 94 (Kouty, 1994)*. de Gruyter, Berlin, 1996, pp. 103–128.

[66] MALÝ, J. Pointwise estimates of nonnegative subsolutions of quasilinear elliptic equations at irregular boundary points. *Comment. Math. Univ. Carolin. 37*, 1 (1996), 23–42.